Volume 38

CRM PROCEEDINGS & LECTURE NOTES

Centre de Recherches Mathématiques
Université de Montréal

Algebraic Structures and Moduli Spaces

CRM Workshop
July 14–20, 2003
Montréal, Canada

Jacques Hurtubise
Eyal Markman
Editors

The Centre de Recherches Mathématiques (CRM) of the Université de Montréal was created in 1968 to promote research in pure and applied mathematics and related disciplines. Among its activities are special theme years, summer schools, workshops, postdoctoral programs, and publishing. The CRM is supported by the Université de Montréal, the Province of Québec (FCAR), and the Natural Sciences and Engineering Research Council of Canada. It is affiliated with the Institut des Sciences Mathématiques (ISM) of Montréal, whose constituent members are Concordia University, McGill University, the Université de Montréal, the Université du Québec à Montréal, and the Ecole Polytechnique. The CRM may be reached on the Web at www.crm.umontreal.ca.

American Mathematical Society
Providence, Rhode Island USA

The production of this volume was supported in part by the Fonds pour la Formation de Chercheurs et l'Aide à la Recherche (Fonds FCAR) and the Natural Sciences and Engineering Research Council of Canada (NSERC).

2000 *Mathematics Subject Classification.* Primary 14Dxx, 14Cxx.

Library of Congress Cataloging-in-Publication Data

Algebraic structures and moduli spaces: CRM Workshop, July 14–20, 2003, Montréal, Canada / Jacques Hurtubise, Eyal Markman, editors.
p. cm. (CRM proceedings & lecture notes, ISSN 1065-8580; v. 38)
Includes bibliographical references.
ISBN 0-8218-3568-8 (soft: alk. paper)
1. Algebra, Abstract. 2. Moduli theory. I. Hurtubise, Jacques. II. Markman, Eyal, 1961– . III. Series

QA162.A42 2004
512′.02—dc22 2004057088

∞ The paper used in this book is acid-free and falls within the guidelines
established to ensure permanence and durability.
This volume was submitted to the American Mathematical Society
in camera ready form by the Centre de Recherches Mathématiques.
Visit the AMS home page at `http://www.ams.org/`

10 9 8 7 6 5 4 3 2 1 09 08 07 06 05 04

Contents

Introduction

Underlying the global structure of any geometric object, one often finds some simple geometric construct. When fully developed into an algebraic machine, surprisingly complete information is revealed about the global structure of the space. Thus, for example, when analyzing loop spaces, the actions of inserting, removing and moving around small disks lead to the structure of an operad, which, when developed fully, gives us back more or less complete information on the homology of the loop space. In some cases, it is useful to think not of one space, but a whole series. An example of this is the disjoint union over n of the Hilbert schemes of length n subschemes on a surface: one again has natural operations of adding and removing points; this, when developed, leads to an action (discovered independently by Nakajima and Grojnowsky) of a Heisenberg algebra on the sum of the cohomologies of these spaces. Subsequent work of Lehn and Li, Qin, Wang, introduced vertex algebra techniques. When taken to their representation theoretic conclusion, these discoveries give an elegant description of the cohomology.

Moduli spaces are particularly suited to such techniques, and the past few years has seen a number of results on these spaces which seem to incorporate the full arsenal of algebraic tools at our disposal and develop along the way deep and important links of the geometry of moduli spaces with a variety of subjects such as representation theory, mathematical physics and combinatorics. It was with this in mind that a conference with the title Algebraic Structures and Moduli Spaces was organized last year at the Centre de Recherches Mathématiques, from July 14th to July 20th, 2003. As is often the case, some of the talks presented spilled over the boundaries defined by the conference title, as preference was given to the quality of the mathematics over an exact fit; the whole, however, remained remarkably faithful to the organizing idea of the conference. The event was a great success, and it was decided to prepare some proceedings.

The conference included two series of introductory lectures as well as individual talks, and we are fortunate in being able to publish notes from both series. The first, by Manfred Lehn, gives a beautiful introduction to the theory of Hilbert schemes of points on surfaces, presenting the basic information, defining the Heisenberg action and discussing some of the ring structure of the cohomology. This series is complemented by a contribution of Wei-Ping Li, Zhenbo Qin and Weiqiang Wang who explain how the Jack polynomials occur in the T-equivariant version of the cohomology of Hilbert schemes.

The second series of lectures, by Hiraku Nakajima and Kōta Yoshioka, describes their recent quite wonderful work on instanton counting (following on work of Nekrasov), which identifies a generating function given in terms of (equivariant cohomology) integrals on the moduli space of instantons over $\mathbb{R}^4$ with a Seiberg-Witten prepotential corresponding to a family of Riemann surfaces. The proof proceeds by showing, in essence, that they satisfy the same differential equation with the same initial conditions; the differential equation on the instanton side is obtained from a relation given by studying how moduli spaces behave under blowing up. Again, there is a complementary contribution, this time by Alexander Braverman, which explains the links of a related counting function, defined in terms of moduli spaces of framed G-bundles on the complex projective plane with a reduction to a parabolic subgroup along a line, with representation-theoretic quantities defined for the affine algebra over the Langlands dual group, and also with a conjectured quantum cohomology of affine flag manifolds.

The category of coherent sheaves on an algebraic variety, and its associated derived category, is a natural invariant, which plays a central role, for example, in Kontsevich's homological interpretation of mirror symmetry. The volume contains several contributions which explore the behavior of these categories: the paper of Yoshinori Namikawa explores the effect of a transformation of a variety by a Mukai flop; the paper of Shinobu Hosono, Bong Lian, Keiji Oguiso and Shing-Tung Yau counts the number of K3 surfaces which have equivalent bounded derived categories to that of a given K3 surface; Justin Sawon explores the role of a twisted version of the derived category in understanding holomorphic symplectic manifolds; and Misha Verbitsky shows how the category of coherent sheaves is constant for the generic (nonalgebraic) compact complex torus.

The variety of techniques exploited in understanding moduli spaces is well represented in the paper of Nan-Kuo Ho and Lisa Jeffrey on the moduli of flat connections for a nonorientable surface, that of Mark Andrea de Cataldo and Luca Migliorini on the Gysin map and mixed Hodge structures, that of Claudio Bartocci and Marcos Jardim on the hyper-Kähler Nahm transform, and that of Mike Roth and Ravi Vakil, which defines a new invariant, the affine stratification number, and applies it to the moduli space of curves.

All in all, a most impressive array. We have quite a few people to thank for their help in compiling it: Hiraku Nakajima for his work as a co-organizer of a most stimulating conference, Josée Laferrière and Louis Pelletier at the CRM for their usual superb handling of the logistical details of the conference, Arthur Greenspoon for his assistance in carefully reading one of the contributions, André Montpetit for his meticulous efforts as technical editor of the volume, and of course the authors, who did most of the work.

Jacques Hurtubise
Eyal Markman

Centre de Recherches Mathématiques
CRM Proceedings and Lecture Notes
Volume **38**, 2004

Lectures on Hilbert Schemes

Manfred Lehn

1. Introduction

These are slightly extended notes of my lectures on the geometry of Hilbert schemes of points that I gave at the workshop on Algebraic Structures and Moduli Spaces at CRM Montreal in July 2003.

The purpose of the lectures was to provide an introduction to some aspects of the theory of Hilbert schemes of points on surfaces. It seems to me that the interest in Hilbert schemes has many sources: There are relations to enumerative geometry, to moduli spaces of sheaves, to the combinatorics of the symmetric group, to orbifold cohomology, to the generalized McKay correspondence, and they provide almost all known examples of irreducible symplectic manifolds (up to deformation).

The scope of these lectures is much narrower. I explain the classical theorems of Fogarty and Briançon based on a technique due to Ellingsrud and Strømme and give a fairly self-contained presentation of Nakajima's theorem on the Heisenberg action on the cohomology of Hilbert schemes with full proofs. I then explain work of Li, Qin and Wang, and of Sorger and myself on the ring structure of the cohomology of Hilbert schemes. In this part proofs are only sketched, or some of the ingredients are explained. The notes end with a brief discussion on the relation to a conjecture of Ruan. Besides the original references, it is worth while also to look at other review articles or lecture notes with the same or a similar subject by Nakajima [**33**], Göttsche [**16**], Göttsche and Ellingsrud [**5**], Wang [**38**], Qin and Wang [**34**].

What is perhaps more interesting is what is not covered here: the relation to the theory of symplectic varieties (cf. the notes [**25**]); the relation to moduli of sheaves [**31**]; the work of Haiman on the $n!$-conjecture and related questions [**20, 21**]; Nakamura's G-Hilbert schemes and the McKay correspondence [**2**]. The reader is invited to look up detailed bibliographies in the quoted articles.

2. Configurations of unordered n-tuples

Configuration spaces of n-tuples of points that move on a given manifold X are interesting topological spaces with fascinating topological and geometric properties. There is a basic difference in whether one is interested in ordered or unordered n-tuples. As a first approach to a proper definition of the right configuration space,

2000 *Mathematics Subject Classification.* 14C05, 14C07.
This is the final form of the paper.

we may consider the product X^n for ordered tuples and its quotient $S^n(X) := X^n/\mathrm{S}_n$ by the symmetric group for unordered tuples. In both spaces, points that correspond to n-tuples of points that are not pairwise distinct are special: In X^n they have nontrivial isotropy groups with respect to the S_n-action, in S^nX they are singular (except for the special case $\dim(X) = 1$). One is therefore interested in other smooth compactifications of the configuration space of n-tuples of distinct points. For ordered n-tuples a very nice compactification was constructed by Fulton and MacPherson [**13**]. For unordered n-tuples a smooth natural compactification for all n exists so far only for curves, where the symmetric product is already smooth, and for surfaces, where such a compactification is provided by the so-called Hilbert schemes of points.

Assume that X is a smooth quasiprojective scheme of finite type over $\mathbb{C}$. Let $Z \subset X$ be a zero-dimensional subscheme. Then $H^0(Z, \mathcal{O}_Z)$ is an Artinian $\mathbb{C}$-algebra. By definition, the length of Z is the length of this algebra, $\ell(Z) := \operatorname{length} H^0(\mathcal{O}_Z) = \dim_{\mathbb{C}} H^0(\mathcal{O}_Z)$. Let $\operatorname{Hilb}^n(X)$ denote the set of all zero-dimensional subschemes $Z \subset X$ of length n. We will see in the next section how $\operatorname{Hilb}^n(X)$ can be given a natural scheme structure. Assume now that X is reduced. If $x \in Z$ is a closed point, the multiplicity of x in Z is defined as $\dim_{\mathbb{C}}(\mathcal{O}_{Z,x})$. We may associate to Z the cycle $|Z|$ corresponding to the underlying set counted with multiplicities. $|Z|$ is a point in the symmetric product:

$$|Z| := \sum_{x\in X} \dim_{\mathbb{C}}(\mathcal{O}_{Z,x}) \cdot x \in S^n(X).$$

Sending Z to $|Z|$ defines the so-called Hilbert–Chow map

$$\rho \colon \operatorname{Hilb}^n(X) \to S^nX.$$

If X is a smooth curve, ρ turns out to be an isomorphism for all n. Essentially, it suffices to understand the case of the affine line: the symmetric group acts on the coordinate ring $\mathbb{C}[x_1, \ldots, x_n]$ of $(\mathbb{C}^1)^n$ by permuting the variables. The invariant ring is generated by the elementary symmetric functions $\sigma_1 = x_1 + \cdots + x_n, \ldots, \sigma_k = \sum_{i_1<\cdots<i_k} x_{i_1}x_{i_2}\cdots x_{i_k}, \ldots, \sigma_n = x_1x_2\cdots x_n$. These are algebraically independent. It follows that

$$S^n(\mathbb{C}) = \operatorname{Spec}(\mathbb{C}[x_1, \ldots, x_n]^{\mathrm{S}_n}) = \operatorname{Spec}(\mathbb{C}[\sigma_1, \ldots, \sigma_n]) = \mathbb{C}^n.$$

On the other hand, any zero-dimensional subscheme in $\mathbb{C}$ is described by a single polynomial $f(T) = T^n + a_1T^{n-1} + \cdots + a_n$ of degree n. The coefficients a_i define a point $a \in \mathbb{C}^n$, and conversely, any such point defines a polynomial and thus a subscheme. If Z consists of n distinct points $p_1, \ldots, p_n \in \mathbb{C}$, the coefficients of f are precisely the elementary symmetric functions in the p_i: $a_i = \sigma_i(p_1, \ldots, p_n)$.

For arbitrary X, the Hilbert–Chow map ρ is still an isomorphism over the open subset of S^nX corresponding to n-tuples of distinct points in X. The opposite of a tuple of distinct points is a subscheme Z with multiplicity n in a single point $x \in X$: consider the fibre $B^n_x(X) := \rho^{-1}(n \cdot x)$. A point in $B^n_x(X)$ corresponds to a subscheme $Z \subset X$ with underlying set $\{x\}$ and length n. Any such subscheme is given by an ideal $I \subset \mathcal{O}_{X,x}$ with $\dim_{\mathbb{C}} \mathcal{O}_{X,x}/I = n$. Let $\mathfrak{m}$ denote the maximal ideal in $\mathcal{O}_{X,x}$. Since $\dim_{\mathbb{C}} \mathcal{O}_{X,x}/I = n$, one must have $\mathfrak{m}^n \subset I$. Thus I is completely determined by the ideal $\bar{I} := I/\mathfrak{m}^n \subset \mathcal{O}_{X,x}/\mathfrak{m}^n$. We conclude that if X is smooth of dimension d, then $B^n_x(X)$ does not depend on either X or x, and that in order to study $B^n_x(X)$ we may as well assume that $X = \mathbb{C}^d$ and that x is the origin O.

Here is a description of B^n_O for the case of surfaces, i.e., $d = 2$ and small n: B^n_O contains all ideals $I \subset \mathbb{C}[x, y]$ with $\dim_{\mathbb{C}} \mathbb{C}[x, y]/I = n$ and $\mathfrak{m}^n \subset I \subset \mathfrak{m} := (x, y)$.

Case $n = 1$. The only choice is $I = \mathfrak{m}$ and B^1_O is a point, as one expects.

Case $n = 2$. Ideals I correspond to one-dimensional linear subspaces $I/\mathfrak{m}^2 \subset \mathfrak{m}/\mathfrak{m}^2 = \mathbb{C}\langle \bar{x}, \bar{y}\rangle$. Thus $B^2_O \cong \mathbb{P}^1 = \mathbb{P}(T_O\mathbb{C}^2)$. This can be geometrically interpreted as follows: if (p_1, p_2) is a pair of distinct points in $\mathbb{C}^2$ that moves toward the origin and collides there then the limiting point in $S^2(\mathbb{C}^2)$ only remembers the number of points that collided and the position where the accident happened whereas the limiting in $\mathrm{Hilb}^n(\mathbb{C}^2)$ also remembers the limiting direction of the line that passes through p_1 and p_2.

Case $n = 3$. Here something interesting happens: there are two qualitatively different types of ideals. One is easy to find: $I_\infty = \mathfrak{m}^2$. It is distinguished from the others by the fact the tangent space $T_O Z_\infty$ of the corresponding subscheme Z_∞ has dimension 2. The other subschemes are contained in the germ of a smooth curve and are therefore called curvilinear. Their ideals have the form $I = (y+\alpha x+\beta x^2, x^3)$ (or x and y exchanged). All these ideals are parametrized by a line bundle over $\mathbb{P}^1$. The subscheme Z_∞ arises as the limiting point of a sequence of triples that approach the origin from three different directions, whereas points of the second type arise as limiting points of triples that move toward the origin along a smooth curve. The affine bundle is compactified by adding the point Z_∞ at infinity. In fact, one can write down an explicit model for B^3_O and a family of schemes parameterized by B^3_O: consider in $\mathbb{P}^4 = \mathrm{Proj}(\mathbb{C}[a, b, c, d, e])$ the projective cone over a rational normal cubic given by the equations $ac - b^2$, $ad - bc$, $bd - c^2$. The family $\mathcal{Z} \subset B^3_O \times \mathbb{C}^2$ is the zero set of the ideal $(ax + by + ex^2, bx + cy + exy, cx + dy + ey^2)$.
For higher n, the general picture is this: call $Z \in B^n_O$ curvilinear, if Z is contained in a germ of a smooth curve, or equivalently, if $\dim_{\mathbb{C}} T_O Z = 1$. Choosing local coordinates in such a way that the line $\{y = 0\}$ is the tangent line to Z at the origin O, we may describe Z by an ideal $I = (y + \alpha_2 x^2 + \cdots + \alpha_{n-1} x^{n-1}, x^n$. The set $B^{n,\mathrm{curv}}_O$ of such I is a bundle over $\mathbb{P}^1$ with fibres $\cong \mathbb{C}^{n-1}$. It is, however, not a vector bundle, since the transition functions are not linear. In any case, $B^{n,\mathrm{curv}}_O$ is irreducible of dimension $n - 1$. By a result of Briançon [1], $B^{n,\mathrm{curv}}_O$ is dense in B^n_O.

If $n \geq 2$ and $\dim(X) \geq 2$, the symmetric product $S^n X$ is singular along the locus of tuples that correspond to non-distinct points. On the other hand there is the following miraculous fact:

THEOREM 2.1 (Fogarty [**10**]). *If X is a smooth surface then for all $n \in \mathbb{N}_0$, the Hilbert scheme* $\mathrm{Hilb}^n(X)$ *is smooth. In particular, ρ:* $\mathrm{Hilb}^n(X) \to S^n(X)$ *is a resolution of the singularities of the symmetric product.*

This fact is at the base of all the interesting features of the Hilbert scheme of surfaces. For higher dimensional varieties the Hilbert scheme is not only not smooth but in general even more singular than the symmetric product. We will prove Fogarty's theorem in the next section.

3. The geometry of Hilbert schemes

3.1. Families of subschemes and the moduli functor. Let X denote a quasiprojective scheme over $\mathbb{C}$. By definition, a flat family of proper subschemes in X parametrized by S is a closed subscheme $Z \subset S \times X$ such that the projection $Z \to S$ is flat and proper. If $s \in S$ is a closed point, we denote the fibre of Z

over s by Z_s. Given such a flat family and a morphism $f\colon S' \to S$, the family $Z' := (f \times \mathrm{id}_X)^{-1}(Z) \subset S' \times X$ is again flat and proper over S'. In this way, we obtain a functor

$$\underline{\mathrm{Hilb}}(X)\colon (\mathrm{Schemes})^{\mathrm{op}} \to (\mathrm{Sets}) \tag{3.1}$$

that associates to S the set of all flat families of proper subschemes in X that are parameterized by S. Let H be an ample Cartier divisor on X. For any proper subscheme $Z \subset X$ the Hilbert polynomial of Z is defined by $P_Z(n) := \chi(\mathcal{O}_Z \otimes \mathcal{O}_X(nH))$. For a flat family $Z \subset S \times X$ the function $S \ni s \mapsto P_{Z_s} \in \mathbb{Q}[T]$ is locally constant. This implies that for a given polynomial P, the functor

$$\underline{\mathrm{Hilb}}^P(X)\colon S \mapsto \{Z \subset S \times X \mid Z \text{ proper and flat}/S, P(Z_s) = P \text{ for all } s \in S\} \tag{3.2}$$

is an open and closed subfunctor of $\underline{\mathrm{Hilb}}(X)$. The dimension of a subscheme can be read off from the degree of its Hilbert polynomial. The functor $\underline{\mathrm{Hilb}}^n(X)$ associated to the constant polynomial $P = n$ parameterizes all zero-dimensional subschemes of length n.

THEOREM 3.1 (Grothendieck [**19**]). *The functor* $\underline{\mathrm{Hilb}}^P(X)$ *is represented by a quasiprojective scheme* $\mathrm{Hilb}^P(X)$. *If X is projective then* $\mathrm{Hilb}^P(X)$ *is also projective.*

The theorem then implies that there exists a universal subscheme

$$\Xi_P \subset \mathrm{Hilb}^P(X) \times X, \tag{3.3}$$

flat over $\mathrm{Hilb}^P(X)$, such that for any $Z \in \underline{\mathrm{Hilb}}^P(X)(S)$ there is a unique classifying morphism $f\colon S \to \mathrm{Hilb}^P(X)$ such that $Z \cong (f \times \mathrm{id}_X)^*(\Xi_P)$.

In the following we will only need the case $P = n = \text{const}$, and the word Hilbert scheme will always refer to Hilbert schemes of points. Note that $\mathrm{Hilb}^n(X)$ does not depend on the choice of the ample line bundle H.

Whenever a scheme or variety represents a functor one gets an intrinsic description of the Zariski tangent space at closed points. For the Hilbert scheme this takes the following form:

THEOREM 3.2 (Grothendieck). *Let* $[Z] \in \mathrm{Hilb}^n(X)$ *be a closed point representing a subscheme* $Z \subset X$ *with ideal sheaf* I_Z *and structure sheaf* $\mathcal{O}_Z$. *Then there is a canonical isomorphism*

$$T_{[Z]}\mathrm{Hilb}^n(X) \cong \mathrm{Hom}_{\mathcal{O}_X}(I_Z, \mathcal{O}_Z). \tag{3.4}$$

PROOF. Let $\mathbb{C}[\varepsilon] = \mathbb{C}[t]/(t^2)$ denote the ring of dual numbers. A tangent vector at $[Z]\colon \mathrm{Spec}\,\mathbb{C} \to \mathrm{Hilb}^n(X)$ corresponds to a morphism $\tau\colon \mathrm{Spec}\,\mathbb{C}[\varepsilon] \to \mathrm{Hilb}^n(X)$ that restricts to $[Z]$ at the closed point. In the modular interpretation τ corresponds to an ideal $\tilde{I} \subset \mathcal{O}_X[\varepsilon]$ such that the quotient sheaf $\tilde{\mathcal{O}} = \mathcal{O}_X[\varepsilon]/\tilde{I}$ is flat over $\mathbb{C}[\varepsilon]$ and restricts to $\mathcal{O}_Z$ at the closed point of $\mathrm{Spec}\,\mathbb{C}[\varepsilon]$. We obtain a commutative diagram with exact columns and rows:

$$
\begin{array}{ccccccccc}
 & & 0 & & 0 & & 0 & & \\
 & & \uparrow & & \uparrow & & \uparrow & & \\
0 & \longrightarrow & I_Z & \longrightarrow & \mathcal{O}_X & \longrightarrow & \mathcal{O}_Z & \longrightarrow & 0 \\
 & & \uparrow & & \uparrow & & \uparrow & & \\
0 & \longrightarrow & \tilde{I} & \longrightarrow & \mathcal{O}_X[\varepsilon] & \longrightarrow & \widetilde{\mathcal{O}} & \longrightarrow & 0 \\
 & & \uparrow & & \uparrow{\scriptstyle \cdot\varepsilon} & & \uparrow & & \\
0 & \longrightarrow & I_Z & \longrightarrow & \mathcal{O}_X & \longrightarrow & \mathcal{O}_Z & \longrightarrow & 0 \\
 & & \uparrow & & \uparrow & & \uparrow & & \\
 & & 0 & & 0 & & 0 & &
\end{array}
\tag{3.5}
$$

That the last row of the diagram is isomorphic to the first row is equivalent to the flatness of $\widetilde{\mathcal{O}}$ and $\tilde{I}$. Thus giving a tangent vector amounts to finding an ideal $\tilde{I}$ that fits into the diagram above. As $\tilde{I}$ always contains εI_Z, it is completely determined by the embedding of the quotient $\tilde{I}/I_Z \cong I_Z$ into $\mathcal{O}_X[\varepsilon]/\varepsilon I_Z = \mathcal{O}_X \oplus \varepsilon\mathcal{O}_Z$. Of the two components of this inclusion, the first is the inclusion $I_Z \to \mathcal{O}_X$, the second a homomorphism $t\colon I_Z \to \mathcal{O}_Z$. In this way, $\tilde{I}$ determines an element $t \in \operatorname{Hom}(I_Z, \mathcal{O}_Z)$, and, conversely, any such t defines an ideal $\tilde{I}$ and hence a tangent vector τ. I leave it to the reader to check that the correspondence $\tau \leftrightarrow t$ is indeed linear. □

THEOREM 3.3 (Fogarty [**10**]). *Let X be a smooth connected quasiprojective surface. Then for each $n \in \mathbb{N}_0$, the Hilbert scheme $\operatorname{Hilb}^n(X)$ is connected and smooth of dimension $2n$.*

PROOF. Let $[Z] \in \operatorname{Hilb}^n(X)$ be a closed point. We need to compute the dimension of $T_{[Z]}\operatorname{Hilb}^n(X) \cong \operatorname{Hom}(I_Z, \mathcal{O}_Z) = \operatorname{Hom}(I_Z/I_Z^2, \mathcal{O}_Z)$. Apparently, this is a completely local datum. We may therefore assume for simplicity that X is projective. From the long exact Ext-sequence associated to

$$0 \to I_Z \to \mathcal{O}_X \to \mathcal{O}_Z \to 0 \tag{3.6}$$

we read off that $\operatorname{Hom}(I_Z, \mathcal{O}_Z) \cong \operatorname{Ext}^1(\mathcal{O}_Z, \mathcal{O}_Z)$. Now

$$\hom(\mathcal{O}_Z, \mathcal{O}_Z) = h^0(\mathcal{O}_Z) = n, \tag{3.7}$$

and, by Serre duality,

$$\operatorname{ext}^2(\mathcal{O}_Z, \mathcal{O}_Z) = \hom(\mathcal{O}_Z, \mathcal{O}_Z \otimes \omega_X) = h^0(\mathcal{O}_Z \otimes \omega_X) = n. \tag{3.8}$$

The theorem of Hirzebruch–Riemann–Roch yields

$$n - \operatorname{ext}^1(\mathcal{O}_Z, \mathcal{O}_Z) + n = \int_X \operatorname{ch}(\mathcal{O}_Z)\operatorname{ch}(\mathcal{O}_Z)^\vee \operatorname{td}(X), \tag{3.9}$$

where $^\vee$ is the automorphism on H^{even} defined as $(-1)^i$ on H^{2i}. The integral is zero for dimension reasons, because Z has codimension 2. We conclude that $\dim_{\mathbb{C}} T_{[Z]}\operatorname{Hilb}^n(X) = 2n$ for all closed points $[Z]$.

Observe that $\operatorname{Hilb}^n(X)$ always contains the smooth $2n$-dimensional configuration space of unordered n-tuples of pairwise disjoint points as an open subscheme. Thus in order to see that $\operatorname{Hilb}^n(X)$ is smooth of dimension $2n$ it suffices to show that $\mathrm{Hilb}^n(X)$ is connected. We will later give a proof (see Lemma 3.7) by a method different from Fogarty's. □

Fogarty's theorem is the basis of all the nice properties of the Hilbert scheme of points on surfaces. For higher-dimensional varieties much less is true:

COROLLARY 3.4. *Let X be a smooth quasiprojective variety of dimension d and let $[Z] \in \mathrm{Hilb}^n(X)$ be a closed point such that $\dim T_x Z \le 2$ for all $x \in Z$. Then $\mathrm{Hilb}^n(X)$ is smooth of dimension dn at $[Z]$. In particular, $\mathrm{Hilb}^n(X)$ is smooth for all d if $n \le 3$.*

PROOF. It suffices to consider the case that Z is supported in a single point $x \in X$. Let $d' := \dim T_x Z \le d$. We may choose local coordinates $z_1, \dots, z_d$ near x such that $z_{d'+1}, \dots, z_d$ are contained in the ideal I of Z. Let $Y \subset X$ be the subvariety defined by $J := (z_{d'+1}, \dots, z_d)$. Then Y is smooth at x and contains Z. The ideal of Z in Y is $\bar{I} := I/J$. There is an isomorphism of $\mathcal{O}_Z$-modules $I/I^2 \cong \mathbb{C}\langle \bar{z}_{d'+1}, \dots, \bar{z}_d\rangle \otimes \mathcal{O}_Z \oplus \bar{I}/\bar{I}^2$. Therefore,

$$T_{[Z]} \mathrm{Hilb}^n(X) = \mathrm{Hom}_{\mathbb{C}}(\mathbb{C}\langle \bar{z}_{d'+1}, \dots, \bar{z}_d\rangle, \mathcal{O}_Z) \oplus T_{[Z]} \mathrm{Hilb}^n(Y). \tag{3.10}$$

By assumption, $d' \le 2$, hence $\mathrm{Hilb}^n(Y)$ is smooth at $[Z]$ of dimension $d'n$. We obtain: $\dim T_{[Z]} \mathrm{Hilb}^n(X) = (d - d') \cdot \ell(\mathcal{O}_Z) + d'n = dn$. This proves the first assertion. In general, one always has $d' \le n - 1$. Hence the assumptions for the first claim are always satisfied if $n \le 3$. This proves the second claim. □

REMARK 3.5. $\mathrm{Hilb}^4(\mathbb{C}^3)$ is singular at $[Z] = V(\mathfrak{m}^2)$, where $\mathfrak{m} = (x, y, z) \subset \mathbb{C}[x, y, z]$. For

$$\begin{aligned} T_{[Z]} \mathrm{Hilb}^n(\mathbb{C}^3) &= \mathrm{Hom}_{\mathbb{C}[x,y,z]}(\mathfrak{m}^2/\mathfrak{m}^4, \mathbb{C}[x,y,z]/\mathfrak{m}^2) \\ &= \mathrm{Hom}_{\mathbb{C}}(\mathfrak{m}^2/\mathfrak{m}^3, \mathfrak{m}/\mathfrak{m}^2) \end{aligned} \tag{3.11}$$

has dimension 18, whereas $\mathrm{Hilb}^4(\mathbb{C}^3)$ is twelve-dimensional.

Let X be a smooth projective surface. The projection $p\colon \Xi_n \to \mathrm{Hilb}^n(X)$ is flat and finite of degree n. This implies that $\mathcal{A} := p_* \mathcal{O}_{\Xi_n}$ is a locally free sheaf of rank n. Moreover, $\mathcal{A}$ carries an algebra structure, and $\Xi_n \cong \underline{\mathrm{Spec}}(\mathcal{A})$.

Let $\mathcal{A} \to \mathcal{O}_{\mathrm{Hilb}^n(X)}$ denote the trace map for the action of $\mathcal{A}$ on itself. The composition with the multiplication $\mathcal{A} \otimes \mathcal{A} \to \mathcal{A}$ is a quadratic form $\mathcal{A} \otimes \mathcal{A} \to \mathcal{O}_{\mathrm{Hilb}^n(X)}$. Its discriminant vanishes in $[\xi]$ if and only if $\mathcal{A}([\xi]) = H^0(\mathcal{O}_\xi)$ is not semisimple, i.e., if ξ has multiplicity ≥ 2 at some of its points. The set

$$\partial \mathrm{Hilb}^n(X) := \{[\xi] \in \mathrm{Hilb}^n(X) \mid \exists x \in X : \mathrm{length}(\mathcal{O}_{\xi,x}) \ge 2\}. \tag{3.12}$$

is called the boundary of the Hilbert scheme. It is the vanishing locus of the homomorphism $\mathcal{A} \to \mathcal{A}^*$. In particular, it is a divisor satisfying

$$[\partial \mathrm{Hilb}^n(X)] = -2c_1(\mathcal{A}). \tag{3.13}$$

3.2. The Hilbert–Chow morphism. Let X be a reduced quasiprojective variety. To any closed point $[Z] \in \mathrm{Hilb}^n(X)$ we may associate a formal sum $\sum_{x \in X} \ell(\mathcal{O}_{Z,x}) \cdot x$, i.e., closed point in the symmetric product $S^n(X)$. This defines the Hilbert–Chow map

$$\rho\colon \mathrm{Hilb}^n(X) \to S^n(X), \tag{3.14}$$

at least set-theoretically. That ρ is indeed a morphism can be seen as follows (cf. [**22**]): suppose $Z \in \mathrm{Hilb}^n(X)(S)$, and let $s_0 \in S$ be a closed point corresponding to a subscheme $Z_0 \subset X$. Since X is quasiprojective, there is an open affine subscheme $U = \operatorname{Spec} A \subset X$ that contains Z_0. As the projection $p\colon Z \to S$ is proper, $p(S \times (X \setminus U) \cap Z)$ is a closed subset in S. Its complement is an open subset V' with the property that $Z_s \subset U$ for all $s \in V'$. Let $V = \operatorname{Spec} B$ be an open affine neighborhood of s_0 in V'. Now $Z_V := Z \cap V \times U$ is affine, say $Z_V \cong \operatorname{Spec} C$, where C is a factor ring of $B \otimes A$. Moreover, C is a projective B-module of rank n, since $p\colon Z_V \to V$ is flat and finite of degree n. The natural surjection $B \otimes A \to C$ induces a ring homomorphism $f\colon A \to \mathrm{End}_B(C)$. There is an induced action of $A^{\otimes n}$ on $C^{\otimes n}$, the tensor products being taken over $\mathbb{C}$ and B, respectively. The subring of invariant tensor $S_n A := (A^{\otimes n})^{\mathrm{S}_n}$ acts on the rank 1 submodule $\Lambda_n C \subset C^{\otimes n}$ of anti-symmetric tensors. This yields a ring homomorphism $\varphi\colon S_n A \to \mathrm{End}_B(\Lambda_n C) = B$ and thus a morphism

$$V = \operatorname{Spec} B \to S^n U := \operatorname{Spec}\big((A^{\otimes n})^{\mathrm{S}_n}\big) \subset S^n X. \tag{3.15}$$

The construction is functorial in V and therefore yields a morphism

$$\rho\colon \mathrm{Hilb}^n(X) \to S^n X. \tag{3.16}$$

Explicitly the ring homomorphism $\varphi\colon S_n A \to B$ can be described as follows: Let $\mathrm{ld}\colon \mathrm{End}_B(C)^{\otimes n} \to B$ be the polar form of the determinant, i.e.,

$$\mathrm{ld}(e_1, \dots, e_n) := \frac{1}{n!} \operatorname{coeff}\big(t_1 \cdots t_n, \det(t_1 e_1 + \cdots + t_n e_n)\big), \tag{3.17}$$

Then $\varphi(a_1 \otimes \cdots \otimes a_n) = \mathrm{ld}\big(f(a_1), \dots, f(a_n)\big)$. We must show that for a $\mathbb{C}$-valued point $[Z]$ of $\mathrm{Hilb}^n(X)$ one has $\rho([Z]) = \sum_{x \in X} \ell(\mathcal{O}_{Z,x}) \cdot x$. The situation above simplifies: $B = \mathbb{C}$ and C is an Artinian factor algebra of A of length n. The ring homomorphism φ factors through $S_n A \to S_n C$ and a homomorphism $S_n C \to \mathrm{End}_{\mathbb{C}}(\Lambda_n C)$. Decompose $C = C_1 \times \cdots \times C_s$ with local Artinian rings C_i of length n_i respectively. Then $\Lambda_n C \cong \bigotimes_i \Lambda_{n_i} C_i$ and $S_n C = \prod_\ell \bigotimes_i S_{\ell_i} C_i$, where the product runs over decompositions $n = \ell_1 + \cdots + \ell_s$. All of these factors act trivially on $\Lambda_n C$ except the one corresponding to the decomposition $\ell_i = n_i$. Thus φ factors as follows:

$$S_n A \to S_n C \to \bigotimes_i S_{n_i} C_i \to \mathrm{End}_{\mathbb{C}}(\Lambda_n C) = \mathbb{C}. \tag{3.18}$$

This shows that the closed point $x_i = (\operatorname{Spec} C_i)_{\mathrm{red}}$ appears in $\rho([Z])$ with multiplicity n_i.

Definition 3.6. Let X be a smooth quasiprojective surface, and let $\Delta := \{nx \mid x \in X\} \subset S^n X$ denote the small diagonal. Let $B^n := \rho^{-1}(\Delta)$ and $B^n_p := \rho^{-1}(np)$ for some $p \in X$, both subsets equipped with the reduced induced subscheme structure.

We will refer to B^n_p as the nth Briançon variety. Any étale morphism $f\colon X \to X'$ induces an isomorphism $B^n_p \to B^n_{f(p)}$. Thus all Briançon varieties are non-canonically isomorphic, regardless which surface X or which point $p \in X$ one considers. In fact one can show that $B \to \Delta$ is a fibre bundle in the Zariski topology.

3.3. An induction scheme. A basic tool to study Hilbert schemes of points on a quasiprojective scheme X is an induction procedure that allows one to compare the geometric properties of $\mathrm{Hilb}^n(X)$ and $\mathrm{Hilb}^{n+1}(X)$.

Let $Z \subset \mathrm{Hilb}^n(X) \times \mathrm{Hilb}^{n+1}(X)$ denote the subscheme whose closed points consist of pairs (ξ, ξ') such that ξ is a subscheme of ξ'. The scheme structure of Z is defined by the vanishing of the homomorphism

$$\mathrm{pr}_{23}^*(I_{\Xi_{n+1}}) \to \mathcal{O}_{\mathrm{Hilb}^n(X)\times\mathrm{Hilb}^{n+1}(X)\times X} \to \mathrm{pr}_{13}^*(\mathcal{O}_{\Xi_n}). \tag{3.19}$$

The subscheme Z comes with two projections:

$$\mathrm{Hilb}^n(X) \xleftarrow{\varphi} Z \xrightarrow{\psi} \mathrm{Hilb}^{n+1}(X). \tag{3.20}$$

In order to make use of this diagram we need a more specific description of φ and ψ. We follow here a method due to Ellingsrud and Strømme [**8**].

Suppose that $\xi \subset \xi' \subset X$ are subschemes of length n and $n+1$, respectively. Then there are short exact sequences

$$0 \to \mathcal{O}_x \to \mathcal{O}_{\xi'} \to \mathcal{O}_\xi \to 0 \tag{3.21}$$

and

$$0 \to I_{\xi'} \to I_\xi \to \mathcal{O}_x \to 0 \tag{3.22}$$

where $\mathcal{O}_x$ denotes the structure sheaf of the reduced point $x \in X$, the point where ξ and ξ' differ. It is apparent from (3.22) that we may pass from ξ to ξ' by choosing a point $x \in X$ and a surjective homomorphism $\lambda\colon I_\xi(x) = I_\xi/\mathfrak{m}_x I_\xi \to \mathbb{C}$. Any ξ' can be obtained in this way, and ξ' determines λ up to a scalar. More formally, let $\Phi = (\phi, \rho)\colon \mathbb{P} := \mathbb{P}(I_{\Xi_n}) \to \mathrm{Hilb}^n(X) \times X$ denote the projectivization of I_{Ξ_n}. We will construct a flat family of subscheme in X of length $n+1$ and parameterized by $\mathbb{P}$ as follows: There is a canonical surjection

$$\alpha\colon \Phi^*(I_{\Xi_n}) \to \mathcal{O}_{\mathbb{P}}(1). \tag{3.23}$$

Let $Y \subset \mathbb{P} \times X$ denote the preimage of the diagonal $\Delta_X \subset X \times X$ under the morphism $\rho \times \mathrm{id}_X : \mathbb{P} \times X \to X$. Consider the surjective homomorphism of sheaves on $\mathbb{P} \times X$:

$$\beta\colon (\phi\times\mathrm{id}_X)^* I_{\Xi_n} \to (\phi\times\mathrm{id}_X)^* I_{\Xi_n}|_Y \cong (\mathrm{pr}_1^*\,\Phi^* I_{\Xi_n})|_Y \xrightarrow{\mathrm{pr}_1^*\,\alpha} (\mathrm{pr}_1^*\,\mathcal{O}_{\mathbb{P}}(1))|_Y. \tag{3.24}$$

Its kernel is the ideal sheaf of a family of subschemes of length $n+1$. Hence there is classifying morphism $\psi : \mathbb{P} \to \mathrm{Hilb}^{n+1}(X)$ such that $\ker(\beta) \cong (\psi \times \mathrm{id}_X)^* I_{\Xi_{n+1}}$. We obtain a short exact sequence

$$0 \to (\psi \times \mathrm{id}_X)^* I_{\Xi_{n+1}} \to (\phi \times \mathrm{id}_X)^* I_{\Xi_n} \to \mathrm{pr}_1^*\,\mathcal{O}_{\mathbb{P}}(1)|_Y \to 0 \tag{3.25}$$

of sheaves on $\mathbb{P} \times X$. This is the global version of sequence (3.22). The morphism $(\phi, \psi)\colon \mathbb{P} \to \mathrm{Hilb}^n(X) \times \mathrm{Hilb}^{n+1}(X)$ is an isomorphism onto the subscheme Z.

We also note that $\mathbb{P}$ comes with a map $\rho\colon \mathbb{P} \to X$ that sends $\xi, \xi')$ to the point x, where ξ is modified. The induction scheme takes the final form

$$\begin{array}{ccc} \mathrm{Hilb}^n(X) & \xleftarrow{\ \phi\ } \mathbb{P} \xrightarrow{\ \psi\ } & \mathrm{Hilb}^{n+1}(X) \\ & \big\downarrow{\scriptstyle\rho} & \\ & X & \end{array} \tag{3.26}$$

Here is a first application of this construction. The following lemma fills the remaining gap in the proof of Fogarty's Theorem 3.3.

LEMMA 3.7. *Let X be a connected quasiprojective scheme. Then* $\mathrm{Hilb}^n(X)$ *is connected for all* $n \geq 0$.

PROOF. $\mathrm{Hilb}^0(X)$ is a point, $\mathrm{Hilb}^1(X) = X_{\mathrm{red}}$ is connected. By induction we may assume that $\mathrm{Hilb}^n(X)$ is known to be connected. The fibres of the map $\Phi := (\phi, \rho)\colon \mathbb{P} \to \mathrm{Hilb}^n(X) \times X$ are projective spaces and hence connected. Thus $\mathbb{P}$ is connected. As ψ is surjective, $\mathrm{Hilb}^{n+1}(X)$ is connected as well. □

The symmetry suggested by diagram (3.26) is not quite apparent from the construction of $\mathbb{P}$ given above. However, if X is a smooth surface the following holds: The dualizing sheaf $\omega_\xi := \mathcal{E}xt^2_{\mathcal{O}_X}(\mathcal{O}_\xi, \omega_X) = \mathcal{E}xt^1(I_\xi, \omega)$ of ξ is an $\mathcal{O}_\xi$-sheaf of length n. Applying $\mathcal{H}om(-, \omega_X)$ to sequence (3.21) we obtain an exact sequence

$$0 \to \omega_\xi \to \omega_{\xi'} \to \mathcal{O}_x \to 0. \tag{3.27}$$

Thus to pass from ξ' to ξ, we have to choose a closed point x in the support of ξ' and a surjective homomorphism $\mu : \omega_{\xi'}(x) \to \mathbb{C}$. This defines a point $[\mu] \in \mathbb{P}(\omega_{\xi'}(x))$. Now we can argue as above and show that the morphism $\Psi\colon (\psi, \rho)\colon \mathbb{P} \to \mathrm{Hilb}^{n+1}(X) \times X$ is isomorphic to the canonical projection $\mathbb{P}(\omega_{\Xi_{n+1}}) \to \mathrm{Hilb}^{n+1}(X) \times X$, where $\omega_{\Xi_{n+1}} = \mathcal{E}xt^2(\mathcal{O}_{\Xi_{n+1}}, \omega_X)$ is the relative dualizing sheaf of the universal family.

From now on let X denote a smooth irreducible projective surface. Let $\mathbb{P}'$ denote the blow-up of $\mathrm{Hilb}^n(X) \times X$ along the subscheme Ξ_n, and let $E \subset \mathbb{P}'$ be the exceptional divisor. The ring epimorphism $S^*(I_{\Xi_n}) \twoheadrightarrow \bigoplus_{\nu \geq 0} I^\nu_{\Xi_n}$ induces a closed embedding $\mathbb{P}' \to \mathbb{P}$.

PROPOSITION 3.8. $\mathbb{P}$ *is irreducible and hence isomorphic to* $\mathbb{P}'$.

PROOF. Let H be an ample divisor on X. For sufficiently large m, the sheaf $B := p^*\big(p_*(\mathcal{O}_{\Xi_n}(mH))\big)(-mH)$ is locally free, and the evaluation map $B \to I_{\Xi_n}$ is surjective. Let A denote its kernel. Since I_{Ξ_n} is flat over $\mathrm{Hilb}^n(X)$, the same is true for A and B. Hence for each $[\xi] \in \mathrm{Hilb}^n(X)$, the sequence

$$0 \to A|_{\{\xi\} \times X} \to B|_{\{\xi\} \times X} \to I_\xi \to 0 \tag{3.28}$$

is still exact. Since X is smooth of dimension 2, the global dimension of $\mathcal{O}_\xi$ is ≤ 2 and that of I_ξ is ≤ 1. It follows that $A|_{\{\xi\} \times X}$ is locally free. By flatness, A is locally free, too. We obtain a global resolution

$$0 \to A \to B \to I_{\Xi_n} \to 0 \tag{3.29}$$

with locally free sheaves A and B of rank a and $b = a + 1$, respectively. The epimorphism $B \to I_{\Xi_n}$ induces a closed embedding $\mathbb{P} \to \mathbb{P}(B)$, and the image is the zero set of the composite map $\Phi^*A \to \Phi^*B \to \mathcal{O}_B(1)$. Thus $\mathbb{P}$ is locally cut out from the smooth variety $\mathbb{P}(B)$ by a equations. It follows that every irreducible component of $\mathbb{P}$ has dimension $\geq \dim(\mathrm{Hilb}^n(X) \times X) + \mathrm{rk}(B) - 1 - \mathrm{rk}(A) = 2n + 2$.

Let $p \in \mathbb{P}$ and let $[\xi] = \phi(p)$, $[\xi'] = \psi(p)$, $x = \rho(p)$ denote the images of p in $\mathrm{Hilb}^n(X)$, $\mathrm{Hilb}^{n+1}(X)$ and X, respectively. The corresponding ideal sheaves, structure sheaves and dualizing sheaves are related by the sequences (3.21), (3.22) and (3.27). The invariants $i := i(\xi, x) := \dim I_\xi(x)$, $i' := i(\xi', x)$ and $j' := j(\xi', x) := \dim \omega_{\xi'}(x)$ are related by

$$|i - i'| \leq 1, \quad j' = i' - 1. \tag{3.30}$$

This can be seen as follows: Tensoring (3.22) by $\mathcal{O}_x$, we obtain

$$\mathbb{C}^2 = \mathrm{Tor}_1(\mathcal{O}_x, \mathcal{O}_x) \to I_{\xi'}(x) \to I_\xi(x) \to \mathbb{C} \to 0 \tag{3.31}$$

and conclude that $-1 \leq i - i' \leq 1$. To prove the second assertion, choose a locally free resolution $0 \to A' \to B' \to I_{\xi'} \to 0$. Then $I_{\xi'}(x) = \mathrm{coker}\big(A'(x) \xrightarrow{m} B'(x)\big)$ and $\omega_{\xi'}(x) = \mathrm{coker}(B'(X)^* \xrightarrow{m^*} A'(X)^*)$. It follows that

$$j' = \mathrm{rk}(A') - \mathrm{rk}(m^*) = (\mathrm{rk}(B') - 1) - \mathrm{rk}(m) = i' - 1. \tag{3.32}$$

We stratify $\mathrm{Hilb}^n(X) \times X$ according to the fibre dimension of Φ:

$$\mathrm{Hilb}^n(X) \times X = \coprod_{i \geq 1} H_{n,i}, \quad \text{with } H_{n,i} := \{(\xi, x) \mid i(\xi, x) = i\} \tag{3.33}$$

Then $\Phi^{-1}(\xi, x) \cong \mathbb{P}^{i-1}$ for $(\xi, x) \in H_{n,i}$, and similarly, $\Psi^{-1}(\xi', x) \cong \mathbb{P}^{j'-1} = \mathbb{P}^{i'-2}$ for $(\xi', x) \in H_{n+1,i'}$. Moreover, because of $|i - i'| \leq 1$, the inclusions

$$\Psi^{-1}(H_{n+1,i'}) \subset \bigcup_{|i-i'|\leq 1} \Phi^{-1}(H_{n,i}) \tag{3.34}$$

hold. This gives the dimension estimate

$$\dim H_{n+1,i'} + (i' - 2) \leq \max\{\dim H_{n,i} + (i - 1) \mid |i - i'| \leq 1\}. \tag{3.35}$$

Using this estimate one can prove the following bound by induction on n:

$$\dim H_{n,i} \leq 2n - 2(i - 2) \quad \text{for all } i \geq 1 \tag{3.36}$$

The assertion is obvious for $i = 1$ and is also true for $i = 2$, since $H_{n,2} \subset \Xi_n$, the latter projecting finitely onto $\mathrm{Hilb}^n(X)$. Assume by induction that the assertion is true for n and consider $H_{n+1,i'}$. It follows from the inequality (3.35) that

$$\begin{aligned} \dim H_{n+1,i'} &\leq (i' - 2) + \max\{2n - 2(i - 2) + (i - 1) \mid -1 \leq i - i' \leq 1\} \\ &= 2(n + 1) - 2(i' - 2). \end{aligned} \tag{3.37}$$

This finishes the induction.

Using (3.36), we obtain $\Phi^{-1}(H_{n,i}) \leq 2n + 2 - (i - 1) < 2n + 2$ for $i \geq 2$. Since any irreducible component of $\mathbb{P}$ must have dimension $\geq 2n + 2$, $\mathbb{P}$ must be contained in the closure of $\Phi^{-1}(H_{n,1})$. This implies that $\mathbb{P}$ is irreducible and therefore isomorphic to the blow-up $\mathbb{P}'$. □

PROPOSITION 3.9. *The divisor $\partial\,\mathrm{Hilb}^n(X) \subset \mathrm{Hilb}^n(X)$ is irreducible.*

PROOF. We will show that $\partial\,\mathrm{Hilb}^n(X)$ is irreducible by induction on n. Clearly, $\partial\,\mathrm{Hilb}^1(X)$ is empty, and $\partial\,\mathrm{Hilb}^2(X) \cong X$. We will make use of the notation and construction introduced in the proof of Proposition 3.8: the exceptional divisor $E \subset \mathbb{P}' = \mathbb{P}$ equals $\bigcup_{i\geq 2} \Phi^{-1}(H_{n,i})$. Every component of E has dimension $\geq 2n+1$. Since $\dim \Phi^{-1}(H_{n,i}) < 2n + 1$ for $i \geq 3$, it follows that $\Phi^{-1}(H_{n,2})$ is dense in E. Since the map $H_{n,2} \mapsto \partial\mathrm{Hilb}^n(X)$ is birational, we may assume by induction that E is irreducible. Since $\partial\,\mathrm{Hilb}^n(X) = \psi(E)$ this proves the claim. □

THEOREM 3.10 (Briançon [1]). *Let X be a smooth projective surface and let $p \in X$ be a closed point. The varieties B_p^n are irreducible and of dimension $n-1$. In particular, they contain the set of curvilinear subschemes as an open dense subset.*

PROOF. We proceed by induction and assume that the theorem holds for n. We may assume that X is irreducible and continue to use the construction and notation introduced above. The stratification (3.33) of $\mathrm{Hilb}^n(X) \times X$ induces a stratification

$$B_p^n \times \{p\} = \coprod_{i \geq 2} B_{n,i}, \quad \text{with } B_{n,i} = B_p^n \times \{p\} \cap H_{n,i}. \tag{3.38}$$

Now $\Phi^{-1}(B_p^n \times \{p\}) = E \cap \Phi^{-1}(B_p^n \times X) =: E'$ is a Cartier divisor in the blow-up of $B_p^n \times X$ along $\Xi_n \cap B_p^n \times X$. Since the blow-up has dimension $n+1$, every irreducible component of E' has dimension $\geq n$. We will prove the following estimate by induction:

$$\dim B_{n,i} \leq (n-1) - (i-1) \quad \text{for } i \geq 3. \tag{3.39}$$

Assuming this estimate for n, it follows that $\dim \Phi^{-1}(B_{n,i}) \leq (n-i)+(i-1) = n-1$ for all $i \geq 3$. This shows that E' is contained in the closure of $\Phi^{-1}(B_{n,2})$ and hence irreducible. It follows that $B_p^{n+1} = \psi(E')$ is irreducible, too. Moreover, since $\psi\colon E' \to B_p^{n+1}$ is generically an isomorphism, $\dim(B_p^{n+1}) \leq \dim(B_p^n) + 1 = n$.

It remains to prove (3.39). We argue as before:

$$\dim B_{n+1,i'} + (i'-2) \leq \max\{\dim B_{n,i} + (i-1) \mid |i-i'| \leq 1\}. \tag{3.40}$$

For $i' \geq 4$, we have always $i \geq 3$ and conclude by induction that $\dim B_{n+1,i'} \leq (n+1) - i'$. If $i' = 3$, we cannot exclude that the case $i = 2$ appears. However, $\Psi^{-1}(B_{n+1,2}) \cap \Phi^{-1}(B_{n,2})$ must be a proper subset in $\Phi^{-1}(B_{n,2})$ and hence have dimension $< n$. We get the desired estimate for $i' = 3$ as well. □

4. The cohomology of $\mathrm{Hilb}^n(X)$

Let X be a smooth irreducible projective surface. By Fogarty's theorem, the Hilbert scheme $\mathrm{Hilb}^n(X)$ is a compact complex manifold of dimension $2n$. This section deals with the computation of the cohomology of the Hilbert scheme and, most important, the additional structure as a module over a certain infinite-dimensional Lie algebra as defined by Nakajima. In the following, all cohomology groups of varieties refer to cohomology with coefficients in $\mathbb{Q}$ (or any other field of characteristic 0).

4.1. Göttsche's formula. Since $\mathrm{Hilb}^n(X)$ is a manifold of complex dimension $2n$, its rational cohomology groups range from degree 0 to degree $4n$. The Betti numbers were computed by Göttsche [**15**] who gave the following formula that expresses the Betti numbers of $\mathrm{Hilb}^n(X)$ in terms of the Betti numbers of X:

THEOREM 4.1 (Göttsche [**15**]).

$$\sum_{n=0}^{\infty} \sum_{i=0}^{4n} b_i(\mathrm{Hilb}^n(X)) p^{i-2n} q^n = \prod_{m=1}^{\infty} \prod_{j=0}^{4} (1-(-1)^j p^{j-2} q^m)^{-(-1)^j b_j(X)}. \tag{4.1}$$

Here t and q are formal parameters. One of the consequences of this formula is that the Betti numbers stabilize, i.e., for fixed i and $n \to \infty$, $b_i(\mathrm{Hilb}^n(X))$ becomes constant.

It also becomes clear that one should not expect to get nice formulas for a single Hilbert scheme, but that one should rather look at all Hilbert schemes simultaneously. Let us introduce the double graded vector space

$$\mathbb{H} := \bigoplus_n \mathbb{H}^n = \bigoplus_{n,i} \mathbb{H}^{n,i}, \quad \mathbb{H}^{n,i} := H^{i-2n}(\mathrm{Hilb}^n(X)). \tag{4.2}$$

The right-hand side in Göttsche's formula is also the Poincaré series of an irreducible module over the so-called Heisenberg Lie algebra. I will explain this in the next section. As a warm-up consider the following toy example.

EXAMPLE 4.2. Consider the Lie algebra $\mathfrak{g}_{\mathrm{even}}$ spanned by elements p, q, c that are subject to the relations

$$[p,q] = c, \quad [p,c] = 0, \quad [q,c] = 0, \tag{4.3}$$

so that c is a central element. $\mathfrak{g}_{\mathrm{even}}$ acts on the vector space $M := \mathbb{C}[x]$ via $p \mapsto \frac{\partial}{\partial x}$, $q \mapsto x\cdot$ and $c \mapsto \mathrm{id}$. The Leibniz rule for the differential implies that this is indeed a representation of $\mathfrak{g}$. Moreover, M is an irreducible representation: If $V \subset M$ is a nonzero submodule, and if $f = f_n x^n + \cdots + f_0 \in V$ a polynomial of degree n, then V also contains $1 = \frac{1}{n! f_n} p^n(f)$. Applying q sufficiently often, we see that V contains all powers of x and therefore coincides with M.

The module M is naturally graded, and its Poincaré series is

$$\sum_{n\geq 0} \dim M_n t^n = \sum_{n\geq 0} t^n = (1-t)^{-1}. \tag{4.4}$$

Next, we modify the example slightly: We now assume that p and q are odd elements. The commutator has to be redefined as $[p,q] = pq + qp$. We keep c even and take the same relations as before. (However, we have to add $[p,p] = 0 = [q,q]$, since these relations are no longer automatic.) The super Lie algebra $\mathfrak{g}_{\mathrm{odd}}$ spanned by p, q and c acts on the vector space underlying the exterior algebra $M = \mathbb{C} \oplus \mathbb{C}x$ of a one dimensional vector space $\mathbb{C}\langle x\rangle$ rather than the symmetric algebra as before. The action is given by $p \mapsto \partial/\partial x$, $q \mapsto x\wedge$ and $c \mapsto \mathrm{id}$. The Poincaré series of M is $(1+t)$.

Both Poincaré series can be uniformly written as $(1-\varepsilon t)^{-\varepsilon}$ with $\varepsilon = \pm 1$ according to whether p and q are even or odd elements.

If we now look at Göttsche's formula again, we realize that the factors on the right-hand side have a very similar structure as the Poincaré series in the example. But instead of one factor there are infinitely many: this indicates that one deals with an infinite tensor product of modules such as M. And instead of one formal parameter t there are two, p and q, indicating that the graded modules are in fact bigraded.

Nakajima constructed an action of an infinite dimensional Lie algebra on $\mathbb{H}$ that realizes all these features. We will first discuss the abstract algebraic construction of the Lie algebra and then Nakajima's geometric representation of it.

4.2. The Heisenberg Lie algebra and an irreducible representation. By definition, a super vector space is a vector space H together with a $\mathbb{Z}/2$-grading $H = H^{\mathrm{even}} \oplus H^{\mathrm{odd}}$. If H and H' are super vector spaces, $\mathrm{Hom}(H, H')$ and $H \otimes H'$ etc. inherit $\mathbb{Z}/2$-gradings in a natural way. The symmetric group S_n acts on $H^{\otimes n}$

by permuting the factors of a product of homogeneous elements with an additional sign:

$$\pi(h_1 \otimes \cdots \otimes h_n) = (-1)^{\varepsilon} h_{\pi^{-1}(1)} \otimes \cdots \otimes h_{\pi^{-1}(n)}, \tag{4.5}$$

where $\varepsilon = \sum_{i<j,\pi(i)>\pi(j)} |h_i| \cdot |h_j|$ and $|h_i| \in \mathbb{Z}/2$ denotes the parity of h_i. This affects the definition of the symmetric and the exterior algebras of a super vector space. As a consequence

$$S^*(H^{\text{even}} \oplus H^{\text{odd}}) = S^*(H^{\text{even}}) \otimes \Lambda^*(\Pi H^{\text{odd}}). \tag{4.6}$$

Here Π is the parity change functor: it makes H^{odd} even and H^{even} odd.

Let H be finite-dimensional super vector space over $\mathbb{Q}$ with a non-degenerate even symmetric bilinear form $\langle -,- \rangle \colon H \times H \to \mathbb{Q}$. Evenness of the form means that $H^{\text{even}} \perp H^{\text{odd}}$, and symmetry means that

$$\langle \alpha, \beta \rangle = (-1)^{|\alpha|\cdot|\beta|} \langle \beta, \alpha \rangle \tag{4.7}$$

for homogeneous elements of degree $|\alpha|$, $|\beta| \in \mathbb{Z}/2$.

A typical example for such a situation is the cohomology ring $H := H^*(X, \mathbb{Q})$ of a compact manifold X of even real dimension. The pairing in this example is the cup product pairing $\int_X \alpha \cup \beta$. The cup product itself is not needed for the moment.

To the datum $(H, \langle -,- \rangle)$ we associate a so-called Heisenberg Lie algebra as follows: Define a Lie bracket on

$$\mathfrak{h} := H[t, t^{-1}] \oplus \mathbb{Q}c \tag{4.8}$$

by $[c, u] := 0$ for all $u \in \mathfrak{h}$ and

$$[\alpha f(t), \beta g(t)] := \langle \alpha, \beta \rangle \operatorname{res}_t f\, dg \cdot c \tag{4.9}$$

for all α, $\beta \in H$, f, $g \in \mathbb{C}[t, t^{-1}]$. This bracket is super skew-symmetric, since the residue of $d(fg)$ vanishes so that the Leibniz rule implies that $\operatorname{res}_t f\, dg = -\operatorname{res}_t g\, df$. Since c is central by definition, the bracket satisfies the Jacobi identity.

It is convenient to introduce the variables $\alpha_n := \alpha\, t^n$ for $\alpha \in H$. Then the relations can be written as $[\alpha_n, c] = 0$ and

$$[\alpha_n, \beta_m] = n\, \delta_{n,-m} \langle \alpha, \beta \rangle\, c. \tag{4.10}$$

Note that here and elsewhere whenever we are concerned with possibly odd elements x and y, the Lie bracket of x and y will be super skew symmetric, i.e.,

$$[x, y] = -(-1)^{|x|\cdot|y|} [y, x]. \tag{4.11}$$

If we choose a homogeneous basis α^i of H with dual basis β^i, then each pair α^i_n, β^i_{-n} spans a Lie algebra isomorphic to either $\mathfrak{g}_{\text{even}}$ or $\mathfrak{g}_{\text{odd}}$ in Example 4.2 depending on whether α is an even or an odd element.

The next step consists in constructing a highest (or rather: lowest) weight representation of $\mathfrak{h}$. Let $I \subset U(\mathfrak{h})$ be the left ideal in the universal enveloping algebra of $\mathfrak{h}$ that is generated by all elements of the form α_n with $n \geq 0$, and the element $c - 1$.

Then the quotient $V := U(\mathfrak{h})/I$ is as a representation of $\mathfrak{h}$. It is generated as a representation by the residue class $\mathbf{1}$ of $1 \in U(\mathfrak{h})$. By construction, we have

$$\alpha_n \mathbf{1} = 0, \text{ for all } n \geq 0, \quad \text{and} \quad c.\mathbf{1} = \mathbf{1}. \tag{4.12}$$

We will refer to $\mathbf{1}$ as the vacuum element. Let $\mathfrak{h}_- = t^{-1}H[t^{-1}]$. The Poincaré–Birkhoff–Witt theorem implies that V is isomorphic as a vector space to $U(\mathfrak{h}_-) \cong S(\mathfrak{h}_-)$.

A good way of looking at V is the following. Using the basis α^i of H and its dual basis b^i as above, V may be considered as the underlying vector space of the (super) polynomial ring in the variables α^i_{-n}, $i = 1, \ldots, \dim(H)$, $n \in \mathbb{N}$. On these monomials the action of $\mathfrak{h}$ is given as follows: c acts as the identity, α^j_{-m} by multiplication with α^j_{-m}, and β^j_m acts as $m\partial/\partial\alpha^j_{-m}$ for $m \in \mathbb{N}_0$. The same reasoning as in Example 4.2 shows:

LEMMA 4.3. *V is an irreducible $\mathfrak{h}$-module*

The Heisenberg Lie algebra can be endowed with an additional $\mathbb{Z}$-grading by *weight*. We declare $\mathrm{wt}(\alpha_n) := -n$ and $\mathrm{wt}(c) = 0$. It is easy to check that the Lie bracket is homogeneous, and also that the ideal $I \subset U(\mathfrak{h})$ is homogeneous. Hence $V = U(\mathfrak{h})/I$ inherits a grading, and the action of $\mathfrak{h}$ on V is homogeneous. In order to compute its Poincaré series, all we need is to remember the fact that $S^*(\bigoplus W_i) = \bigotimes S^*(W_i)$ for any family of vector spaces. Therefore there are isomorphisms of graded vector spaces, where I have put in a formal variable to indicate the weight of the appearing vector spaces:

$$\bigoplus_{n\geq 0} V_n q^n = S^*\left(\bigoplus_{m>0} Hq^m\right) = \bigotimes_{m>0} S^*(H^{\mathrm{even}}q^m) \otimes \bigotimes_{m>0} \Lambda^*(H^{\mathrm{odd}}q^m). \tag{4.13}$$

Passing to dimensions, we obtain the following formula for the Poincaré series:

$$\sum_{n\geq 0} \dim(V^n)q^n = \prod_{m>0} \frac{(1+q^m)^{\dim H^{\mathrm{odd}}}}{(1-q^m)^{\dim H^{\mathrm{even}}}}. \tag{4.14}$$

Assume now in addition that the $\mathbb{Z}/2$-grading of H comes from a $\mathbb{Z}$-grading $H = \bigoplus_i H^i$. We will denote the degree of a homogeneous element $\alpha \in H$ as $|\alpha| \in \mathbb{Z}$. Note that this reinterpretation of the symbol $|\alpha|$ does not affect any of the formulas above. Then V becomes a bigraded vector space, $V = \bigoplus_{n,i} V^{n,i}$, the bidegree of α_{-n} being $(n, |\alpha|)$.

A straightforward extension of the calculation just completed gives

$$\sum_{n\geq 0}\sum_i \dim(V^{n,i})q^n p^i = \prod_{m>0}\prod_j (1-(-1)^j p^j q^m)^{-(-1)^j \dim H^j}. \tag{4.15}$$

Apply the construction above to the following setting: Let X be a smooth projective surface and let $H := H^*(X,\mathbb{Q})[-2]$. Let $\langle\alpha,\beta\rangle := -\int_X \alpha\cup\beta$. (By definition, the pairing is zero if the degrees of α and β don't match. The additional sign was thrown in in order to keep future formulas free from signs.) Let $\mathfrak{h}$ be the Heisenberg Lie algebra constructed from $(H, \langle -,-\rangle)$ and let $\mathbb{V}$ denote the representation constructed above. Then $\mathbb{V}$ has the same Poincaré series as the joint cohomology $\mathbb{H}$ of all Hilbert schemes. In particular, $\mathbb{V}$ and $\mathbb{H}$ are isomorphic as bigraded vector spaces. Of course, this in itself doesn't mean much. The question arises: Is there a natural isomorphism $\mathbb{V} \xrightarrow{\cong} \mathbb{H}$? This was answered positively by Nakajima [**32**] and Grojnowski [**18**].

4.3. Nakajima's operators. As before X is a smooth irreducible projective surface, $H := H^*(X,\mathbb{Q})[-2]$ and $\mathbb{H} := \bigoplus H^*(\mathrm{Hilb}^n(X,\mathbb{Q})[-2n])$. To keep formulas and diagrams readable, we will denote the Hilbert scheme by $X^{[n]} := \mathrm{Hilb}^n(X)$.

Let n, $\ell \geq 0$. Inside the triple product $X^{[\ell+n]} \times X \times X^{[\ell]}$ consider the incidence variety

$$Z = Z^{\ell,n+\ell} = \{(\xi', x, \xi) \mid \xi \subset \xi', \rho(\xi') = \rho(\xi) + nx\}. \tag{4.16}$$

Thus Z is made up from all triples where ξ is a subscheme of ξ', and ξ and ξ' differ only at the point x. Apparently, Z generalizes the incidence variety from Section 3.3.

We may decompose Z into strata $Z_0 \cup \cdots \cup Z_\ell$ with Z_i consisting of all triples such that $h^0(\mathcal{O}_{\xi,x}) = i$.

If x and ξ have disjoint support, $\xi' = \xi \cup \eta$ for some $\eta \in X^{[n]}$ with support in x. In fact, Z_0 is isomorphic to an open subscheme in $X^{[\ell]} \times B^n$, where B^n is the $(n+1)$-dimensional Briançon scheme introduced in Section 3.2. Similarly, Z_1 is isomorphic to an open subscheme in $X^{[\ell-1]} \times B^{n+1}$. It follows that $\dim(Z_0) = 2\ell + n + 1$ and $\dim(Z_1) = 2\ell+n$. With more care one can show that all other strata have dimension $< 2\ell+n$. To the best of my knowledge, we still do not know whether Z is irreducible or not. However, Z_1 is in the closure of Z_0, and certainly Z has no other component of dimension $\geq 2\ell+n$. This suffices for the present purpose. The fundamental class $[Z]$ is an element of degree $2\ell + n + 1$ in the Chow group of $X^{[\ell+n]} \times X \times X^{[\ell]}$.

Following Nakajima [**32**], we define an operator α_{-n} on $\mathbb{H}$ with $\alpha_{-n} \colon \mathbb{H}^\ell \to \mathbb{H}^{\ell+n}$ for any $\alpha \in H$ and $n \in \mathbb{N}$ as follows. Let p_i denote the projection from $X^{[\ell]} \times X \times X^{[\ell+n]}$ onto its i-th factor, $i = 1, 2, 3$. For $y \in \mathbb{H}^\ell = H^*(X^{[\ell]})[-2\ell]$ let

$$\alpha_{-n}(y) := \mathrm{PD}^{-1} p_{1*}((p_2^*(\alpha) \cup p_3^*(y)) \cap [Z]), \tag{4.17}$$

where PD denotes Poincaré duality. It is clear from the construction that α_{-n} is homogeneous of bidegree $(n, |\alpha|)$.

For positive indices α_n can be defined by reading the same diagram backward. I prefer to define α_n as the operator adjoint to α_{-n} with respect to the pairings on $\mathbb{H}^\ell$ given by $\langle x, y\rangle = (-1)^\ell \int_{X^{[\ell]}} x \cup y$. Note that this convention tallies with our definition for the pairing on the cohomology of $X = X^{[1]}$. Finally α_0 is set zero. Then Nakajima's main result is

THEOREM 4.4 (Nakajima [**32**]). *For all α, $\beta \in H$ and n, $m \in \mathbb{Z}$,*

$$[\alpha_n, \beta_m] = n\delta_{n,m}\langle\alpha, \beta\rangle \mathrm{id}_{\mathbb{H}}. \tag{4.18}$$

We give a proof of this theorem in the next section.

We can rephrase the theorem by saying that the Heisenberg Lie algebra $\mathfrak{h}$ constructed in the previous section acts on $\mathbb{H}$ via the geometrically defined operators α_n. As this action is also faithful our otherwise irresponsible double usage of the symbol α_n as an element in $\mathfrak{h}$ and as an endomorphism on $\mathbb{H}$ is justified.

PROPOSITION 4.5. *$\mathbb{V} \cong \mathbb{H}$ as representations of $\mathfrak{h}$.*

PROOF. The zeroth Hilbert scheme is a point representing the empty subscheme of X. Its cohomology is generated by the unique element $\mathbf{1} := 1 \in H^0(X^{[0]})$. It is clear for dimension reasons that $\alpha_n\mathbf{1} = 0$ for all $n \geq 0$. Sending the generator $\mathbf{1} \in \mathbb{V}$ to $\mathbf{1} \in \mathbb{H}$ defines a homomorphism $f \colon \mathbb{V} \to \mathbb{H}$. Since $\mathbb{V}$ is irreducible f must be injective. It follows that the dimensions of the homogeneous components of V give lower bounds for the Betti numbers of the Hilbert schemes. But we have already seen that by Göttsche's formula these numbers are equal. It follows that f is an isomorphism. □

As I learned from Marc de Cataldo in this conference, the use of Göttsche's formula can be avoided altogether by the following argument. Consider the Leray spectral sequence for the constant sheaf $\mathbb{Q}$ and the Hilbert–Chow morphism ρ: $\mathrm{Hilb}^n(X) \to S^n X$:

$$E_2^{pq} \Longrightarrow H^{p+q}(\mathrm{Hilb}^n(X)), \tag{4.19}$$

where the E_2 term is given by $\bigoplus_\alpha \bigotimes_{i\in\mathbb{N}} S^{\alpha_i}(H)$, and α runs over all partitions $\alpha = (1^{\alpha_1}2^{\alpha_2}\cdots)$ of n, i.e., $n = \sum_i \alpha_i i$. Göttsche's formula is equivalent to the degeneracy of this spectral sequence. However, and this is de Cataldo's point: we need not know that the spectral sequence degenerates. The Betti numbers of the Hilbert schemes can only become *smaller* than the Betti numbers of E_2. On the other hand, Nakajima's theorem shows that the same numbers provide lower bounds. Thus we must have degeneracy of the sequence and $\mathbb{V} \cong \mathbb{H}$.

One important consequence of the isomorphism $\mathbb{V} \cong \mathbb{H}$ is that we can write down a natural basis for the cohomology of the Hilbert schemes that also has a geometrical interpretation. Let α^i be a basis of homogeneous element in H. Then $\mathbb{H}$ has a basis of elements

$$\alpha^{i_1}_{-n_1}\cdots\alpha^{i_s}_{-n_s}\mathbf{1}, \tag{4.20}$$

where a factor $\alpha^{i_j}_{-n_j}$ should not show up twice if the degree of α^{i_j} is odd.

4.4. Proof of the Nakajima relations. We will sketch a proof for the commutator relations (4.18). The most interesting case is of course the bracket for operators α_n and β_{-n}. The cases where $n+m\neq 0$ are treated essentially in the same way, except that they are simpler: one never needs seriously compute an intersection product, dimension arguments suffice.

We want to calculate the two operators $\alpha_n\beta_{-n}$ and $\beta_{-n}\alpha_n$ on $\mathbb{H}$, or rather their difference.

Geometrically speaking, the operator $\alpha_n\beta_{-n}$ first adds an n-fold point somewhere and afterward subtracts an n-fold point, perhaps somewhere else. In contrast, the operator $\beta_{-n}\alpha_n$ first subtracts an n-fold point, and then adds an n-fold point, possibly a different one and possibly somewhere else. Heuristically, it is evident that the second procedure is more difficult to realize geometrically: in order to subtract something of length n from a subscheme $\xi\subset X$, the local ring of ξ should first of all have length $\geq n$ at some point x of its support. This is not so in the first case: we can always at least subtract what we added in first.

Consider the product variety

$$Y_+ := X^{[\ell]}\times X\times X^{[\ell+n]}\times X\times X^{[\ell]} \tag{4.21}$$

and let p_{123} etc. denote the projection onto the product of the first three factors etc. The cycle

$$w_+ := p_{1245*}(p_{123}^*[Z^{\ell,\ell+n}].p_{345}^*[Z^{\ell,\ell+n}]) \in A_{2\ell+2}(X^{[\ell]}\times X\times X\times X^{[\ell]}) \tag{4.22}$$

defines the operator $\alpha_n\beta_{-n}$:

$$\alpha_n\beta_{-n}(y) = (-1)^n\,\mathrm{PD}\big(p_{1*}(p_2^*(\alpha)\cup p_3^*(\beta)\cup p_4^*(y)\cap w_+)\big). \tag{4.23}$$

Geometrically, w_+ is supported on the scheme

$$W_+ := p_{1245}\big(p_{123}^{-1}(Z^{\ell,\ell+n})\cap p_{345}^{-1}(Z^{\ell,\ell+n})\big). \tag{4.24}$$

Set-theoretically, W_+ is easy to describe:

$$W_+ := \{(\xi, x, y, \zeta) \mid \exists \eta : \xi \subset \eta \supset \zeta, \rho(\eta) = \rho(\xi) + nx = \rho(\zeta) + ny\}. \tag{4.25}$$

We can immediately identify two irreducible components of W_+ of dimension $2\ell+2$: the first is the closure W' of the image of the rational map

$$X^{[\ell-n]} \times B^n \times B^n \to X^{[\ell]} \times X \times X \times X^{[\ell]} \tag{4.26}$$

given by $(\eta, \eta', \eta'') \mapsto (\eta \cup \eta', \rho(\eta'), \rho(\eta''), \eta \cup \eta'')$ and defined on all pairwise disjoint triples (η, η', η''). The second is the diagonal

$$\Delta = \{(\xi, x, x, \xi) \mid \xi \in X^{[\ell]}, x \in X\} \subset X^{[\ell]} \times X \times X \times X^{[\ell]}. \tag{4.27}$$

A dimension argument essentially based on Briançon's theorem shows that any other component of W_+, if there is any at all, must be of dimension strictly less than $2\ell + 2$ and therefore cannot support a cycle class of this degree.

The cycle w_+ is used to compute the composite operator $\alpha_n \beta_{-n}$. For the composition in the opposite order we proceed analogously.

Consider the product variety

$$Y_- := X^\ell \times X \times X^{[\ell-n]} \times X \times X^{[\ell]} \tag{4.28}$$

and the cycle

$$w_- := p_{1245*}(p_{123}^*[Z^{\ell,\ell-n}].p_{345}^*[Z^{\ell,\ell-n}]) \in A_{2\ell+2}(X^{[\ell]} \times X \times X \times X^{[\ell]}). \tag{4.29}$$

Geometrically, w_- is supported by the variety

$$W_- := p_{1245}\big(p_{123}^{-1}(Z^{\ell,\ell-n}) \cap p_{345}^{-1}(Z^{\ell,\ell-n})\big). \tag{4.30}$$

The latter has the following set theoretic description:

$$W_- := \{(\xi, x, y, \zeta) \mid \exists \theta : \xi \supset \theta \subset \zeta, \rho(\xi) = \rho(\theta) + nx, \rho(\zeta) = \rho(\theta) + ny\}. \tag{4.31}$$

As predicted by our heuristic statement at the very beginning, there is only one component of dimension $2\ell+2$, namely the variety W' defined above, and any other component of W_-, if there is any, must be of dimension strictly less than $2\ell + 2$.

Thus we must have $w_+ = a_+[W'] + N[\Delta]$ and $w_- = a_-[W']$ for some integers a_-, a_+, N. To compute these coefficients, it suffices to localize the geometric setting at a generic point of W' and Δ.

A generic point of W' has the form $(\eta \cup \eta', x, y, \eta \cup \eta'')$ where η, η' and η'' are disjoint, and η' and η'' have length n and are supported at disjoint point x and y, respectively. It is not difficult to check that the intersection of $p_{123}^{-1}Z^{\ell,\ell+n}$ and $p_{345}^{-1}Z^{\ell,\ell+n}$ in Y_+ is transverse above $(\eta \cup \eta', x, y, \eta \cup \eta'')$ and yields $a_+ = 1$, and that $a_- = 1$ for similar reasons.

This shows that $w_+ - w_- = N[\Delta]$. Note that the structure of this equation already proves (4.18) except for the precise value of the coefficient that appears on the right-hand side in (4.18). This factor was first computed by Ellingsrud and Strømme [**8**] and then by a quite different and very interesting argument by Nakajima [**33**].

It remains to compute N: again we localize the geometric setting at a generic point of the diagonal, i.e., a point of the form (ξ, x, x, ξ) where x is disjoint from ξ. Consider a point (ξ, x, η, x, ξ) in Y_+ above (ξ, x, x, ξ). Up to an étale map, we may replace $X^{[\ell+n]}$ near (ξ, x, η, x, ξ) by $X^{[\ell]} \times X^{[n]}$, and $Z^{\ell,n+\ell}$ by $X^{[\ell]} \times B^n$ etc. In this way the whole situation becomes a product of $X^{[\ell]}$ with a much simpler scenario corresponding to the case $\ell = 0$. In other words, N does not depend on ℓ, and we may assume without loss of generality that $\ell = 0$.

The problem is reduced to the following: Let $i\colon B^n \to X^{[n]}$ denote the natural inclusion and $\rho\colon B^n \to X$ the map that sends ξ to its support. Moreover, let $j = (i,\rho)\colon B^{[n]} \to X^{[n]} \times X$ be the embedding as graph of ρ. We need to compute the intersection of $j(B^n) \times X$ and $X \times j(B^n)$ in $X \times X^{[n]} \times X$, or more precisely, the coefficient N in

$$p_{13*}(j_*[j(B^n)] \times [X].[X] \times j_*[B^n]) = N[\Delta], \tag{4.32}$$

where $[\Delta]$ is the class of the diagonal in $X \times X$ and p_{13} is the projection from $X \times X^{[n]} \times X$ to $X \times X$. Intersect with $\mathrm{pr}_3^{-1}(p)$ for some point $p \in X$ and get

$$p_{1*}(j_*[B^n].[X] \times [B^n_p]) = N[p], \tag{4.33}$$

where p_1 is the projection from $X \times X^{[n]}$ to X. This shows that N is the intersection number of $[B^n]$ and $[B^n_p]$ in $X^{[n]}$.

THEOREM 4.6 (Ellingsrud–Strømme [**8**]).

$$[B^n].[B^n_p] = (-1)^{n-1}n. \tag{4.34}$$

PROOF. Recall the maps $\mathrm{Hilb}^n(X) \xleftarrow{\varphi} \mathbb{P} \xrightarrow{\psi} \mathrm{Hilb}^{n+1}(X)$. As before, let $E \subset \mathbb{P}$ denote the exceptional divisor of the blow-up $\Phi = (\phi,\rho) : \mathbb{P} \to \mathrm{Hilb}^n(X) \times X$. Moreover, let $B' := \psi^{-1}(B^{n+1})$ and $B'_p = \psi^{-1}(B^{n+1}_p)$ for some closed point $p \in X$. Then one has

$$\psi^*[B^{n+1}] = (n+1)[B'], \qquad \psi^*[B^{n+1}_p] = (n+1)[B'_p], \tag{4.35}$$

$$[E].\phi^*[B^n] = n[B'], \qquad [E].\phi^*[B^n_p] = n[B'_p]. \tag{4.36}$$

The first of these identities follows easily from the fact that $\psi\colon \mathbb{P} \to \mathrm{Hilb}^{n+1}$ is a generically finite map of degree $n+1$. Moreover, it is clear that $E \cap \phi^{-1}(B^n) = B'$. Hence one needs to determine the factor λ in $[E].\phi^*[B^n] = \lambda[B']$. This can be done by an explicit calculation in local coordinates for $\mathrm{Hilb}^n(\mathbb{C}^2)$ near the point $\xi = V\big((y,x^n)\big)$ in B^n.

We can now prove the theorem by induction on n and assume that it has been shown for n.

$$\begin{aligned} \frac{1}{n+1}[B^{n+1}].[B^{n+1}_p] &= \frac{1}{(n+1)^2}\psi^*[B^{n+1}].\psi^*[B^{n+1}_p] = [B'].[B'_p] \\ &= \frac{1}{n^2}[E].\phi^*[B^n].[E].\phi^*[B^n_p] = \frac{(-1)^n n}{n^2}[E]^2.[F], \end{aligned} \tag{4.37}$$

where in the last line we have used the induction hypothesis and $[F]$ denotes a generic fibre of $\phi\colon \mathbb{P} \to \mathrm{Hilb}^n$. Such a generic fibre equals the surface X blown up in n distinct points. Therefore $[E]^2.[F] = -n$. It follows that $[B^{n+1}].[B^{n+1}_p] = (-1)^{n+1}(n+1)$. □

5. Vertex algebras

The irreducible $\mathfrak{h}$-representation $\mathbb{V}$ that we constructed in Section 4.2 carries naturally the structure of a vertex algebra. In fact, it is one of easiest nontrivial vertex algebras around. By Nakajima's isomorphism $\mathbb{V} \to \mathbb{H}$, this vertex algebra structure is transplanted onto the cohomology of the Hilbert schemes. The important point is of course, that this is not just an empty formal procedure, but that certain features of the vertex algebra are reflected in the geometry, and conversely.

In the little time and space available to me in these lectures I cannot give more than a very superficial introduction to vertex algebras. I recommend the text books of Kac [**23**] and Frenkel and Ben-Zvi [**11**]. However, I would like to introduce just as much of some basic notions as to give some idea why the language of fields might be a good way of organizing the algebraic structures that have appeared in the previous section and will be used in the next.

Vertex algebras are (usually infinite dimensional) vector spaces with a distinguished element $\mathbf{1}$ (the so-called vacuum element) and a countable number of products $-_{(n)}-\colon V\otimes V\to V$ referred to as the nth product, $n\in\mathbb{Z}$. None of these products is commutative or associative, and the failure is measured in terms of all the other products. A good way to organize these data is in terms of endomorphism valued power series: For a fixed $a\in V$, there is an endomorphism $a_{(n)}-\colon V\to V$ for each $n\in\mathbb{Z}$, and we put all of them into formal sum

$$a(z) := \sum_{n\in\mathbb{Z}} a_{(n)} z^{-n-1} \in \operatorname{End}(V)[\![z,z^{-1}]\!]. \tag{5.1}$$

The unusual indexing is introduced so as to give $a_{(n)} = \operatorname{res}_z z^n a(z)$. In this way, we get a map, the so-called state-field correspondence,

$$Y\colon V\to \operatorname{End}(V)[\![z,z^{-1}]\!],\quad a\mapsto a(z). \tag{5.2}$$

The first vertex algebra axiom on Y requires that for any a and b in V the formal V-valued power series $a(z)b = \sum_{n\in\mathbb{Z}} a_{(n)}bz^{-n-1}$ should have a finite pole order, i.e., we require that $a_{(n)}b=0$ for $n\gg 0$. In this case, the power series $a(z)$ is called a field. The set of all fields is a linear subspace $\operatorname{QEnd}(V)\subset \operatorname{End}(V)[\![z,z^{-1}]\!]$.

Here is an example: let $\mathbb{V}$ denote the representation of $\mathfrak{h}$ that we discussed in the previous sections. For each $a\in H$, the Lie algebra elements a_n, $n\in\mathbb{Z}$, act on V, and may be organized into an $\operatorname{End}(V)$ valued formal power series. The grading of V and homogeneity of the a_n imply that $\sum_{n\in\mathbb{Z}} a_n z^{-n-1} =: a(z)$ is a field for each $a\in H\subset V$.

We must resist the temptation to multiply two fields

$$a(z)b(z) \stackrel{?}{=} \sum_{p\in\mathbb{Z}}\Big(\sum_{m+n+1=p} a_{(m)}b_{(n)}\Big) z^{-p-1}, \tag{5.3}$$

for the inner sum is infinite, and the expression does not make sense. However, because of the field property of $a(z)$ and $b(z)$ the term $a_{(p-n-1)}b_{(n)}c$ vanishes for any given c and $n\gg 0$. This is not so when n becomes very negative. In this case $a_{(p-n-1)}$ becomes zero when evaluated on a fixed vector, except that in the present situation its argument changes as well, and we are lost. In quantum field theory there is a radical cure for this: Simply let the a_n's act first. The normal ordered product is defined as follows

$$:a(z)b(z): := a(z)_- b(z) - (-1)^{|a|\cdot|b|} b(z)a(z)_+, \tag{5.4}$$

where $a(z)_+ = \sum_{n\geq 0} a_{(n)}z^{-n-1}$ and $a(z)_- := \sum_{n<0} a_{(n)}z^{-n-1}$ are the principal and the holomorphic part of $a(z)$, respectively. It is not difficult to see that this definition makes sense and that the normal ordered product of two fields is again a field. It is less obvious that to make such a definition really is a reasonable thing to do. More generally, we define an n-th product of two fields $a, b \in \mathrm{QEnd}(V)$ for any $n \in \mathbb{Z}$ by

$$(5.5)\quad a(z)_{(n)}b(z)$$
$$:= \mathrm{res}_w\left(w^n\left(1-\frac{z}{w}\right)^n a(w)b(z) - (-1)^{|a|\cdot|b|}(-z)^n\left(1-\frac{w}{z}\right)^n b(z)a(w)\right),$$

where $(1-w/z)^n = 1 - nw/z + \binom{n}{2}(w/z)^2 + \cdots$ etc. Again this is well-defined, and $a(z)_{(n)}b(z)$ is again a field. One can check that the (-1)-product of two fields is the normal ordered product. If $(V, Y, \mathbf{1})$ is a vertex algebra then $(a_{(n)}b)(z) = a(z)_{(n)}b(z)$.

Let us go back to the $\mathfrak{h}$-representation on $\mathbb{V}$. Choose a basis a^i of H and a dual basis b^i with respect to the symmetric bilinear form $\langle -, - \rangle$. To each element $a \in H$ we have associated a field $a(z)$. We may therefore consider the field

$$(5.6)\qquad L(z) := \sum_{n\in\mathbb{Z}} L_n z^{-n-2} := \frac{1}{2}\sum_i {:}a^i(z)b^i(z){:}.$$

If one is not accustomed to this kind of calculation it might take a while to verify that the coefficients L_n of the field $L(z)$ satisfy the following commutator relations

$$(5.7)\qquad [L_n, L_m] = (n-m)L_{n+m} + \delta_{n,-m}\frac{n^3-n}{12}\,\mathrm{sdim}(H)\mathrm{id}_{\mathbb{V}},$$

where $\mathrm{sdim}(H) := \dim H^{\mathrm{even}} - \dim H^{\mathrm{odd}}$ is the super dimension of H. The commutator relations (5.7) are characteristic for the Virasoro algebra.

The Virasoro algebra is defined as follows: $\mathrm{Witt} := \mathrm{Der}(\mathbb{Q}[t, t^{-1}])$ is the Lie algebra of vector fields on a punctured disc. It has a basis $\ell_n := t^{-n-1}d/dt$ that satisfies the Lie bracket relations $[\ell_n, \ell_m] = (n-m)\ell_{n+m}$. Apparently, Witt is a perfect Lie algebra and therefore has a universal central extension

$$(5.8)\qquad 0 \to Z \to \mathrm{Vir} \to \mathrm{Witt} \to 0.$$

It turns out that the center of Vir is one-dimensional, say spanned by an element c, and that one may choose lifts $L_n \in \mathrm{Vir}$ of $\ell_n \in \mathrm{Witt}$ in such a way that the Lie brackets for the L_n have the normalized form

$$(5.9)\qquad [L_n, L_m] = (n-m)L_{n+m} + \delta_{n,-m}\frac{n^3-n}{12}c.$$

Comparing this with (5.7), we may express the meaning of (5.7) by saying that there is a natural representation of Vir on $\mathbb{V}$ where the element c acts as a scalar. This scalar is often referred to as the central charge of the representation. In the present case it is $\mathrm{sdim}(H)$.

Assume now that H has in addition to its grading and its symmetric pairing a ring structure $\cup\colon H \otimes H \to H$ with unit 1 and a trace map $t\colon H \to \mathbb{C}$ such that $\langle\alpha, \beta\rangle = t(\alpha \cup \beta)$. Let $\Delta : H \to H \otimes H$ be the map adjoint to the cup product with respect to the pairing $\langle -, - \rangle$. For instance, $\Delta(1) = \sum a^i \otimes b^i$, where (α^i) and (β^i) are dual bases as above. The element $e := \sum_i \alpha^i \cup \beta^i \in H$ is called the Euler class of H. It is the image of 1 under the composite map $H \xrightarrow{\Delta} H \otimes H \xrightarrow{\cup} H$. Furthermore we have $t(e) = \mathrm{sdim}(H)$.

The following construction generalizes the definition of the field $L(z)$ above: For any $\gamma \subset H$ write $\Delta(\gamma) = \sum \gamma' \otimes \gamma''$

$$L(\gamma)(z) := \sum_{n\in\mathbb{Z}} L_n(\gamma) z^{-n-2} := \frac{1}{2}\sum_i {:}\gamma'(z)\gamma''(z){:}. \tag{5.10}$$

This defines endomorphisms $L_n(\gamma) \in \mathrm{End}(\mathbb{V})$ that satisfy the twisted Virasoro relations

$$[L_n(\alpha), L_m(\beta)] = (n-m)L_{n+m}(\alpha\cup\beta) + \delta_{n,-m}\frac{n^3-n}{12}t(\alpha\cup\beta\cup e). \tag{5.11}$$

For $\alpha = \beta = 1$, one gets back the old relations (5.9). We will need these endomorphisms in the next section. Explicitly, one has

$$L_n(\gamma) = \frac{1}{2}\sum_i\sum_{\nu\in\mathbb{Z}} \gamma'_\nu \gamma''_{n-\nu} \tag{5.12}$$

for $n \neq 0$ and

$$L_0(\gamma) = \sum_i\sum_{\nu\in\mathbb{N}} \gamma'^{i}_\nu \gamma''_{-\nu}. \tag{5.13}$$

I leave it as an exercise to check that

$$[L_n(\alpha), \beta_m] = -m(\alpha\cup\beta)_{n+m}. \tag{5.14}$$

6. The ring structure

Nakajima's beautiful theorem gives us a little bit more than just an explicit isomorphism of vector spaces $\mathbb{V} \to \mathbb{H}$. It also allows one to effectively compute the cup product pairing on $\mathbb{H}^n$. Since α_{-n} is adjoint to α_n, one has

$$\begin{aligned}
&\langle \alpha^{i_1}_{-n_1}\cdots\alpha^{i_s}_{-n_s}\mathbf{1}, \alpha^{j_1}_{-m_1}\cdots\alpha^{j_t}_{-m_t}\mathbf{1}\rangle \\
&\quad= \langle \alpha^{i_2}_{-n_2}\cdots\alpha^{i_s}_{-n_s}\mathbf{1}, \alpha^{i_1}_{+n_1}\alpha^{j_1}_{-m_1}\cdots\alpha^{j_t}_{-m_t}\mathbf{1}\rangle \\
&\quad= \sum_{\tau=1}^{t} n_1\delta_{n_1,m_\tau}\langle\alpha^{i_1},\alpha^{j_\tau}\rangle\langle \alpha^{i_2}_{-n_2}\cdots\alpha^{i_s}_{-n_s}\mathbf{1}, \alpha^{i_1}_{+n_1}\alpha^{j_1}_{-m_1}\cdots\hat{\alpha}^{j_\tau}_{n_\tau}\cdots\alpha^{j_t}_{-m_t}\mathbf{1}\rangle.
\end{aligned}$$

Inductively, we may reduce everything to $\langle\mathbf{1},\mathbf{1}\rangle = 1$.

However, so far there is no relation between the representation of $\mathfrak{h}$ on $\mathbb{H}$ and the additional ring structure on $\mathbb{H}^n$, given by the cup product. To understand this ring structure is the goal of this section.

6.1. Multiplication operators. The basic idea is to think of the cup product in terms of multiplication operators and to relate these with Nakajima's operators.

Recall that $\mathrm{Hilb}^n(X)$ parameterizes a universal family

$$\begin{array}{ccc}
\Xi_n \subset \mathrm{Hilb}^n(X)\times X & \xrightarrow{q} & X \\
\downarrow{\scriptstyle p} & & \\
\mathrm{Hilb}^n(X) & &
\end{array} \tag{6.1}$$

If F is a locally free sheaf on X, then $F^{[n]} := p_*(\mathcal{O}_{\Xi_n}\otimes q^*F)$ is locally free on $\mathrm{Hilb}^n(X)$ of rank $n\cdot\mathrm{rk}(F)$. Sheaves of this type are called tautological sheaves. Applying this construction to the trivial sheaf we obtain the sheaf of algebras $\mathcal{A} = \mathcal{O}^{[n]}$ encountered before.

Following Li, Qin and Wang [**29**] one can generalize this construction from elements in K-theory to cohomology classes. For each $\alpha \in H$ let

$$\alpha^{[n]} := p_*\big(\mathrm{ch}(\mathcal{O}_{\Xi_n}) \cup q^*(\mathrm{td}(X) \cup \alpha)\big). \tag{6.2}$$

With this notation the Grothendieck–Riemann–Roch theorem says that

$$\mathrm{ch}(F^{[n]}) = \big(\mathrm{ch}(F)\big)^{[n]}. \tag{6.3}$$

Define operators $m(\alpha) \in \mathrm{End}(\mathbb{H})$ by

$$m(\alpha)|_{\mathbb{H}^n}(y) = \alpha^{[n]} \cup y. \tag{6.4}$$

Note that even if α is homogeneous, $m(\alpha)$ is not so. As a consequence, the operator has weight 0 but is not homogeneous in the cohomological direction. Let the homogeneous component of bidegree $(0, j)$ be denoted by $m(\alpha)_j$. For example, $m(1)_0|_{\mathbb{H}_n}$ is the multiplication by $\mathrm{rk}(\mathcal{A}) = n$, and $m(1)_2|_{\mathbb{H}^n}$ is the cup product with $c_1(\mathcal{A}) = -\frac{1}{2}[\partial \operatorname{Hilb}^n(X)]$. This operator is of particular importance and deserves its own symbol

$$\partial := m(1)_2 = c_1(\mathcal{O}^{[n]}). \tag{6.5}$$

Up to a factor $-\frac{1}{2}$, ∂ describes the intersection of a cohomology class with the divisor $\partial \operatorname{Hilb}^n(X)$.

Finally, let $c(F) \in \mathrm{End}(\mathbb{H})$ be the endomorphism with $c(F)|_{\mathbb{H}^n} = c(F^{[n]}) \cup -$ for a locally free sheaf F on X.

Every element in $\mathbb{H}$ is a linear combination of elements of the form

$$x = \alpha^{i_1}_{-n_1} \cdots \alpha^{i_s}_{-n_s} \mathbf{1}. \tag{6.6}$$

Then

$$m(\beta)(x) = \sum_{j=1^s} \alpha^{i_1}_{-n_1} \cdots [m(\beta), \alpha^{i_j}_{-n_j}] \cdots \alpha^{i_s}_{-n_s} \mathbf{1} + \alpha^{i_1}_{-n_1} \cdots \alpha^{i_s}_{-n_s} m(\beta)\mathbf{1}. \tag{6.7}$$

The last summand is zero, since $\Xi_0 = \varnothing$ so that $m(\beta)\mathbf{1} = 0$. We can compute $m(\beta)(x)$ if we understand the commutators $[m(\beta), \alpha_{-n}]$.

THEOREM 6.1.

$$[m(\beta), \alpha_{-1}] = \sum_{\nu \geq 0} \frac{1}{\nu!} (\mathrm{ad}\, \partial)^\nu (\beta \cup \alpha)_{-1}. \tag{6.8}$$

For the special case $\beta = ch(F)$ this theorem can be found in [**24**], for arbitrary β it is due to Li, Qin and Wang [**29**]. Theorem 6.1 reduces the calculation of $[m(\beta), -]$ to the special case $[\partial, -]$. The proof of the theorem is in fact quite easy:

PROOF. Recall the incidence scheme $\mathbb{P}$ studied in Section 3.3 with the maps $\psi \colon \mathbb{P} \to \operatorname{Hilb}^{n+1}(X)$, $\phi : \mathbb{P} \to \operatorname{Hilb}^n(X)$ and $\rho \colon \mathbb{P} \to X$. Passing from the sequence (3.25) of ideal sheaves to structure sheaves we obtain the global version of sequence (3.21):

$$0 \to \mathrm{pr}_1^* \mathcal{O}_{\mathbb{P}}(1)|_Y \to (\psi \times \mathrm{id}_X)^* \mathcal{O}_{\Xi_{n+1}} \to (\phi \times \mathrm{id}_X)^* \mathcal{O}_{\Xi_n} \to 0. \tag{6.9}$$

Applying $p_*\big(\mathrm{ch}(-) \cup \mathrm{pr}_X^*(\mathrm{td}(X) \cup \beta)\big)$ to this sequence we obtain the identity

$$\psi^* \beta^{[n+1]} - \phi^* \beta^{[n]} = \mathrm{ch}(\mathcal{O}_{\mathbb{P}}(1)) \cup \rho^* \beta \tag{6.10}$$

in $H^*(\mathbb{P})$. Now, by definition, $\alpha_{-1}(y) = \psi_*\big(\rho^*\alpha \cup \phi^*(y)\big)$. It follows that

$$\begin{aligned}(6.11)\quad [m(\beta), \alpha_{-1}](y) \\ &= \beta^{[n+1]} \cup \psi_*(\rho^*\alpha \cup \phi^* y) - (-1)^{|\alpha|\,|\beta|}\psi_*\big(\rho^*\alpha \cup \phi^*(\beta^{[n]} \cup y)\big) \\ &= \psi_*\big((\psi^*\beta^{[n+1]} - \phi^*\beta^{[n]}) \cup \rho^*\alpha \cup \phi^* y\big) \\ &= \psi_*\big(\mathrm{ch}\,\mathcal{O}_{\mathbb{P}}(1) \cup \rho^*(\beta \cup \alpha) \cup \phi^* y\big),\end{aligned}$$

where we have used the identity (6.10) in the last line. From this one can already deduce what Li, Qin and Wang called the transfer property:

$$(6.12)\qquad [m(\beta), \alpha_{-1}] = [m(1), (\beta \cup \alpha)_{-1}].$$

Moreover, if in the calculation above we take $\beta = 1$ and the degree 2 part of $m(1)$, we get

$$(6.13)\qquad [\partial, \alpha_{-1}](y) = \psi_*(c_1(\mathcal{O}_{\mathbb{P}}(1)) \cup \rho^*\alpha \cup \phi^* y),$$

and more generally,

$$(6.14)\qquad (\mathrm{ad}\,\partial)^\nu(\alpha_{-1})(y) = \psi_*\big(c_1\big(\mathcal{O}_{\mathbb{P}}(1)\big)^\nu \cup \rho^*\alpha \cup \phi^* y\big),$$

If we reinsert this expression into (6.11), it follows that

$$(6.15)\qquad [m(\beta), \alpha_{-1}] = \sum_{\nu \geq 0} \frac{1}{\nu!}(\mathrm{ad}\,\partial)^\nu(\beta \cup \alpha)_{-1}. \qquad \square$$

Let L be a line bundle on X. Let $c(L) \in \mathrm{End}(\mathbb{H})$ denote the operator, that is the multiplication with the total Chern class $c(L^{[n]})$ of the tautological rank n bundle $L^{[n]}$ on $\mathbb{H}^n$. Note that $c(L)$ is invertible.

THEOREM 6.2 ([**24**]).

$$(6.16)\qquad c(L)\alpha_{-1}c(L)^{-1} = (c(L) \cup \alpha)_{-1} + [\partial, \alpha_{-1}].$$

The proof is very similar to the proof of Theorem 6.1 and will be omitted.

The following theorem is the key to the ring structure of the Hilbert scheme. The proof is too long to reproduce it here. In the theorem, $K \in H^0 = H^2(X)$ is the canonical class of the surface. Moreover, $L_n(\alpha)$ is the twisted Virasoro operator as defined by equation (5.10)

THEOREM 6.3 ([**24**]). *Let $n \in \mathbb{N}$. Then*

$$(6.17)\qquad [\partial, \alpha_{-n}] = -nL_n(\alpha) + \frac{n(n-1)}{2}(\alpha \cup K)_{-n}.$$

Observe that due to the relations (5.12) and (5.13), the right-hand side in the theorem is a quadratic expression in Nakajima operators. In particular, iterated applications of the theorem will compute all expressions of the form $(\mathrm{ad}\,\partial)^\nu(\alpha_n)$.

Here are some immediate consequences of Theorem 6.3:

COROLLARY 6.4. (1) $[\partial, \alpha_{-1}] = -L_{-1}(\alpha)$.
(2) *For all $n \in \mathbb{N}$:* $(\mathrm{ad}[\partial, 1_{-1}])^n(\alpha_{-1}) = (-1)^n n!\, \alpha_{-n-1}$.

PROOF. The first statement is only a special case of the theorem, and the second follows from iterated applications of the identity (5.14). $\square$

This corollary has in turn important applications: since any Nakajima operator α_{-n} can be written as a linear combination of words in the symbols ∂ and β_{-1}, Theorem 6.1 is, in principle, strong enough to compute any of the commutators $[m(\beta), \alpha_{-n}]$: firstly, one has to rewrite α_{-n} as such a linear combination. Secondly, using the fact that $m(\beta)$ must commute with ∂, since both are multiplication operators in the graded commutative ring $\mathbb{H}^n$, express the commutator with $m(\beta)$ in terms of commutators with ∂. Use Theorem 6.3 to compute these. In practice, this is difficult to carry out. However, one can turn these instructions into a rigid algorithm that allows to compute any cup product in $\mathbb{H}^n$.

COROLLARY 6.5. $\mathbb{H} = \partial(\mathbb{H}) + H_{-1}\mathbb{H}$.

COROLLARY 6.6. *Let L be a line bundle on X. Then*

$$\sum_{n\geq 0} c(L^{[n]})q^n = \exp\Big(\sum_{m>0} \frac{(-1)^{m-1}}{m} c(L)_{-m}q^m\Big)\mathbf{1}. \tag{6.18}$$

PROOF. The unit $1_{\mathrm{Hilb}^n(X)} \in \mathbb{H}^n$ equals $(1/n!)(1_X)_1^n\mathbf{1}$. Hence the left-hand side in (6.18) equals

$$\sum_{n\geq 0} c(L^{[n]})q^n = c(L)\exp\big((1_X)_{-1}q\big)\mathbf{1} = \exp(c(L)(1_X)_1 c(L)^{-1}q)\mathbf{1}. \tag{6.19}$$

By Theorems 6.2 and 6.3,

$$c(L)(1_X)_1c(L)^{-1} = c(L)_{-1} + [\partial, (1_X)_{-1}] = c(L)_{-1} + L(1_X)_{-1} =: A. \tag{6.20}$$

We must show that

$$\exp(Aq)\mathbf{1} = \exp\Big(\sum_{m>0} \frac{(-1)^{m-1}}{m} c(L)_{-m}q^m\Big)\mathbf{1}. \tag{6.21}$$

It suffices to show that the right-hand side satisfies the differential equation

$$\frac{\partial \mathfrak{X}}{\partial q} = A\mathfrak{X} \tag{6.22}$$

with initial condition $\mathfrak{X}(0) = \mathbf{1}$. I leave the verification to the reader. □

6.2. The affine plane. The affine plane is an ideal example for the study of the cohomology ring of the Hilbert scheme because it has trivial cohomology. Therefore, there are no complications arising from the topology of $\mathbb{C}^2$, and $H^*\big(\mathrm{Hilb}^n(\mathbb{C}^2)\big)$ becomes an essentially combinatorial object.

REMARK 6.7. However, there is one problem with $\mathbb{C}^2$: it is not projective. Therefore, we have to modify our presentation of Nakajima's work. I will only briefly indicate the necessary changes: if X is a smooth but not necessarily projective surface then the projection from the incidence variety Z in (4.16) to the factor $X^{[\ell+n]}$ is still proper. Therefore the operators α_{-n} are well-defined for cohomology classes $\alpha \in H^*(X)$. However, the projection to $X^{[\ell]}$ is not proper. In order to define α_n for $n \in \mathbb{N}$ one needs to take α from cohomology with compact support. In this way, one can build up a Heisenberg Lie algebra $\mathfrak{h} = t^{-1}H^*(X)[t^{-1}] \oplus tH_c^*(x)[t]$ that acts on $\mathbb{H}$. It remains true that $\mathbb{H}$ is isomorphic to $S^*(t^{-1}H^*(X)[t^{-1}])$.

Let $X = \mathbb{C}^2$. Then $H = H^*(\mathbb{C}^2)[2]$ is is one-dimensional. For better readability, we write $p_n := (1_X)_{-n}$. Thus

$$\mathbb{H} = \bigoplus_{n\geq 0} H^*\big(\mathrm{Hilb}^n(\mathbb{C}^2)\big)[2n] = \mathbb{Q}[p_1, p_2, p_3, \ldots].$$

Here p_n has weight n and degree -2. For example,

$$H^*\big(\mathrm{Hilb}^4(\mathbb{C}^2)\big)[8] = \mathbb{Q}\langle p_1^4, p_1^2 p_2, p_1 p_3, p_2^2, p_4\rangle,$$

and $\dim H^*\big(\mathrm{Hilb}^n(\mathbb{C}^2)\big)$ equals the number of partitions of n.

Let $\mathbb{Q}[\mathrm{S}_n]$ denote the group ring of the symmetric group. Every permutation π admits a decomposition $\pi = z_1 \cdots z_s$ into disjoint cycles z_i of length λ_i with $\lambda_1 \geq \cdots \geq \lambda_s$. We refer to the partition $\lambda = (\lambda_1, \ldots, \lambda_s)$ of n as the cycle type of π.

The center $Z\mathbb{Q}[\mathrm{S}_n]$ of $\mathbb{Q}[\mathrm{S}_n]$ consists of all elements that are invariant under the conjugation action of S_n. It has a $\mathbb{Q}$-basis consisting of elements $e_\lambda := \sum_{\pi\vdash\lambda} \pi$, where the sum runs over all permutations π of cycle type λ.

Consider the map $\Phi\colon \mathbb{Q}[\mathrm{S}_n] \to \mathbb{Q}[p_1, p_2, \ldots]$ which sends a permutation π of cycle type λ to the monomial

$$\Phi(\pi) = \frac{1}{n!} p_{\lambda_1} \cdots p_{\lambda_s}.$$

Then Φ induces an isomorphism of vector spaces

$$\Phi\colon Z\mathbb{Q}[\mathrm{S}_n] \to \mathbb{H}^n = \mathbb{Q}[p_1, p_2, \ldots]_n, \quad e_\lambda \mapsto \prod_{i\in\mathbb{N}} \frac{1}{m_i(\lambda)!}\left(\frac{p_i}{i}\right)^{m_i(\lambda)},$$

where $m_i(\lambda) = |\{j \mid \lambda_j = i\}|$.

Both $Z\mathbb{Q}[\mathrm{S}_n]$ and $\mathbb{H}^n = H^*\big(\mathrm{Hilb}^n(\mathbb{C}^2)\big)[2n]$ are rings. However, they cannot be isomorphic for trivial reasons: $Z\mathbb{Q}[\mathrm{S}_n]$ has a $\mathbb{Q}$-basis of idempotent elements, whereas $\mathbb{H}^n$ has only one idempotent: $1_{\mathrm{Hilb}^n(\mathbb{C}^2)}$. Nevertheless, these two rings are curiously related:

Proposition 6.8.

$$\Phi\Big(\sum_{\pi\in\mathrm{S}_n} \mathrm{sgn}(\pi)\pi\Big) = c(\mathcal{O}^{[n]}).$$

In other words, Φ sends the alternating character to the total Chern class of the tautological sheaf $\mathcal{O}^{[n]} = \mathcal{A}$ on $\mathrm{Hilb}^n(\mathbb{C}^2)$.

Proof. It is a combinatorial exercise to see that

$$\Phi\Big(\sum_{n\geq 0}\sum_{\pi\in\mathrm{S}_n} \mathrm{sgn}(\pi)\pi q^n\Big) = \exp\Big(\sum_{m>0} \frac{(-1)^{m-1}}{m} p_m q^m\Big).$$

By Corollary 6.6, the right-hand side equals $\sum_{n\geq 0} c(\mathcal{O}^{[n]}) q^n$. □

The action of $\partial = c_1(\mathcal{O}^{[n]}) \cup -$ on $\mathbb{H} = \mathbb{C}[p_1, p_2, \ldots]$ can be described as a differential operator. The following proposition is a consequence of Theorem 6.3:

Proposition 6.9 ([**24**]).

$$\partial = \frac{-1}{2} \sum_{n,m>0} nm p_{n+m} \frac{\partial}{\partial p_n}\frac{\partial}{\partial p_m}.$$

As Frenkel and Wang [**12**] pointed out, the differential operator on the right-hand side is very similar to the following operator introduced by I. Goulden [**17**]:

$$G := \frac{1}{2}\sum_{n,m>0} nm p_{n+m}\frac{\partial}{\partial p_n}\frac{\partial}{\partial p_m} + \frac{1}{2}\sum_{n,m>0}(n+m)p_n p_m\frac{\partial}{\partial p_{n+m}}.$$

Let $\tau \in Z\mathbb{Q}[\mathrm{S}_n]$ denote the sum of all transpositions. Then Goulden's operator describes the multiplication with τ in $Z\mathbb{Q}[\mathrm{S}_n]$ as differential operator the same way as ∂ describes the multiplication with $c_1(\mathcal{O}^{[n]})$, namely:

$$\Phi(\tau\cdot y) = G\big(\Phi(y)\big), \quad c_1(\mathcal{O}^{[n]})\cup\Phi(y) = \partial\big(\Phi(y)\big) \tag{6.23}$$

To see this observe the following: if a transposition (ij) is composed with a permutation $\pi = z_1\cdots z_s$ two things can happen: either i and j belong to two different cycles z_k, then as a result of the multiplication the two cycles are merged. For instance

$$(14)\cdot(1\,2\,3)(4\,5) = (1\,2\,3\,4\,5). \tag{6.24}$$

Thus two cycles of length n and m are replaced by a cycle of length $n+m$. This is accounted for with the correct multiplicities by the first half of G. If on the other hand i and j belong to the same cycle, then this cycle will be split, as in

$$(14)\cdot(1\,2\,3\,4\,5) = (1\,2\,3)(4\,5). \tag{6.25}$$

Thus one cycle of length $n+m$ is split into two cycles of length n and m. The second half of G takes care of this possibility.

As we already know that $\Phi(-\tau) = c_1(\mathcal{O}^{[n]})$, the similarity of the differential operators ∂ and G allows only the following conclusion: in order to obtain the correct cup product in $\mathbb{H}^n$, the multiplication in $Z\mathbb{Q}[\mathrm{S}_n]$ has to be modified in such a way that only multiplications as in (6.24) remain nonzero. This can be achieved by introducing an appropriate grading.

Define the age of $\pi\in\mathrm{S}_n$ as the minimal number k needed to write π as a product of k transpositions. Clearly, $\mathrm{age}(\pi) = n-s$, where s is the number of disjoint cycles in the cycle decomposition of π. Moreover, it if follows directly from the definition that $\mathrm{age}(\pi\sigma)\leq\mathrm{age}(\pi)+\mathrm{age}(\sigma)$. Let $F_i^{\mathrm{age}}Z\mathbb{Q}[\mathrm{S}_n]$ be generated by all elements of age $\leq i$. Then $F_i\cdot F_j\subset F_{i+j}$. Hence, the subspaces F_i^{age} define an ascending filtration of $Z\mathbb{Q}[\mathrm{S}_n]$, and we may pass to the associated graded ring $\mathrm{gr}^{\mathrm{age}}Z\mathbb{Q}[\mathrm{S}_n]$. The discussion above makes the following theorem look very plausible. The same result was obtained independently and by a quite different method by Vasserot [**37**]:

THEOREM 6.10. *The map*

$$\Phi\colon \mathrm{gr}^{\mathrm{age}}Z\mathbb{Q}[\mathrm{S}_n]\cong H^*\big(\mathrm{Hilb}^n(\mathbb{C}^2)\big)$$

is an isomorphism of rings that is degree preserving if the age is doubled.

For a proof within the given framework I refer to the joint paper with Sorger [**26**]. The key point is that due to Proposition 6.8 we have $\Phi(\sum_{\pi\in\mathrm{S}_n,\mathrm{age}(\pi)=i}) = c_i(\mathcal{O}^{[n]})$ and that these elements generate the rings $\mathrm{gr}^{\mathrm{age}}Z\mathbb{Q}[\mathrm{S}_n]$ and $H^*(\mathrm{Hilb}^n(\mathbb{C}^2))$, respectively.

6.3. Orbifold cohomology. In order to generalize Theorem 6.10 for other smooth surfaces, it is best to use the notion of orbifold cohomology rings. Orbifold cohomology itself has a longer history, but apparently the ring structure was first introduced by Chen and Ruan [**4, 35**]. We will need here only the case of global group quotients. For this case we refer also to the paper of Fantechi and Göttsche [**9**].

Let X be a smooth projective variety on which a finite group G acts in such a way, that ω_X is locally trivial as a G-linearized sheaf. Equivalently, the stabilizer group G_x of every point $x \in X$ is supposed to lie in $\mathrm{SL}(T_xX)$.

Let $g \in G$ be an element of order n. The fixed point set X^g is a smooth subvariety of X. The isomorphism class of the tangent space T_xX considered as a representation of the cyclic group $\langle g\rangle$ is locally constant on X^g. Let $\exp(2\pi i k_j)$, $j = 1, \dots, \dim(X)$, denote the eigenvalues of $g|_{T_xX}$. The k_j are normalized by the requirement $k_j \in \mathbb{Q} \cap [0, 1)$. Then the age of g is the locally constant function $\mathrm{age}(g)\colon X^g \to \mathbb{N}_0$ with $\mathrm{age}(g)(x) = \sum_j k_j$.

REMARK 6.11. In the previous subsection we have already defined the age of a permutation π. The connection with the present definition is this: The symmetric group acts on $(\mathbb{C}^2)^n$ by permuting the factors. Let π be a permutation with disjoint cycle decomposition $\pi = z_1 \cdots z_s$, and let $\ell(z_i) =: \ell_i$. The fixed point set of π is isomorphic to $(\mathbb{C}^2)^s$, and the eigenvalues of π are $\exp(2\pi i k/\ell_p)$ with $k = 0, \dots, \ell_p - 1$ and $p = 1, \dots, s$, each of them counted twice. The sum of all these k/ℓ_p gives the age of π in the new sense, namely

$$\mathrm{age}(\pi) = 2\sum_{p=1}^{s}\sum_{k=0}^{\ell_p-1} k/\ell_p = \sum_{p=1}^{s}(\ell_p - 1) = n - s.$$

This agrees with the previous definition.

Consider $(m, \mathrm{pr}_2)\colon G \times X \to X \times X$, where m is the action of G on X, and define the inertia subvariety $I \subset G \times X$ by means of the Cartesian diagram

$$\begin{array}{ccc} I & \longrightarrow & G \times X \\ \downarrow & & \downarrow \\ X & \xrightarrow{\Delta} & X \times X. \end{array}$$

In other words, $I = \bigcup_{g\in G}\{g\} \times X^g$. We may consider age as a locally constant function on I. The group G acts naturally on $G \times X$ (by conjugation on the first factor) and I is a G-invariant subspace. Let

$$H^*(X, G) := H^*(I)[-2\,\mathrm{age}] = \bigoplus_{g\in G} H^*(X^g)[-2\,\mathrm{age}(g)]$$

and

$$H^*_{\mathrm{orb}}(X/G) := H^*(X, G)^G = \bigoplus_{[g]\in G/\sim} H^*(X^g)^{C_g}[-2\,\mathrm{age}(g)],$$

where in the last sum g runs through a set of representatives for the conjugacy classes of G, and $C_g \subset G$ is the centralizer of g.

This defines the orbifold cohomology of X/G as a vector space. It contains the ordinary cohomology $H^*(X/G)$ as the direct summand corresponding to the neutral element of G. The other direct summands are called twisted sectors in the physically inspired terminology.

In order to define a product on $H^*_{\mathrm{orb}}(X/G)$ one first defines a product on $H^*(X,G)$ and shows then that this product is G-equivariant. Define a map

$$m\colon H^*(X^g)[-2\,\mathrm{age}(g)] \times H^*(X^h)[-2\,\mathrm{age}(h)] \to H^*(X^{gh})[-2\,\mathrm{age}(gh)]$$

by

$$m(\alpha,\beta) = i_*(\alpha|_{X^{g,h}} \cup \beta|_{X^{g,h}} \cup c_{g,h}),$$

where $X^{g,h} = X^g \cap X^h$ is the fixed point set of the subgroup $H \subset G$ generated by g and h, $i\colon X^{g,h} \to X^{gh}$ is the inclusion map, and $c_{g,h} \in H^*(X^{g,h})$ is a certain cohomology class still to be defined:

There is a unique ramified cover $C \to \mathbb{P}^1$ with Galois group H, ramified over $\{0,1,\infty\}$, such that the inertia groups over the points 0, 1 and ∞ are generated by g, h, and gh, respectively. Now

$$c_{g,h} := c_{\mathrm{top}}\Big(\big(TX|_{X^{g,h}} \otimes_{\mathbb{C}} H^1(C,\mathcal{O}_C)\big)^H\Big).$$

On the right-hand side, $TX|_{X^{g,h}}$ is an H-bundle that is twisted with the H-representation $H^1(C,\mathcal{O}_C)$, and $c_{g,h}$ is the top Chern class of the H-invariant part of this bundle. The following theorem is by no means obvious [**9**].

THEOREM 6.12 (Fantechi–Göttsche). *m preserves the grading and is associative.*

So finally, $H^*(X,G)$ has got a ring structure. It is not difficult to check that this ring structure is G-equivariant and therefore gives rise to a well-defined ring structure on the orbifold cohomology $H^*_{\mathrm{orb}}(X/G)$.

THEOREM 6.13. *Let X be a smooth projective surface with numerically trivial canonical divisor. Then there is a ring isomorphism*

$$H^*(\mathrm{Hilb}^n(X)) \cong H^*_{\mathrm{orb}}(X^n/\mathrm{S}_n).$$

I will restrict myself to a few comments on the logical dependence of the various ingredients for the proof:

(1) The theorem is not literally true unless one modifies the definition of the orbifold cohomology by making certain sign twists; see [**9**].

(2) The isomorphism of the underlying graded vector spaces follows directly from Göttsche's formula.

(3) Li, Qin and Wang [**28, 29**] have shown that the homogeneous components of the elements $\alpha^{[n]}$, $\alpha \in H^*(X)$, generate $H^*\big(\mathrm{Hilb}^n(X)\big)$ as a ring.

(4) Sorger and the author [**27**] constructed an abstract ring $H^*(X)^{[n]}$ and showed that $H^*\big(\mathrm{Hilb}^n(X)\big) \cong H^*(X)^{[n]}$. The key point being again the explicit formula of $c(\mathcal{O}_X^{[n]})$.

(5) Fantechi and Göttsche [**9**], and independently, Uribe [**36**], computed the orbifold cohomology ring $H^*_{\mathrm{orb}}(X^n/\mathrm{S}_n)$ and showed that $H^*_{\mathrm{orb}}(X^n/\mathrm{S}_n) \cong H^*(X)^{[n]}$.

Ruan proposed the following more general conjecture, of which the theorem above then verifies a special case:

CONJECTURE 6.14 (Ruan). *Let Y be a symplectic complex manifold and assume that the group G acts on Y preserving the symplectic structure. Assume furthermore that the quotient Y/G admits a crepant resolution $Z \to Y/G$. Then there is a ring isomorphism $H^*Z \cong H^*_{\mathrm{orb}}(Y/G)$.*

Besides the case of the Hilbert–Chow map $\rho\colon \mathrm{Hilb}^n(X) \to X^n/\mathrm{S}_n$ for $K3$ and abelian surfaces that is covered by Theorem 6.13 the following instances of Ruan's conjecture have been proved:

• The Kummer varieties $K_{n-1}(A)$ associated to an abelian surface A resolve the quotient A^{n-1}/S_n for the action of S_n on the subvariety $\cong A^{n-1}$ of all points in A^n of sum 0. This case was done by Britze [**3**] in his Ph.D. thesis, using calculations of Fantechi and Göttsche [**9**].

• Let $\Gamma \subset \mathrm{SL}_2(\mathbb{C})$ be a finite subgroup. The wreath product $\Gamma_n = \Gamma \wr \mathrm{S}_n$ acts on $(\mathbb{C}^2)^n$, and the n-th Hilbert scheme of a minimal resolution of $\mathbb{C}^2/\Gamma$ provides a crepant resolution of the quotient $\mathbb{C}^{2n}/\Gamma_n = S^n(\mathbb{C}^2/\Gamma)$. The conjecture was proved for this case by Wang [**39**].

• More recently, Ginzburg and Kaledin [**14**] proved the conjecture for the linear case, i.e., a symplectic action of a finite group G on a symplectic vector space V. The methods are very interesting and quite different from those discussed above.

References

1. J. Briançon, *Description de* $\mathrm{Hilb}^n\, \mathbb{C}\{x, y\}$, Invent. Math. **41** (1977), no. 1, 45–89.
2. T. Bridgeland, A. King, and M. Reid, *The McKay correspondence as an equivalence of derived categories*, J. Amer. Math. Soc. **14** (2001), no. 3, 535–554.
3. M. Britze, *On the cohomology of generalised Kummer varieties*, Dissertation, Universität zu Köln, 2002.
4. W. Chen, Y. Ruan, *Orbifold Gromov–Witten theory*, Orbifolds in Mathematics and Physics (Madison, WI, 2001), Contemp. Math., vol. 310, Amer. Math. Soc., Providence, RI, 2002, pp. 25–85.
5. G. Ellingsrud and L. Göttsche, *Hilbert schemes of points and Heisenberg algebras*, School on Algebraic Geometry (Trieste, 1999), ICTP Lect. Notes, vol. 1, Abdus Salam Int. Cent. Theoret. Phys., Trieste, 2000, pp. 59–100.
6. G. Ellingsrud and S. A. Strømme, *On the homology of the Hilbert scheme of points in the plane*. Invent. Math. **87** (1987), no. 2, 343–352.
7. ———, *Towards the Chow ring of the Hilbert scheme of* $\mathbb{P}^2$, J. Reine Angew. Math. **441** (1993), 33–44.
8. ———, *An intersection number for the punctual Hilbert scheme*, Trans. Amer. Math. Soc. **350** (1998), no. 6, 2547–2552.
9. B. Fantechi and L. Göttsche, *Orbifold cohomology for global quotients*, Duke Math. J. **117** (2003), no. 2, 197–227.
10. J. Fogarty, *Algebraic families on an algebraic surface*, Amer. J. Math. **90** (1968), 511–521.
11. E. Frenkel and D. Ben-Zvi, *Vertex Algebras and Algebraic Curves*, Math. Surveys Monogr., vol. 88, Amer. Math. Soc., Providence, RI, 2001.
12. I. Frenkel and W. Wang, *Virasoro algebra and wreath product convolution*, J. Algebra **242** (2001), no. 2, 656–671.
13. W. Fulton and R. MacPherson, *A compactification of configuration spaces*, Ann. of Math. (2) **139** (1994), no. 1, 183–225.
14. V. Ginzburg and D. Kaledin, *Poisson deformations of symplectic quotient singularities*, math.AG/0212279.
15. L. Göttsche, *The Betti numbers of the Hilbert scheme of points on a smooth projective surface*, Math. Ann. **286** (1990), no. 1-3, 193–207.
16. ———, *Hilbert schemes of zero-dimensional subschemes of smooth varieties*, Lecture Notes in Math., vol. 1572. Springer-Verlag, Berlin, 1994.
17. I. P. Goulden, *A differential operator for symmetric functions and the combinatorics of multiplying transpositions*, Trans. Amer. Math. Soc. **344** (1994), no. 1, 421–440.
18. I. Grojnowski, *Instantons and affine algebras. I. The Hilbert scheme and vertex operators*, Math. Res. Lett. **3** (1996), no. 2, 275–291.
19. A. Grothendieck, *Techniques de construction et théorèmes d'existence en géométrie algébrique. IV. Les schéas de Hilbert*, Séminaire Bourbaki, Vol. 6, Exp. No. 221, Soc. Math. France, Paris, 1995, pp. 249–276.

20. M. Haiman,*Hilbert schemes, polygraphs and the Macdonald positivity conjecture*, J. Amer. Math. Soc. **14** (2001), no. 4, 941–1006.
21. ———, *Vanishing theorems and character formulas for the Hilbert scheme of points in the plane*, Invent. Math. **149** (2002), no. 2, 371–407.
22. B. Iversen, *Linear determinants with applications to the Picard scheme of a family of algebraic curves*, Lectures Notes in Math., vol. 174, Springer-Verlag, Berlin–New York, 1970.
23. V. Kac, *Vertex algebras for beginners*, Univ. Lecture Ser., vol. 10, Amer. Math. Soc., Providence, RI, 1997.
24. M. Lehn, *Chern classes of tautological sheaves on Hilbert schemes of points on surfaces*, Invent. Math. **136** (1999), no. 1, 157–207.
25. ———, *Symplectic moduli spaces*, School and Conference on Intersection Theory and Moduli, ICTP Trieste, September 2002.
26. M. Lehn and C. Sorger, *Symmetric groups and the cup product on the cohomology of Hilbert schemes*, Duke Math. J. **110** (2001), no. 2, 345–357.
27. ———, *The cup product of Hilbert schemes for* K3 *surfaces*,. Invent. Math. **152** (2003), no. 2, 305–329.
28. W.-P. Li, Z. Qin, and W. Wang, *Generators for the cohomology ring of Hilbert schemes of points on surfaces*, Internat. Math. Res. Notices (2001), no. 20, 1057–1074.
29. ———, *Vertex algebras and the cohomology ring structure of Hilbert schemes of points on surfaces*, Math. Ann. **324** (2002), no. 1, 105–133.
30. ———, *Stability of the cohomology rings of Hilbert schemes of points on surfaces*, J. Reine Angew. Math. **554** (2003), 217–234.
31. E. Markman, *Generators of the cohomology ring of moduli spaces of sheaves on symplectic surfaces*, J. Reine Angew. Math. **544** (2002), 61–82.
32. H. Nakajima, *Heisenberg algebra and Hilbert schemes of points on projective surfaces*, Ann. of Math. (2) **145** (1997), no. 2, 379–388.
33. ———, *Lectures on Hilbert schemes of points on surfaces*, Univ. Lecture Ser., vol. 18, Amer. Math. Soc., Providence, RI, 1999.
34. Z. Qin and W. Wang, *Hilbert schemes and symmetric products: a dictionary*, Orbifolds in Mathematics and Physics (Madison, WI, 2001), Contemp. Math., vol. 310, Amer. Math. Soc., Providence, RI, 2002, pp. 233–257.
35. Y. Ruan, *Stringy geometry and topology of orbifolds.* Symposium in Honor of C. H. Clemens (Salt Lake City, UT, 2000), Contemp. Math., vol. 312, Amer. Math. Soc., Providence, RI, 2002, pp. 187–233.
36. B. Uribe, *Orbifold cohomology of the symmetric product*, math.AT/0109125.
37. E. Vasserot, *L'anneau de cohomologie du schéma de Hilbert de* $\mathbb{C}^2$, C. R. Acad. Sci. Paris Sér. I Math. **332** (2001), no. 1, 7–12.
38. W. Wang, *Algebraic structures behind Hilbert schemes and wreath products*, Recent Developments in Infinite-Dimensional Lie Algebras and Conformal Field Theory (Charlottesville, VA, 2000), Contemp. Math., vol. 297, Amer. Math. Soc., Providence, RI, 2002, pp. 271–295.
39. ———, *The Farahat–Higman ring of wreath products and Hilbert schemes*, math.QA/0205071.

Fachbereich Mathematik und Informatik, Johannes Gutenberg-Universität Mainz, 55099 Mainz, Germany

E-mail address: `lehn@mathematik.uni-mainz.de`

Centre de Recherches Mathématiques
CRM Proceedings and Lecture Notes
Volume **38**, 2004

Lectures on Instanton Counting

Hiraku Nakajima and Kōta Yoshioka

Dedicated to Professor Akihiro Tsuchiya on his sixtieth birthday

ABSTRACT. These notes have two parts. The first is a study of Nekrasov's deformed partition functions $Z(\varepsilon_1, \varepsilon_2, \vec{a}; \mathfrak{q}, \vec{\tau})$ of $N = 2$ SUSY Yang–Mills theories, which are generating functions of the integration in equivariant cohomology over the moduli spaces of instantons on $\mathbb{R}^4$. The second is a review of the geometry of the Seiberg–Witten curves and the geometric engineering of the gauge theory, which are the physical backgrounds of Nekrasov's partition functions.

The first part is a continuation of our previous paper [**69**], where we identified the Seiberg–Witten prepotential with $Z(0, 0, \vec{a}; \mathfrak{q}, 0)$. We associate higher Casimir operators to the partition function and clarify their relation to the Seiberg–Witten u-plane. We also determine the coefficients of $\varepsilon_1\varepsilon_2$ and $(\varepsilon_1^2 + \varepsilon_2^2)/3$ (the genus 1 part) of the partition function, which coincide with two measure factors A, B appearing in the u-plane integral. The proof is based on the blowup equation which we derived in [**69**].

1. Introduction

In this long introduction, we review the history of the Donaldson invariants and Seiberg–Witten geometry, which leads to Nekrasov's deformed partition function. This section contains no mathematically rigorous results, but provides the motivation for our study in later sections.

1.1. Donaldson invariants: a mathematical definition. Let X be a smooth, compact, oriented, 4-manifold with a Riemannian metric g with $b^+ \geq 1$ and odd. We also assume $\pi_1(X) = 1$ for simplicity. Let $P \to X$ be an SO(3)-bundle over X. Let $M(P)$ be the moduli space of irreducible anti-self-dual connections

2000 *Mathematics Subject Classification.* Primary 14D21; Secondary 57R57, 81T13, 81T60.

The first author is supported by the Grant-in-aid for Scientific Research (No. 15540023), JSPS.

The authors are grateful to N. Nekrasov for discussions, especially on perturbation terms and the genus 1 correction.

The authors would like to express their sincere gratitude to Professor Akihiro Tsuchiya for his encouragement to us and the organization of many workshops over the years. This work could not have been done without his enthusiasm for understanding physics.

This is the final form of the paper.

on P. This is a manifold of dimension $-2p_1(P) - 3(1 + b^+)$ for a generic metric g. Let $\mathcal{P} \to X \times M(P)$ be the universal bundle. Then the Donaldson invariant is a polynomial on $H_0(X) \oplus H_2(X)$ defined by

$$D_P(p^a S^b) = \int_{M(P)} \mu(p)^a \mu(S)^b, \quad p \in H_0(X),\ S \in H_2(X),$$

where $\mu\colon H_i(X) \to H^{4-i}(M(P))$ is given by the slant product $\mu(\,\cdot\,) = -\frac{1}{4}p_1(\mathcal{P})/\,\cdot$. Since $M(P)$ is not compact, we must justify the definition of the integration, and this can be done by using the Uhlenbeck compactification, as one can find in textbooks on Donaldson theory [**14, 26**].

We then formulate a generating function

$$D_\xi(p, S) = \sum_P \sum_{m,n \geq 0} D_P\left(\frac{S^m}{m!}\frac{p^n}{n!}\right),$$

where $\xi = w_2(P)$ is fixed. When $b^+ > 1$, D_ξ is independent of the metric and defines invariants of the differentiable structure of X. When $b^+ = 1$, it is piecewise constant as a function of g.

Although the invariants D_ξ can be defined, their calculation was difficult in general. This was because it is difficult to describe the moduli spaces $M(P)$ explicitly. The situation changed when Kronheimer and Mrowka [**40, 41**] proved a structure theorem for D_ξ in 1994: Although D_ξ involves infinitely many moduli spaces, it is determined by a finite amount of data if D_ξ satisfies a so-called *simple type* condition.

Soon afterward, Fintushel and Stern obtained the "blowup formula" which describes the relation between D_ξ on X and that on the blowup $\widehat{X}$ [**24**]. The formula involves an elliptic function. The underlying elliptic curve is related to the structure theorem so that the simple type condition means that it degenerates to a rational curve. The blowup formula will play a fundamental role in this paper.

1.2. Seiberg–Witten geometry. In 1988, Witten described Donaldson invariants as correlation functions of certain operators in a twisted version of $\mathcal{N} = 2$ SUSY (supersymmetric) Yang–Mills theory [**78**]. We do not explain what this statement means here, but mention that it is an infinite dimensional analogue of the Chern–Weil formula [**5**].

Shortly after [**40, 41**] appeared, Seiberg and Witten analyzed the original $\mathcal{N} = 2$ SUSY Yang–Mills theory with gauge group SU(2) [**74**]. The original theory is formulated on $\mathbb{R}^4$, and there was no mathematically rigorous definition of the "prepotential," which they calculated, at that time. Giving such a definition is one of the main purposes of these notes. (See Theorem 5.3.) But we present an "informal" definition here.

Let $H^*_{\mathrm{SU}(2)}(\mathrm{pt})$ be the SU(2)-equivariant cohomology of a point with complex coefficients. It is naturally identified with the Weyl group (in this case $\{\pm 1\}$) invariant part of the symmetric product of the dual of the (complexified) Cartan subalgebra $\mathfrak{h}$ (in this case $\mathbb{C}$). It is the coordinate ring of $\mathfrak{h}/W$. This space $\mathfrak{h}/W$ is the classical limit of the so-called *u-plane*, a family of "vacuum states," which plays a most important role in the Seiberg–Witten geometry.

The coordinate ring $A(\mathfrak{h}/W) = H^*_{\mathrm{SU}(2)}(\mathrm{pt})$ has a generator $-\frac{1}{2}\operatorname{tr}\left(\begin{smallmatrix}-a & 0\\ 0 & a\end{smallmatrix}\right)^2 = -a^2$, where a is considered as a coordinate on $\mathfrak{h}$. Let us denote it by u_{cl} since it is a coordinate of the classical limit of the u-plane. We make a "quantum correction" u

of the function $u_{\rm cl}$ by using the framed moduli space $M(2,n)$ of instantons on S^4. The precise definition will be given below, but it is roughly given by

$$u = -\sum_{n\geq 0} \Lambda^{4n} \int_{M(2,n)} \mu(p) \Big/ \sum_{n\geq 0} \Lambda^{4n} \int_{M(2,n)} 1. \tag{1.1}$$

Here Λ is a formal variable, the integration is done in the equivariant homology group, and μ is defined by the same formula as for Donaldson invariants. The moduli space $M(2,n)$ has an SU(2)-action given by the change of the framing. The classical part is the term $n = 0$; then $M(2,0)$ is a single point, so the integration is just an identity operator. In this case, $\mu(p) \in H^*_{\mathrm{SU}(2)}(\mathrm{pt})$ is nothing but the generator $-u_{\rm cl}$. Thus $u_{\rm cl}$ is the classical limit of u, as we explained.

When $n > 0$, the moduli space $M(2,n)$ is *noncompact* and we need to justify the integration. Here the problem is not a technical one, and has a very different nature from the noncompactness appearing in the definition of Donaldson invariants, which was overcome by Uhlenbeck compactification. In fact, if $M(2,n)$ had a suitable compactification, the integration of 1 would be 0 for degree reasons. The integration will be defined via the localization theorem in the equivariant homology group. The precise formulation will be given in §4.1. As the upshot, the integral does not take values in $H^*_{\mathrm{SU}(2)}(\mathrm{pt})$, but in its *fraction field.* (In fact, we need to consider an extra two-dimensional torus as below. Or we should consider u as an operator as in §6.2.) Thus u is a rational function on $\mathfrak{h}/W$. In the Seiberg–Witten geometry, the roles of u and $u_{\rm cl}$ are reversed. We define the *u-plane* as the parameter space for u, i.e., u is the coordinate of the u-plane. Then we consider $u_{\rm cl}$ (and a) as a rational function on the u-plane.

Other than the function u, there are several important geometric objects on the u-plane. They are defined via integration over the instanton moduli spaces. One of the most important objects is the *prepotential*, which has the form:

$$\mathcal{F}_0 = \mathcal{F}_0^{\rm pert} + \mathcal{F}_0^{\rm inst}, \tag{1.2}$$

where $\mathcal{F}_0^{\rm pert}$ is the *perturbative part* of the prepotential, which is an explicit rational function on $\mathfrak{h}/W$. The part $\mathcal{F}_0^{\rm inst}$ is the *instanton part*, and is a power series in Λ^4. The coefficient of Λ^{4n} is given by integration over $M(2,n)$. The u-plane is a *special Kähler* manifold, where the prepotential is included in its definition. For example, the Kähler metric is the imaginary part of the second derivative of the prepotential. See [**25**] for more details.

The main result of [**74**] is the determination of the u-plane and the prepotential $\mathcal{F}_0$. As a result, the u-plane is the parameter space for elliptic curves:

$$y^2 = (z^2 + u - 2\Lambda^2)(z^2 + u + 2\Lambda^2).$$

The prepotential $\mathcal{F}_0$ is given by using certain elliptic integrals. The original method used for the determination was a highly nontrivial physical argument. One of the most essential ingredients is understanding its behavior under the "duality" transformation $\tau \mapsto -1/\tau$, where τ is the period of the above elliptic curve, which is given by the second derivative of $\mathcal{F}_0$ with respect to the coordinate a. This is a rather mysterious transformation in view of the definition (1.2). In our approach, we will see theta functions quite naturally. So the duality will come from the Poisson summation formula, but we do not really understand its geometric origin.

Note that this picture is very similar to that of *mirror symmetry.* The prepotential above is the counterpart of the Gromov–Witten invariants and is on the "symplectic" side. The elliptic curves (Seiberg–Witten curves) are on the "complex" side. In fact, this is not just an analogy. The geometric engineering which will be reviewed in §7.5 explains the result as a special case of the mirror symmetry.

For later purposes, we give some functions explicitly. Let τ be the period of the Seiberg–Witten elliptic curve. Then

$$(1.3)\qquad u = -\frac{\theta_{00}^4+\theta_{10}^4}{\theta_{00}^2\theta_{10}^2}\Lambda^2, \qquad \frac{du}{da} = \frac{2\sqrt{-1}}{\theta_{00}\theta_{10}}\Lambda, \qquad a = \sqrt{-1}\frac{2E_2+\theta_{00}^4+\theta_{10}^4}{3\theta_{00}\theta_{10}}\Lambda.$$

Here $\theta_* = \theta_*(0|\tau)$ is the theta function and $E_2 = E_2(\tau)$ is the (normalized) second Eisenstein series. The reader should be careful when he/she compares these with the formulas in [**59**]. Our u (resp. a) is multiplied by -2 (resp. $2\sqrt{-1}$).

Finally note that the elliptic curve becomes singular at $u = \pm 2\Lambda^2$. In the classical limit $\Lambda \to 0$, these merge into a single point 0, which is the singular point in the classical u-plane $\mathfrak{h}/W$.

1.3. The u-plane integral. We return back to Donaldson invariants. Witten [**79**] explained that D_ξ has three contributions:

$$D_\xi(p,S) = Z_u(p,S) + Z_+(p,S) + Z_-(p,S).$$

The parts $Z_\pm(p,S)$ come from the measure supported on the singularity $\pm 2\Lambda^2$ of the u-plane. These are given by invariants defined via the moduli spaces of monopoles, called Seiberg–Witten invariants. As for application to topology, Z_u is irrelevant as it depends only on $H^2(X,\mathbb{Z})$. Furthermore, Z_u vanishes when $b_+ > 1$. But we are interested in structures of instanton moduli spaces which are reflected in Z_u.

When $b_+ = 1$, a more precise description of D_ξ was given by Moore and Witten [**59**]. (See also [**50, 51**] for similar results.) We briefly recall their description, since some parts are closely related to our study. The parts $Z_\pm(p,S)$ are written in terms of the Seiberg–Witten invariants summed over various choices of Spin^c structures. See [**59**, §7] for the explicit expression. The remaining part Z_u is the integration with respect to a smooth volume form. It is called the *u-plane integral.* We choose and fix a harmonic self-dual two form ω with $\int_X \omega\wedge\omega = 1$. This is unique up to sign, and the choice of ω is related to the orientation of the moduli space. We also put $\Lambda = 1$. Then

$$(1.4)\qquad Z_u(p,S) = \int_{u\text{-plane}} da\, d\bar{a}\, A(u)^\chi B(u)^\sigma e^{pu+S^2T}\Psi,$$

with

$$A(u) = \alpha\left(\frac{du}{da}\right)^{1/2}, \qquad B(u) = \beta(u^2-4)^{1/8}, \qquad T = \frac{1}{24}\left(\frac{du}{da}\right)^2 E_2(\tau) - \frac{1}{6}u,$$

$$\Psi = -\frac{\sqrt{-2}}{4y^{1/2}}\frac{d\bar\tau}{d\bar a}\exp\left[\frac{1}{8\pi y}\left(\frac{du}{da}\right)^2 S_+^2\right] e^{2\pi\sqrt{-1}\lambda_0^2}$$
$$\times \sum_{\lambda\in H^2+\frac12\xi} (-1)^{(\lambda-\lambda_0)\cdot w_2(X)}\left[(\lambda,\omega) - \frac{1}{4\pi y}\frac{du}{da}(S,\omega)\right]$$
$$\times \exp\left[-\sqrt{-1}\pi\bar\tau\lambda_+^2 - \sqrt{-1}\pi\tau\lambda_-^2 + \frac{du}{da}(S,\lambda_-)\right].$$

Here χ (resp. σ) is the Euler number (resp. signature) of X, α, β are universal constants independent of X, $\tau = x+iy$, λ_0 is a fixed element in $\frac12\xi + H^2(X,\mathbb{Z})$, and $(\,\cdot\,)_\pm$ denotes the self-dual and anti-self-dual part of $\cdot$ respectively.

Since this is a divergent integral, we must regularize it. See the original paper [59] for how it is done.

The term T is called a *contact term*. Its determination can be done in several ways. In [50], by equating the answers given by various methods, a nontrivial equation was derived. This is the *contact term equation*, which will be important for our study. See Theorems 2.4 and 5.3. The terms A, B come from a Riemannian metric g.

Let us analyze the effect of the blowup $\widehat{X} \to X$ on the u-plane integral since it is closely related to our study. Let C be the exceptional curve. We want to evaluate $Z_u(p, S+tC)$, where $S \in H_2(X)$ is considered as a class of $H_2(\widehat{X})$ via the projection.

Since $\chi(\widehat{X}) = \chi(X)+1$, $\sigma(\widehat{X}) = \sigma(X)-1$, the factor $A(u)^\chi B(u)^\sigma$ is multiplied by

$$\frac{A(u)}{B(u)} = \frac{\alpha}{\beta}(u^2-4)^{-1/8}\left(\frac{du}{da}\right)^{1/2} = \frac{1}{\theta_{01}} \quad \text{(up to a constant)}.$$

We work in the chamber $C_+ = 0$, so we have

$$\frac{\Psi_{\widehat{X}}}{\Psi_X} = \sum_{n \in \mathbb{Z}+\frac{1}{2}w_2(\widetilde{P})\cdot C} (-1)^n \exp\left[\sqrt{-1}\pi\tau n^2 - nt\frac{du}{da}\right] = \theta_*\left(\frac{t\sqrt{-1}}{2\pi}\frac{du}{da}\middle|\tau\right),$$

where $* = 01$ or 11 according as $w_2(\widetilde{P}) \cdot C = 0$ or 1. Therefore we get

$$\frac{Z_u(p, S+tC)}{Z_u(p,S)} = \exp(-Tt^2)\frac{\theta_*\left((t\sqrt{-1}/2\pi)du/da|\tau\right)}{\theta_{01}(0|\tau)}$$

up to a constant multiple. The constant turns out to be 1 as the left-hand side is 1 at $t = 0$ when $* = 01$.

1.4. Nekrasov's deformed partition function. As we explained, the prepotential $\mathcal{F}_0$ was given as integration over instanton moduli spaces. Before Nekrasov gave an explicit expression [70], it was written in terms of differential forms on moduli spaces. So it was difficult to calculate, understand its meaning, etc. (See [15].) Nekrasov's idea was to use an extra two-dimensional torus action and apply the localization theorem in equivariant homology. Technically it was also important that the Uhlenbeck (partial) compactification of the moduli space has a nice resolution of singularities introduced by the first author [66]. (The latter space will be denoted by $M(2,n)$ in the main body of the paper.) Let ε_1, ε_2 be two generators of $H^*_{T^2}(\mathrm{pt})$. Then we define

$$F = \varepsilon_1\varepsilon_2 F^{\mathrm{pert}} + \varepsilon_1\varepsilon_2 \log\left(\sum_{n\geq 0}\Lambda^{4n}\int_{M(2,n)} 1\right), \tag{1.5}$$

where F^{pert} is a certain two-parameter deformation of $\mathcal{F}_0^{\mathrm{pert}}$. Each coefficient of Λ^{4n} is a rational function in ε_1, ε_2, and is mathematically rigorously defined. Nekrasov conjectured that $F|_{\varepsilon_1,\varepsilon_2=0}$ is equal to $\mathcal{F}_0$, given by the Seiberg–Witten curve. This is mathematically meaningful statement. This conjecture was proved in [69] and [71] by totally different methods.

The method used in [71] was geometric and a standard technique in the study of Donaldson invariants. We consider the instanton moduli spaces $\widehat{M}(2, c_1, n)$ on the blowup, introduce an operator $\mu(C)$ in this equivariant setting, and compute this equivariant analog of Donaldson invariants. From the explicit expression given by the localization theorem, it is very easy to derive the blowup formula in a combinatorial form. On the other hand, a simple dimension counting argument shows that $\int_{\widehat{M}(2,0,n)} \mu(C)^2 = 0$. This vanishing gives a differential equation satisfied by the original F. We call it *the blowup equation.* (See (5.1).) It characterizes F. When we put $\varepsilon_1 = \varepsilon_2 = 0$, this equation turns out to be the contact term equation, which we mentioned. Since the contact term equation can be derived from the Seiberg–Witten curve in a mathematically rigorous way (see §2), this gives a proof of Nekrasov's conjecture.

1.5. Gravitational corrections. After identifying $F|_{\varepsilon_1,\varepsilon_2=0}$ with the Seiberg–Witten prepotential (2.4), it becomes natural to ask for the meaning of higher-order terms in the expansion

$$F = F_0 + (\varepsilon_1 + \varepsilon_2)H + \varepsilon_1\varepsilon_2 A + \frac{\varepsilon_1^2 + \varepsilon_2^2}{3} B + \cdots .$$

Nekrasov asserted that these are *gravitational corrections* to the gauge theory [**70**, §4]. Using the differential equation mentioned above, we prove that these A, B coincides with those A, B appearing in the u-plane integrand. (H turns out to be a simple function.) (The calculation was done jointly with N. Nekrasov.)

Moreover, by geometric engineering [**39**] (see §7.5), we can expect that these terms are certain limits of higher genus Gromov–Witten invariants for a noncompact Calabi–Yau 3-fold, in this case the canonical bundle of $\mathbb{P}^1 \times \mathbb{P}^1$. More precisely, we put $\varepsilon_1 = -\varepsilon_2 = \hbar$ and consider

$$F = F_0 + F_1\hbar^2 + F_2\hbar^4 + \cdots .$$

Then F_g is a limit of the genus g Gromov–Witten invariants. Since $(1/\varepsilon_1\varepsilon_2)F$ is more fundamental (see (1.5)), we should write this as

$$\frac{1}{\hbar^2}F = \sum_{g=0}^{\infty} \hbar^{2g-2}F_g.$$

This is more natural as $2 - 2g$ is the Euler number of a genus g Riemann surface. It probably explains the singularity $1/\hbar^2$.

Recently many Gromov–Witten invariants for noncompact toric Calabi–Yau have been calculated (see [**1**] and the references therein). These are identified with the Jones–Witten invariants via the geometric transition (called "large N duality"), as first proposed by Gopakumar and Vafa [**28**]. The first of such examples is the identification of Gromov–Witten for the resolved conifold and the SU(N) Jones–Witten invariant for S^3. These identifications have been proved in a mathematical rigorous way in a number of examples (see [**73, 81**]).

In the case of $K_{\mathbb{P}^1\times\mathbb{P}^1}$, the Jones–Witten side is the SU(N)-invariant for the Hopf link. Using the calculation by Morton and Lukac [**61**], Iqbal and Kashani-Poor show that the invariant of the Hopf link has the same combinatorial expression as that of F given by the localization formula [**37, 38**]. (See also [**19**].)

Note that these results identify the n-instanton correction with the Gromov–Witten invariants of degree n (with respect to one of the factors of $\mathbb{P}^1 \times \mathbb{P}^1$) for each n. Thus they do not say much about the structure of the generating function F, which is studied in this paper. Therefore it is interesting to understand the blowup equation from the Gromov–Witten side.

2. Seiberg–Witten curves

In this section we introduce the Seiberg–Witten curves, give the definition of the prepotential, and derive the renormalization equation and the contact term equation, which will characterize the prepotential.

We give some details, though one can find most of them in the physics literature. The reason is that we must carefully choose cycles on the Seiberg–Witten curve to determine the characteristic of the theta function in a mathematically rigorous way. It is a standard exercise but we cannot find the argument in the literature.

The material discussed here is a minimal amount of Seiberg–Witten geometry. We omit many things, such as monodromies, Picard–Fuchs equations, relations to integrable systems, etc. Even for the differential equations satisfied by the prepotential, our treatment is minimal. The Whitham hierarchy underlying these equations will not be discussed. The reader may wonder where these equations come from, though the authors' approach through the instanton moduli spaces will be explained in §5. For the original approaches, see [**55, 57**] and the references therein.

There is a nice survey article [**12**] for mathematicians which describes the relation between integrable systems and the Seiberg–Witten geometry, as well as background in physics. We recommend it to our readers since it has no overlap with this paper.

Note that we multiply a_α by $-\sqrt{-1}$ from the conventional usage in order to match with its appearance in instanton counting.

2.1. Definition of the Seiberg–Witten prepotential. We consider a family of curves (Riemann surfaces) parametrized by $\vec{u} = (u_2, \dots, u_r)$:

$$C_{\vec{u}} : \Lambda^r \left(w + \frac{1}{w} \right) = P(z) = z^r + u_2 z^{r-2} + u_3 z^{r-3} + \cdots + u_r.$$

We call them *Seiberg–Witten curves.* The projection $C_{\vec{u}} \ni (w, z) \mapsto z \in \mathbb{P}^1$ gives the structure of hyperelliptic curves. The hyperelliptic involution ι is given by $\iota(w) = 1/w$.

If we introduce $y = \Lambda^r(w - 1/w)$, we have

$$y^2 = P(z)^2 - 4\Lambda^{2r} = (P(z) - 2\Lambda^r)(P(z) + 2\Lambda^r).$$

This special form of the right-hand side will play a crucial role later.

The parameter space $\{\vec{u} \in \mathbb{C}^{r-1}\}$ is called the *u-plane.* Here Λ is also a parameter, but we treat it separately from $\vec{u}$. The parameter Λ is called the *renormalization scale* in physics. When $\Lambda = 0$, the theory goes to the *classical limit.* We consider $\vec{u} = (u_2, \dots, u_r)$ as a coordinate system on the u-plane. This is a global coordinate.

Let $z_1, \dots, z_r$ be the solutions of $P(z) = 0$. We will work on a region of the u-plane where $|z_\alpha - z_\beta|$, $|z_\alpha|$ are much larger than $|\Lambda|$, and then analytically continue. In particular the z_α's are distinct. The vector $\vec{z} = (z_1, \dots, z_r)$ ($\sum_\alpha z_\alpha = 0$) is a local coordinate on the u-plane. The relation between $\vec{z}$ and $\vec{u}$ is very simple. The former is a coordinate on $\mathbb{C}^{r-1}$ while the latter is on $\mathbb{C}^{r-1}/S_r \approx \mathbb{C}^{r-1}$, where S_r is the symmetric group on r letters. In other words, $(-1)^p u_p$ is the pth elementary symmetric function in $z_1, \dots, z_r$. It is better to keep this simple relation in mind, since this coordinate system $\vec{z}$ is a *quantum correction* of another coordinate system $\vec{a}$ introduced below.

We can find $z_\alpha^\pm$ near z_α such that $P(z_\alpha^\pm) = \pm 2\Lambda^r$ when $|u| \gg |\Lambda|$. These are the $2r$-branch points of the projection $C_{\vec{u}} \to \mathbb{P}^1$. Infinity is not a branch point, and its inverse image consists of ∞_+ $(w = \infty)$ and ∞_- $(w = 0)$. The genus of $C_{\vec{u}}$ is $r-1$. In the classical limit $\Lambda \to 0$, both $z_\alpha^\pm$ go to z_α, and the curves develop singularities.

Let us define the *quantum discriminant* by

$$\Delta = (4\Lambda^r)^{2r} \prod_{\alpha<\beta} (z_\alpha^+ - z_\beta^+)^2 (z_\alpha^- - z_\beta^-)^2. \tag{2.1}$$

On the locus $\Delta = 0$, the Seiberg–Witten curves develop singularities. As we mentioned, we study a region away from this locus.

The hyperelliptic curve $C_{\vec{u}}$ is made up of two copies of the Riemann sphere, glued along the r-cuts between z_α^- and z_α^+ $(\alpha = 1, \dots, r)$, as usual. Let A_α be the cycle encircling the cut between z_α^- and z_α^+. We have $\sum_\alpha A_\alpha = 0$. We draw $C_{\vec{u}}$ as in Figure 1. The hyperelliptic involution ι is the rotation by π about the axis passing through the branch points $z_\alpha^\pm$. Then we choose cycles B_α $(\alpha = 2, \dots, r)$ as in Figure 1 so that $\{A_\alpha, B_\alpha \mid \alpha = 2, \dots, r\}$ form a symplectic basis of $H_1(C_{\vec{u}}, \mathbb{Z})$, i.e., $A_\alpha \cdot A_\beta = 0 = B_\alpha \cdot B_\beta$, $A_\alpha \cdot B_\beta = \delta_{\alpha\beta}$ for α, $\beta = 2, \dots, r$. (The cycle A_1 is omitted.) In the figure the branch points are lined up as z_1^+, z_1^-, z_2^-, $z_2^+, \dots$, from the left. That is z_α^+ is on the left (resp. right) of z_α^- for α odd (resp. even).

Note that we *cannot* take the A, B-cycles *globally* on the u-plane. The cycles are transformed by monodromies around the locus $\Delta = 0$. In fact, the study of monodromies is important as it has been used for consistency checks of the Seiberg–Witten curves related to some physically expected properties of the prepotential (introduced below). However, we do not study the monodromy behavior here except that around $\Lambda = 0$. We first fix a small region in the u-plane and then analytically continue. We choose a region containing the part in which z_α's are real and satisfy $z_1 > z_2 > \cdots > z_r$. We also assume Λ is a positive real number. Since we assume Λ small, we have $z_1^+ > z_1 > z_1^- > z_2^- > z_2 > z_2^+ > \cdots$. This choice determines A, B-cycles as in Figure 1. Note that the branch points are lined up from the right by the order in Figure 1. Thus the choice is natural in this region. Note also that we choose the inverse image of the region with respect to the projection $\mathbb{C}^{r-1} \to \mathbb{C}^{r-1}/S_r$. The permutation ambiguity is less important than the monodromies, but we use this choice as we want to specify what is a_α.

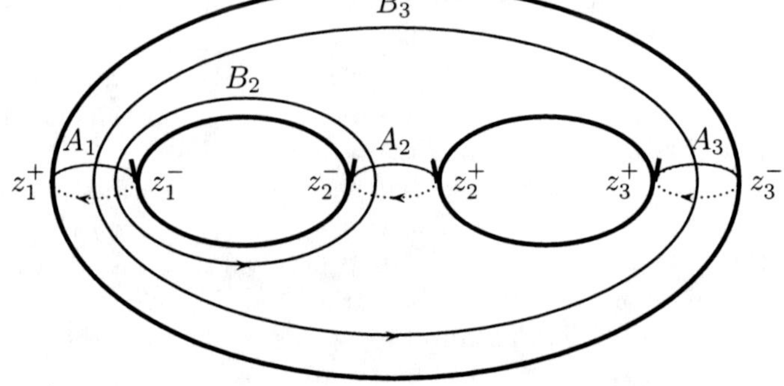

FIGURE 1. Seiberg–Witten curve and cycles $(r = 3)$

Let us define the *Seiberg–Witten differential* by

$$dS = -\frac{1}{2\pi} z \frac{dw}{w} = -\frac{1}{2\pi} \frac{zP'(z)\,dz}{\sqrt{P(z)^2 - 4\Lambda^{2r}}} = -\frac{1}{2\pi} \frac{zP'(z)\,dz}{y}.$$

It is a meromorphic differential having poles at $\infty_\pm$. We define functions a_α, a_β^D on the u-plane ($|u| \gg |\Lambda|$) by

$$a_\alpha = \int_{A_\alpha} dS, \quad \alpha = 1, \dots, r, \qquad a_\beta^D = 2\pi\sqrt{-1} \int_{B_\beta} dS, \quad \beta = 2, \dots, r. \tag{2.2}$$

Let us study the behavior of the function a_α around $\Lambda = 0$. We move the cycle A_α so that $P(z)$ and $1/P(z)$ are bounded there. In particular, we are in a sheet where $\sqrt{P(z)^2 - 4\Lambda^{2r}}$ is single-valued. We choose the sheet so that it is approximated by $P(z)$ on the A_α-cycle. We suppose A_α has counterclockwise rotation in the sheet. In Figure 1, the sheet is the part lower than the plane containing the A_α's. (See also the proof of Proposition 2.2 below.) Then we have the following expansion:

$$a_\alpha = -\frac{1}{2\pi} \int_{A_\alpha} dz\, z \frac{P'(z)}{P(z)} \sum_{n=0}^{\infty} \binom{\frac{1}{2}}{n} \left(\frac{4\Lambda^{2r}}{P(z)^2}\right)^n = -\sqrt{-1} z_\alpha + O(\Lambda^{2r}), \tag{2.3}$$

where $\binom{a}{n}$ is the binomial coefficient. In particular, $\vec{a} = (a_1, \dots, a_r)$ ($\sum_\alpha a_\alpha = 0$) is a local coordinate system for small Λ. As we mentioned before, the coordinate $\vec{z}$ is the quantum correction of $\vec{a}$. a_α is a function of u_p, but conversely we consider u_p as a function of a_α (and also of Λ).

We differentiate the Seiberg–Witten differential dS by setting w to be constant:

$$\frac{\partial}{\partial u_p} dS \Big|_{w=\text{const}} = \frac{1}{2\pi} \frac{z^{r-p}}{P'(z)} \frac{dw}{w} = \frac{1}{2\pi} \frac{z^{r-p}\,dz}{y}.$$

It is well-known that these form a basis of holomorphic differentials on $C_{\vec{u}}$ for $p = 2, \dots, r$ (see, e.g., [**33**, §2.3]). In other words, the Seiberg–Witten differential is a "*potential*" for holomorphic differentials. Let $(\sigma_{\alpha p})$ be the matrix given by

$$\sigma_{\alpha p} = \frac{\partial a_\alpha}{\partial u_p} = \frac{1}{2\pi} \int_{A_\alpha} \frac{z^{r-p}}{P'(z)} \frac{dw}{w}, \quad \alpha, p = 2, \dots, r.$$

If $(\sigma^{p\alpha})$ is the inverse matrix, the normalized holomorphic 1-forms

$$\omega_\beta = \frac{1}{2\pi} \sum_p \sigma^{p\beta} \frac{z^{r-p}}{P'(z)} \frac{dw}{w} = \frac{\partial}{\partial a_\beta} dS \Big|_{w=\text{const}}$$

satisfy $\int_{A_\alpha} \omega_\beta = \delta_{\alpha\beta}$. Therefore the *period matrix* $\tau = (\tau_{\alpha\beta})$ of the curve $C_{\vec{u}}$ is given by

$$\tau_{\alpha\beta} = \int_{B_\alpha} \omega_\beta = \frac{1}{2\pi\sqrt{-1}} \frac{\partial a_\alpha^D}{\partial a_\beta}.$$

Since $(\tau_{\alpha\beta})$ is symmetric (see, e.g., [**33**, §2.2]), there exists a locally defined function $\mathcal{F}_0$ on the u-plane such that

$$a_\alpha^D = -\frac{\partial \mathcal{F}_0}{\partial a_\alpha}. \tag{2.4}$$

It is unique up to a constant independent of a_α. This function $\mathcal{F}_0$ is called the *Seiberg–Witten prepotential.* We may also write $\mathcal{F}_0(\vec{a})$ or $\mathcal{F}_0(\vec{a};\Lambda)$. From the definition we have

$$\tau_{\alpha\beta} = -\frac{1}{2\pi\sqrt{-1}}\frac{\partial^2\mathcal{F}_0}{\partial a_\alpha \partial a_\beta}. \tag{2.5}$$

We use the subscript 0 because this will be identified with the genus 0 part of Nekrasov's deformed partition function.

2.2. Homogeneity. We regard u_p, a_α^D as functions of a_β and Λ. Then

LEMMA 2.1.

$$\begin{aligned}\sum_\alpha a_\alpha \frac{\partial u_p}{\partial a_\alpha} + \Lambda\frac{\partial u_p}{\partial \Lambda} &= p u_p,\\ \sum_\alpha a_\alpha \frac{\partial a_\beta^D}{\partial a_\alpha} + \Lambda\frac{\partial a_\beta^D}{\partial \Lambda} &= a_\beta^D.\end{aligned}$$

PROOF. We differentiate the following equation by setting w to be constant,

$$\Lambda^r\left(w + \frac{1}{w}\right) = P(z),$$

to get

$$rP(z) = \Lambda\frac{\partial z}{\partial\Lambda}P'(z) + \sum_p \Lambda\frac{\partial u_p}{\partial\Lambda}z^{r-p}.$$

Thus

$$\Lambda\frac{\partial z}{\partial\Lambda} = \frac{rP(z)}{P'(z)} - \sum_p \Lambda\frac{\partial u_p}{\partial\Lambda}\frac{z^{r-p}}{P'(z)}.$$

Since $rP(z) - zP'(z) = \sum_p p u_p z^{r-p}$,

$$\frac{rP(z)}{P'(z)} = z + \sum_p p u_p \frac{z^{r-p}}{P'(z)}.$$

We substitute this into the above equation to get

$$\Lambda\frac{\partial}{\partial\Lambda}dS\bigg|_{w=\text{const}} = dS - \sum_p\left(p u_p - \Lambda\frac{\partial u_p}{\partial\Lambda}\right)\frac{\partial}{\partial u_p}dS\bigg|_{w=\text{const}}. \tag{2.6}$$

By integrating over A_α, we get

$$0 = \Lambda\frac{\partial a_\alpha}{\partial\Lambda} = a_\alpha - \sum_p\left(p u_p - \Lambda\frac{\partial u_p}{\partial\Lambda}\right)\frac{\partial a_\alpha}{\partial u_p}.$$

Multiplying by the matrix $(\partial u_p/\partial a_\alpha) = (\partial a_\alpha/\partial u_p)^{-1}$, we obtain

$$0 = \sum_\alpha a_\alpha\frac{\partial u_p}{\partial a_\alpha} - \left(p u_p - \Lambda\frac{\partial u_p}{\partial\Lambda}\right). \tag{2.7}$$

This is nothing but the first equation.

On the other hand, from (2.6), we get

$$\Lambda \frac{\partial a^D_\alpha}{\partial \Lambda} - a^D_\alpha \quad \sum_p \left(p u_p - \Lambda \frac{\partial u_p}{\partial \Lambda} \right) \frac{\partial a^D_\alpha}{\partial u_p}$$
$$= a^D_\alpha - \sum_p \sum_\alpha a_\alpha \frac{\partial u_p}{\partial a_\alpha} \frac{\partial a^D_\alpha}{\partial u_p} = a^D_\alpha - \sum_\alpha a_\alpha \frac{\partial a^D_\alpha}{\partial a_\alpha},$$

where we have used (2.7) in the second equality. This is nothing but the second equation. □

As a corollary, we get

$$\frac{\partial}{\partial a_\beta} \left(\sum_\alpha a_\alpha \frac{\partial \mathcal{F}_0}{\partial a_\alpha} + \Lambda \frac{\partial \mathcal{F}_0}{\partial \Lambda} - 2\mathcal{F}_0 \right) = 0$$

for all β. Hence we can choose $\mathcal{F}_0$ so that

$$\sum_\alpha a_\alpha \frac{\partial \mathcal{F}_0}{\partial a_\alpha} + \Lambda \frac{\partial \mathcal{F}_0}{\partial \Lambda} = 2\mathcal{F}_0. \tag{2.8}$$

This together with (2.4) determines the prepotential $\mathcal{F}_0$ up to (constant) $\cdot\, \Lambda^2$. We finally fix the constant so that the instanton part (see below) is a power series of Λ^{2r}, and has no quadratic term. Thus the prepotential is uniquely determined.

2.3. The logarithmic singularities of the Seiberg–Witten prepotential. Let us study the behavior of $\mathcal{F}_0$ as a function of Λ, following [**11**]. Our aim is to show

PROPOSITION 2.2. *We have*

$$\mathcal{F}_0 = \sum_{\alpha \neq \beta} \gamma_0(a_\alpha - a_\beta; \Lambda) + O(\Lambda^{2r})$$
$$= \sum_{\alpha \neq \beta} \left[\frac{1}{2}(a_\alpha - a_\beta)^2 \log\left(\frac{a_\alpha - a_\beta}{\Lambda} \right) - \frac{3}{4}(a_\alpha - a_\beta)^2 \right] + O(\Lambda^{2r}),$$

where $\gamma_0(x; \Lambda) = \frac{1}{2}x^2 \log(x/\Lambda) - \frac{3}{4}x^2$ *is the coefficient of* $1/\varepsilon_1\varepsilon_2$ *in* $-\gamma_{\varepsilon_1,\varepsilon_2}(x;\Lambda)$ *as in* (E.3).

The part $\sum_{\alpha\neq\beta} \gamma_0(a_\alpha - a_\beta; \Lambda)$ is called the *perturbative part* of the prepotential $\mathcal{F}_0$, and is denoted by $\mathcal{F}_0^{\text{pert}}$. The remaining part is called the *instanton part*, and is denoted by $\mathcal{F}_0^{\text{inst}}$. It is a power series in Λ^{2r}: $\mathcal{F}_0^{\text{inst}} = f_1\Lambda^{2r} + f_2\Lambda^{4r} + \cdots + f_n\Lambda^{2rn} + \cdots$. The coefficient f_n is called the *nth instanton correction* to the prepotential. This is because we will identify f_n with something defined via the n-instanton moduli space.

The choice of the branch of log is as follows. Suppose that z_α, $z_\alpha^\pm \in \mathbb{R}$, $\Lambda \in \mathbb{R}_{>0}$ and $z_1 > z_2 > \cdots > z_r$ as above. We choose a path A_α encircling z_α^- and z_α^+ so that it is invariant under the complex conjugation $z \mapsto \bar{z}$. Then a_α is purely imaginary. We have $\sqrt{-1}a_1 > \sqrt{-1}a_2 > \cdots > \sqrt{-1}a_r$. We choose the branch of log so that

$$\frac{1}{2}\log\left(\frac{a_\alpha - a_\beta}{\Lambda}\right) + \frac{1}{2}\log\left(\frac{a_\beta - a_\alpha}{\Lambda}\right) = \log\left(\frac{\sqrt{-1}(a_\alpha - a_\beta)}{\Lambda}\right)$$

is real for $\alpha < \beta$. In what follows, we assume this choice of z_α, etc. It is enough to consider this case, by analytic continuation.

PROOF OF PROPOSITION 2.2. First we study

$$a_\alpha^D = 2\pi\sqrt{-1}\int_{B_\alpha} dS.$$

Locally, this is a function of Λ^{2r}. But it is multivalued, as the cycle B_α transforms to $B_\alpha + A_\alpha - A_1$ when we analytically continue from Λ^{2r} to $e^{2\pi\sqrt{-1}}\Lambda^{2r}$. Therefore $a_\alpha^D + \sqrt{-1}(z_\alpha - z_1)\log\Lambda^{2r}$ is a single-valued function of Λ. This kind of monodromy behavior is quite important in the conventional arguments.

For $\beta = 2, \dots, r$, let C'_β be the straight line from $z^{\pm}_{\beta-1}$ to $z^{\pm}_\beta$ (+ for β odd, − for β even) in one sheet and define the cycle C_β as C'_β followed by $-\iota C'_\beta$. This is a cycle encircling the hole in Figure 1. We have

$$B_\alpha = \sum_{\beta=2}^{\alpha} C_\beta,$$

and

$$-2\pi\int_{C_\beta} dS = 2\int_{z^{\pm}_{\beta-1}}^{z^{\pm}_\beta} \frac{zP'(z)dz}{\sqrt{P(z)^2 - 4\Lambda^{2r}}}.$$

In the last expression, z is real. But we should be careful in the choice of the branch of $\sqrt{P(z)^2 - 4\Lambda^{2r}}$. This is not necessarily ≥ 0, contrary to the usual convention for a real function. It is determined by analytic continuation. We choose the sheet so that $\sqrt{P(z)^2 - 4\Lambda^{2r}}$ has the same sign as $P(z)$ on each interval $[z^{\pm}_{\beta-1}, z^{\pm}_\beta]$. This is the same sheet used in (2.3) (i.e., the lower half) and we have the right orientation so that $A_\alpha \cdot B_\alpha = 1$. Note also that a_α^D is pure imaginary as the integrals are real.

Fix $\delta > 0$ small with $|\Lambda| \ll |\delta|$ and rewrite the integral as

$$-\pi\int_{C_\beta} dS = \left(\int_{z_{\beta-1}+\delta}^{z_\beta-\delta} + \int_{z_\beta-\delta}^{z^{\pm}_\beta} - \int_{z_{\beta-1}+\delta}^{z^{\pm}_{\beta-1}}\right)\frac{zP'(z)dz}{\sqrt{P(z)^2 - 4\Lambda^{2r}}}. \tag{2.9}$$

The first integral is regular at $\Lambda = 0$:

$$\begin{aligned}\int_{z_{\beta-1}+\delta}^{z_\beta-\delta}\frac{zP'(z)dz}{P(z)} + O(\delta) &= \int_{z_{\beta-1}+\delta}^{z_\beta-\delta}\sum_{\gamma=1}^{r}\left(1 + \frac{z_\gamma}{z - z_\gamma}\right)dz + O(\delta)\\ &= z_\beta(r + \log\delta) - z_{\beta-1}(r + \log\delta) + \sum_{\gamma\neq\beta} z_\gamma\log|z_\beta - z_\gamma|\\ &\qquad - \sum_{\gamma\neq\beta-1} z_\gamma\log|z_{\beta-1} - z_\gamma| + O(\delta).\end{aligned}$$

Here we choose the branch of log so that all the above expressions are real.

The second integral of (2.9) is equal to

$$z_\beta\int_{z_\beta-\delta}^{z^{\pm}_\beta}\frac{P'(z)\,dz}{\sqrt{P(z)^2 - 4\Lambda^{2r}}} + \int_{z_\beta-\delta}^{z^{\pm}_\beta}\frac{(z - z_\beta)P'(z)\,dz}{\sqrt{P(z)^2 - 4\Lambda^{2r}}}. \tag{2.10}$$

The first term is

$$z_\beta\int_{w=w_\beta(\delta)}^{w=\pm 1}\frac{dw}{w} = -z_\beta\log|w_\beta(\delta)|,$$

where

$$w_\beta(\delta) = \frac{1}{2\Lambda^r}\left(P(z_\beta - \delta) + \sqrt{P(z_\beta - \delta)^2 - 4\Lambda^{2r}}\right).$$

We have

$$\log|w_\beta(\delta)| = \log\frac{|P(z_\beta-\delta)|}{2\Lambda^r} + \log\left(1+\sqrt{1-\frac{4\Lambda^{2r}}{P(z_\beta-\delta)^2}}\right)$$
$$= \log\left(\frac{|\delta\prod_{\gamma\neq\beta}(z_\beta-z_\gamma)|}{\Lambda^r}\right) + O(\delta).$$

Let us consider the second term of (2.10). Let $z-z_\beta = \prod_{\gamma\neq\beta}(z_\beta-z_\gamma)^{-1}P(z)+E(z)$. We have $E(z)=O(\delta^2)$ in the range of integration. But the above calculation of the first part shows that the integral of $E(z)$ yields $O(\delta^2)O(\log\delta)=O(\delta)$. Therefore the second term is

$$\frac{1}{\prod_{\gamma\neq\beta}(z_\beta-z_\gamma)}\int_{z_\beta^\pm}^{z_\beta-\delta}\frac{P(z)P'(z)\,dz}{\sqrt{P(z)^2-4\Lambda^{2r}}} + O(\delta)$$
$$= \frac{1}{\prod_{\gamma\neq\beta}(z_\beta-z_\gamma)}\left[\sqrt{P(z)^2-4\Lambda^{2r}}\right]_{z_\beta^\pm}^{z_\beta-\delta} + O(\delta).$$

But as $P(z_\beta^\pm)=2\Lambda^r$, $P(z_\beta-\delta)=O(\delta)$, the contribution is $O(\delta)$.

The third integral has a similar expression with z_β, δ replaced by $z_{\beta-1}$, $-\delta$ respectively. Altogether we get

$$-\pi\int_{C_\beta}dS - r(z_\beta-z_{\beta-1})(1+\log\Lambda) + \sum_{\gamma\neq\beta}(z_\beta-z_\gamma)\log|z_\beta-z_\gamma|$$
$$- \sum_{\gamma\neq\beta-1}(z_{\beta-1}-z_\gamma)\log|z_{\beta-1}-z_\gamma| = O(\delta).$$

But the left-hand side is independent of δ. This means that the left-hand side is, in fact, $O(\Lambda^{2r})$. Combining with (2.3), we have

$$\frac{1}{2}a_\alpha^D = r(a_\alpha-a_1)(1+\log\Lambda) - \sum_{\beta\neq\alpha}(a_\alpha-a_\beta)\log|\sqrt{-1}(a_\alpha-a_\beta)|$$
$$+\sum_{\beta\neq 1}(a_1-a_\beta)\log\sqrt{-1}(a_1-a_\beta) + O(\Lambda^{2r}).$$

In the last part, we do not take the absolute value of $\sqrt{-1}(a_1-a_\beta)$ since $\sqrt{-1}a_1 > \sqrt{-1}a_\beta$.

Now let us differentiate $\mathcal{F}_0^{\text{pert}}$ in the statement. Let

$$\overline{\gamma}_0(x;\Lambda) = x^2\log\left(\frac{\sqrt{-1}x}{\Lambda}\right) - \frac{3}{2}x^2.$$

(Remember our choice of the branch of log.) We have the following

$$-\frac{\partial\mathcal{F}_0^{\text{pert}}}{\partial a_\alpha} = -\sum_{\beta<\gamma}\frac{\partial}{\partial a_\alpha}\overline{\gamma}_0(a_\beta-a_\gamma;\Lambda)$$
$$= -\sum_{\alpha<\beta}\overline{\gamma}_0'(a_\alpha-a_\beta;\Lambda) + \sum_{\beta<\alpha}\overline{\gamma}_0'(a_\beta-a_\alpha;\Lambda) + \sum_{1\neq\beta}\overline{\gamma}_0'(a_1-a_\beta;\Lambda).$$

We substitute $\overline{\gamma}_0'(x) = 2x\log(\sqrt{-1}x/\Lambda) - 2x$ to get

$$-\frac{1}{2}\frac{\partial \mathcal{F}_0^{\text{pert}}}{\partial a_\alpha} = r(a_\alpha - a_1) - \sum_{\beta\neq\alpha}(a_\alpha - a_\beta)\log\frac{|\sqrt{-1}(a_\alpha - a_\beta)|}{\Lambda} + \sum_{\beta\neq 1}(a_1 - a_\beta)\log\frac{\sqrt{-1}(a_1 - a_\beta)}{\Lambda}.$$

This coincides with the above expression. Finally $\mathcal{F}_0^{\text{pert}}$ satisfies the homogeneity property required in §2.2. □

2.4. A renormalization group equation. We prove the so-called "renormalization group equation" following [**76**] in this subsection:

PROPOSITION 2.3.

$$\Lambda\frac{\partial}{\partial\Lambda}\mathcal{F}_0 = -2ru_2.$$

This equation was found earlier by [**58**] for SU(2), and independently by [**20**]. See also [**10**].

PROOF. We differentiate the Euler equation (2.8):

$$\begin{aligned}
\frac{\partial}{\partial u_p}\left(\Lambda\frac{\partial}{\partial\Lambda}\mathcal{F}_0\right) &= 2\frac{\partial\mathcal{F}_0}{\partial u_p} - \sum_\alpha \frac{\partial}{\partial u_p}\left(a_\alpha\frac{\partial\mathcal{F}_0}{\partial a_\alpha}\right)\\
&= -\sum_\alpha\left(\frac{\partial a_\alpha}{\partial u_p}a_\alpha^D - a_\alpha\frac{\partial a_\alpha^D}{\partial u_p}\right)\\
&= -2\pi\sqrt{-1}\sum_\alpha\left[\int_{A_\alpha}\frac{\partial}{\partial u_p}dS\int_{B_\alpha}dS - \int_{A_\alpha}dS\int_{B_\alpha}\frac{\partial}{\partial u_p}dS\right].
\end{aligned}$$

Let us make a change of variable $x = 1/z$. We expand the Seiberg–Witten differential and its differential around $x = 0$:

$$\begin{aligned}
dS &= -\frac{1}{2\pi}\frac{zP'(z)\,dz}{\sqrt{P(z)^2 - 4\Lambda^{2r}}} = (s_{-2}x^{-2} + s_0 + s_1x + \cdots)\,dx,\\
\frac{\partial}{\partial u_p}dS &= \frac{1}{2\pi}\frac{z^{r-p}dz}{\sqrt{P(z)^2 - 4\Lambda^{2r}}} = (\omega_0^p + \omega_1^p x + \cdots)\,dx.
\end{aligned}$$

(Recall that dS is a meromorphic differential having poles only at $\infty_\pm$.) By the Riemann bilinear relation (see, e.g., [**33**, §2.3]), we have

$$\frac{\partial}{\partial u_p}\left(\Lambda\frac{\partial}{\partial\Lambda}\mathcal{F}_0\right) = 8\pi^2\sum_n\frac{s_{-n}\omega_{n-2}^p}{n-1} = 8\pi^2 s_{-2}\omega_0^p.$$

Since $s_{-2} = r/(2\pi)$, $\omega_0^p = -1/(2\pi)\delta_{p2}$, we get

$$\frac{\partial}{\partial u_p}\left(\Lambda\frac{\partial}{\partial\Lambda}\mathcal{F}_0\right) = -2r\delta_{p2}.$$

Integrating out, we get the assertion. Here the integration constant is zero thanks to the homogeneity of $\mathcal{F}_0$. □

2.5. The contact term equation. In this subsection we show the following partial differential equation:

THEOREM 2.4. *We have*

$$\Lambda\frac{\partial}{\partial\Lambda}u_p = \frac{2r}{\pi\sqrt{-1}}\sum_{\alpha,\beta=2}^{r}\frac{\partial u_p}{\partial a_\alpha}\frac{\partial u_2}{\partial a_\beta}\frac{\partial}{\partial\tau_{\alpha\beta}}\log\Theta_E(0|\tau)$$

for $p = 2, 3, \dots, r$. *Here* E *is the even half-integer characteristic given by* $\left[\begin{smallmatrix}\vec{0}\\ \vec{\Delta}\end{smallmatrix}\right]$ *in* (B.1).

This equation was first derived by Losev, Nekrasov and Shatashvili [**50, 51**] during their study of the topologically twisted version of $\mathcal{N} = 2$ SUSY Yang–Mills theory, i.e., the physical counterpart of Donaldson theory. More precisely, they derived the equation by studying the effect of the blowup on the so-called "contact terms." So we call the equation the *contact term* equation. Later Gorsky, Marshakov, Mironov and Morozov [**29**] derived the contact term equation in the framework of the Seiberg–Witten curve. We give the proof following their approach in this subsection. In fact, they did not determine the characteristic. It was determined in [**50, 51**], but the argument involves physical intuition. Here we can give a mathematically rigorous proof thanks to our precise definition of the B-cycles used in the definition of the prepotential.

For a later purposes, we make a remark. Recall that $(-1)^p u_p$ is the pth elementary symmetric function in the variables $z_1, \dots, z_r$. Let c_p be the pth power sum multiplied by $(-\sqrt{-1})^p/p!$:

$$c_p = \frac{(-\sqrt{-1})^p}{p!}\sum_{\alpha=1}^{r} z_\alpha^p, \qquad c_1 = 0, \quad c_2 = u_2, \dots . \tag{2.11}$$

Then $(c_2, c_3, \dots, c_r)$ is another coordinate system on the $\vec{u}$-plane. Since c_p is a polynomial in u_q's, it is also a solution of the contact term equation:

$$\Lambda\frac{\partial}{\partial\Lambda}c_p = \frac{2r}{\pi\sqrt{-1}}\sum_{\alpha,\beta=2}^{r}\frac{\partial c_p}{\partial a_\alpha}\frac{\partial u_2}{\partial a_\beta}\frac{\partial}{\partial\tau_{\alpha\beta}}\log\Theta_E(0|\tau).$$

Before giving the proof of Theorem 2.4, we give a corollary which will play an important role later.

COROLLARY 2.5.

$$\left(\Lambda\frac{\partial}{\partial\Lambda}\right)^2\mathcal{F}_0 = \frac{-1}{\pi\sqrt{-1}}\sum_{\alpha,\beta=2}^{r}\frac{\partial}{\partial a_\alpha}\left(\Lambda\frac{\partial}{\partial\Lambda}\mathcal{F}_0\right)\frac{\partial}{\partial a_\beta}\left(\Lambda\frac{\partial}{\partial\Lambda}\mathcal{F}_0\right)\frac{\partial}{\partial\tau_{\alpha\beta}}\log\Theta_E(0|\tau).$$

This equation together with the description of the perturbative part (Proposition 2.2) completely determines the prepotential $\mathcal{F}_0$. See the proof of Theorem 5.3 and §5.2 below. This observation is due to [**17**]. (See also [**58**] for an earlier result for SU(2).)

PROOF OF THEOREM 2.4. Recall that we consider u_p as functions of a_α, Λ. We differentiate (2.2) by $\log\Lambda$ to get

$$\sum_p \frac{\partial u_p}{\partial\log\Lambda}\int_{A_\alpha}\frac{\partial}{\partial u_p}dS + \int_{A_\alpha}\frac{\partial}{\partial\log\Lambda}dS = 0.$$

Therefore

$$\sum_p \frac{\partial u_p}{\partial \log \Lambda} \frac{\partial a_\alpha}{\partial u_p} = -\int_{A_\alpha} \frac{\partial}{\partial \log \Lambda} dS = \frac{r}{2\pi} \int_{A_\alpha} \frac{P(z)}{P'(z)} \frac{dw}{w}$$
$$= \frac{r}{2\pi} \int_{A_\alpha} \frac{P(z)\, dz}{y} = \frac{r}{2\pi} \int_{A_\alpha} \frac{(P(z)+y)\, dz}{y}.$$

The last expression can be given by the Szegö kernel (see (B.7)) as

$$\Psi_E^2(z_1, \infty_\pm) = -\frac{P(z_1) \pm y(z_1)}{2y(z_1)}\, dz_1\, d\left(\frac{1}{z_2}\right)\bigg|_{z_2=\infty_\pm}.$$

Here we choose the leftmost point z_1^+ as the base point for the Abel–Jacobi map. And the even half-integer characteristic E corresponds to the partition of the branch points into

$$\{z_\alpha^+ \mid \alpha = 1, \dots, r\} \sqcup \{z_\alpha^- \mid \alpha = 1, \dots, r\}.$$

On the other hand, we have

$$\omega_\beta|_{z=\infty_\pm} = -\frac{1}{2\pi} \sum_p \frac{\partial u_p}{\partial a_\beta} \frac{z^{r-p} dz}{y}\bigg|_{z=\infty_\pm} = \frac{1}{2\pi} \frac{\partial u_2}{\partial a_\beta}\, d\left(\frac{1}{z}\right)\bigg|_{z=\infty_\pm},$$

where we have used $y \sim z^r$ at $z = \infty_\pm$ in the second equality. Therefore by Fay's identity (B.6) we have

$$\frac{r}{2\pi} \int_{A_\alpha} \frac{(P(z)+y)\, dz}{y} = -\frac{r}{2\pi^2} \frac{\partial u_2}{\partial a_\beta} \frac{\partial^2}{\partial \xi_\alpha \partial \xi_\beta} \log \Theta_E(0|\tau)$$
$$= \frac{2r}{\pi\sqrt{-1}} \frac{\partial u_2}{\partial a_\beta} \frac{\partial}{\partial \tau_{\alpha\beta}} \log \Theta_E(0|\tau).$$

In the second equality we have used the heat equation and the fact that E is an even half-integer characteristic, and hence the derivative of $\Theta_E(\vec{\xi}|\tau)$ at $\vec{\xi} = 0$ vanishes. Multiplying by the matrix $(\partial u_p / \partial a_\alpha)$ both sides, we get the differential equation as in the assertion.

Finally we determine the even half-integer characteristic E explicitly. Looking at Figure 1, we find that the partition corresponding to characteristic 0 is

$$\{z_1^+, z_2^-, z_3^+, \cdots\} \sqcup \{z_1^-, z_2^+, z_3^-, \cdots\}.$$

Namely z_α^+ and z_α^- for α even are interchanged from E. Since $\int_{z_\alpha^-}^{z_\alpha^+} \omega_\beta = \frac{1}{2} \int_{A_\alpha} \omega_\beta = \frac{1}{2}\delta_{\alpha\beta}$, we find that the characteristic E is $\left[\begin{smallmatrix} \vec{0} \\ \vec{\Delta} \end{smallmatrix}\right]$ in (B.1). □

REMARK 2.6. In **[29]** "time variables" $T_1, T_2, \dots, T_{r-1}$ are introduced in the framework of the Whitham hierarchy. Then the contact term equations are the specialization of certain equations at $T_2 = T_3 = \cdots = 0$ (T_1 is essentially $\log \Lambda$). On the other hand, we will introduce *infinitely many* variables $\tau_1, \tau_2, \dots$ in §4. We will show $\partial / \partial T_p = \partial / \partial \tau_p$ for $p = 1, 2, \dots, r-1$ when it is restricted to $\tau_1 = \tau_2 = \cdots = 0$ in Theorem 5.3(2). However we do not know the equations for the τ_p-derivatives, which should different from the equations for T_p-derivatives outside this subspace.

2.6. Rank 2 case. When $r = 2$, i.e., the Seiberg–Witten curve is an elliptic curve, we have the expressions (1.3) for u, a in terms of theta functions and Eisenstein series. Here we write $u = u_2$, $a = a_2$. The derivations of the expressions are left to the reader as an exercise. One can prove Theorem 2.4 using the expressions. See [**29**, Appendix].

Let us record the following formula for later purposes.

$$\begin{aligned}(2.12)\qquad \Lambda\frac{\partial u}{\partial \Lambda} &= 2u - a\frac{du}{da} = 2u + \sqrt{-1}\frac{2E_2 + \theta_{00}^4 + \theta_{10}^4}{3\theta_{00}\theta_{10}}\Lambda\frac{du}{da} \\ &= 2u - \frac{1}{3}E_2\left(\frac{du}{da}\right)^2 + \frac{2}{3}\frac{\theta_{00}^4 + \theta_{10}^4}{\theta_{00}^2\theta_{10}^2}\Lambda^2 = -\frac{1}{3}E_2\left(\frac{du}{da}\right)^2 + \frac{4}{3}u,\end{aligned}$$

where the first equality follows from the homogeneity of u.

For the reader who wants to compare the formulas with ones in the literature, we record how parameters differ. Let us make a change of variable by

$$w - \frac{u}{3} = -\Lambda^2\frac{z - \sqrt{-u + 2\Lambda^2}}{z + \sqrt{-u + 2\Lambda^2}}.$$

The branched points $z = \sqrt{-u + 2\Lambda^2}, -\sqrt{-u + 2\Lambda^2}, \sqrt{-u - 2\Lambda^2}, -\sqrt{-u - 2\Lambda^2}$ are mapped to $w = u/3, \infty, (-u - 3\sqrt{u^2 - 4\Lambda^4})/6, (-u + 3\sqrt{u^2 - 4\Lambda^4})/6$ respectively. And the curve $y^2 = (z^2 + u)^2 - 4\Lambda^4$ is isomorphic to a Weierstrass form $y^2 = 4w^3 - g_2 w - g_3$ with

$$g_2 = 4(\tfrac{1}{3}u^2 - \Lambda^4), \qquad\qquad g_3 = -\frac{1}{27}u(8u^2 - 36\Lambda^4).$$

This is the form of curves appearing in [**24**] with u replaced by $-x$ therein. If we make a further change of variable by $w = 2x + u/3$, we get

$$\frac{1}{32}y^2 = x\left(x^2 + \frac{u}{2}x + \frac{\Lambda^4}{4}\right).$$

This is the form of the curves in [**59**] after the replacement $u \mapsto -2u$, $y \mapsto 4\sqrt{2}y$.

3. Instanton moduli spaces

3.1. Basic definitions. In this and the next subsections we briefly recall properties of framed moduli spaces of instantons (resp. torsion-free sheaves) on S^4 (resp. $\mathbb{P}^2$) and the corresponding moduli spaces on the blowup. For more details, see [**69**, §1] and [**68**, Chapters 2, 3] and the references therein.

Let $M(r,n)$ be the framed moduli space of torsion free sheaves on $\mathbb{P}^2$ with rank r and $c_2 = n$, which parametrizes isomorphism classes of (E, Φ) such that

(1) E is a torsion free sheaf of rank $E = r$, $\langle c_2(E), [\mathbb{P}^2]\rangle = n$ which is locally free in a neighborhood of ℓ_∞,

(2) $\Phi\colon E|_{\ell_\infty} \xrightarrow{\sim} \mathcal{O}_{\ell_\infty}^{\oplus r}$ is an isomorphism called the "framing at infinity."

Here $\ell_\infty = \{[0{:}z_1{:}z_2] \in \mathbb{P}^2\} \subset \mathbb{P}^2$ is the line at infinity. Notice that the existence of a framing Φ implies $c_1(E) = 0$.

This space is known to be nonsingular of dimension $2nr$.

Let $M_0^{\mathrm{reg}}(r,n)$ be the open subset consisting of locally free sheaves. By a result of Donaldson [**13**] it can be identified with the framed moduli space of instantons on S^4, which parametrize anti-self-dual connections A on a principal $\mathrm{SU}(r)$-bundle P with $\langle c_2(P), [S^4]\rangle = n$ modulo gauge transformations γ with $\gamma_\infty = \mathrm{id}$.

Let $M_0(r,n)$ be the Uhlenbeck (partial) compactification of $M_0^{\text{reg}}(r,n)$. Set theoretically it is defined by

$$M_0(r,n) = \bigsqcup_{k=0}^{n} M_0^{\text{reg}}(r,n-k) \times S^k\mathbb{C}^2,$$

where $S^k\mathbb{C}^2$ is the kth symmetric product of $\mathbb{C}^2$. We can endow this space with the structure of an affine algebraic variety so that there is a projective morphism

$$\pi\colon M(r,n) \to M_0(r,n).$$

The corresponding map between closed points can be identified with

$$(3.1) \qquad (E,\Phi) \mapsto ((E^{\vee\vee},\Phi), \operatorname{Supp}(E^{\vee\vee}/E)) \in M_0^{\text{reg}}(r,n') \times S^{n-n'}\mathbb{C}^2.$$

where $E^{\vee\vee}$ is the double dual of E and $\operatorname{Supp}(E^{\vee\vee}/E)$ is the support of $E^{\vee\vee}/E$ counted with multiplicities. Note that $E^{\vee\vee}$ is a locally free sheaf. For moduli spaces on general projective surfaces, such morphisms from moduli spaces of sheaves to Uhlenbeck compactifications were constructed by J. Li [**43**] and Morgan [**60**].

Let T be the maximal torus of $\operatorname{GL}_r(\mathbb{C})$ consisting of diagonal matrices and let $\widetilde{T} = \mathbb{C}^* \times \mathbb{C}^* \times T$. We define an action of $\widetilde{T}$ on $M(r,n)$ as follows: For $(t_1,t_2) \in \mathbb{C}^* \times \mathbb{C}^*$, let F_{t_1,t_2} be the automorphism of $\mathbb{P}^2$ defined by

$$F_{t_1,t_2}([z_0{:}z_1{:}z_2]) = [z_0{:}t_1z_1{:}t_2z_2].$$

For $\operatorname{diag}(e_1,\dots,e_r) \in T$ let $G_{e_1,\dots,e_r}$ denote the isomorphism of $\mathcal{O}_{\ell_\infty}^{\oplus r}$ given by

$$\mathcal{O}_{\ell_\infty}^{\oplus r} \ni (s_1,\dots,s_r) \mapsto (e_1s_1,\dots,e_rs_r).$$

Then for $(E,\Phi) \in M(r,n)$, we define

$$(3.2) \qquad (t_1,t_2,e_1,\dots,e_r)\cdot(E,\Phi) = ((F_{t_1,t_2}^{-1})^*E, \Phi'),$$

where Φ' is the composite of the homomorphisms

$$(F_{t_1,t_2}^{-1})^*E|_{\ell_\infty} \xrightarrow{(F_{t_1,t_2}^{-1})^*\Phi} (F_{t_1,t_2}^{-1})^*\mathcal{O}_{\ell_\infty}^{\oplus r} \longrightarrow \mathcal{O}_{\ell_\infty}^{\oplus r} \xrightarrow{G_{e_1,\dots,e_r}} \mathcal{O}_{\ell_\infty}^{\oplus r}.$$

Here the middle arrow is the homomorphism given by the action.

In a similar way, we have a $\widetilde{T}$-action on $M_0(r,n)$. The map $\pi\colon M(r,n) \to M_0(r,n)$ is equivariant.

The fixed points $M(r,n)^{\widetilde{T}}$ consist of $(E,\Phi) = (I_1,\Phi_1)\oplus\cdots\oplus(I_r,\Phi_r)$ such that

(a) I_α is the ideal sheaf of a 0-dimensional subscheme Z_α contained in $\mathbb{C}^2 = \mathbb{P}^2 \setminus \ell_\infty$.
(b) Φ_α is an isomorphism from $(I_\alpha)_{\ell_\infty}$ to the αth factor of $\mathcal{O}_{\ell_\infty}^{\oplus r}$.
(c) I_α is fixed by the action of $\mathbb{C}^* \times \mathbb{C}^*$, coming from that on $\mathbb{P}^2$.

On the other hand, the fixed points $M_0(r,n)^{\widetilde{T}}$ consist of the single point $n[0] \in S^n\mathbb{C}^2 \subset M_0(r,n)$.

We parametrize the fixed point set $M(r,n)^{\widetilde{T}}$ by an r-tuple of Young diagrams $\vec{Y} = (Y_1,\dots,Y_r)$ so that the ideal I_α is spanned by monomials x^iy^j placed at $(i-1,j-1)$ outside Y_α as illustrated in Figure 2. The constraint is that the total number of boxes $|\vec{Y}| \overset{\text{def}}{=} \sum_\alpha |Y_\alpha|$ is equal to n.

Let $Y = (\lambda_1 \geq \lambda_2 \geq \cdots)$ be a Young diagram, where λ_i is the length of the ith column. Let $Y' = (\lambda'_1 \geq \lambda'_2 \geq \dots)$ be the transpose of Y. Thus λ'_j is the length of the jth row of Y. Let $l(Y)$ denote the number of columns of Y, i.e., $l(Y) = \lambda'_1$. Let

$$a_Y(i,j) = \lambda_i - j, \qquad a'(i,j) = j - 1$$
$$l_Y(i,j) = \lambda'_j - i, \qquad l'(i,j) = i - 1.$$

Here we set $\lambda_i = 0$ when $i > l(Y)$. Similarly $\lambda'_j = 0$ when $j > l(Y')$. When the square $s = (i,j)$ lies in Y, these are called the *arm-length, arm-colength, leg-length, leg-colength* respectively, and we usually consider this case. But our formula below involves these also for squares outside Y. So these take negative values in general. Note that a' and l' do not depend on the diagram, and we do not write the subscript Y.

THEOREM 3.1. *Let (E, Φ) be a fixed point of the $\widetilde{T}$-action corresponding to $\vec{Y} = (Y_1, \dots, Y_r)$. Then the $\widetilde{T}$-module structure of $T_{(E,\Phi)}M(r,n)$ is given by*

$$\sum_{\alpha,\beta=1}^{r} N_{\alpha,\beta}^{\vec{Y}}(t_1,t_2),$$

where

$$N_{\alpha,\beta}^{\vec{Y}}(t_1,t_2) = e_\beta e_\alpha^{-1} \times \left\{ \sum_{s \in Y_\alpha} \left(t_1^{-l_{Y_\beta}(s)} t_2^{a_{Y_\alpha}(s)+1}\right) + \sum_{t \in Y_\beta} \left(t_1^{l_{Y_\alpha}(t)+1} t_2^{-a_{Y_\beta}(t)}\right) \right\}.$$

Here we have used the following notation.

NOTATION 3.2. We denote by e_α ($\alpha = 1, \dots, r$) the one-dimensional $\widetilde{T}$-module given by

$$\widetilde{T} \ni (t_1, t_2, e_1, \dots, e_r) \mapsto e_\alpha.$$

Similarly, t_1, t_2 denote one-dimensional $\widetilde{T}$-modules. Thus the representation ring $R(\widetilde{T})$ is isomorphic to $\mathbb{Z}[t_1^{\pm}, t_2^{\pm}, e_1^{\pm}, \dots, e_r^{\pm}]$, where e_α^{-1} is the dual of e_α.

3.2. Moduli spaces on the blowup. Let $\widehat{\mathbb{P}}^2$ be the blowup of $\mathbb{P}^2$ at $[1{:}0{:}0]$. Let $p\colon \widehat{\mathbb{P}}^2 \to \mathbb{P}^2$ denote the projection. The manifold $\widehat{\mathbb{P}}^2$ is the closed subvariety of $\mathbb{P}^2 \times \mathbb{P}^1$ defined by

$$\{([z_0{:}z_1{:}z_2],[z{:}w]) \in \mathbb{P}^2 \times \mathbb{P}^1 \mid z_1 w = z_2 z\},$$

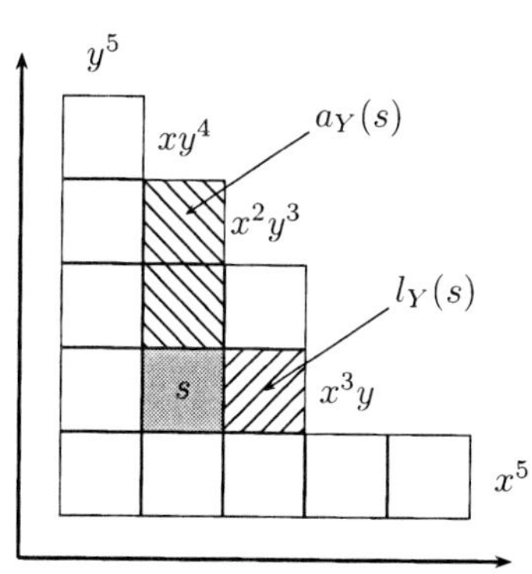

FIGURE 2. Young diagram and ideal

where the map $p\colon \widehat{\mathbb{P}}^2 \to \mathbb{P}^2$ is the projection to the first factor. Let us denote the inverse image of ℓ_∞ under $\widehat{\mathbb{P}}^2 \to \mathbb{P}^2$ also by ℓ_∞ for brevity. It is given by the equation $z_0 = 0$. The complement $\widehat{\mathbb{P}}^2 \setminus \ell_\infty$ is the blowup $\widehat{\mathbb{C}}^2$ of $\mathbb{C}^2$ at the origin. Let C denote the exceptional set. It is given by $z_1 = z_2 = 0$.

In this subsection, $\mathcal{O}$ denotes the structure sheaf of $\widehat{\mathbb{P}}^2$, $\mathcal{O}(C)$ the line bundle associated with the divisor C, and $\mathcal{O}(mC)$ its mth tensor product.

Let $\widehat{M}(r,k,n)$ be the framed moduli space of torsion free sheaves (E,Φ) on $\widehat{\mathbb{P}}^2$ with rank r, $\langle c_1(E), [C]\rangle = -k$ and $\langle c_2(E) - (r-1)c_1(E)^2/(2r), [\widehat{\mathbb{P}}^2]\rangle = n$. This is also nonsingular of dimension $2nr$. (Note that n may not be an integer in general.)

THEOREM 3.3. *There is a projective morphism*

$$\widehat{\pi}\colon \widehat{M}(r,k,n) \to M_0\left(r, n - \frac{k(r-k)}{2r}\right) \qquad (0 \le k < r)$$

defined by

$$(E,\Phi) \mapsto \Big(\big((p_*E)^{\vee\vee}, \Phi\big), \operatorname{Supp}(p_*E^{\vee\vee}/p_*E) + \operatorname{Supp}(R^1p_*E)\Big).$$

The proof of this theorem was not given in [**69**] and will be given in Appendix F. In fact, we prove the corresponding result for arbitrary projective surfaces.

Let us define an action of the $(r+2)$-dimensional torus $\widetilde{T} = \mathbb{C}^* \times \mathbb{C}^* \times T$ on $\widehat{M}(r,k,n)$ by modifying the action on $M(r,n)$ as follows. For $(t_1,t_2) \in \mathbb{C}^* \times \mathbb{C}^*$, let F'_{t_1,t_2} be the automorphism of $\widehat{\mathbb{P}}^2$ defined by

$$F'_{t_1,t_2}([z_0{:}z_1{:}z_2],[z{:}w]) = ([z_0{:}t_1z_1{:}t_2z_2],[t_1z{:}t_2w]).$$

Then we define the action by replacing F_{t_1,t_2} by F'_{t_1,t_2} in (3.2). The action of the factor T is exactly the same as before. The morphism $\widehat{\pi}$ is equivariant.

Note that the fixed point set of $\mathbb{C}^* \times \mathbb{C}^*$ in $\widehat{\mathbb{C}}^2 = \widehat{\mathbb{P}}^2 \setminus \ell_\infty$ consists of the two points $([1{:}0{:}0],[1{:}0])$, $([1{:}0{:}0],[0{:}1])$. Let us denote them by p_1 and p_2.

The fixed points $\widehat{M}(r,k,n)^{\widetilde{T}}$ consist of $(E,\Phi) = (I_1(k_1C),\Phi_1) \oplus \cdots \oplus (I_r(k_rC),\Phi_r)$ such that

(a) $I_\alpha(k_\alpha C)$ is the tensor product $I_\alpha \otimes \mathcal{O}(k_\alpha C)$, where $k_\alpha \in \mathbb{Z}$ and I_α is the ideal sheaf of a 0-dimensional subscheme Z_α contained in $\widehat{\mathbb{C}}^2 = \widehat{\mathbb{P}}^2 \setminus \ell_\infty$.
(b) Φ_α is an isomorphism from $(I_\alpha)_{\ell_\infty}$ to the αth factor of $\mathcal{O}_{\ell_\infty}^{\oplus r}$.
(c) I_α is fixed by the action of $\mathbb{C}^* \times \mathbb{C}^*$, coming from that on $\widehat{\mathbb{P}}^2$.

The support of Z_α must be contained in the fixed point set in $\widehat{\mathbb{C}}^2$, i.e., $\{p_1,p_2\}$. Thus Z_α is the union of Z_α^1 and Z_α^2, subschemes supported at p_1 and p_2 respectively. If we take a coordinate system $(x,y) = (z_1/z_0, w/z)$ (resp. $= (z/w, z_2/z_0)$) around p_1 (resp. p_2), then Z_α^1 (resp. Z_α^2) is generated by monomials x^iy^j. (See Figure 3.) Then Z_α^1 (resp. Z_α^2) corresponds to a Young diagram Y_α^1 (resp. Y_α^2) as before. Therefore the fixed point set is parametrized by r-tuples $(\vec{k}, \vec{Y}^1, \vec{Y}^2) = ((k_1,Y_1^1,Y_1^2),\dots,(k_r,Y_r^1,Y_r^2))$, where $k_\alpha \in \mathbb{Z}$ and Y_α^1, Y_α^2 are Young diagrams. The constraint is

$$\sum_\alpha k_\alpha = k, \qquad |\vec{Y}^1| + |\vec{Y}^2| + \frac{1}{2r}\sum_{\alpha<\beta}|k_\alpha - k_\beta|^2 = n. \tag{3.3}$$

We will use the convention for $\vec{k}$ given in Appendix A.

Note that the fixed point data have three parts, $\vec{k}$, $\vec{Y}^1$ and $\vec{Y}^2$. This will be reflected in the blowup formula (4.6) below. Also, the appearance of $\vec{k} \in \mathbb{Z}^r$ explains the reason why many formulas below contain the theta function.

THEOREM 3.4. *Let (E, Φ) be a fixed point of the $\widetilde{T}$-action corresponding to $(\vec{k}, \vec{Y}^1, \vec{Y}^2)$. Then the $\widetilde{T}$-module structure of $T_{(E,\Phi)}\widehat{M}(r,k,n)$ is given by*

$$\sum_{\alpha,\beta=1}^{r} L^{\vec{k}}_{\alpha,\beta}(t_1,t_2) + t_1^{k_\beta - k_\alpha} N^{\vec{Y}^1}_{\alpha,\beta}(t_1, t_2/t_1) + t_2^{k_\beta - k_\alpha} N^{\vec{Y}^2}_{\alpha,\beta}(t_1/t_2, t_2),$$

where

$$L^{\vec{k}}_{\alpha,\beta}(t_1,t_2) = e_\beta\, e_\alpha^{-1} \times \begin{cases} \sum\limits_{\substack{i,j\ge 0\\ i+j\le k_\alpha - k_\beta - 1}} t_1^{-i} t_2^{-j} & \text{if } k_\alpha > k_\beta, \\ \sum\limits_{\substack{i,j\ge 0\\ i+j\le k_\beta - k_\alpha - 2}} t_1^{i+1} t_2^{j+1} & \text{if } k_\alpha + 1 < k_\beta, \\ 0 & \text{otherwise.} \end{cases}$$

The reason for the change of weights $(t_1, t_2/t_1)$, $(t_1/t_2, t_2)$ is clear. It comes from the action on the coordinate system around p_1 and p_2.

3.3. Topology of moduli spaces. Thanks to the existence of the torus action, the homology groups of $M(r,n)$, $\widehat{M}(r,k,n)$ enjoy nice properties. In particular, we can calculate their Betti numbers, whose generating functions have beautiful formulas. The results of this and the next subsections will not be used in the other parts of this paper. The reader in a hurry may skip this and the next subsections.

THEOREM 3.5. (1) *$\pi^{-1}(n[0])$ is isomorphic to the punctual quot-scheme parametrizing zero dimensional quotients $\mathcal{O}_{\mathbb{P}^2}^{\oplus r} \to Q$ with* $\mathrm{Supp}(Q) = n[0]$.

(2) *$M(r,n)$ is homotopy equivalent to $\pi^{-1}(n[0])$.*

(3) *Both $M(r,n)$ and $\pi^{-1}(n[0])$ have α-partitions into affine spaces.*

(4) *$H_{\mathrm{odd}}(M(r,n),\mathbb{Z}) = 0$ and $H_{\mathrm{even}}(M(r,n),\mathbb{Z})$ are free abelian groups. The cycle map $A_*(M(r,n)) \to H_{\mathrm{even}}(M(r,n),\mathbb{Z})$ is an isomorphism. The same assertions hold for $\pi^{-1}(n[0])$.*

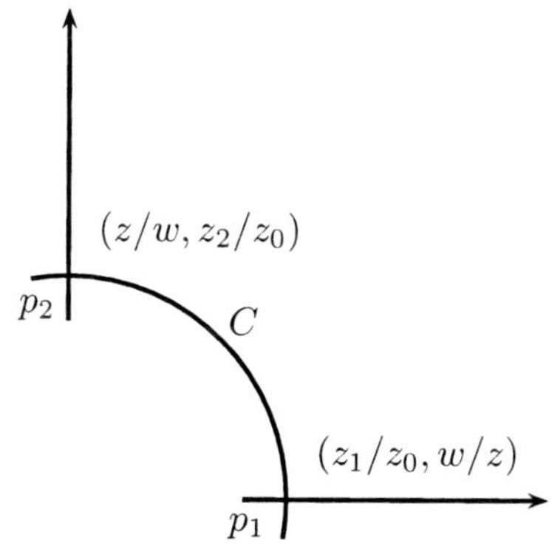

FIGURE 3. Blowup and fixed points

Recall that a finite partition of a variety X into locally closed subvarieties is said to be an α-partition if the subvarieties in the partition can be indexed $X_1, \dots, X_n$ in such a way that $X_1 \cup X_2 \cup \dots \cup X_i$ is closed in X for $i = 1, \dots, n$.

Here $H_*(\cdot, \mathbb{Z})$, $A_*(\cdot)$ denote the Borel–Moore homology group and the Chow group. See Appendix C.

PROOF. (1) By the geometric description (3.1) of the map π, the fiber $\pi^{-1}(n[0])$ consists of (E, Φ) such that $E^{\vee\vee} = \mathcal{O}_{\mathbb{P}^2}^{\oplus r}$ and $\operatorname{Supp}(\mathcal{O}_{\mathbb{P}^2}^{\oplus r}/E) = n[0]$. Thus the quotient $\mathcal{O}_{\mathbb{P}^2}^{\oplus r}/E$ is a point in the punctual quot-scheme.

(2) A similar result was proved in [**65**, 5.5] for quiver varieties by using a $\mathbb{C}^*$-action. The proof can be adapted to our situation as follows.

Let us consider a one-parameter subgroup $\mathbb{C}^* \ni t \mapsto \lambda(t) = (t, t, 1, \dots, 1) \in \widetilde{T}$. When $t \to 0$, $\lambda(t) \cdot x$ goes to 0 for any x in $M_0(r, n)$. Now apply an argument of Slodowy [**75**, 4.3] to $\pi\colon M(r, n) \to M_0(r, n)$.

(3) Choose a generic one-parameter subgroup $\lambda\colon \mathbb{C}^* \to \widetilde{T}$ so that the fixed point set is unchanged: $M(r, n)^{\lambda(\mathbb{C}^*)} = M(r, n)^{\widetilde{T}}$. Moreover we take it so that $\lim_{t\to 0} \lambda(t) = 0$. Then points fixed by $\lambda(\mathbb{C}^*)$ are given as above. In particular, they form a finite set. For each fixed point w, we consider the $(\pm)$-attracting set:

$$S_w = \left\{ x \in M(r, n) \;\middle|\; \lim_{t\to 0} \lambda(t) \cdot x = w \right\}, \quad U_w = \left\{ x \in M(r, n) \;\middle|\; \lim_{t\to \infty} \lambda(t) \cdot x = w \right\}.$$

These are affine spaces by [**7**]. Moreover, there exists an order on the fixed points so that $\bigcup_{y\le w} S_y$ (resp. $\bigcup_{y\le w} U_y$) is closed in $\bigcup_x S_x$ (resp. $\bigcup_x U_x$) for each w. (See, e.g., [**3**, §1].) We claim that $\bigcup S_w = M(r, n)$, $\bigcup U_w = \pi^{-1}(n[0])$. Consider the corresponding action on $M_0(r, n)$. For this purpose, we recall the ADHM description: $M_0(r, n)$ is an affine algebro-geometric quotient

$$\{(B_1, B_2, i, j) \mid [B_1, B_2] + ij = 0\} /\!\!/ \operatorname{GL}_n(\mathbb{C}),$$

where B_1, B_2 are $n \times n$ complex matrices, and $i \in \operatorname{Hom}(\mathbb{C}^r, \mathbb{C}^n)$, $j \in \operatorname{Hom}(\mathbb{C}^n, \mathbb{C}^r)$. (See [**68**, Chapter 2] and [**69**].) The action of $\operatorname{GL}_n(\mathbb{C})$ is given by $g \cdot (B_1, B_2, i, j) = (gB_1g^{-1}, gB_2g^{-1}, gi, jg^{-1})$. The $\widetilde{T}$-action is given by

$$(B_1, B_2, i, j) \mapsto (t_1B_1, t_2B_2, ie^{-1}, t_1t_2ej),$$
$$\text{for } t_1, t_2 \in \mathbb{C}^*,\ e = \operatorname{diag}(e_1, \dots, e_r) \in (\mathbb{C}^*)^r.$$

By [**53**] the coordinate ring of $M_0(r, n)$ is generated by the following two types of functions:

(a) $\operatorname{tr}(B_{\alpha_N} B_{\alpha_{N-1}} \cdots B_{\alpha_1}\colon \mathbb{C}^n \to \mathbb{C}^n)$, where $\alpha_i = 1$ or 2.
(b) $\chi(jB_{\alpha_N} B_{\alpha_{N-1}} \cdots B_{\alpha_1} i)$, where $\alpha_i = 1$ or 2, and χ is a linear form on $\operatorname{End}(\mathbb{C}^r)$.

Both types of functions have positive weights with respect to λ. Therefore every point in $M_0(r, n)$ converges to 0 as $t \to 0$, and every point except 0 goes to infinity as $t \to \infty$. Since π is proper, we get the claim. (We use the fact that the orbit of a one-parameter subgroup has a limit if it is contained in a compact set.) Thus $\bigcup S_w$ (resp. $\bigcup U_w$) gives us an α-partition of $M(r, n)$ (resp. $\pi^{-1}(0)$) into affine spaces.

(4) is a consequence of (3) by [**8**, Lemma 1.8]. □

THEOREM 3.6. *The Poincaré polynomial of the punctual quot-scheme* $\pi^{-1}(n[0])$ *is given by*

$$P_t(\pi^{-1}(n[0])) = \sum_{(Y_1,\dots,Y_r)} \prod_{\alpha=1}^{r} t^{2(r|Y_\alpha|-\alpha l(Y_\alpha))}, \tag{3.4}$$

where the summation runs over the r*-tuples of Young diagrams* $\vec{Y} = (Y_1, \dots, Y_r)$ *with* $|\vec{Y}| = n$.

As in the calculation in [**68**, the end of Chapter 6], we get the following nice expression for the generating function.

COROLLARY 3.7. *The generating function of* $P_t(\pi^{-1}(n[0]))$ *is given by*

$$\sum_n P_t(\pi^{-1}(n[0]))q^n = \prod_{\alpha=1}^{r}\prod_{d=1}^{\infty} \frac{1}{1 - t^{2(rd-\alpha)}q^d}.$$

PROOF OF THEOREM 3.6. Since $\pi^{-1}(n[0])$ is homotopic to $M(r,n)$, it is enough to calculate the dimensions of (ordinary) homology groups. For this calculation, we use the α-partition given in Theorem 3.5.

We need to specify the one-parameter subgroup $\lambda\colon \mathbb{C}^* \to \widetilde{T}$ in the proof of Theorem 3.5. Let

$$\lambda(t) = (t^{m_1}, t^{m_2}, t^{n_1}, \dots, t^{n_r}). \tag{3.5}$$

If we choose the weights m_1, m_2, n_α generic, the Zariski closure of $\lambda(\mathbb{C}^*)$ is equal to the whole $\widetilde{T}$, and the fixed point set of $\lambda(\mathbb{C}^*)$ coincides with that of $\widetilde{T}$.

Furthermore, we assume

$$m_2 \gg n_1 > n_2 > \dots > n_r \gg m_1 > 0 \tag{3.6}$$

in order to make the calculation of the index simpler. (This choice of weights is due to T. Gocho, and was given as an answer to an exercise in a preliminary version of [**68**].)

By the proof of Theorem 3.5 the corresponding $(-)$-attracting set is an α-partition of $\pi^{-1}(0)$.

Our remaining task is to calculate indices of critical points. Let (E, Φ) be a critical point, i.e., a fixed point of the $\widetilde{T}$-action. Then $T_{(E,\Phi)}M(r,n)$ has a $\widetilde{T}$-module structure, and has an induced $\mathbb{C}^*$-module structure via λ. By our choice of weights (3.6), negative weight spaces for the $\mathbb{C}^*$-action are direct sums of weight spaces for the $\widetilde{T}$-action such that one of the following holds:

(1) the weight of t_2 is negative,
(2) the weight of t_2 is zero and the weight of e_1 is negative,
(3) the weights of t_2, e_1 are zero and the weight of e_2 is negative,
(4) the weights of t_2, e_1, e_2 are zero and the weight of e_3 is negative,

⋮

$(r+1)$ the weights of t_2, e_1, $e_2, \dots, e_{r-1}$ are zero and the weight of e_r is negative,
$(r+2)$ the weights of t_2, e_1, $e_2, \dots, e_r$ are zero and the weight of t_1 is negative.

Recall that we decompose the tangent space into $\sum_{\alpha,\beta} N^{\vec{Y}}_{\alpha,\beta}(t_1,t_2)$ in Theorem 3.1. We calculate the sum of dimensions of weight spaces with the above condition in each summand separately, and then sum up the contribution from each summand. In the summand $\alpha = \beta$, the contribution is

$$|Y_\alpha| - l(Y_\alpha).$$

If $\alpha < \beta$, the above condition is equivalent to the weight of t_2 being nonpositive. Hence the contribution is equal to the number of terms in [**69**, (1.18)], i.e.,

$$|Y_\beta|.$$

If $\alpha > \beta$, the above condition is equivalent to the weight of t_2 being negative. If we look at [**69**, (1.18)], we find that the contribution is

$$|Y_\beta| - l(Y_\beta).$$

Thus the total contribution is

$$\sum_{\beta=1}^{r} r|Y_\beta| - (r - \beta + 1)l(Y_\beta).$$

Changing the variable as $\alpha = r - \beta + 1$, we get formula (3.4). □

We now turn to the moduli spaces on the blowup. The proof of the following is the same as that of Theorem 3.5 and hence omitted.

THEOREM 3.8. (1) *$\widehat{M}(r,k,n)$ is homotopy equivalent to $\widehat{\pi}^{-1}(n[0])$.*
(2) *Both $\widehat{M}(r,k,n)$ and $\widehat{\pi}^{-1}(n[0])$ have α-partitions into affine spaces.*
(3) *$H_{\text{odd}}(\widehat{M}(r,k,n),\mathbb{Z}) = 0$ and $H_{\text{even}}(\widehat{M}(r,k,n),\mathbb{Z})$ are free abelian groups. The cycle map $A_*(\widehat{M}(r,k,n)) \to H_{\text{even}}(\widehat{M}(r,k,n),\mathbb{Z})$ is an isomorphism. The same holds for $\widehat{\pi}^{-1}(n[0])$.*

The following theorem is proved in a similar way as Theorem 3.6. The details are left to the reader.

THEOREM 3.9. *The Poincaré polynomial of $\widehat{\pi}^{-1}(n[0])$ is given by*

$$P_t(\widehat{\pi}^{-1}(n[0])) = \sum \prod_{\alpha=1}^{r} t^{2(r|Y_\alpha^1|+r|Y_\alpha^2|-\alpha l(Y_\alpha^1))} \prod_{\alpha<\beta} t^{(k_\alpha-k_\beta)(k_\alpha-k_\beta+1)},$$

where the summation runs over the set $\big((k_1,Y_1^1,Y_1^2),\dots,(k_r,Y_r^1,Y_r^2)\big)$ with (3.3).

COROLLARY 3.10. *The generating function of the Poincaré polynomials of $\widehat{\pi}^{-1}(n[0])$ is given by*

$$\sum_n P_t(\widehat{\pi}^{-1}(n[0]))q^n = \left(\prod_{\alpha=1}^{r}\prod_{d=1}^{\infty}\frac{1}{1-t^{2(rd-\alpha)}q^d}\right)\left(\prod_{d=1}^{\infty}\frac{1}{1-t^{2rd}q^d}\right)^r \times \sum_{\{\vec{k}\}=-k/r} t^{2\langle\vec{k},\rho\rangle}(t^{2r}q)^{(\vec{k},\vec{k})/2}.$$

3.4. A different choice of the one-parameter subgroup. This subsection is an interesting detour. We compute the Betti numbers of $\widehat{\pi}^{-1}(n[0])$ in a different way. A comparison with the formula in the previous subsection gives us a nontrivial combinatorial identity.

Let us choose weights for the one-parameter subgroup λ in (3.5) so that

$$m_1 = m_2 \gg n_1 > n_2 > \cdots > n_r > 0$$

and m_1, n_α are generic.

Since this λ is not generic, the fixed points are different from those for $\widetilde{T}$. But they are described similarly as $(E, \Phi) = (I_1(k_1C), \Phi_1) \oplus \cdots \oplus (I_r(k_rC), \Phi_r)$ such that

(a) $I_\alpha(k_\alpha C)$ is the tensor product $I_\alpha \otimes \mathcal{O}(k_\alpha C)$, where $k_\alpha \in \mathbb{Z}$ and I_α is the ideal sheaf of a 0-dimensional subscheme Z_α contained in $\widehat{\mathbb{C}}^2 = \widehat{\mathbb{P}}^2 \setminus \ell_\infty$.
(b) Φ_α is an isomorphism from $(I_\alpha)_{\ell_\infty}$ to the αth factor of $\mathcal{O}_{\ell_\infty}^{\oplus r}$.
(c) I_α is fixed by the diagonal subgroup $\Delta\mathbb{C}^*$ of $\mathbb{C}^* \times \mathbb{C}^*$, coming from that on $\widehat{\mathbb{P}}^2$.

Furthermore, we can parametrize the components of the fixed point set by $(\vec{k}, \vec{Y}) = \left((k_1, Y_1), \dots, (k_r, Y_r)\right)$ with k_α as above and Y_α is a Young diagram. Here the constraint is

$$\sum_\alpha k_\alpha = k, \qquad |\vec{Y}| + \frac{1}{2r}\sum_{\alpha<\beta} |k_\alpha - k_\beta|^2 = n. \tag{3.7}$$

Since this can be proved by the method in [**68**, Chapter 7], we explain it only briefly. A general point in the component $(\vec{k}, \vec{Y})$ is $(E, \Phi) = (I_1(k_1C), \Phi_1) \oplus \cdots \oplus (I_r(k_rC), \Phi_r)$ such that

(a) the support of I_α consists of P_1, $P_2, \dots, P_{l(Y_\alpha)}$, contained in the exceptional curve C,
(b) if ξ is the inhomogeneous coordinate of $C = \mathbb{P}^1$ and η is the coordinate of the fiber $\widehat{\mathbb{C}}^2 \cong \mathcal{O}(-1) \to C$,
$$I_\alpha = (\xi - \xi_1, \eta^{\lambda_1^\alpha}) \cap (\xi - \xi_2, \eta^{\lambda_2^\alpha}) \cap \cdots$$
with $\xi_l = \xi(P_l)$.

See [**68**, Figure 7.4]. The points P_1, $P_2, \dots,$ move in $\mathbb{P}^1$, but their order is irrelevant when the values λ_l^α are the same. Therefore the component is isomorphic to

$$S^{Y_1}\mathbb{P}^1 \times \cdots \times S^{Y_r}\mathbb{P}^1,$$

with the following notation: For a Young diagram $Y = (\lambda_1 \geq \lambda_2 \geq \cdots)$, we define $m_i = \#\{l \mid \lambda_l = i\}$. We denote $Y = (1^{m_1}2^{m_2}\cdots)$ in this case. We set

$$S^Y\mathbb{P}^1 = S^{m_1}\mathbb{P}^1 \times S^{m_2}\mathbb{P}^1 \times \cdots = \mathbb{P}^{m_1} \times \mathbb{P}^{m_2} \times \cdots,$$

where $S^m\mathbb{P}^1$ is the mth symmetric product of $\mathbb{P}^1$.

Let us suppose that (E, Φ) is a fixed point in the component corresponding to $\left((a_1, Y_1), \dots, (a_r, Y_r)\right)$. Then the tangent space $T_{(E,\Phi)}\widehat{M}(r, k, n)$ is a $\Delta\mathbb{C}^* \times T^r$-module. The $\Delta\mathbb{C}^* \times T^r$-module structure is independent of the choice of point; we take the $\widetilde{T}$-fixed point corresponding to $((k_1, \varnothing, Y_1), \dots, (k_r, \varnothing, Y_r))$. By the formula in Theorem 3.4 we have

$$T_{(E,\Phi)}\widehat{M}(r, k, n) = \sum_{\alpha,\beta}(L_{\alpha,\beta}^{\vec{k}}(t_1, t_1) + t_1^{a_\beta - a_\alpha} N_{\alpha,\beta}^{\vec{Y}'}(t_1)),$$

where $N^{\vec{Y}\prime}_{\alpha,\beta}(t_1) = N^{\vec{Y}}_{\alpha,\beta}(1, t_1)$. By Theorem 3.1 we have

$$N^{\vec{Y}\prime}_{\alpha,\beta}(t_1) = \left(\sum_{s\in Y_\alpha} t_1^{a_{Y_\alpha}(s)+1} + \sum_{t\in Y_\beta} t_1^{-a_{Y_\beta}(t)}\right) e_\beta e_\alpha^{-1}. \tag{3.8}$$

The following theorem is proved in a similar way as Theorem 3.6. The details are left to the reader.

THEOREM 3.11. *The Poincaré polynomial of $\widehat{\pi}^{-1}(n[0])$ is given by*

$$P_t(\widehat{\pi}^{-1}(n[0])) = \sum \prod_{\alpha=1}^{r} t^{2(|Y_\alpha|-l(Y_\alpha))} P_t(S^{Y_\alpha}\mathbb{P}^1) \prod_{\alpha<\beta} t^{2(l'_{\alpha,\beta}+|Y_\alpha|+|Y_\beta|-n'_{\alpha,\beta})}$$

where the summation runs over the set $(\vec{k}, \vec{Y})$ satisfying (3.7), *and*

$$l'_{\alpha,\beta} = \begin{cases} \frac{1}{2}(k_\alpha - k_\beta + 1)(k_\alpha - k_\beta) & \text{if } k_\alpha \geq k_\beta, \\ \frac{1}{2}(k_\beta - k_\alpha + 1)(k_\beta - k_\alpha) - 1 & \text{otherwise,} \end{cases}$$

$$n'_{\alpha,\beta} = \begin{cases} (\#\text{ of columns of } Y_\alpha \text{ which are longer than } k_\alpha - k_\beta) & \text{if } k_\alpha \geq k_\beta, \\ (\#\text{ of columns of } Y_\beta \text{ which are longer than } k_\beta - k_\alpha - 1) & \text{otherwise.} \end{cases}$$

Let us consider the generating function of the Poincaré polynomials. We consider the simplest case.

COROLLARY 3.12. *Assume $r = 2$ and $c_1 = 0$. The generating functions of the Poincaré polynomial is*

$$\left(\prod_{d=1}^{\infty} \frac{1}{(1-t^{4d}q^d)(1-t^{4d-2}q^d)^2(1-t^{4d-4}q^d)}\right) \times \left[\sum_{k\geq 0}\prod_{d=1}^{2k} \frac{1-t^{4d-4}q^d}{1-t^{4d}q^d} t^{2k(2k+1)} q^{k^2} + \sum_{k>0}\prod_{d=1}^{2k-1} \frac{1-t^{4d-4}q^d}{1-t^{4d}q^d} t^{2k(2k+1)-2} q^{k^2}\right].$$

Comparing with the formula in Corollary 3.10, we get the following identity

$$\begin{aligned} \sum_{k\geq 0}\prod_{d=1}^{2k} \frac{1-t^{4d-4}q^d}{1-t^{4d}q^d} t^{2k(2k+1)} q^{k^2} + \sum_{k>0}\prod_{d=1}^{2k-1} \frac{1-t^{4d-4}q^d}{1-t^{4d}q^d} t^{2k(2k+1)-2} q^{k^2} \\ = \prod_{d=1}^{\infty} \frac{1-t^{4d-2}q^d}{1-t^{4d}q^d} \sum_{k=-\infty}^{\infty} t^{2k(2k+1)} q^{k^2}. \end{aligned} \tag{3.9}$$

Since this identity does not involve any geometric information, it is natural to expect it to have a direct proof. Such a proof was provided for us by Hiroyuki Ochiai. See Appendix D.

Finally let us remark that the results of this and the previous subsections give the blowup formula for the virtual Hodge polynomials of moduli spaces for an arbitrary projective surface X. Let H be an ample line bundle over X. For $c_1 \in H^2(X, \mathbb{Z})$, $n \in \mathbb{Q}$, let $M_H(r, c_1, n)$ be the moduli space of H-stable sheaves E on X with $c_1(E) = c_1$, $c_2(E) - (r-1)c_1(E)^2/(2r) = n$. We assume $\gcd(r, \langle c_1, H\rangle) = 1$.

Let $\widehat{M}_H(r, c_1 + kC, n)$ be the moduli space of $(H - \varepsilon C)$-stable sheaves E on $\widehat{X}$ with $c_1(E) = p^*c_1 + kC$, $c_2(E) - (r-1)c_1(E)^2/(2r) \equiv \Delta(E) = n$, where c_1, n are as above, $k \in \mathbb{Z}$, and $\varepsilon > 0$ is sufficiently small.

Let $e(Y; x, y)$ denote the virtual Hodge polynomial of Y introduced in [**9**].

THEOREM 3.13. *The ratio*

$$\sum_n e(\widehat{M}_H(r, c_1 + kC, n); x, y)q^n \Big/ \sum_n e(M_H(r, c_1, n); x, y)q^n$$

is independent of the surface X and is given by

$$\left(\prod_{d=1}^{\infty} \frac{1}{1 - (xy)^{rd}q^d}\right)^r \sum_{\{\vec{k}\}=-k/r} (xy)^{\langle \vec{k},\rho\rangle} \left((xy)^r q\right)^{(\vec{k},\vec{k})/2}.$$

This follows from Corollary F.22 together with Corollaries 3.7 and 3.10.

Finally, note that the above holds in the Grothendieck group of varieties, if we replace $(xy)^n$ by $[\mathbb{C}^n]$. This generalizes [**31**] from rank 2 to higher ranks.

REMARK 3.14. This result was obtained by the second author [**80**]. In fact, he assumed that the moduli spaces are nonsingular and used the Weil conjecture to count numbers of rational points over finite fields. He did not use the framed moduli spaces nor the morphism $\widehat{\pi}$ as is done here. The above proof, under the same assumption, was obtained in July, 1997. The authors then noticed that W-P. Li and Z. Qin [**45–47**] obtained the above result for the rank 2 case, where the universal function is given in the form corresponding to Corollary 3.12. The authors then learned that the virtual Hodge polynomials are the natural language here.

4. Nekrasov's deformed partition function

Nekrasov's deformed partition function [**70**], more precisely, the one with *higher-order Casimir operators* turned on, can be considered as the generating function of the *equivariant homology* version of Donaldson invariants on $\mathbb{C}^2$. We give its definition and also the one for the blowup $\widehat{\mathbb{C}}^2$ at the origin in this section. We then study their relation by using the localization theorem in the equivariant homology groups. (See Appendix C.)

As for the calculation of the *original* Donaldson invariants, the localization technique was not so useful even if we assume that the base manifold has large symmetry (say $X = \mathbb{P}^2$). This was because the fixed point sets are not isolated in general, and are still difficult to study. The crucial difference here is the existence of the framing: A point in $M(r,n)$ is fixed by the action given by the change of the framing if and only if it is a direct sum of rank 1 sheaves. The rank 1 sheaves are easy to study.

4.1. Equivariant integration.

Before giving the definition, we explain a general setting for "equivariant integration" via the localization theorem.

Let T be a torus acting on an algebraic variety N. Suppose that the fixed point set N^T consists of a single point o. Then the push-forward homomorphism for the inclusion $\iota_0 \colon \{o\} \to N$ induces an isomorphism between localized equivariant homology groups

$$(\iota_0)_* \colon \mathcal{S} \cong H_*^T(o) \otimes_S \mathcal{S} \xrightarrow{\cong} H_*^T(N) \otimes_S \mathcal{S},$$

where $S = H_T^*(\mathrm{pt})$ and $\mathcal{S}$ is its quotient field. Furthermore, if $f \colon M \to N$ is a T-equivariant proper morphism, we can define $H_*^T(M) \to \mathcal{S}$ by

$$\alpha \mapsto (\iota_0)_*^{-1} f_* \alpha.$$

We denote this by $\int_M$. This makes sense even when M is not necessarily compact. But it takes values in the rational function field $\mathcal{S}$. When M is compact, it coincides with the usual integration and has values in S.

For equivariant K-homology groups, we have a similar homomorphism defined by the same formula:

$$K^G(N) \ni \alpha \mapsto (\iota_0)_*^{-1} f_* \alpha \in \mathcal{R},$$

where $\mathcal{R}$ is the quotient field of the representation ring of G. This has a relation with equivariant Hilbert polynomials. See [**69**, §3].

Suppose M is nonsingular. Let $M^T = \bigsqcup_i F_i$ be the decomposition of the fixed point set M^T into irreducible components. Let N_i be the normal bundle. Note that we only have finitely many components, and each F_i is compact, as f is proper and $N^T = \{o\}$. By the functoriality of the push-forward homomorphism, we have

$$\int_M \alpha = \sum_i \int_{F_i} \frac{1}{e_T(F_i)} \iota_i^* \alpha,$$

where $e_T(N_i)$ is the equivariant Euler class and ι_i^* is the pull-back homomorphism for the inclusion $\iota_i \colon F_i \to M$ defined via the Poincaré duality homomorphism. Here $\int_{F_i}$ is the usual integration as F_i is compact. When M is compact, the fractional parts of each summand of the right hand side cancel out, and the final answer is in S. But this does not happen when M is noncompact in general.

4.2. Universal sheaves. Since we need a higher-rank generalization of Donaldson's μ-map, we begin with the description of universal sheaves on the moduli spaces.

Over the moduli space $M(r,n)$, we have a natural vector bundle V, whose fiber at (E,φ) is $H^1(E(-\ell_\infty))$. In the ADHM description in [**68**, Chapter 2], this is the bundle associated with the natural principal $\mathrm{GL}_n(\mathbb{C})$-bundle, coming from the construction of $M(r,n)$ as a quotient space. If $\mathcal{E}$ denotes a universal sheaf on $\mathbb{P}^2 \times M(r,n)$, we have $V = R^1 p_{2*}(\mathcal{E} \otimes p_1^*(\mathcal{O}(-\ell_\infty)))$.

We also have another natural vector bundle W, given by the fiber at infinity: $W = H^0(\mathcal{E}|_{\ell_\infty})$. This is a trivial bundle, but nontrivial as an equivariant bundle. We also consider bundles $S^+ = \mathcal{O}_{\mathbb{P}^2}(-1) \oplus \mathcal{O}_{\mathbb{P}^2}(1)$ and $S^- = \mathcal{O}_{\mathbb{P}^2} \oplus \mathcal{O}_{\mathbb{P}^2}$ over $\mathbb{P}^2$ with the T^2-action such that $\operatorname{ch} S_0^+ = 1 + t_1^{-1} t_2^{-1}$, $\operatorname{ch} S_0^- = t_1^{-1} + t_2^{-1}$ on the fibers at the origin. (They are positive and negative spinor bundles S^+, S^- when restricted on $\mathbb{R}^4$.) We then form a virtual equivariant vector bundle on $\mathbb{P}^2 \times M(r,n)$ by

$$\mathcal{O}_{\mathbb{P}^2} \boxtimes W + (S^- - S^+) \boxtimes V.$$

By [**69**, Lemma 1.8], this virtual equivariant bundle is isomorphic to the universal sheaf $\mathcal{E}$ in the equivariant K-cohomology group $K_{\tilde{T}}(\mathbb{P}^2 \times M(r,n))$. We denote the virtual bundle by $\mathcal{E}$ hereafter.

The character of the fiber of $\mathcal{E}$ at the fixed point $(0, \vec{Y})$ is given by

$$\iota_{(0,\vec{Y})}^* \operatorname{ch}(\mathcal{E}) = \sum_{\alpha=1}^r e^{a_\alpha} \left(1 - (1 - e^{\varepsilon_1})(1 - e^{\varepsilon_2}) \sum_{s \in Y_\alpha} e^{l'(s)\varepsilon_1 + a'(s)\varepsilon_2} \right), \tag{4.1}$$

where we set $e_\alpha = e^{-a_\alpha}$, $t_1 = e^{-\varepsilon_1}$, $t_2 = e^{-\varepsilon_2}$ as usual. Let $q_1 = [0{:}1{:}0]$, $q_2 = [0{:}0{:}1]$ be two other fixed points in $\mathbb{P}^2$. We have

$$\iota_{(q_1,\vec{Y})}^* \operatorname{ch}(\mathcal{E}) = \iota_{(q_2,\vec{Y})}^* \operatorname{ch}(\mathcal{E}) = \sum_{\alpha=1}^r e^{a_\alpha}.$$

We define

$$\operatorname{ch}(\mathcal{E})/[\mathbb{C}^2] = \frac{1}{\varepsilon_1\varepsilon_2}\iota^*_{\{0\}\times M(r,n)} \operatorname{ch}(\mathcal{E}).$$

Since $\mathbb{C}^2$ is noncompact, the slant product $/[\mathbb{C}^2]$ is not defined in the usual sense. So we define it by formally applying Bott's formula. The homogeneous degree part of this is an element of the localized equivariant cohomology group $H^*_{\widetilde{T}}(M(r,n))\otimes_S \mathcal{S}$, but its fractional part is a constant in the following sense:

$$\begin{aligned}
&\iota^*_{\vec{Y}}(\operatorname{ch}(\mathcal{E})/[\mathbb{C}^2]) \\
&\quad = \iota^*_{\vec{Y}}(\operatorname{ch}(\mathcal{E})/[\mathbb{P}^2]) + \frac{1}{\varepsilon_1(\varepsilon_2-\varepsilon_1)}\iota^*_{(q_1,\vec{Y})}\operatorname{ch}(\mathcal{E}) + \frac{1}{(\varepsilon_1-\varepsilon_2)\varepsilon_2}\iota^*_{(q_2,\vec{Y})}\operatorname{ch}(\mathcal{E}) \\
&\quad = \iota^*_{\vec{Y}}(\operatorname{ch}(\mathcal{E})/[\mathbb{P}^2]) + \frac{1}{\varepsilon_1\varepsilon_2}\sum_{\alpha=1}^{r} e^{a_\alpha},
\end{aligned}$$

and each degree part of

$$\iota^*_{\vec{Y}}(\operatorname{ch}(\mathcal{E})/[\mathbb{P}^2]) = \frac{(1-e^{\varepsilon_1})(1-e^{\varepsilon_2})}{\varepsilon_1\varepsilon_2}\sum_{\alpha=1}^{r} e^{a_\alpha}\sum_{s\in Y_\alpha} e^{l'(s)\varepsilon_1 + a'(s)\varepsilon_2}$$

is a polynomial. We note that

$$\operatorname{ch}(\mathcal{E})/[\mathbb{P}^2] = \operatorname{ch}\big((S^- - S^+)\boxtimes V\big)/[\mathbb{P}^2]$$

since $\operatorname{ch}(W\boxtimes\mathcal{O}_{\mathbb{P}^2})/[\mathbb{P}^2] = 0$. Now the slant product $/[\mathbb{P}^2]$ on the right-hand side can be replaced by $/[\mathbb{C}^2]$ since $S^- - S^+$ is zero on $\ell_\infty = \mathbb{P}^2\setminus\mathbb{C}^2$. This observation also matches with the above formula for $\iota^*_{\vec{Y}}(\operatorname{ch}(\mathcal{E})/[\mathbb{P}^2])$.

Let us define the instanton part of the *partition function* by

$$\begin{aligned}
&Z^{\mathrm{inst}}(\varepsilon_1,\varepsilon_2,\vec{a};\mathfrak{q},\vec{\tau}) \\
&\quad = \sum_{n=0}^{\infty}\mathfrak{q}^n\int_{M(r,n)}\exp\Big(\sum_{p=1}^{\infty}\tau_p\operatorname{ch}_{p+1}(\mathcal{E})/[\mathbb{C}^2]\Big) \\
&\quad = \sum_{\vec{Y}}\frac{\mathfrak{q}^{|\vec{Y}|}}{\prod_{\alpha,\beta} n^{\vec{Y}}_{\alpha,\beta}(\varepsilon_1,\varepsilon_2,\vec{a})} \\
&\qquad \times\exp\Bigg(\sum_{p=1}^{\infty}\sum_{\alpha=1}^{r}\tau_p\Bigg[\frac{e^{a_\alpha}}{\varepsilon_1\varepsilon_2}\Big(1-(1-e^{\varepsilon_1})(1-e^{\varepsilon_2})\sum_{s\in Y_\alpha}e^{l'(s)\varepsilon_1+a'(s)\varepsilon_2}\Big)\Bigg]_{p-1}\Bigg),
\end{aligned}$$

where ch_{p+1} is the degree $(p+1)$-part of the Chern character, $[\,\cdot\,]_{p-1}$ denotes the degree $(p-1)$-part of $\cdot$, and $\int_{M(r,n)}\cdot$ means $(\iota_{0*})^{-1}\pi_*(\cdot\cap[M(r,n)])$ with $\iota_0\colon\{0\}\to M_0(r,n)$ the inclusion of the unique fixed point $0\in M_0(r,n)^{\widetilde{T}}$. Furthermore, $n^{\vec{Y}}_{\alpha,\beta}(\varepsilon_1,\varepsilon_2,\vec{a})$ is the equivariant Euler class of the tangent space at the fixed point $\vec{Y}$. It is given by the explicit formula

$$\begin{aligned}
n^{\vec{Y}}_{\alpha,\beta}(\varepsilon_1,\varepsilon_2,\vec{a}) = &\prod_{s\in Y_\alpha}\big(-l_{Y_\beta}(s)\varepsilon_1 + (a_{Y_\alpha}(s)+1)\varepsilon_2 + a_\beta - a_\alpha\big) \\
&\times\prod_{t\in Y_\beta}\big((l_{Y_\alpha}(t)+1)\varepsilon_1 - a_{Y_\beta}(t)\varepsilon_2 + a_\beta - a_\alpha\big).
\end{aligned}$$

The indeterminate $\mathfrak{q}$ should be distinguished from $q = e^{\pi\sqrt{-1}\tau}$ which will appear later.

Note that each coefficient of $\mathfrak{q}^n$ has a complicated, but explicit expression. For small n, it is easy to compute. (The authors wrote a MAPLE program, which was very useful in finding our main result.) But when n increases, the number of Young diagrams becomes large, and it becomes difficult to compute. We are interested not in individual coefficients, but in the generating function.

By (4.1) we have

$$\iota_{\vec{Y}}^*(\mathrm{ch}_1(\mathcal{E})/[\mathbb{C}^2]) = \frac{1}{\varepsilon_1\varepsilon_2}\sum_\alpha a_\alpha = 0, \quad \iota_{\vec{Y}}^*(\mathrm{ch}_2(\mathcal{E})/[\mathbb{C}^2]) = \frac{1}{2\varepsilon_1\varepsilon_2}\sum_\alpha a_\alpha^2 - n.$$

By the first equality, we did not include τ_0 in $\vec{\tau}$. Also, the second equation means that τ_1 is essentially equal to $-\log\mathfrak{q}$ (see also §4.5), but we use both of them to simplify the blowup formula below.

If we set $\vec{\tau} = 0$, we get the partition function studied in [**69**], which was denoted by $Z(\varepsilon_1, \varepsilon_2, \vec{a}; \mathfrak{q})$ there. But we emphasize that this is only the *instanton part* of the partition function. It is more natural to include the perturbative part also. This will be done in §4.4. This is the reason why we change the notation.

If we expand the exponential, $Z^{\mathrm{inst}}(\varepsilon_1, \varepsilon_2, \vec{a}; \mathfrak{q}, \vec{\tau})$ becomes

$$\sum_{n=0}^\infty \mathfrak{q}^n \sum_\mu \prod_i \tau_{\mu_i} \int_{M(r,n)} \prod_i \mathrm{ch}_{\mu_i+1}(\mathcal{E})/[\mathbb{C}^2],$$

where $\mu = (\mu_1 \ge \mu_2 \ge \cdots)$ is a partition. Therefore $Z^{\mathrm{inst}}(\varepsilon_1, \varepsilon_2, \vec{a}; \mathfrak{q}, \vec{\tau})$ is the generating function of *all* intersection numbers of Chern classes of universal sheaves slanted by the fundamental cycle $[\mathbb{C}^2]$.

The following formula will be useful later:

$$(4.2)\quad \left(\prod_i \frac{\partial}{\partial\tau_{\mu_i}}\right) Z^{\mathrm{inst}}(\varepsilon_1, \varepsilon_2, \vec{a}; \mathfrak{q}, \vec{\tau})$$
$$= \sum_{n=0}^\infty \mathfrak{q}^n \int_{M(r,n)} \left(\prod_i \mathrm{ch}_{\mu_i+1}(\mathcal{E})/[\mathbb{C}^2]\right) \cap \exp\left(\sum_{p=1}^\infty \tau_p \,\mathrm{ch}_{p+1}(\mathcal{E})/[\mathbb{C}^2]\right)$$

for a partition $\mu = (\mu_1 \ge \mu_2 \ge \cdots)$.

4.3. Partition function on the blowup. We consider similar partition functions on the blowup:

$$\widehat{Z}^{\mathrm{inst}}_{c_1=k}(\varepsilon_1, \varepsilon_2, \vec{a}; \mathfrak{q}, \vec{\tau}, \vec{t})$$
$$= \sum_n \mathfrak{q}^n \int_{\widehat{M}(r,k,n)} \exp\left(\sum_{p=1}^\infty \{t_p(\mathrm{ch}_{p+1}(\widehat{\mathcal{E}})/[C]) + \tau_p(\mathrm{ch}_{p+1}(\widehat{\mathcal{E}})/[\widehat{\mathbb{C}}^2])\}\right),$$

where $\widehat{\mathcal{E}}$ is a universal sheaf over $\widehat{\mathbb{P}}^2 \times \widehat{M}(r,k,n)$ and $\mathrm{ch}(\widehat{\mathcal{E}})/[\widehat{\mathbb{C}}^2]$ is defined as above via the localization (see below). The summation runs over $n \in \mathbb{Z}_{\ge 0} + k(r-k)/(2r)$. Here we do not include t_0 though $\mathrm{ch}_1(\widehat{\mathcal{E}})/[C]$ is not 0 in general. In fact, it is the constant $-k$.

We calculate this by using the localization formula. Recall that

$$\widehat{\mathbb{P}}^2 = \{([z_0{:}z_1{:}z_2], [z{:}w]) \in \mathbb{P}^2 \times \mathbb{P}^1 \mid z_1 w = z_2 z\}.$$

Let $p_1 = ([1{:}0{:}0], [1{:}0])$, $p_2 = ([1{:}0{:}0], [0{:}1])$, $q_1 = ([0{:}1{:}0], [1{:}0])$, $q_2 = ([0{:}0{:}1], [0{:}1])$ be the fixed points in $\widehat{\mathbb{P}}^2$. We use the same notation for the latter two points as fixed points in $\mathbb{P}^2$, since they are mapped to the corresponding points under the projection $p\colon \widehat{\mathbb{P}}^2 \to \mathbb{P}^2$. The characters of the fibers over the fixed points are given by

$$\iota^*_{(p_1,\vec{k},\vec{Y}^1,\vec{Y}^2)}\left(\mathrm{ch}(\widehat{\mathcal{E}})\right) = \iota^*_{(0,\vec{Y}^1)}(\mathrm{ch}(\mathcal{E}))\Big|_{\substack{\varepsilon_1\to\varepsilon_1\\ \varepsilon_2\to\varepsilon_2-\varepsilon_1\\ \vec{a}\to\vec{a}+\varepsilon_1\vec{k}}},$$

$$\iota^*_{(p_2,\vec{k},\vec{Y}^1,\vec{Y}^2)}\left(\mathrm{ch}(\widehat{\mathcal{E}})\right) = \iota^*_{(0,\vec{Y}^2)}(\mathrm{ch}(\mathcal{E}))\Big|_{\substack{\varepsilon_1\to\varepsilon_1-\varepsilon_2\\ \varepsilon_2\to\varepsilon_2\\ \vec{a}\to\vec{a}+\varepsilon_2\vec{k}}},$$

$$\iota^*_{(q_1,\vec{k},\vec{Y}^1,\vec{Y}^2)}\left(\mathrm{ch}(\widehat{\mathcal{E}})\right) = \iota^*_{(q_2,\vec{k},\vec{Y}^1,\vec{Y}^2)}\left(\mathrm{ch}(\widehat{\mathcal{E}})\right) = \sum_{\alpha=1}^{r} e^{-a_\alpha}.$$

For the first two equalities, see the proof of [**69**, 2.4]. The last two equalities are obvious since p is isomorphism outside the exceptional set. Therefore $\mathrm{ch}(\widehat{\mathcal{E}})/[\widehat{\mathbb{C}}^2]$ has the same constant fractional part as $\mathrm{ch}(\mathcal{E})/[\mathbb{C}^2]$.

Therefore we have

$$\begin{aligned}
&\iota^*_{(\vec{k},\vec{Y}^1,\vec{Y}^2)}(\mathrm{ch}(\widehat{\mathcal{E}})/[\mathbb{C}])\\
&= \sum_{\alpha=1}^{r}\left[\frac{e^{a_\alpha+\varepsilon_1 k_\alpha}\varepsilon_1}{\varepsilon_1(\varepsilon_2-\varepsilon_1)}\left(1-(1-e^{\varepsilon_1})(1-e^{\varepsilon_2-\varepsilon_1})\sum_{s\in Y^1_\alpha} e^{l'(s)\varepsilon_1+a'(s)(\varepsilon_2-\varepsilon_1)}\right)\right.\\
&\qquad\left.+\frac{e^{a_\alpha+\varepsilon_2 k_\alpha}\varepsilon_2}{(\varepsilon_1-\varepsilon_2)\varepsilon_2}\left(1-(1-e^{\varepsilon_1-\varepsilon_2})(1-e^{\varepsilon_2})\sum_{s\in Y^2_\alpha} e^{l'(s)(\varepsilon_1-\varepsilon_2)+a'(s)\varepsilon_2}\right)\right]\\
&= \varepsilon_1\iota^*_{\vec{Y}^1}(\mathrm{ch}(\mathcal{E})/[\mathbb{C}^2])\Big|_{\substack{\varepsilon_1\to\varepsilon_1\\ \varepsilon_2\to\varepsilon_2-\varepsilon_1\\ \vec{a}\to\vec{a}+\varepsilon_1\vec{k}}} + \varepsilon_2\iota^*_{\vec{Y}^2}(\mathrm{ch}(\mathcal{E})/[\mathbb{C}^2])\Big|_{\substack{\varepsilon_1\to\varepsilon_1-\varepsilon_2\\ \varepsilon_2\to\varepsilon_2\\ \vec{a}\to\vec{a}+\varepsilon_2\vec{k}}}
\end{aligned}$$

and

$$\iota^*_{(\vec{k},\vec{Y}^1,\vec{Y}^2)}\left(\mathrm{ch}(\widehat{\mathcal{E}})/[\widehat{\mathbb{C}}^2]\right) = \iota^*_{\vec{Y}^1}(\mathrm{ch}(\mathcal{E})/[\mathbb{C}^2])\Big|_{\substack{\varepsilon_1\to\varepsilon_1\\ \varepsilon_2\to\varepsilon_2-\varepsilon_1\\ \vec{a}\to\vec{a}+\varepsilon_1\vec{k}}} + \iota^*_{\vec{Y}^2}(\mathrm{ch}(\mathcal{E})/[\mathbb{C}^2])\Big|_{\substack{\varepsilon_1\to\varepsilon_1-\varepsilon_2\\ \varepsilon_2\to\varepsilon_2\\ \vec{a}\to\vec{a}+\varepsilon_2\vec{k}}}.$$

Note also that the equivariant Euler class of the tangent space $(\vec{k},\vec{Y}^1,\vec{Y}^2)$ is given by

$$\prod_{\alpha,\beta} l^{\vec{k}}_{\alpha,\beta}(\varepsilon_1,\varepsilon_2,\vec{a})\, n^{\vec{Y}^1}_{\alpha,\beta}(\varepsilon_1,\varepsilon_2-\varepsilon_1,\vec{a}+\varepsilon_1\vec{k})\, n^{\vec{Y}^2}_{\alpha,\beta}(\varepsilon_1-\varepsilon_2,\varepsilon_2,\vec{a}+\varepsilon_2\vec{k}),$$

where

$$l^{\vec{k}}_{\alpha,\beta}(\varepsilon_1,\varepsilon_2,\vec{a}) = s^{k_\alpha-k_\beta}(\varepsilon_1,\varepsilon_2,a_\beta-a_\alpha),$$

$$s^k(\varepsilon_1,\varepsilon_2,x) = \begin{cases}\prod_{\substack{i,j\ge 0\\ i+j\le k-1}}(-i\varepsilon_1-j\varepsilon_2+x) & \text{if } k>0,\\ \prod_{\substack{i,j\ge 0\\ i+j\le -k-2}}((i+1)\varepsilon_1+(j+1)\varepsilon_2+x) & \text{if } k<-1,\\ 1 & k=0 \text{ or } -1.\end{cases} \tag{4.3}$$

Therefore

$$(4.4)\quad \widehat{Z}^{\text{inst}}_{c_1=k}(\varepsilon_1,\varepsilon_2,\vec{a};\mathfrak{q},\vec{\tau},\vec{t}) = \sum_{\{\vec{k}\}=-k/r} \frac{\mathfrak{q}^{(\vec{k},\vec{k})/2}}{\prod_{\alpha,\beta} l^{\vec{k}}_{\alpha,\beta}(\varepsilon_1,\varepsilon_2,\vec{a})} Z^{\text{inst}}(\varepsilon_1,\varepsilon_2-\varepsilon_1,\vec{a}+\varepsilon_1\vec{k};\mathfrak{q},\vec{\tau}+\varepsilon_1\vec{t}) \times Z^{\text{inst}}(\varepsilon_1-\varepsilon_2,\varepsilon_2,\vec{a}+\varepsilon_2\vec{k};\mathfrak{q},\vec{\tau}+\varepsilon_2\vec{t}),$$

where $\vec{\tau}+\varepsilon_1\vec{t}$ means $(\tau_1+\varepsilon_1 t_1,\tau_2+\varepsilon_1 t_2,\dots)$. This is a generalization of the blowup formula. (In our previous paper [**69**], we only considered the case $\vec{t}=(t,0,0,\dots)$.)

4.4. Adding the perturbation term. We define the *full* partition function by

$$Z(\varepsilon_1,\varepsilon_2,\vec{a};\mathfrak{q},\vec{\tau}) = \exp\left[-\sum_{\alpha\neq\beta}\gamma_{\varepsilon_1,\varepsilon_2}(a_\alpha-a_\beta;\Lambda)\right] Z^{\text{inst}}(\varepsilon_1,\varepsilon_2,\vec{a};\mathfrak{q},\vec{\tau}),$$

where $\gamma_{\varepsilon_1,\varepsilon_2}$ is as in Appendix E and $\Lambda=\mathfrak{q}^{1/2r}$ as usual. The first term is the perturbation term of the partition function.

We choose the branch of log in the perturbative term as explained in the paragraph after Proposition 2.2. See (E.5).

We have

$$\begin{aligned}&\sum_{\alpha\neq\beta}\gamma_{\varepsilon_1,\varepsilon_2-\varepsilon_1}(a_\alpha-a_\beta+\varepsilon_1(k_\alpha-k_\beta);\Lambda)+\gamma_{\varepsilon_1-\varepsilon_2,\varepsilon_2}(a_\alpha-a_\beta+\varepsilon_2(k_\alpha-k_\beta);\Lambda)\\ &\quad=\sum_{\alpha\neq\beta}\left[\gamma_{\varepsilon_1,\varepsilon_2}(a_\alpha-a_\beta;\Lambda)+\log s^{k_\beta-k_\alpha}(\varepsilon_1,\varepsilon_2,a_\alpha-a_\beta)-\frac{(k_\alpha-k_\beta)^2}{2}\log\Lambda\right]\end{aligned}$$

by (E.2). Therefore

$$\begin{aligned}(4.5)\quad &Z(\varepsilon_1,\varepsilon_2-\varepsilon_1,\vec{a}+\varepsilon_1\vec{k};\mathfrak{q},\vec{\tau}+\varepsilon_1\vec{t})Z(\varepsilon_1-\varepsilon_2,\varepsilon_2,\vec{a}+\varepsilon_2\vec{k};\mathfrak{q},\vec{\tau}+\varepsilon_2\vec{t})\\ &=\exp\left[-\sum_{\alpha\neq\beta}\gamma_{\varepsilon_1,\varepsilon_2}(a_\alpha-a_\beta;\Lambda)\right]\prod_{\alpha,\beta}\frac{\Lambda^{(k_\beta-k_\alpha)^2/2}}{l^{\vec{k}}_{\beta,\alpha}(\varepsilon_1,\varepsilon_2,\vec{a})}\\ &\qquad\times Z^{\text{inst}}(\varepsilon_1,\varepsilon_2-\varepsilon_1,\vec{a}+\varepsilon_1\vec{k};\mathfrak{q},\vec{\tau}+\varepsilon_1\vec{t})\\ &\qquad\times Z^{\text{inst}}(\varepsilon_1-\varepsilon_2,\varepsilon_2,\vec{a}+\varepsilon_2\vec{k};\mathfrak{q},\vec{\tau}+\varepsilon_2\vec{t}).\end{aligned}$$

Then the blowup formula (4.4) is simplified to

$$(4.6)\quad \widehat{Z}^{c_1=k}(\varepsilon_1,\varepsilon_2,\vec{a};\mathfrak{q},\vec{\tau},\vec{t}) = \sum_{\{\vec{k}\}=-k/r} Z(\varepsilon_1,\varepsilon_2-\varepsilon_1,\vec{a}+\varepsilon_1\vec{k};\mathfrak{q},\vec{\tau}+\varepsilon_1\vec{t}) \times Z(\varepsilon_1-\varepsilon_2,\varepsilon_2,\vec{a}+\varepsilon_2\vec{k};\mathfrak{q},\vec{\tau}+\varepsilon_2\vec{t}),$$

where

$$\widehat{Z}^{c_1=k}(\varepsilon_1,\varepsilon_2,\vec{a};\mathfrak{q},\vec{\tau},\vec{t}) = \exp\left[-\sum_{\alpha\neq\beta}\gamma_{\varepsilon_1,\varepsilon_2}(a_\alpha-a_\beta;\Lambda)\right]\widehat{Z}^{\text{inst}}_k(\varepsilon_1,\varepsilon_2,\vec{a};\mathfrak{q},\vec{\tau},\vec{t}).$$

REMARK 4.1. As we saw, the blowup equations are simplified if we include the perturbation part. This was pointed out to us by N. Nekrasov. Probably this is already enough reason for the perturbation term. But it has a geometric meaning as a regularization of the Euler class of an "infinite rank" vector bundle. (See [**70**, §3.10] and §7.2 below.)

4.5. τ_1 versus $\log \mathfrak{q}$. Let $\vec{\tau}_1 = (\tau_1, 0, 0, \dots)$ be a vector with only the first entry nonzero. We have

$$Z^{\text{inst}}(\varepsilon_1, \varepsilon_2, \vec{a}; \mathfrak{q}, \vec{\tau} + \vec{\tau}_1) = \exp\left[\frac{\tau_1}{2\varepsilon_1\varepsilon_2} \sum_\alpha a_\alpha^2\right] Z^{\text{inst}}(\varepsilon_1, \varepsilon_2, \vec{a}; \mathfrak{q}e^{-\tau_1}, \vec{\tau}).$$

On the other hand, we have

$$\begin{aligned}
&\exp\left[-\sum_{\alpha\neq\beta} \gamma_{\varepsilon_1,\varepsilon_2}(a_\alpha - a_\beta; \Lambda\exp(-\frac{\tau_1}{2r}))\right] \Big/ \exp\left[-\sum_{\alpha\neq\beta} \gamma_{\varepsilon_1,\varepsilon_2}(a_\alpha - a_\beta; \Lambda)\right] \\
&\quad = \exp\left[-\sum_{\alpha\neq\beta} \frac{\tau_1}{2r}\left\{\frac{(a_\alpha - a_\beta)^2}{2\varepsilon_1\varepsilon_2} + \frac{(a_\alpha - a_\beta)(\varepsilon_1+\varepsilon_2)}{2\varepsilon_1\varepsilon_2} + \frac{\varepsilon_1^2 + \varepsilon_2^2 + 3\varepsilon_1\varepsilon_2}{12\varepsilon_1\varepsilon_2}\right\}\right] \\
&\quad = \exp\left[-\frac{\tau_1}{2\varepsilon_1\varepsilon_2}\sum_\alpha a_\alpha^2\right] \exp\left[-\frac{\tau_1(r-1)(\varepsilon_1^2+\varepsilon_2^2+3\varepsilon_1\varepsilon_2)}{24\varepsilon_1\varepsilon_2}\right]
\end{aligned}$$

by (E.4). Therefore

$$Z(\varepsilon_1, \varepsilon_2, \vec{a}; \mathfrak{q}, \vec{\tau} + \vec{\tau}_1) = \exp\left[-\frac{\tau_1(r-1)(\varepsilon_1^2+\varepsilon_2^2+3\varepsilon_1\varepsilon_2)}{24\varepsilon_1\varepsilon_2}\right] Z(\varepsilon_1, \varepsilon_2, \vec{a}; \mathfrak{q}e^{-\tau_1}, \vec{\tau}).$$

In particular, we have

$$\begin{aligned}
(4.7)\quad &\left(\frac{\partial}{\partial \tau_1}\right)^N Z(\varepsilon_1, \varepsilon_2, \vec{a}; \mathfrak{q}, \vec{\tau}) \\
&\qquad = \left(-\frac{(r-1)(\varepsilon_1^2+\varepsilon_2^2+3\varepsilon_1\varepsilon_2)}{24\varepsilon_1\varepsilon_2} - \mathfrak{q}\frac{\partial}{\partial\mathfrak{q}}\right)^N Z(\varepsilon_1, \varepsilon_2, \vec{a}; \mathfrak{q}, \vec{\tau})
\end{aligned}$$

for $N \in \mathbb{Z}_{\geq 0}$.

5. The blowup equation and Nekrasov's conjecture

The blowup formula (4.6) equates the unknown function $\widehat{Z}^{c_1=k}$ to the unknown Z. It is useless unless we know either or their independent relation. We do not have such knowledge so far in general, but we do know something when we restrict to the subspace $\vec{\tau} = 0$. This will be given in this section. An application is a solution of Nekrasov's conjecture: $\varepsilon_1\varepsilon_2 \log Z(\varepsilon_1, \varepsilon_2, \vec{a}; \mathfrak{q}, 0)|_{\varepsilon_1,\varepsilon_2=0}$ is equal to the Seiberg–Witten prepotential $\mathcal{F}_0(\vec{a}; \Lambda)$ introduced in §2.

Let

$$F(\varepsilon_1, \varepsilon_2, \vec{a}; \mathfrak{q}, \vec{\tau}) = \varepsilon_1\varepsilon_2 \log Z(\varepsilon_1, \varepsilon_2, \vec{a}; \mathfrak{q}, \vec{\tau}).$$

Here the logarithm is defined as follows: We first separate this into the perturbative part and the instanton part

$$F(\varepsilon_1, \varepsilon_2, \vec{a}; \mathfrak{q}, \vec{\tau}) = -\varepsilon_1\varepsilon_2 \sum_{\alpha\neq\beta} \gamma_{\varepsilon_1,\varepsilon_2}(a_\alpha - a_\beta; \Lambda) + \varepsilon_1\varepsilon_2 \log Z^{\text{inst}}(\varepsilon_1, \varepsilon_2, \vec{a}; \mathfrak{q}, \vec{\tau}).$$

We denote the second part by $F^{\text{inst}}(\varepsilon_1, \varepsilon_2, \vec{a}; \mathfrak{q}, \vec{\tau})$. It has the form

$$\sum_{p=1}^\infty \sum_{\alpha=1}^r \tau_p [e^{a_\alpha}]_{p+1} + \varepsilon_1\varepsilon_2 \log\left[\sum_{n=0}^\infty \mathfrak{q}^n \int_{M(r,n)} \exp\left(\sum_{p=1}^\infty \tau_p \operatorname{ch}_{p+1}(\mathcal{E})/[\mathbb{C}^2]\right)\right].$$

Since the summation in the last part starts with 1, its logarithm makes sense as a formal power series in $\mathfrak{q}$.

5.1. Gap from the dimension counting. In this subsection we derive a differential equation from a simple geometric consideration, which is well-known in the context of Donaldson invariants.

LEMMA 5.1. *Let $0 < k < r$. Then*

$$\widehat{Z}^{\text{inst}}_{c_1=k}(\varepsilon_1, \varepsilon_2, \vec{a}; \mathfrak{q}, \vec{\tau}, \vec{t}) = O\big((\vec{\tau}, \vec{t})^{k(r-k)}\big).$$

Here we put $\deg \tau_p = p - 1$, $\deg t_p = p$ and $O\big((\vec{\tau}, \vec{t})^N\big)$ means that it is a sum of monomials of degree greater than or equal to N. When it is a function only of $\vec{\tau}$ (or $\vec{t}$), we simply denote by $O(\vec{\tau}^N)$ (or $O(\vec{t}^N)$). This convention will be used in what follows.

PROOF. Consider the projective morphism

$$\widehat{\pi}\colon \widehat{M}(r, k, n) \to M_0\left(r, n - \frac{1}{2r}k(r-k)\right).$$

If $x \in H^{2d}_{\widetilde{T}}(\widehat{M}(r, k, n))$, we have

$$\widehat{\pi}_*(x \cap [\widehat{M}(r, k, n)]) \in H^{\widetilde{T}}_{2\dim \widehat{M}(r,k,n) - 2d}\left(M_0\left(r, n - \frac{1}{2r}k(r-k)\right)\right),$$

and this is 0 if

$$\dim \widehat{M}(r, k, n) - d > \dim M_0\left(r, n - \frac{1}{2r}k(r-k)\right) \iff k(r-k) > d.$$

In the definition of the partition function on the blowup, we have $\operatorname{ch}_{p+1}(\widehat{\mathcal{E}})/[C] \in H^{2p}_{\widetilde{T}}$, $\operatorname{ch}_{p+1}(\widehat{\mathcal{E}})/[\widehat{\mathbb{P}}^2] \in H^{2(p-1)}_{\widetilde{T}}$. The degrees exactly match with the above definition of the degrees of t_p and τ_p. □

Similarly we consider the $k = 0$ case. The morphism $\widehat{\pi}\colon \widehat{M}(r, 0, n) \to M_0(r, n)$ is an isomorphism outside the inverse image of the closure of $\{0\} \times M_0^{\text{reg}}(r, n-1)$. Furthermore we have $\widehat{\pi}^*(\mathcal{E}) \cong \widehat{\mathcal{E}}$ there. Since $\operatorname{codim}(\{0\} \times M_0^{\text{reg}}(r, n-1)) = 2r$, the same argument as above shows

$$\widehat{Z}^{\text{inst}}_{c_1=0}(\varepsilon_1, \varepsilon_2, \vec{a}; \mathfrak{q}, \vec{\tau}, \vec{t}) = Z^{\text{inst}}(\varepsilon_1, \varepsilon_2, \vec{a}; \mathfrak{q}, \vec{\tau}) + O\big((\vec{\tau}, \vec{t})^{2r}\big).$$

Combined with the blowup formula in the previous subsection, we get

$$\begin{aligned}(5.1)\quad \sum_{\{\vec{k}\}=0} Z(\varepsilon_1, \varepsilon_2 - \varepsilon_1, \vec{a} + \varepsilon_1\vec{k}; \mathfrak{q}, \vec{\tau} + \varepsilon_1\vec{t}) Z(\varepsilon_1 - \varepsilon_2, \varepsilon_2, \vec{a} + \varepsilon_2\vec{k}; \mathfrak{q}, \vec{\tau} + \varepsilon_2\vec{t}) \\ = Z(\varepsilon_1, \varepsilon_2, \vec{a}; \mathfrak{q}, \vec{\tau}) + O\big((\vec{\tau}, \vec{t})^{2r}\big),\end{aligned}$$

$$\begin{aligned}(5.2)\quad \sum_{\{\vec{k}\}=-k/r} Z(\varepsilon_1, \varepsilon_2 - \varepsilon_1, \vec{a} + \varepsilon_1\vec{k}; \mathfrak{q}, \vec{\tau} + \varepsilon_1\vec{t}) Z(\varepsilon_1 - \varepsilon_2, \varepsilon_2, \vec{a} + \varepsilon_2\vec{k}; \mathfrak{q}, \vec{\tau} + \varepsilon_2\vec{t}) \\ = O\big((\vec{\tau}, \vec{t})^{k(r-k)}\big) \quad (0 < k < r).\end{aligned}$$

We call these *blowup equations.*

5.2. Recursive structure. In this subsection we illustrate the power of the blowup equations: They determine $Z(\varepsilon_1, \varepsilon_2, \vec{a}; \mathfrak{q}, \vec{\tau})$ up to $O(\vec{\tau}^{2r-3})$.

We introduce two auxiliary functions:

$$F_a(\vec{a}) = F(\varepsilon_1, \varepsilon_2 - \varepsilon_1, \vec{a}; \mathfrak{q}, \vec{\tau}), \qquad F_b(\vec{a}) = F(\varepsilon_1 - \varepsilon_2, \varepsilon_2, \vec{a}; \mathfrak{q}, \vec{\tau}).$$

We suppress the ε_1, ε_2-dependence in the notation.

We divide (5.1) for $c_1 = k = 0$ by $Z(\varepsilon_1, \varepsilon_2 - \varepsilon_1, \vec{a}; \mathfrak{q}, \vec{\tau})Z(\varepsilon_1 - \varepsilon_2, \varepsilon_2, \vec{a}; \mathfrak{q}, \vec{\tau})$, expand with respect to the variables $\vec{t}$, and take the coefficients of t_1 and t_1^2:

$$\sum_{\{\vec{k}\}=0} \frac{1}{\varepsilon_2 - \varepsilon_1}\left(\frac{\partial}{\partial \tau_1} F_a(\vec{a} + \varepsilon_1 \vec{k}) - \frac{\partial}{\partial \tau_1} F_b(\vec{a} + \varepsilon_2 \vec{k})\right)$$
$$\times \exp\left[\frac{1}{\varepsilon_2 - \varepsilon_1}\left(\frac{F_a(\vec{a} + \varepsilon_1 \vec{k}) - F_a(\vec{a})}{\varepsilon_1} - \frac{F_b(\vec{a} + \varepsilon_2 \vec{k}) - F_b(\vec{a})}{\varepsilon_2}\right)\right] = O(\vec{\tau}^{2r-1}),$$

$$(5.3)\quad \sum_{\{\vec{k}\}=0}\left[\left\{\frac{1}{\varepsilon_2 - \varepsilon_1}\left(\frac{\partial}{\partial \tau_1} F_a(\vec{a} + \varepsilon_1 \vec{k}) - \frac{\partial}{\partial \tau_1} F_b(\vec{a} + \varepsilon_2 \vec{k})\right)\right\}^2 + \frac{1}{\varepsilon_2 - \varepsilon_1}\left(\varepsilon_1 \frac{\partial^2}{\partial \tau_1^2} F_a(\vec{a} + \varepsilon_1 \vec{k}) - \varepsilon_2 \frac{\partial^2}{\partial \tau_1^2} F_b(\vec{a} + \varepsilon_2 \vec{k})\right)\right]$$
$$\times \exp\left[\frac{1}{\varepsilon_2 - \varepsilon_1}\left(\frac{F_a(\vec{a} + \varepsilon_1 \vec{k}) - F_a(\vec{a})}{\varepsilon_1} - \frac{F_b(\vec{a} + \varepsilon_2 \vec{k}) - F_b(\vec{a})}{\varepsilon_2}\right)\right] = O(\vec{\tau}^{2r-2}).$$

THEOREM 5.2. (1) *The solution $F(\varepsilon_1, \varepsilon_2, \vec{a}; \mathfrak{q}, \vec{\tau})$ of the equations (5.3) is unique up to $O(\vec{\tau}^{2r-2})$. In fact, (5.3) determine the coefficients of $\mathfrak{q}^n$ in $F^{\text{inst}}(\varepsilon_1, \varepsilon_2, \vec{a}; \mathfrak{q}, \vec{\tau})$ recursively up to $O(\vec{\tau}^{2r-2})$.*

(2) *The coefficients of monomials in $\vec{\tau}$ of degree $< 2r - 2$ in $F(\varepsilon_1, \varepsilon_2, \vec{a}; \mathfrak{q}, \tau)$ are regular at $\varepsilon_1 = \varepsilon_2 = 0$.*

PROOF. We prove the assertions by induction on the power of $\mathfrak{q}$. Suppose that the coefficients of $\mathfrak{q}^m$ in $F(\varepsilon_1, \varepsilon_2, \vec{a}; \mathfrak{q}, \vec{\tau})$ are determined for $m < n$. We show that the coefficients of $\mathfrak{q}^n$ in F_a, F_b, and hence in $F(\varepsilon_1, \varepsilon_2, \vec{a}; \mathfrak{q}, \vec{\tau})$ are determined from (5.3).

Let us separate the terms with $\vec{k} = 0$. The remaining terms with $\vec{k} \neq 0$ are divisible by $\mathfrak{q}$ by (4.5) (recall $\prod_{\alpha,\beta} \Lambda^{(k_\beta - k_\alpha)^2/2} = \mathfrak{q}^{(\vec{k},\vec{k})/2}$). Then the equations are written as

$$\frac{\partial}{\partial \tau_1} F_a(\vec{a}) - \frac{\partial}{\partial \tau_1} F_b(\vec{a}) = \mathfrak{q} \times \text{known up to order } n-1,$$
$$\varepsilon_1 \frac{\partial^2}{\partial \tau_1^2} F_a(\vec{a}) - \varepsilon_2 \frac{\partial^2}{\partial \tau_1^2} F_b(\vec{a}) = \mathfrak{q} \times \text{known up to order } n-1.$$

After noticing that $\partial/\partial\tau_1$ is essentially equal to $\mathfrak{q}\partial/\partial\mathfrak{q}$ as in §4.5, the above equations give a system of linear equations on the coefficients of $\mathfrak{q}^n$ in $F_a(\vec{a}, \varepsilon_1 \vec{t}')$, $F_b(\vec{a}, \varepsilon_2 \vec{t}')$. This system is uniquely solvable since the determinant of $\begin{pmatrix} n & -n \\ \varepsilon_1 n^2 & -\varepsilon_2 n^2 \end{pmatrix}$ is nonzero.

Furthermore, the right-hand sides divided by $\varepsilon_1 - \varepsilon_2$ are regular at ε_1, $\varepsilon_2 = 0$ if $F(\varepsilon_1, \varepsilon_2, \vec{a}; \mathfrak{q}, \tau)$ is regular. Again by induction, we get the second assertion. □

5.3. Contact term equations as limit of blowup equations. In this subsection we study the specialization of the differential equation (5.1) at $\varepsilon_1 = \varepsilon_2 = 0$.

Let

$$(5.4)\quad F(\varepsilon_1, \varepsilon_2, \vec{a}; \mathfrak{q}, \vec{\tau}) = F_0(\vec{a}; \mathfrak{q}, \vec{\tau}) + (\varepsilon_1 + \varepsilon_2) H(\vec{a}; \mathfrak{q}, \vec{\tau}) + (\varepsilon_1 + \varepsilon_2)^2 G(\vec{a}; \mathfrak{q}, \vec{\tau}) + \varepsilon_1 \varepsilon_2 F_1(\vec{a}; \mathfrak{q}, \vec{\tau}) + \cdots,$$

where we consider terms up to $O(\vec{\tau}^{2r-2})$.

By exactly the same argument as in [**69**, 6.1], we have $Z^{\text{inst}}(\varepsilon_1, -2\varepsilon_1, \vec{a}; \mathfrak{q}, \vec{\tau}) = Z^{\text{inst}}(2\varepsilon_1, -\varepsilon_1, \vec{a}; \mathfrak{q}, \vec{\tau})$ up to $O(\vec{\tau}^{2r-2})$. In particular, $H(\vec{a}; \mathfrak{q}, \vec{\tau})$ up to $O(\vec{\tau}^{2r-2})$ comes only from the perturbative term:

$$H(\vec{a}; \mathfrak{q}, \vec{\tau}) = \frac{1}{2}\sum_{\alpha<\beta}(a_\alpha - a_\beta)\log(-1) = \pi\sqrt{-1}\langle\vec{a}, \rho\rangle.$$

See (E.5).

The first part of the following was proved by [**69**] and independently by [**71**].

THEOREM 5.3. (1) *$F_0(\vec{a}; \mathfrak{q}, 0)$ is equal to the Seiberg–Witten prepotential $\mathcal{F}_0(\vec{a}; \Lambda)$ with $\mathfrak{q} = \Lambda^{2r}$.*

(2) *For $p = 2, \dots, r$, let c_p be the pth power sum in $z_1, \dots, z_r$ multiplied by $(-\sqrt{-1})^p/p!$ given in (2.11). We have*

$$\left.\frac{\partial F_0}{\partial \tau_{p-1}}\right|_{\vec{\tau}=0} = c_p.$$

PROOF. As the heading of this subsection suggests, we prove the assertion by studying the limit of the blowup equations.

We have

$$\frac{F_0(\vec{a} + \varepsilon_1\vec{k}; \mathfrak{q}, \varepsilon_1\vec{t})}{\varepsilon_1(\varepsilon_2 - \varepsilon_1)} + \frac{F_0(\vec{a} + \varepsilon_2\vec{k}; \mathfrak{q}, \varepsilon_2\vec{t})}{(\varepsilon_1 - \varepsilon_2)\varepsilon_2}$$
$$= \frac{1}{\varepsilon_1\varepsilon_2}F_0 - \left[\frac{\partial^2 F_0}{\partial\tau_p\partial\tau_q}\frac{t_p t_q}{2} + \frac{\partial^2 F_0}{\partial\tau_p\partial a^l}t_p k^l + \frac{\partial^2 F_0}{\partial a^l\partial a^m}\frac{k^l k^m}{2}\right] + \cdots,$$

$$\frac{\varepsilon_2 H(\vec{a} + \varepsilon_1\vec{k}; \mathfrak{q}, \varepsilon_1\vec{t})}{\varepsilon_1(\varepsilon_2 - \varepsilon_1)} + \frac{\varepsilon_1 H(\vec{a} + \varepsilon_2\vec{k}; \mathfrak{q}, \varepsilon_2\vec{t})}{(\varepsilon_1 - \varepsilon_2)\varepsilon_2}$$
$$= \frac{\varepsilon_1 + \varepsilon_2}{\varepsilon_1\varepsilon_2}H + \left[\frac{\partial H}{\partial\tau_p}t_p + \frac{\partial H}{\partial a^l}k^l\right] + \cdots = \frac{\varepsilon_1 + \varepsilon_2}{\varepsilon_1\varepsilon_2}H + \pi\sqrt{-1}\langle\vec{k}, \rho\rangle + \cdots,$$

$$\frac{\varepsilon_2^2 G(\vec{a} + \varepsilon_1\vec{k}; \mathfrak{q}, \varepsilon_1\vec{t})}{\varepsilon_1(\varepsilon_2 - \varepsilon_1)} + \frac{\varepsilon_1^2 G(\vec{a} + \varepsilon_2\vec{k}; \mathfrak{q}, \varepsilon_2\vec{t})}{(\varepsilon_1 - \varepsilon_2)\varepsilon_2} = \frac{(\varepsilon_1 + \varepsilon_2)^2 - \varepsilon_1\varepsilon_2}{\varepsilon_1\varepsilon_2}G + \cdots,$$
$$F_1(\vec{a} + \varepsilon_1\vec{k}; \mathfrak{q}, \varepsilon_1\vec{t}) + F_1(\vec{a} + \varepsilon_2\vec{k}; \mathfrak{q}, \varepsilon_2\vec{t}) = 2F_1 + \cdots.$$

Here and throughout the proof, F_0, H, G, F_1 and their derivatives on the right-hand side are all restrictions to $\vec{\tau} = 0$.

We divide both sides of (5.1) by $Z(\varepsilon_1, \varepsilon_2, \vec{a}; \mathfrak{q}, \vec{\tau})$, set $\vec{\tau} = 0$, and take the limit ε_1, $\varepsilon_2 \to 0$:

$$(5.5)\quad \sum_{\{\vec{k}\}=0}\exp\left[-\frac{\partial^2 F_0}{\partial\tau_p\partial\tau_q}\frac{t_p t_q}{2} - \frac{\partial^2 F_0}{\partial\tau_p\partial a^l}t_p k^l - \frac{\partial^2 F_0}{\partial a^l\partial a^m}\frac{k^l k^m}{2}\right](-1)^{\langle\vec{k},\rho\rangle}\exp(F_1 - G)$$
$$= 1 + O(\vec{t}^{2r}),$$

where the summation symbols over p, q, l, m are omitted. Logically speaking, we only show F is regular up to $O(\vec{t}^{2r-2})$ at this moment. Thus the higher-order terms may diverge in the limit $\varepsilon_1 = \varepsilon_2 = 0$. Therefore this equation should be understood that the left-hand side is equal to 1 if we set all higher-order terms to be zero.

Let

$$(5.6)\qquad \tau_{kl} = -\frac{1}{2\pi\sqrt{-1}}\frac{\partial^2 F_0}{\partial a^k\partial a^l}.$$

This is symmetric and positive definite for $\mathfrak{q}$ small. So we consider the corresponding theta function Θ_E with the characteristic $E = {}^t\left(\frac{1}{2}\ \frac{1}{2}\ \cdots\right)$ in the notation for the root system of type A_{r-1}.

Comparing with the constant term of (5.5), we get

$$\exp(G - F_1) = \Theta_E(0|\tau). \tag{5.7}$$

Comparing the coefficients of $t_p t_q$ with $p + q \leq 2r - 1$, we get

$$0 = -\frac{\partial^2 F_0}{\partial\tau_p\partial\tau_q} + \frac{1}{\pi\sqrt{-1}}\sum_{l,m}\frac{\partial^2 F_0}{\partial\tau_p\partial a^l}\frac{\partial^2 F_0}{\partial\tau_q\partial a^m}\frac{\partial}{\partial\tau_{lm}}\log\Theta_E(0|\tau). \tag{5.8}$$

By (4.7) we have

$$\frac{\partial}{\partial\tau_1}F_0 = -\mathfrak{q}\frac{\partial}{\partial\mathfrak{q}}F_0 = -\frac{1}{2r}\Lambda\frac{\partial}{\partial\Lambda}F_0 \quad\text{and}\quad \frac{\partial^2}{\partial\tau_1^2}F_0 = \frac{1}{4r^2}\left(\Lambda\frac{\partial}{\partial\Lambda}\right)^2 F_0.$$

Therefore the equation with $p = q = 1$ is nothing but the contact term equation in Corollary 2.5.

When we consider the contact term equation as the differential equation for F_0, it has a similar recursive structure as the blowup equation studied in §5.2. The coefficients of $\mathfrak{q}^n$ in the instanton part of F_0 are determined from lower coefficients. In particular, the solution is unique if the perturbative part is given. Since F_0 and the Seiberg–Witten prepotential $\mathcal{F}_0$ have the same perturbative part, we have (1) of Theorem 5.3.

Let us prove the second assertion. The case $p = 2$ is nothing but the renormalization group equation in Proposition 2.3. We substitute $p = p - 1$, $q = 2$ in (5.8):

$$\Lambda\frac{\partial}{\partial\Lambda}\left(\frac{\partial F_0}{\partial\tau_{p-1}}\right) = \frac{2r}{\pi\sqrt{-1}}\sum_{l,m}\frac{\partial u_2}{\partial a^m}\frac{\partial}{\partial a^l}\left(\frac{\partial F_0}{\partial\tau_{p-1}}\right)\frac{\partial}{\partial\tau_{lm}}\log\Theta_E(0|\tau).$$

If we expand the both sides into power series in $\mathfrak{q}$ (plus the perturbative part), the equation determines the coefficients recursively. The point here is the observation that $\partial\log\Theta_E(0|\tau)/\partial\tau_{lm}$ is divisible by $\mathfrak{q}$. In particular, the solution is unique if the perturbative part is given.

The perturbative part of $\partial F_0/\partial\tau_{p-1}$ is given by

$$[e^{a_\alpha}]_p = \sum_\alpha \frac{a_\alpha^p}{p!}.$$

This is equal to the perturbative part of c_p. This shows our assertion. □

6. Fintushel–Stern's blowup formula

6.1. In this subsection and the next one we assume that derivatives of $F(\varepsilon_1, \varepsilon_2, \vec{a}; \mathfrak{q}, \vec{\tau})$ up to the second order are regular at $\varepsilon_1 = \varepsilon_2 = 0$, when restricted to $\vec{\tau} = 0$.

As we did in (5.5) we can derive the limit of the blowup formula (4.6) as

$$(6.1)\quad \lim_{\varepsilon_1,\varepsilon_2\to 0} \frac{\widehat{Z}^{c_1=k}(\varepsilon_1,\varepsilon_2,\vec{a};\mathfrak{q},0,\vec{t})}{Z(\varepsilon_1,\varepsilon_2,\vec{a};\mathfrak{q},0)} = \sum_{\{\vec{k}\}=-k/r} \exp\left[-\frac{\partial^2 F_0}{\partial\tau_p\partial\tau_q}\frac{t_p t_q}{2} - \frac{\partial^2 F_0}{\partial\tau_p\partial a^l}t_p k^l - \frac{\partial^2 F_0}{\partial a^l\partial a^m}\frac{k^l k^m}{2}\right] \times (\sqrt{-1})^{2\langle\vec{k},\rho\rangle}\exp(F_1 - G),$$

where the summation symbol is omitted as before, and we restrict functions to the subspace $\vec{\tau}=0$ also as before.

We define the *contact term* by

$$T_{p,q}(\vec{a};\mathfrak{q}) = \frac{1}{2}\frac{\partial^2 F_0}{\partial\tau_p\partial\tau_q}(\vec{a};\mathfrak{q},0).$$

We also set $t_r = t_{r+1} = \cdots = 0$. Then the right-hand side of (6.1) is rewritten as

$$(6.2)\quad \exp\left(-\sum_{p,q=1}^{r-1} T_{p,q}t_p t_q\right)\frac{\Theta_{E_k}\left((\sqrt{-1}/2\pi)\sum_{p=1}^{r-1}(dc_{p+1}/d\vec{a})t_p\middle|\tau\right)}{\Theta_E(0|\tau)}$$

where

$$\frac{dc_{p+1}}{d\vec{a}} = {}^t\!\left(\frac{\partial c_{p+1}}{\partial a^1}\ \cdots\ \frac{\partial c_{p+1}}{\partial a^{r-1}}\right),$$

and E_k is the characteristic given in §B.1. We have used Theorem 5.3(2) and (5.7).

This is a generalization of [**69**, 7.1]. Note that this expression for $r=2$ coincides with the blowup formula derived from the u-plane integral in §1.3. The identification of the contact term follows from (2.12).

For a physical derivation of a higher-rank generalization, see [**50**].

6.2. A reformulation. We reformulate the blowup formula in the previous subsection in a form which does not involve the limit. It also provides an interpretation of Theorem 5.3(2) which does not involve the limit. The goal is to express the formula in H_*^T instead of $H_*^{\widetilde{T}}$. (Recall $\widetilde{T} = \mathbb{C}^*\times\mathbb{C}^*\times T$.)

Let $j^!\colon H_*^{\widetilde{T}}(M_0(r,n)) \to H_*^T(M_0(r,n))$ be the homomorphism given by the restriction of the action. This can be defined via the pull-back homomorphism with respect to a locally trivial fibration with fiber $\widetilde{T}/T$

$$j\colon M_0(r,n)\times_T U \to M_0(r,n)\times_{\widetilde{T}} U,$$

where U is a $\widetilde{T}$-variety as in Appendix C.

Recall that we have made an identification

$$\prod_n (H_*^{\widetilde{T}}(M_0(r,n))\otimes_S \mathcal{S})\Lambda^{2rn} \cong \prod_n \mathcal{S}\Lambda^{2rn}$$

via $(\iota_0)_*$. The multiplication by Λ on the right-hand side is identified with the push-forward homomorphism $i_{n,n+1*}$ of the natural embedding $i_{n,n+1}\colon M_0(r,n)\to$

$M_0(r,n+1)$. The identification follows from the commutativity of the diagram

$$\begin{array}{ccc} H_*^{\widetilde{T}}(\mathrm{pt}) & \xrightarrow{\iota_{0*}} & H_*^{\widetilde{T}}(M_0(r,n)) \\ \| & & \downarrow i_{n,n+1*} \\ H_*^{\widetilde{T}}(\mathrm{pt}) & \xrightarrow{\iota_{0*}} & H_*^{\widetilde{T}}(M_0(r,n+1)). \end{array}$$

In particular, we note that the multiplication of Λ^{2r} makes sense as operators on $\prod_n H_*^{\widetilde{T}}(M_0(r,n))\Lambda^{2rn}$. It also makes sense on $\prod_n H_*^{T}(M_0(r,n))\Lambda^{2rn}$, and two homomorphisms commute with $j^!$.

We consider

$$Z^{\mathrm{inst}}(\vec{a};\Lambda) = \sum_n \Lambda^{2rn}\pi_*[M(r,n)],$$
$$\widehat{Z}^{\mathrm{inst}}_{c_1=k}(\vec{a};\Lambda,\vec{t}) = \sum_n \Lambda^{2rn}\widehat{\pi}_*\left[\exp\left(\sum_{p=1}^\infty t_p\left(\mathrm{ch}_{p+1}(\widehat{\mathcal{E}})/[C]\right)\right)\cap[\widehat{M}(r,k,n)]\right].$$

These are formal sums of elements in $H_*^T(M_0(r,n))$. They are the pull-backs of the corresponding elements in $H_*^{\widetilde{T}}(M_0(r,n))$ via $j^!$.

LEMMA 6.1. *Let $R = \{f(\vec{a},\varepsilon_1,\varepsilon_2) \mid f(\vec{a},0,0)\neq 0\}$. Then $H_*^{\widetilde{T}}(M_0(r,n))_R :=$ $H_*^{\widetilde{T}}(M_0(r,n))\otimes_{S(T)} S(T)_R$ is a torsion-free $S(\widetilde{T})_R$-module.*

PROOF. By the localization theorem, we have

$$H_*^{\widetilde{T}}(M_0(r,n))_R \cong H_*^{\widetilde{T}}(M_0(r,n)^T)_R \cong H_*^{\mathbb{C}^*\times\mathbb{C}^*}(M_0(r,n)^T)\otimes_{\mathbb{C}} S(T)_R.$$

Since $M_0(r,n)^T = S^n\mathbb{C}^2$, $H_*^{\mathbb{C}^*\times\mathbb{C}^*}(M_0(r,n)^T)$ is a torsion-free $\mathbb{C}[\varepsilon_1,\varepsilon_2]$-module. □

Therefore the blowup formula in the previous section can be restricted:

$$\widehat{Z}^{\mathrm{inst}}_{c_1=k}(\vec{a};\Lambda,\vec{t}) = (6.2)\times Z^{\mathrm{inst}}(\vec{a};\Lambda). \tag{6.3}$$

This is an equality in formal power series in Λ^{2r} and $\vec{t}$ with values in $H_*^T(M_0(r,n))$ $\otimes_{S(T)} \mathcal{S}(T)$. Let us emphasize again that the multiplication by Λ^{2r} is $i_{n,n+1*}$.

Recall that $Z^{\mathrm{inst}}(\varepsilon_1,\varepsilon_2,\vec{a};\mathfrak{q},\vec{\tau}) = \exp(F^{\mathrm{inst}}(\varepsilon_1,\varepsilon_2,\vec{a};\mathfrak{q},\vec{\tau})/\varepsilon_1\varepsilon_2)$. By (4.2) we have

$$\begin{aligned}\sum_{n=0}^\infty \Lambda^{2rn}\int_{M(r,n)}\prod_i(\mathrm{ch}_{\mu_i+1}(\mathcal{E})/[0]) &= \left(\prod_i \varepsilon_1\varepsilon_2\frac{\partial}{\partial\tau_{\mu_i}}\right)Z^{\mathrm{inst}}(\varepsilon_1,\varepsilon_2,\vec{a};\Lambda^{2r},\vec{\tau})\Big|_{\vec{\tau}=0}\\ &= \left(\prod_i\frac{\partial F^{\mathrm{inst}}}{\partial\tau_{\mu_i}}(\varepsilon_1,\varepsilon_2,\vec{a};\Lambda^{2r},0)\right)Z^{\mathrm{inst}}(\varepsilon_1,\varepsilon_2,\vec{a};\Lambda^{2r},0),\end{aligned}$$

where $[0]$ is the fundamental class of the origin. We can make a restriction:

$$\sum_{n=0}^\infty \Lambda^{2rn}\pi_*\left(\prod_i(\mathrm{ch}_{\mu_i+1}(\mathcal{E})/[0])\cap[M(r,n)]\right) = \left(\prod_i c_{\mu_i+1}\right)Z^{\mathrm{inst}}(\vec{a};\Lambda^{2r}) \tag{6.4}$$

thanks to Theorem 5.3(2), where we assume all $\mu_i\le r-1$.

Note that this formula explains the meaning of $c_p = \partial F_0/\partial\tau_{p-1}$ without taking limits. It is just multiplication by $\mathrm{ch}_p(\mathcal{E})/[0]$. Also it formally looks like (1.1) when $r = 2$, $\mu_i = 1$.

We assume

(1) The factor (6.2) is in $\mathbb{C}[c_2, \dots, c_r, \Lambda^{2r}][[t_1, \dots, t_{r-1}]]$.

Let us denote it by $B^{c_1=k}(\vec{c}, \Lambda, \vec{t})$. Note that this is a conjecture on theta functions. It seems that this was proved in [**16**]. But we have not quite checked the details. Then (6.3) becomes

$$\text{(6.5)}\quad \sum_n \Lambda^{2rn}\widehat{\pi}_*\left[\exp\Bigl(\sum_{p=1}^{\infty} t_p(\mathrm{ch}_{p+1}(\widehat{\mathcal{E}})/[C])\Bigr)\cap[\widehat{M}(r,k,n)]\right] = \sum_n \Lambda^{2rn}\pi_*(B^{c_1=k}(\mathrm{ch}(\mathcal{E})/[0], \Lambda, \vec{t})\cap[M(r,n)]).$$

We conjecture

(2) (6.5) holds for the moduli spaces $M_H(r, c_1, n)$, $\widehat{M}_H(r, c_1 + kC, n)$ for an arbitrary projective surface X.

Note also that (6.4) explains the meaning of Kronheimer–Mrowka's simple type condition if we have the same formula for an arbitrary surface:

$$\sum_{n=0}^{\infty}\Lambda^{4n}\pi_*\bigl((\mathrm{ch}_2(\mathcal{E})/[0])^2\cap[M_H(2,c_1,n)]\bigr) = 4\Lambda^2\sum_{n=0}^{\infty}\Lambda^{4n}\pi_*[M_H(2,c_1,n)].$$

It is equivalent to $c_2^2 = u_2^2 = 4\Lambda^2$. It means that the Seiberg–Witten curve is singular.

6.3. Rank 2 case. We assume $r = 2$. In this subsection we give an explicit expression for $B^{c_1=k}(u, \Lambda, t)$ in terms of Weierstrass σ-functions as in [**24**]. This is an exercise in elliptic functions and involves no geometry.

The regularity assumption made in the previous subsections is true as we consider the derivative with respect to τ_1.

6.3.1. *Weierstrass functions.* Let $\wp(z)$ be the Weierstrass $\wp$-function with the period $\mathbb{Z}\omega + \mathbb{Z}\omega'$, where $\omega'/\omega = \tau$, $\Im\tau > 0$. Then the associated elliptic curve is given by

$$y^2 = 4(x - e_1)(x - e_2)(x - e_3) = 4x^3 - g_2 x - g_3,$$

where $e_1 = \wp(\omega/2)$, $e_2 = \wp(-\omega/2 - \omega'/2)$ and $e_3 = \wp(\omega'/2)$.

Let $\sigma(z)$ be the Weierstrass σ-function. We have the following expansion:

$$\wp(z) = \sum_{n\geq 0} c_n(g_2, g_3) z^{2n-2} = \frac{1}{z^2} + \frac{g_2}{2^2\cdot 5}z^2 + \frac{g_3}{2^2\cdot 7}z^4 + \cdots$$

$$\sigma(z) = \sum_{n\geq 0} c'_n(g_2, g_3) z^{2n+1} = z - \frac{g_2}{2^4\cdot 3\cdot 5}z^5 - \frac{g_3}{2^3\cdot 3\cdot 5\cdot 7}z^7 - \cdots,$$

where c_n, c'_n are weighted homogeneous polynomials of degree n with $\deg g_2 = 2$ and $\deg g_3 = 3$. Let $\sigma_i(z)$, $i = 1, 2, 3$ be three more sigma functions associated to e_i. We assign $\deg e_i := 1$. Since $\sigma_i(z)^2 = \sigma(z)^2(\wp(z) - e_i)$, $\sigma_i(z)$ also has an expansion

$$\text{(6.6)}\qquad \sigma_i(z) = 1 - \frac{e_i}{2}z^2 + \sum_{n\geq 2} c''_n(e_i, g_2, g_3) z^{2n}$$

where c''_n are weighted homogeneous polynomials of degree n.

In terms of modular forms, we have the following expressions:

$$(6.7)\quad \begin{aligned}\sigma(\omega z) &= \omega e^{\omega\eta z^2}\frac{\theta_{11}(z|\tau)}{\theta'_{11}(0|\tau)} = -\frac{\omega}{\pi\theta_{10}(0|\tau)\theta_{00}(0|\tau)}e^{\omega\eta z^2}\frac{\theta_{11}(z|\tau)}{\theta_{01}(0|\tau)}\\ \sigma_3(\omega z) &= e^{\omega\eta z^2}\frac{\theta_{01}(z|\tau)}{\theta_{01}(0|\tau)},\end{aligned}$$

where $\eta = \zeta(\omega/2)$ is given by

$$\eta\omega = \frac{\pi^2}{6}E_2(\tau).$$

We also have

$$(6.8)\quad \begin{gathered} e_1 = \frac{1}{3}\left(\frac{\pi}{\omega}\right)^2(\theta_{00}^4+\theta_{01}^4), \qquad e_2 = \frac{1}{3}\left(\frac{\pi}{\omega}\right)^2(\theta_{10}^4-\theta_{01}^4),\\ e_3 = -\frac{1}{3}\left(\frac{\pi}{\omega}\right)^2(\theta_{10}^4+\theta_{00}^4).\end{gathered}$$

$$(6.9)\quad \begin{aligned} g_2 &= -4\left(\frac{\pi}{\omega}\right)^4\left(-\tfrac{1}{3}(\theta_{10}^4+\theta_{00}^4)^2+(\theta_{10}^4\theta_{00}^4)\right),\\ g_3 &= \frac{1}{27}\left(\frac{\pi}{\omega}\right)^6(\theta_{10}^4+\theta_{00}^4)\left(8(\theta_{10}^4+\theta_{00}^4)^2-36(\theta_{10}^4\theta_{00}^4)\right).\end{aligned}$$

6.3.2. We write $u = u_2$, $a = a_2$ as in §2.6. We set $\omega = \pi\theta_{00}\theta_{10}/\Lambda$. Then

$$g_2 = 4(\tfrac{1}{3}u^2-\Lambda^4), \qquad g_3 = -\frac{1}{27}u(8u^2-36\Lambda^4), \qquad e_3 = \frac{u}{3}.$$

Hence we get

$$(6.10)\quad \begin{aligned} e^{-T_{1,1}t^2}\theta_{01}(0|\tau)^{-1}\theta_{01}\left(\frac{\sqrt{-1}}{2\pi}\frac{du}{da}t\middle|\tau\right) &= e^{ut^2/6}\sigma_3(t),\\ e^{-T_{1,1}t^2}\theta_{01}(0|\tau)^{-1}\theta_{11}\left(\frac{\sqrt{-1}}{2\pi}\frac{du}{da}t\middle|\tau\right) &= e^{ut^2/6}\sigma(t)\Lambda.\end{aligned}$$

This checks the conjecture (1) in the previous subsection.

Since $\mathrm{ch}_2(\mathcal{E}) = -c_2(\mathcal{E})$, we put $x = -u$. Then x corresponds to the insertion of the point class $\mu(p) = c_2(\mathcal{E})/[0]$ in (6.5). The above (6.10) exactly coincides with the functions $B(x,t)$, $S(x,t)$ appearing in Fintushel–Stern's blowup formula for Donaldson invariants [**24**].

This checks the conjecture (2) in a weak sense, i.e., if we multiply products of $\mu(S)$ and integrate, then the equality holds. Conversely if we can prove (2) directly, it gives a new proof of Fintushel–Stern's blowup formula. The proof of (2) probably requires a more detailed study of the map $\widehat{\pi}$.

7. Gravitational corrections

As we mentioned in the introduction, Nekrasov asserts that higher-order terms in (5.4) are *gravitational corrections* to the gauge theory [**70**, §4]. In some cases these are some known quantities, which are really related genus g curves, e.g., Gromov–Witten invariants with domain of genus g. We are still far away from verifying this conjecture in full generality. But we have some nontrivial examples, which we review in this section.

7.1. Genus 1 part. The result of this subsection is based on discussions with N. Nekrasov.

We determine the coefficients G, F_1 in the expansion (5.4) for $\vec{\tau}=0$. These terms are considered as genus 1 gravitational corrections, as we said. Since we are only interested in the $\vec{\tau}=0$ case, we omit $\vec{\tau}$ from the notation.

We consider the blowup equation (5.2) for $c_1 = kC$ ($k\neq 0$) with $\vec{\tau}=\vec{t}=0$:

$$0 = \sum_{\{\vec{k}\}=-k/r} Z(\varepsilon_1, \varepsilon_2-\varepsilon_1, \vec{a}+\varepsilon_1\vec{k};\mathfrak{q}) Z(\varepsilon_1-\varepsilon_2,\varepsilon_2,\vec{a}+\varepsilon_2\vec{k};\mathfrak{q}).$$

As in the derivation of (5.5) we have

$$\begin{aligned}
0 &= \sum_{\{\vec{k}\}=-k/r} \exp\Bigg[-\frac{\partial^2 F_0}{\partial a^l \partial a^m}\frac{k^l k^m}{2} + \frac{\partial H}{\partial a^l}k^l \\
&\qquad + (\varepsilon_1+\varepsilon_2)\left\{-\frac{\partial^3 F_0}{\partial a^l\partial a^m\partial a^n}\frac{k^l k^m k^n}{3!} + \frac{\partial(G+F_1)}{\partial a^l}k^l\right\} + \cdots\Bigg] \\
&= \sum_{\{\vec{k}\}=-k/r} \sqrt{-1}^{2\langle\vec{k},\rho\rangle}\exp\Bigg[\pi\sqrt{-1}\tau_{lm}k^l k^m \\
&\qquad + (\varepsilon_1+\varepsilon_2)\left\{-\frac{\partial^3 F_0}{\partial a^l\partial a^m\partial a^n}\frac{k^l k^m k^n}{3!} + \frac{\partial(G+F_1)}{\partial a^l}k^l\right\} + \cdots\Bigg],
\end{aligned}$$

where τ_{lm} is the period of the Seiberg–Witten curve as before (2.5).

Setting $\varepsilon_1=\varepsilon_2=0$, we get

$$0 = \sum_{\{\vec{k}\}=-k/r} \sqrt{-1}^{2\langle\vec{k},\rho\rangle}\exp(\pi\sqrt{-1}\tau_{lm}k^l k^m) = \Theta_{E_k}(\vec{0}|\tau).$$

Next we take the coefficient of $\varepsilon_1+\varepsilon_2$ in the above to get

$$\begin{aligned}
0 = &\sum_{\{\vec{k}\}=-k/r} \sqrt{-1}^{2\langle\vec{k},\rho\rangle}\exp[\pi\sqrt{-1}\tau_{lm}k^l k^m] \\
&\qquad\times\left\{2\pi\sqrt{-1}\frac{\partial\tau_{mn}}{\partial a^l}\frac{k^l k^m k^n}{3!} + \frac{\partial(G+F_1)}{\partial a^l}k^l\right\},
\end{aligned}$$

i.e.,

$$\sum_l \frac{\partial(G+F_1)}{\partial a^l}\frac{\partial}{\partial\xi^l}\Theta_{E_k}(\vec{0}|\tau) + \frac{1}{3}\frac{\partial^2}{\partial a^l\partial\xi^l}\Theta_{E_k}(\vec{0}|\tau) = 0.$$

We believe this equation with $k=1,\dots,r-1$ determines $\partial(G+F_1)/\partial a^l$ for $l=2,\dots,r$. But we do not know the required identities for the theta functions. So we assume $r=2$ from now on. Then the equation is

$$\frac{\partial(G+F_1)}{\partial a} = -\frac{1}{3}\frac{\partial}{\partial a}\log\left(\frac{\partial}{\partial\xi}\Theta_{E_1}(\vec{0}|\tau)\right). \tag{7.1}$$

We now switch to the notation in §B.2. From (5.7), (7.1), we get

$$\exp(G-F_1) = \theta_{01}(0,\tau), \qquad \exp(G+F_1) = C\theta'_{11}(0,\tau)^{-1/3}$$

for some constant C independent of a. Therefore

$$\exp(2F_1) = C\theta'_{11}(0,\tau)^{-1/3}\theta_{01}(0,\tau)^{-1}.$$

By Jacobi's triple product identity (see, e.g., [**63**, Chapter I, §14]) we have

$$\theta_{01}(0,\tau) = \prod_{d=1}^{\infty}[(1-q^{2d})(1-q^{2d-1})^2].$$

By Jacobi's derivative formula (see [**63**, Chapter I, §13]) we have

$$\theta'_{11}(0,\tau) = -2\pi q^{1/4}\prod_{d=1}^{\infty}(1-q^{2d})^3.$$

Therefore we get

$$\exp F_1 = C' q^{-1/24}\prod_{d=1}^{\infty}(1-q^d)^{-1} = \frac{C'}{\eta(\tau/2)}$$

for some constant C' independent of a. A priori, C' may depend on $\mathfrak{q}$ (or Λ), but in fact, it does not, as follows. Let us define the degrees of variables by

$$\deg\varepsilon_1 = \deg\varepsilon_2 = \deg a_\alpha = 1, \qquad \deg\mathfrak{q} = 2r \quad (\deg\Lambda = 1).$$

(This definition applies for arbitrary r, not necessarily 2.) By definition, Z has degree 0. Therefore F_0 has degree 2, while G and F_1 have degree 0. Then τ has degree 0, and hence so does $\eta(\tau/2)$. Therefore C' has degree 0. Since it is independent of a, this means that it is also independent of $\mathfrak{q}$ (or Λ). Therefore C' can be computed by studying the expansions of F_1 and η in a/Λ:

$$F_1 = \frac{1}{12}\left\{\log\left(\frac{2a}{\Lambda}\right) + \log\left(\frac{-2a}{\Lambda}\right)\right\} + \cdots = \frac{1}{6}\log\left(\frac{2\sqrt{-1}a}{\Lambda}\right) + \cdots,$$

$$\log\frac{1}{\eta(\tau/2)} = -\frac{1}{24}\log q - \sum_{d=1}^{\infty}\log(1-q^d) = -\frac{1}{24}\pi\sqrt{-1}\tau + \cdots.$$

Furthermore, we have

$$\tau = \frac{\sqrt{-1}}{\pi}4\log\left(\frac{2\sqrt{-1}a}{\Lambda}\right) + \cdots.$$

Therefore we get $C' = 1$, and hence

$$F_1 = -\log\eta\left(\frac{\tau}{2}\right), \qquad G = \log\left[q^{-1/24}\prod_{d=1}^{\infty}(1-q^{2d-1})\right].$$

It is better to form the following combination:

$$F_1 = A - \tfrac{2}{3}B, \qquad G = \tfrac{1}{3}B.$$

Then

$$\exp A = \exp(F_1 + 2G) = q^{-1/8}\prod_{d=1}^{\infty}\frac{1-q^{2d-1}}{1-q^{2d}} = \left(-\frac{2\pi\theta_{01}}{\theta'_{11}}\right)^{1/2} = \left(\frac{2}{\theta_{00}\theta_{10}}\right)^{1/2},$$

$$\exp B = \exp(3G) = q^{-1/8}\prod_{d=1}^{\infty}(1-q^{2d-1})^3 = \left(\frac{-2\pi\theta_{01}^3}{\theta'_{11}}\right)^{1/2} = \left(\frac{2\theta_{01}^2}{\theta_{00}\theta_{10}}\right)^{1/2}$$

$$= \left(\frac{4\theta_{01}^4}{\theta_{00}^2\theta_{10}^2}\right)^{1/4} = \left(\frac{4(\theta_{00}^4 - \theta_{10}^4)}{\theta_{00}^2\theta_{10}^2}\right)^{1/4}.$$

Comparing with (1.3), we find

$$\exp A = \left(\frac{\sqrt{-1}}{\Lambda}\frac{du}{da}\right)^{1/2}, \qquad \exp B = \left(\frac{4(u^2 - 4\Lambda^4)}{\Lambda^4}\right)^{1/8}. \tag{7.2}$$

Note that the last expression is given by the quantum discriminant (2.1):

$$\Delta = 2^{12}\Lambda^8(u^2 - 4\Lambda^4).$$

Therefore

$$\varepsilon_1\varepsilon_2 F_1 + (\varepsilon_1+\varepsilon_2)^2 G = \varepsilon_1\varepsilon_2 \log\left(\frac{\sqrt{-1}}{\Lambda}\frac{du}{da}\right)^{1/2} + \frac{\varepsilon_1^2+\varepsilon_2^2}{3}\log\left(\frac{\Delta}{2^{10}\Lambda^{12}}\right)^{1/8}.$$

Comparing with (1.4), this suggests the following formula for the equivariant Euler number and signature for $\mathbb{C}^2$:

$$\chi(\mathbb{C}^2) = \varepsilon_1\varepsilon_2, \qquad \sigma(\mathbb{C}^2) = \frac{\varepsilon_1^2+\varepsilon_2^2}{3}.$$

This is natural from the following formal computation:

$$\chi(\mathbb{C}^2) = c_2(\mathbb{C}^2), \qquad \sigma(\mathbb{C}^2) = \tfrac{1}{3}\left(c_1(\mathbb{C}^2)^2 - 2c_2(\mathbb{C}^2)\right),$$
$$c_1(\mathbb{C}^2) = \varepsilon_1+\varepsilon_2, \qquad c_2(\mathbb{C}^2) = \varepsilon_1\varepsilon_2.$$

Nekrasov conjectures that (7.2) holds also in the higher-rank case if we replace du/da by $\det(\partial u_p/\partial a^i)$.

7.2. Coordinate rings of symmetric products. The next example is related to the *perturbative part* of the K-theory version of the partition function in §4. (See [**71**, §A.0.3].) It fits with geometric engineering quite well (see below), but it looks like an accident if we understand it as a purely mathematical statement. The authors learned the result from [**71**, §A.0.3].

Let us consider the nth symmetric product $S^n(\mathbb{C}^2)$ of the affine plane $\mathbb{C}^2$. We define an action of the two-torus $T^2 = \mathbb{C}^*\times\mathbb{C}^*$ on $\mathbb{C}^2$ by

$$(x,y)\mapsto(t_1x, t_2y), \quad (t_1,t_2)\in T^2.$$

We also have an induced action on $S^n(\mathbb{C}^2)$.

The coordinate ring $H^0(S^n(\mathbb{C}^2),\mathcal{O})$, that is, the ring of polynomial functions on $S^n(\mathbb{C}^2)$, is a T^2-module. We consider its character

$$\operatorname{ch} H^0(S^n(\mathbb{C}^2),\mathcal{O}) = \sum_{m,n\ge 0} t_1^m t_2^n \dim H^0(S^n(\mathbb{C}^2),\mathcal{O})_{m,n},$$

where

$$H^0(S^n(\mathbb{C}^2),\mathcal{O})_{m,n} = \{f\in H^0(S^n(\mathbb{C}^2),\mathcal{O}) \mid (t_1,t_2)\cdot f = t_1^m t_2^n f\}$$

is a simultaneous eigenspace, i.e., a weight space. The character is sometimes called the *Hilbert series.* It is standard in algebraic geometry to show that

(1) each weight space is finite-dimensional, and hence the character is well-defined as a formal sum,
(2) the character is a rational function in $t_1^{\pm}$, $t_2^{\pm}$.

In fact, in this case, an explicit answer can be written down:

PROPOSITION 7.1. *The generating function of the character is given by*

$$\sum_{n=0}^{\infty} \mathfrak{q}^n \operatorname{ch} H^0(S^n(\mathbb{C}^2),\mathcal{O}) = \exp\left(\sum_{d=1}^{\infty}\frac{\mathfrak{q}^d}{d(1-t_1^d)(1-t_2^d)}\right). \tag{7.3}$$

We refer [**69**, §3] for the proof. But we recommend the reader to write down the proof himself/herself since it is a nice exercise on a treatment of generating functions.

The generating function (7.3) is the rank 1 version of Nekrasov's deformed partition function (for the K-theory version). Let us expand it into a formal power series in $\hbar$, after putting $t_1 = \exp\hbar$, $t_2 = \exp(-\hbar)$:

$$\begin{aligned}(7.4)\quad &\log\left(\sum_{n=0}^{\infty} \mathfrak{q}^n \operatorname{ch} H^0(S^n(\mathbb{C}^2), \mathcal{O})\right)\Bigg|_{\substack{t_1=e^{\hbar}\\ t_2=e^{-\hbar}}} \\ &= \sum_{d=1}^{\infty} \frac{\mathfrak{q}^d}{d(1-e^{\hbar d})(1-e^{-\hbar d})} \\ &= -\sum_{d=1}^{\infty} \frac{\mathfrak{q}^d}{d^3}\hbar^{-2} - \frac{1}{12}\log(1-\mathfrak{q})\hbar^0 + \sum_{g\geq 2} \frac{B_{2g}}{2g(2g-2)!}\sum_{d=1}^{\infty} d^{2g-3}\mathfrak{q}^d\hbar^{2g-2},\end{aligned}$$

where B_{2g} is the $2g$th Bernoulli number as in Appendix E. The series start with $\hbar^{-2}$ and has only even powers of $\hbar$. In the next subsection we will see that (7.4) is equal to the generating function of certain Gromov–Witten invariants. Then g will be identified with the genus of the domain curve, and hence $2-2g$ with the Euler number.

7.3. Gromov–Witten invariants of the resolved conifold. Let X be the total space of the rank 2 vector bundle $E = \mathcal{O}(-1) \oplus \mathcal{O}(-1)$ over $\mathbb{P}^1$. This space is called the *resolved conifold* in physics. The local Gromov–Witten invariants for target X are defined as follows: Let $M_{g,n}(\mathbb{P}^1, d)$ be the moduli space of stable maps of degree d with target $\mathbb{P}^1$ from a genus g curve with n marked points. We consider the $d > 0$ case only, that is, the stable map is not constant. We have the diagram

$$M_{g,0}(\mathbb{P}^1, d) \xleftarrow{\text{forget}} M_{g,1}(\mathbb{P}^1, d) \xrightarrow{\text{eval}} \mathbb{P}^1,$$

where forget is the map given by forgetting the marked point, and eval is the map given by taking the image of the marked point under the stable map. We consider the vector bundle $\mathbb{E} = R^1 \operatorname{forget}_* \operatorname{eval}^* E$. Let $c_{\text{top}}(\mathbb{E})$ be its top Chern class. Then the local Gromov–Witten invariant is defined by

$$C(g,d) = \int_{M_{g,0}(\mathbb{P}^1,d)^{\text{vir}}} c_{\text{top}}(\mathbb{E}),$$

where $M_{g,0}(\mathbb{P}^1, d)^{\text{vir}}$ is the virtual fundamental class. This is a rational number. This is a local contribution to the *global* Gromov–Witten invariant of a Calabi–Yau 3-fold of multiple covers of a fixed rational curve with normal bundle $E = \mathcal{O}(-1) \oplus \mathcal{O}(-1)$. Thus the formula for $C(g,d)$ is important in Gromov–Witten theory. The genus $g = 0$ case is known as the Aspinwall–Morrison formula. The complete answer is given by

THEOREM 7.2 ([**2, 54, 77**] for $g = 0$, [**32**] for $g = 1$, [**21**] for $g \geq 2$).

$$C(0,d) = \frac{1}{d^3}, \qquad C(1,d) = \frac{1}{12d}, \qquad C(g,d) = \frac{(-1)^{g-1}B_{2g}}{2g(2g-2)!}d^{2g-3}.$$

Comparing with (7.4), we get

$$(7.5)\qquad \log\left(\sum_{n=0}^{\infty} \mathfrak{q}^n \operatorname{ch} H^0(S^n(\mathbb{C}^2), \mathcal{O})\right)\Bigg|_{\substack{t_1=e^{\hbar}\\ t_2=e^{-\hbar}}} = \sum_{d=1}^{\infty}\sum_{g=0}^{\infty} C(g,d)\mathfrak{q}^d(i\hbar)^{2g-2}.$$

This is our first example of the assertion that the gauge theory partition function is identified with generating functions of Gromov–Witten invariants.

It is worthwhile mentioning that the exponential of the right-hand side is the generating function of Gromov–Witten invariants whose domain curves are not necessarily *connected.* This does not make sense on the gauge theory side, but is somehow related to the recursive structure among the symmetric products $S^n\mathbb{C}^2$ for various n.

7.4. The $r = 1$ case and Gromov–Witten invariants for $\mathbb{P}^1$. In the main body of the paper, the case $r = 1$ was excluded as the Seiberg–Witten geometry does not make sense. However Nekrasov's partition function does make sense for $r = 1$ as well. The K-theory version with $\vec{\tau} = 0$ is what we already saw in §7.2. This is because the moduli space $M(1,n)$ is nothing but the Hilbert scheme of points, and it is known that the higher direct image sheaves for $M(1,n) \to S^n\mathbb{C}^2$ vanish and $\operatorname{ch} H^0(S^n\mathbb{C}^2, \mathcal{O})$ can be given by the Atiyah–Bott formula for $M(1,n)$. (See [**69**, §3] for more details.) When $\vec{\tau} \neq 0$, it was studied in [**34**]. (His main result is a positivity property, which is not studied in this paper. But it is natural to conjecture a similar property for the higher-rank cases also.)

The homology version of the partition function, i.e., our $Z(\varepsilon_1, \varepsilon_2; \mathfrak{q}, \vec{\tau})$, has the presentation by the Fock space when it is restricted to $\varepsilon_2 = -\varepsilon_1$. Then comparing with a presentation for the Gromov-Witten invariants for $\mathbb{P}^1$ [**72**], one gets

$$\log Z(\varepsilon_1, -\varepsilon_1; \mathfrak{q}, \vec{\tau}) \\ = \text{the generating function of the Gromov–Witten invariants for } \mathbb{P}^1,$$

where ε_1 is mapped to an indeterminate for the domain genus, and $\vec{\tau}$ to those for gravitational descendants. This remarkable observation was done by [**48**] and [**49**] independently. We refer for the precise statement and the proof to the original papers.

7.5. Geometric engineering. The geometric engineering of Katz–Klemm–Vafa [**39**] realizes 4-dimensional $\mathcal{N} = 2$ supersymmetric gauge theories as limits of type IIA string theory compactified on certain noncompact Calabi–Yau 3-folds. Mathematically it poses the following conjecture:

(7.6) | Partition functions in 4-dimensional $\mathcal{N} = 2$ supersymmetric gauge theories are equal to limits of generating functions of local Gromov–Witten invariants for certain noncompact Calabi–Yau 3-folds. |

The noncompact Calabi–Yau 3-fold is chosen according to the gauge theory to be realized. Typically it is an ALE space fibration over $\mathbb{P}^1$. Recall that an ALE space is the minimal resolution of a simple singularity and contains a configuration of $\mathbb{P}^1$'s intersecting as one of the ADE Dynkin diagrams. The group of the gauge theory is the corresponding ADE group.

This statement is rather striking since it seems difficult to compare the two types of moduli spaces directly, i.e., moduli spaces of stable maps and instanton moduli spaces. Moreover, we must sum up over degrees for Gromov–Witten invariants as we will see below.

For simplicity, we restrict ourselves to the rank 2 case. (In higher-rank cases, the above naive definition of the local Gromov–Witten invariants must be modified as the base space become singular.) As we have already seen in §7.3, the K-theory version of the partition function is more natural here as we do not need to take a limit. We define

$$Z_K(\hbar, a; \beta) = \exp\left[-\gamma_\hbar^K(2a;\beta) - \gamma_\hbar^K(-2a;\beta)\right] Z_K^{\text{inst}}(\hbar, a; \beta),$$

$$Z_K^{\text{inst}}(\hbar, a; \beta) = \sum_{n=0}^{\infty} \beta^{2rn} \sum_i (-1)^i \operatorname{ch} H^i(M(r,n), \mathcal{O}),$$

$$t_1 = e^{\beta\hbar}, \quad t_2 = e^{-\beta\hbar}, \quad e_1 = e^{-\beta a}, \quad e_2 = e^{\beta a},$$

where ch is the character of a $\widetilde{T}$-module and the one-dimensional $\widetilde{T}$-modules (in the notation 3.2) are replaced as indicated. The characters of the cohomologies have expressions in terms of Young diagrams $\vec{Y}$ by the localization formula for K-theory:

$$Z_K^{\text{inst}}(\hbar, a; \beta) = \sum_{\vec{Y}} \beta^{2r|\vec{Y}|} \frac{1}{\operatorname{ch}(\bigwedge_{-1} T_{\vec{Y}}^* M(r,n))}.$$

(See [**69**, §3] for details.) The perturbative term is given by

$$\gamma_\hbar^K(x;\beta) = \sum_{d=1}^{\infty} \frac{e^{-\beta d x}}{d(1-e^{\hbar d})(1-e^{-\hbar d})}.$$

This is equal to the one in [**71**, §A.0.3] up to polynomials in a.

Note that we do not include higher Casimir operators. This is not for brevity. We (at least the authors) do not know what are their counterparts in Gromov–Witten theory.

In the rank 2 case, the noncompact Calabi-Yau 3-fold is assumed to be the canonical bundle $K_{\mathbf{F}_n}$ of the Hirzebruch surface $\mathbf{F}_n$. And it is conjectured that the Gromov–Witten invariants are *essentially* independent of n. So we further restrict to the case $n = 0$, i.e., the $K_{\mathbf{F}_0}$ of $\mathbf{F}_0 = \mathbb{P}^1 \times \mathbb{P}^1$. (For the actual calculation, it is necessary to assume $\mathbf{F}_n$ is a toric variety in order to apply the localization technique.)

We define the local Gromov–Witten invariants as in the case of the resolved conifold. The degree of a map is a pair of integers (n, d) corresponding to the base and fiber respectively. (Although we have a symmetry exchanging the two factors, we break it and consider one as the base and the other as the fiber. This is automatic for the other $\mathbf{F}_n$.) Let us introduce two parameters q_{b}, q_{f} respectively.

Now geometric engineering asserts that $\log Z_K(\hbar, a; \beta)$ is equal to the generating function of Gromov–Witten invariants, under a suitable identification of parameters. The parameter $\hbar$ should count the genus of domain curves as in §7.3.

Next we match the degree n for the base with the instanton number n. Although this identification is quite natural, we do not have any mathematically rigorous justification of this statement. Anyway we should identify q_{b} with β^{2r}. In fact, the analysis below gives the exact answer:

$$q_{\mathrm{b}} = \left(\frac{\beta}{2}\right)^{2r}.$$

Let us consider the Gromov–Witten invariants for $n = 0$. In this case, we only have multiple covers of the fiber. It is given by (7.5) multiplied by -2. Since $n = 0$

means zero instanton number on the gauge theory side, it should be equal to the perturbation term of the gauge theory, namely:

$$-\gamma_\hbar^K(2a|\beta) - \gamma_\hbar^K(-2a|\beta) \overset{?}{=} -2\sum_{d=1}^{\infty} \frac{q_{\mathrm{f}}^d}{d(1-e^{\hbar d})(1-e^{-\hbar d})}$$

Since the left-hand side is equal to $-2\gamma_\hbar^K(2a|\beta)$ up to a polynomial in a, thanks to the inversion formula for polylogarithms, this (up to a polynomial in a) follows from what we observed in §7.3 when we equate the parameters as

$$q_{\mathrm{f}} = e^{-2\beta a}.$$

Note that for this identification, we must sum up the Gromov–Witten invariants for various degrees on the fibers (and various genera). This is also true for n-instanton corrections.

For $n > 0$, the genus 0 Gromov–Witten invariants were calculated using the local mirror symmetry for $X = K_{\mathbf{F}_0}$ in [**39**]. And the limit of their generating function was identified with the Seiberg–Witten prepotential F_0. In fact, they identified the limit of the local mirror of X with the Seiberg–Witten curve. Note that the genus 0 case is enough to identify the parameters.

Recently Iqbal and Kashani-Poor identified $\log Z_K$ with the generating function of all genus Gromov-Witten invariants by using the large N duality [**37, 38**]. In fact, they identified the expression of $\log Z_K$ via Young diagrams $\vec{Y}$ with Jones-Witten invariants for the Hopf link. (They assume certain combinatorial identities which are proved in more recent paper [**19**].)

Appendix A. The root system of type A_{r-1}

Let Q be the coroot lattice of type A_{r-1}:

$$Q = \left\{\vec{k} = (k_1,\dots,k_r) \in \mathbb{Z}^r \;\middle|\; \sum_\alpha k_\alpha = 0\right\}.$$

We take simple coroots

$$\alpha_i^\vee = (0,\dots,0,\overset{i}{1},\overset{i+1}{-1},0,\dots,0) \qquad (i = 1,\dots,r-1).$$

We can write

$$Q \ni \vec{k} = \sum_i k^i \alpha_i^\vee.$$

For a given $k \in \mathbb{Z}$, elements $\vec{k} \in \mathbb{Z}^r$ with $\sum_\alpha k_\alpha = k$ are identified

$$\left\{\vec{l} = (l_1,\dots,l_r) \in \mathbb{Q}^r \;\middle|\; \sum_\alpha l_\alpha = 0, \forall\alpha\; l_\alpha \equiv -\frac{k}{r} \bmod \mathbb{Z}\right\}.$$

This is a subset of the coweight lattice $P = \{\vec{l} = (l_1,\dots,l_r) \in \mathbb{Q}^r \mid \sum_\alpha l_\alpha = 0, \exists k \in \mathbb{Z}\; \forall\alpha\; l_\alpha \equiv -k/r \bmod \mathbb{Z}\}$. There exists a homomorphism $P \to \mathbb{Z}/r\mathbb{Z}$ by taking the fractional part of l_α. It can be identified with the natural quotient homomorphism $P \to P/Q$. We denote it by $\vec{l} \mapsto \{\vec{l}\}$. Hereafter we identify $\vec{l}$ with $\vec{k}$ and denote both by $\vec{k}$. We write $\vec{k} = \sum_i k^i\alpha_i^\vee$ in both the cases $k = 0, \neq 0$. But k^i may be rational in the latter case. Let $(\ ,\ )$ be the standard inner product on Q.

The Killing form $B_{\mathrm{SU}(r)}$ of $\mathrm{SU}(r)$ satisfies $B_{\mathrm{SU}(r)} = 2r(\ ,\ )$. The following formulas will be useful later:

$$\text{(A.1)}\qquad \begin{gathered} \frac{1}{2r}\sum_{\alpha,\beta}(k_\alpha - k_\beta)(a_\alpha - a_\beta) = (\vec{k},\vec{a}) = \sum_{i,j} C_{ij}a^i k^j, \\ \frac{1}{2r}\sum_{\alpha,\beta}(k_\alpha - k_\beta)^2 = (\vec{k},\vec{k}) = \sum_{i,j} C_{ij}k^i k^j, \\ \sum_{\alpha<\beta}\frac{k_\alpha - k_\beta}{2} = \langle \vec{k},\rho\rangle = \sum_i k^i. \end{gathered}$$

Here C_{ij} is the Cartan matrix, and ρ is the half of the sum of the positive roots, as usual.

Appendix B. Theta functions

We give definitions and some properties of Riemann theta functions.

B.1. Riemann theta functions. Let $Q = \mathbb{Z}^g$. Let $\tau = (\tau_{\alpha\beta})$ be a symmetric $g \times g$ complex matrix whose imaginary part is positive definite. For $\vec{\mu}$, $\vec{\nu} \in \mathbb{C}^g$, we define the theta function with characteristic $\left[\begin{smallmatrix}\vec{\mu}\\ \vec{\nu}\end{smallmatrix}\right]$ by

$$\begin{aligned}\Theta&\begin{bmatrix}\vec{\mu}\\ \vec{\nu}\end{bmatrix}(\vec{\xi}|\tau) \\ &= \sum_{\vec{k}\in Q}\exp\Big(\pi\sqrt{-1}\sum_{\alpha,\beta}\tau_{\alpha\beta}(k_\alpha+\mu_\alpha)(k_\beta+\mu_\beta) + 2\pi\sqrt{-1}\sum_\alpha(k_\alpha+\mu_\alpha)(\xi_\alpha+\nu_\alpha)\Big).\end{aligned}$$

When $\left[\begin{smallmatrix}\vec{\mu}\\ \vec{\nu}\end{smallmatrix}\right] = \left[\begin{smallmatrix}0\\ 0\end{smallmatrix}\right]$, we simply denote it by Θ. We have

$$\Theta\begin{bmatrix}\vec{\mu}\\ \vec{\nu}\end{bmatrix}(\vec{\xi}|\tau) = \exp\big(\pi\sqrt{-1}\,{}^t\vec{\mu}\tau\vec{\mu} + 2\pi\sqrt{-1}\,{}^t\vec{\mu}(\vec{\xi}+\vec{\nu})\big)\Theta(\vec{\xi}+\tau\vec{\mu}+\vec{\nu}|\tau).$$

The theta function is quasi-periodic with respect to the lattice $\mathbb{Z}^g \oplus \tau\mathbb{Z}^g$. It satisfies the heat equation:

$$\frac{\partial^2}{\partial\xi_\alpha\partial\xi_\beta}\Theta\begin{bmatrix}\vec{\mu}\\ \vec{\nu}\end{bmatrix}(\vec{\xi}|\tau) = 4\pi\sqrt{-1}\frac{\partial}{\partial\tau_{\alpha\beta}}\Theta\begin{bmatrix}\vec{\mu}\\ \vec{\nu}\end{bmatrix}(\vec{\xi}|\tau).$$

When $\vec{\mu}$, $\vec{\nu} \in \frac{1}{2}\mathbb{Z}^g$, $\left[\begin{smallmatrix}\vec{\mu}\\ \vec{\nu}\end{smallmatrix}\right]$ is called a *half-integer characteristic*. The set of half-integer characteristics are divided into two, *odd* or *even*, according to whether $\Theta\left[\begin{smallmatrix}\vec{\mu}\\ \vec{\nu}\end{smallmatrix}\right](\vec{\xi}|\tau)$ is an odd or even function.

In the main body of the paper, we use the A_{r-1}-lattice

$$\Big\{\vec{k} = (k_1,\dots,k_r) \in \mathbb{Z}^r \,\Big|\, \sum_\alpha k_\alpha = 0\Big\}.$$

(See Appendix A.) This is identified with $\mathbb{Z}^{r-1}$ by taking $(k_2,\dots,k_r)$. Then we apply the above convention, i.e., the suffix runs over $\alpha = 2,\dots,r$.

A theta function with a particular half-integer even characteristic appears often in this paper:

$$\text{(B.1)}\qquad \Theta_E(\vec{\xi}|\tau) = \Theta\begin{bmatrix}\vec{0}\\ \vec{\nu}\end{bmatrix}(\vec{\xi}|\tau) \qquad \nu_2 = \frac{1}{2}, \nu_3 = 0, \nu_4 = \frac{1}{2}, \nu_5 = 0, \dots.$$

We denote this characteristic by E and the corresponding theta function by Θ_E.

We also use the notation for the root system of the Lie algebra of type A_{r-1}. Then

$$\begin{aligned}\frac{1}{2}k_2 + \frac{1}{2}k_4 + \cdots &\equiv -\frac{1}{2}k_2 - \frac{2}{2}k_3 - \cdots - \frac{r-1}{2}k_r \\ &= \frac{r-1}{4}k_1 + \frac{r-3}{4}k_2 + \cdots + \frac{1-r}{4}k_r = \sum_{\alpha<\beta}\frac{k_\alpha - k_\beta}{4} = \frac{1}{2}\sum_i k^i,\end{aligned}$$

where $\equiv$ means equality modulo $\mathbb{Z}^{r-1}$. The last equality is (A.1). Therefore the characteristic is

$$^t\begin{pmatrix}\frac{1}{2} & \frac{1}{2} & \cdots\end{pmatrix}$$

in this notation.

We also use the theta function where the summation range is replaced by

$$\left\{\vec{k} = (k_1, \ldots, k_r) \in \mathbb{Q}^r \;\middle|\; \sum_\alpha k_\alpha = 0, \forall\alpha\; l_\alpha \equiv -\frac{k}{r} \bmod \mathbb{Z}\right\}$$

for a fixed $k \in \mathbb{Z}$. It is $\Theta\left[\begin{smallmatrix}\vec{a}_k\\ \vec{\Delta}\end{smallmatrix}\right]$ with $\vec{a}_k = (k/r)^t\begin{pmatrix}1 & 2 & \cdots & r-1\end{pmatrix}$. We denote this characteristic by E_k.

B.2. When $g = 1$, we use the following notation for the theta functions:

$$\theta_{00}(z,\tau) = \sum_{n\in\mathbb{Z}} q^{n^2} w^{2n}, \qquad \theta_{01}(z,\tau) = \sum_{n\in\mathbb{Z}} (-1)^n q^{n^2} w^{2n},$$

$$\theta_{10}(z,\tau) = \sum_{n\in\mathbb{Z}} q^{(n+1/2)^2} w^{2n+1}, \quad \theta_{11}(z,\tau) = \sqrt{-1}\sum_{n\in\mathbb{Z}} (-1)^n q^{(n+1/2)^2} w^{2n+1},$$

where

$$q = \exp(\pi\sqrt{-1}\tau), \qquad w = \exp(\pi\sqrt{-1}z).$$

This is the same as [**63**]. We have

$$\Theta_E = \theta_{01}, \qquad \Theta_{E_1} = \theta_{11}.$$

B.3. Riemann surfaces and theta functions. Let C be a compact Riemann surface of genus g. Let K_C be its canonical bundle. We choose and fix a symplectic basis $A_1, \ldots, A_g$, $B_1, \ldots, B_g$ of $H_1(C,\mathbb{Z})$ so that $A_\alpha \cdot A_\beta = 0 = B_\alpha \cdot B_\beta$, $A_\alpha \cdot B_\beta = \delta_{\alpha\beta}$ for α, $\beta = 1, \ldots, g$. We then have a basis $\omega_1, \ldots, \omega_g$ of holomorphic differentials $H^0(C, K_C)$ such that $\int_{A_\alpha} \omega_\beta = \delta_{\alpha\beta}$. The period matrix of C is defined by

$$\tau_{\alpha\beta} = \int_{B_\beta} \omega_\alpha.$$

It is symmetric and its imaginary part is positive-definite.

Using the period matrix $(\tau_{\alpha\beta})$ of the Riemann surface C, we consider the associated theta function $\Theta\left[\begin{smallmatrix}\vec{\mu}\\ \vec{\nu}\end{smallmatrix}\right](\vec{\xi}|\tau)$ as in the previous section. We consider it as a multivalued function (or a section of a line bundle) on the Jacobian variety $J(C) = H^0(C,K_C)^*/H_1(C,\mathbb{Z})$. Here $H^0(C,K_C)$ is identified with $\mathbb{C}^g$ by the basis $\omega_1, \ldots, \omega_g$.

We choose a base point P_0 in C. Then we have the *Abel–Jacobi map*

$$C \ni P \mapsto \int_{P_0}^P \omega \in J(C); \quad \omega \in H^0(C, K_C).$$

We denote it by $\mathbf{A}$. It extends to a map from $J_g(C)$, the Picard variety of divisor classes (linear equivalence classes) of degree g divisors. When $g = 0$, it is independent of the base point and we have an isomorphism $J_0(C) \cong J(C)$.

Riemann's theorem ([**23**, Theorem 1.1], [**63**, Chapter II, 3.1]) says that there exists a vector $\overrightarrow{K} \in \mathbb{C}^g$ such that for all $\vec{e} \in \mathbb{C}^g$ the composition $\Theta \circ (\mathbf{A} + \vec{e})$ either vanishes identically or has g zeroes $Q_1, \dots Q_g$ such that $\sum_{k=1}^{g} \mathbf{A}(Q_k) + \vec{e} \equiv \overrightarrow{K}$ mod $H_1(C, \mathbb{Z})$. The vector is called the *Riemann constant*. The vector $\overrightarrow{K}$ depends on the choice of the symplectic basis of $H_1(C, \mathbb{Z})$ and the base point P_0. However, if we denote it by $\overrightarrow{K}_{P_0}$, then $\Delta = (g-1)P_0 + \overrightarrow{K}_{P_0} \in J_{g-1}(C)$ is independent of P_0. See [**22**, VI.3.7], [**63**, Chapter II, 3.11, 3.18]. In fact, we have $\Theta = W_{g-1} - \Delta$, where Θ is considered as a divisor in $J(C)$ as zero set $\Theta = 0$, and $W_{g-1} = \{x_1 + \cdots + x_{g-1} \mid x_\alpha \in C\}$.

The set Σ of divisor classes $D \in J_{g-1}(C)$ such that $2D = K_C$ is called the set of *theta characteristics*. The above Δ is an example. The set Σ is bijective to $\frac{1}{2}H_1(C, \mathbb{Z})/H_1(C, \mathbb{Z}) \cong (\mathbb{Z}/2\mathbb{Z})^{2g}$ (considered as a subset of $J(C)$) via $D \mapsto D - \Delta$. We identify Σ with a characteristic for the theta function by the further identification $J(C) \cong \mathbb{C}^g/\mathbb{Z}^g \oplus \tau\mathbb{Z}^g$ given by the choice of cycles.

B.4. Green functions. Let $\phi\colon \widetilde{C} \to C$ be the universal covering of C. We take a nonsingular odd theta characteristic $D \in J_{g-1}(C)$ (i.e., the theta divisor is smooth at D [**63**, Chapter IIIb, Lemma 1]) and let $\delta = \left[\begin{smallmatrix}\vec{\mu}\\ \vec{\nu}\end{smallmatrix}\right]$ be the corresponding half integer characteristic. By Riemann's theorem,

$$\zeta(x) = \sum_{\alpha=1}^{g} \frac{\partial \Theta_\delta}{\partial \xi_\alpha}(0)\omega^\alpha(x) \tag{B.2}$$

is a section of K_C which vanishes on D. Since D is a nonsingular odd characteristic, $H^0\big(C, K_C(-D)\big) \cong H^0\big(C, \mathcal{O}_C(D)\big) \cong \mathbb{C}$, and hence $\sqrt{\zeta(x)}$ is a section of $\mathcal{O}_C(D)$. We define the *prime form* by

$$E(x,y) := \frac{\Theta_\delta(\int_x^y \vec{\omega}|\tau)}{\sqrt{\zeta(x)}\sqrt{\zeta(y)}}. \tag{B.3}$$

This is a holomorphic differential form on $\widetilde{C} \times \widetilde{C}$. It is also regarded as a holomorphic section of a line bundle on $C \times C$: Let $\pi_i\colon C \times C \to C$, $i = 1, 2$ be the projections and $\mu\colon C \times C \to J(C)$ be the map $(x, y) \mapsto y - x$. Then $E(x, y)$ is a section of $\pi_1^*\mathcal{O}_C(-\Delta) \otimes \pi_2^*\mathcal{O}_C(-\Delta) \otimes \mu^*\big(\mathcal{O}_{J(C)}(\Theta)\big)$, where Θ is the theta divisor.

We list some properties of $E(x, y)$.

(1) $E(x, y) = 0 \iff \phi(x) = \phi(y)$.

(2) $E(x, y)$ has a first-order zero along the diagonal $\Delta_C \subset C \times C$ and locally $E(x,y) = (y-x)/(\sqrt{dx}\sqrt{dy})(1 + O\big((x-y)^2\big))$.

(3) $E(x, y) = -E(y, x)$.

Let

$$W(z_1, z_2) = \partial_{z_1}\partial_{z_2} \log E(z_1, z_2). \tag{B.4}$$

This is a well-defined meromorphic 2-form on $C \times C$ and it is used to construct differentials of the 2nd kind. For $c \in \mathbb{C}^g$ with $\Theta(c) \neq 0$, we set

$$\Psi_c(z_1, z_2) = \frac{\Theta_c(\int_{z_2}^{z_1} \vec{\omega}|\tau)}{\Theta_c(0)E(z_1, z_2)}. \tag{B.5}$$

It is called the *Szegö kernel*.

By Fay's trisecant identity ([**23**, p. 34, formula 45] or [**63**, Chapter IIIb, §2]), we have

$$\Psi_c^2(z_1, z_2) = W(z_1, z_2) + \sum_{\alpha,\beta} \omega^\alpha(z_1)\omega^\beta(z_2)\frac{\partial^2}{\partial\xi_\alpha\partial\xi_\beta}\log\Theta_c(0|\tau) \tag{B.6}$$

for a half-integer characteristic c ([**23**, Corollary 2.12, formula 38], [**63**, Chapter IIIb, §3(2)]).

B.5. Hyperelliptic curves. Let $Q(z)$ be a polynomial of degree $2g+2$. Let $C = \{y^2 = Q(z)\}$ be the corresponding hyperelliptic curve of genus g. We denote by $\iota\colon C \to C$ the involution. Let $\{Q_1, \dots, Q_{2g+2}\}$ be the set of branched points, i.e., the roots of $Q(z) = 0$. We choose cycles A_α, B_α as in Figure 1 where we make replacements as

$$z_1^+ \to Q_1, \quad z_1^- \to Q_2, \quad z_2^- \to Q_3, \quad z_2^+ \to Q_4, \quad \cdots.$$

Then the choice of cycles is exactly the same as [**23**, p. 12, Example] with a shift of the numbering by 1, i.e., $A_2 \to A_1$, $B_2 \to B_1$, etc. If we choose Q_1 for the base point of the Abel–Jacobi map, we have ([**23**, p. 14], [**63**, Chapter IIIa.5])

$$\overrightarrow{K} = \tau\,{}^t\!\begin{pmatrix}\frac{1}{2} & \frac{1}{2} & \cdots & \frac{1}{2}\end{pmatrix} + {}^t\!\begin{pmatrix}\frac{1}{2} & \frac{2}{2} & \cdots & \frac{g}{2}\end{pmatrix}.$$

Let L be the divisor class of degree 2 containing $P + \iota P$ for $P \in C$. Then the set of theta characteristics is in bijection with the set of subsets $T \subset \{Q_1, \dots, Q_{2g+2}\}$ with $\#T \equiv (g+1) \bmod 2$ modulo the equivalence relation $T \sim T^c$ by

$$T \mapsto \sum_{P \in T} P + \frac{g - 1 - \#T}{2} L.$$

Under this correspondence, the vector $\Delta = (g-1)Q_1 + \overrightarrow{k}$ is mapped to $\{Q_1, Q_3, \dots, Q_{2g+1}\}$.

When $\#T = (g+1)$, the corresponding Szegö kernel is given explicitly by

$$\Psi_c(z_1, z_2) = \frac{1}{2}\left(\sqrt[4]{\frac{\psi(z_1)}{\psi(z_2)}} + \sqrt[4]{\frac{\psi(z_2)}{\psi(z_1)}}\right)\frac{\sqrt{dz_1 dz_2}}{z_1 - z_2}, \tag{B.7}$$

where $\psi(z) = \prod_{Q_\alpha \in T}(z - Q_\alpha) \times \prod_{Q_\beta \in T^c}(z - Q_\beta)^{-1}$. See [**23**, p. 12, Example].

Appendix C. Equivariant Borel–Moore homology

We use equivariant Borel-Moore homology in this paper. For the usual Borel–Moore homology, see, e.g., [**27**, §B.2]. As we only use the Borel–Moore homology, we denote it by $H_*(\,\cdot\,)$. If we do not specify the coefficients, we mean the complex coefficients.

The following properties are crucial.

(a) If X is nonsingular, $H_i(X)$ is isomorphic to the ordinary cohomology group $H^{2\dim X - i}(X)$.
(b) For an irreducible algebraic variety X, its fundamental class $[X] \in H_{2\dim X}(X)$ is defined.
(c) For a proper continuous map $f\colon X \to Y$, the push-forward homomorphism $f_*\colon H_*(X) \to H_*(Y)$ is defined.

(d) If $U \subset X$ is open with complement $Y = X \setminus U$, we have the long exact sequence

$$\cdots \to H_i(Y) \xrightarrow{\iota_*} H_i(X) \xrightarrow{j^*} H_i(U) \to H_{i-1}(Y) \to \cdots,$$

where $\iota\colon Y \to X$, $j\colon U \to X$ are inclusions, and j^* is the restriction homomorphism.

For an equivariant Borel–Moore homology, we use the one given in [**52**], but we shift the degree so that the fundamental class $[X]$ has degree $2\dim X$. This definition is the same as [**18**].

Let us recall the definition briefly. Let G be a linear algebraic group acting on an algebraic variety X. (Everything is over $\mathbb{C}$.) We have a finite-dimensional approximation of the classifying space $EG \to BG$, i.e., for any n, there exists a smooth irreducible variety U with G-action such that

(a) The quotient $U \to U/G$ exists and is a principal G-bundle.
(b) $H^i(U) = 0$ for $i = 1, \dots, n$.

We then define

$$H_n^G(X) = H_{n-2\dim G+2\dim U}(X \times_G U).$$

Here U is smooth, in particular $\dim U$ makes sense. One can show that this is independent of the choice of U, using the double fibration argument.

Note that $H_n^G(X) = 0$ if $n > 2\dim X$, but $H_n^G(X)$ may be nonzero for $n < 0$. (X is pure dimensional.)

On the other hand, we define the equivariant *co*-homology as

$$H_G^n(X) = H^n(X \times_G U),$$

where $H^n(\,\cdot\,)$ is the ordinary cohomology. This coincides with the usual definition. It is a graded ring. We have the Poincaré duality isomorphism

$$H_G^n(X) \cong H^G_{2\dim X-n}(X)$$

when X is nonsingular.

As the projection $X \times_G U \to U/G$ is flat, $H_*^G(X)$ has the structure of an $H_G^*(\mathrm{pt})$-module.

Suppose that G is reductive. Then $H_G^*(\mathrm{pt})$ is isomorphic to $S^*(\mathfrak{h}^*)^W$, where $\mathfrak{h}$ is the Cartan subalgebra, $S^*(\mathfrak{h}^*)$ is the symmetric algebra of its dual, and W is the Weyl group. We denote this by S or $S(G)$.

Let T be a torus acting on X. Let X^T be the fixed point set and $\iota\colon X^T \to X$ be the inclusion. We have the push-forward homomorphism $\iota_*\colon H_*^T(X^T) \to H_*^T(X)$. Since T acts trivially on X^T, we have $H_*^T(X^T) = H_*(X^T) \otimes_{\mathbb{C}} S$. The localization theorem (see [**4**]) says that ι_* becomes an isomorphism after tensoring by the quotient field $\mathcal{S}$ of S.

When X is nonsingular, the inverse of ι_* can be explicitly given. Let $X^T = \bigsqcup F_i$ be the decomposition into irreducible components. Each F_i is nonsingular. Let N_i be the normal bundle of F_i in X. Then we have

$$(\iota_*)^{-1} = \sum_i \frac{1}{e_T(N_i)} \iota_i^*,$$

where $e_T(N_i)$ is the equivariant Euler class and ι_i^* is the pull-back homomorphism for the inclusion $\iota_i\colon F_i \to X$ defined via the Poincaré duality homomorphism.

Appendix D. The proof of (3.9) by Hiroyuki Ochiai

Let

$$(a)_k = (a,q)_k = (1-a)(1-aq)\cdots(1-aq^{k-1}), \qquad (a)_\infty = \prod_{d=0}^{\infty}(1-aq^d).$$

We start with the formula, [**6**, p. 16, (7.2)].
We substitute $d = aq/c$, then we have

$$\sum_{k=0}^{\infty} \frac{(a)_k(e)_k(f)_k(1-aq^{2k})(-a/qef)^k q^{k(k+3)/2}}{(aq/e)_k(aq/f)_k(q)_k(1-a)} = \frac{(aq)_\infty(aq/ef)_\infty}{(aq/e)_\infty(aq/f)_\infty}.$$

(the q-hypergeometric part vanishes since $(1)_k = 0$ for $k \geq 1$.) We put $e = -(aq)^{1/2}$. Then we have

$$\sum_{k=0}^{\infty} \frac{(a)_k(f)_k(1-aq^{2k})(a^{1/2}q^{-3/2}/f)^k q^{k(k+3)/2}}{(aq/f)_k(q)_k(1-a)} = \frac{(aq)_\infty(-(aq)^{1/2}/f)_\infty}{(-(aq)^{1/2})_\infty(aq/f)_\infty}.$$

Finally, we put $f = aq^{-1/2}$. Then we have

$$\sum_{k=0}^{\infty} \frac{(a)_k(aq^{-1/2})_k(1-aq^{2k})(a^{-1/2}q^{-1})^k q^{k(k+3)/2}}{(q^{3/2})_k(q)_k} = \frac{(a)_\infty(-a^{-1/2}q)_\infty}{(-(aq)^{1/2})_\infty(q^{3/2})_\infty}.$$

Using the Jacobi triple product identity

$$(q)_\infty(-a^{1/2})_\infty(-a^{-1/2}q)_\infty = \sum_{l=-\infty}^{\infty}(q/a)^{l/2}q^{l^2/2},$$

the right-hand side is

$$\frac{(a)_\infty}{(-a^{1/2}q^{1/2})_\infty(-a^{1/2})_\infty(q)_\infty(q^{3/2})_\infty}\sum_{l=-\infty}^{\infty}(q/a)^{l/2}q^{l^2/2}.$$

Using $(b)_\infty(bq^{1/2})_\infty = (b,q^{1/2})_\infty$, and $(b^2)_\infty = (b,q^{1/2})_\infty(-b,q^{1/2})_\infty$, we get

$$\sum_{k=0}^{\infty} \frac{(aq^{-1/2},q^{1/2})_{2k}(1-aq^{2k})(a^{-1/2}q^{-1})^k q^{k(k+3)/2}}{(q,q^{1/2})_{2k}} = \frac{(a^{1/2},q^{1/2})_\infty}{(q,q^{1/2})_\infty}\sum_{l=-\infty}^{\infty}(q/a)^{l/2}q^{l^2/2}.$$

Also

$$\sum_{k=0}^{\infty} \frac{(aq^{-1/2},q^{1/2})_{2k}(1-aq^{2k})(a^{-1/2}q^{-1})^k q^{k(k+3)/2}}{(q^{1/2},q^{1/2})_{2k+1}} = \frac{(a^{1/2},q^{1/2})_\infty}{(q^{1/2},q^{1/2})_\infty}\sum_{l=-\infty}^{\infty}(q/a)^{l/2}q^{l^2/2}.$$

Now we substitute $a \mapsto q^2t^2$ and $q \mapsto q^4t^2$. Then

$$\sum_{k=0}^{\infty} \frac{(t,q^2t)_{2k}(1-q^{8k+2}t^{4k+2})q^{k(2k+1)}t^{k^2}}{(q^2t,q^2t)_{2k+1}} = \frac{(qt,q^2t)_\infty}{(q^2t,q^2t)_\infty}\sum_{l=-\infty}^{\infty}q^{l(2l+1)}t^{l^2}.$$

Using the identity

$$1-q^{8k+2}t^{4k+2} = (1-q^{4k}t^{2k+1})q^{4k+2}t^{2k+1} + (1-q^{4k+2}t^{2k+1}),$$

we see that the left-hand side is

$$\sum_{k=0}^{\infty}\left(\frac{(t,q^2t)_{2k}}{(q^2t,q^2t)_{2k}}q^{k(2k+1)}t^{k^2}+\frac{(t,q^2t)_{2k+1}}{(q^2t,q^2t)_{2k+1}}q^{(k+1)(2k+3)-1}t^{(k+1)^2}\right).$$

This is the end of the proof. □

Appendix E. Perturbation term

E.1. One-parameter version. Let

$$\gamma_\hbar(x;\Lambda)=\frac{d}{ds}\bigg|_{s=0}\frac{\Lambda^s}{\Gamma(s)}\int_0^\infty\frac{dt}{t}t^s\frac{e^{-tx}}{(e^{\hbar t}-1)(e^{-\hbar t}-1)},$$

where $\Gamma(s)$ is the Gamma function

$$\Gamma(s)=\int_0^\infty\frac{dt}{t}t^se^{-t}.$$

The integral on the right-hand side converges when $\Re(s)>2$. The analytic continuation can be done by the standard procedure using the Taylor expansion of the integrand. (See below.)

If we formally expand as

$$\frac{1}{(e^{\hbar t}-1)(e^{-\hbar t}-1)}=\sum_{m,n\geq 0}e^{\hbar(m-n)t},$$

we get

$$\gamma_\hbar(x;\Lambda)\overset{\text{formally}}{=}\sum_{m,n\geq 0}\log\left(\frac{x-\hbar(m-n)}{\Lambda}\right).$$

Thus $\gamma_\hbar(x;\Lambda)$ is a regularization of the right-hand side.

We introduce Bernoulli numbers by

$$\frac{t}{e^t-1}=\sum_{n=0}^\infty\frac{B_n}{n!}t^n.$$

We have $B_0=1$, $B_1=-\frac{1}{2}$, $B_2=\frac{1}{6}$, $B_{2k+1}=0$ for $k\geq 1$. Note

$$\frac{1}{(e^t-1)(e^{-t}-1)}=\frac{d}{dt}\frac{1}{e^t-1}=-\frac{1}{t^2}+\sum_{g=1}^\infty\frac{B_{2g}}{2g(2g-2)!}t^{2g-2}.$$

Then

(E.1) $$\begin{aligned}\gamma_\hbar(x;\Lambda)&=\frac{d}{ds}\bigg|_{s=0}\Bigg[-\left(\frac{x}{\hbar}\right)^2\left(\frac{\Lambda}{x}\right)^s\frac{\Gamma(s-2)}{\Gamma(s)}+\frac{B_2}{2}\left(\frac{\Lambda}{x}\right)^s\\&\qquad+\sum_{g=2}^\infty\frac{B_{2g}}{2g(2g-2)!}\left(\frac{\hbar}{x}\right)^{2g-2}\frac{\Gamma(s+2g-2)}{\Gamma(s)}\Bigg]\\&=\hbar^{-2}\left\{\frac{1}{2}x^2\log\left(\frac{x}{\Lambda}\right)-\frac{3}{4}x^2\right\}-\frac{1}{12}\log\left(\frac{x}{\Lambda}\right)+\sum_{g=2}^\infty\frac{B_{2g}}{2g(2g-2)}\left(\frac{\hbar}{x}\right)^{2g-2}.\end{aligned}$$

We have the difference equation

$$\gamma_\hbar(x+\hbar;\Lambda)+\gamma_\hbar(x-\hbar;\Lambda)-2\gamma_\hbar(x;\Lambda)=\log\left(\frac{x}{\Lambda}\right).$$

In fact, the left-hand side is equal to

$$\frac{d}{ds}\bigg|_{s=0}\frac{\Lambda^s}{\Gamma(s)}\int_0^\infty \frac{dt}{t}t^s e^{-tx}\frac{e^{t\hbar}+e^{-t\hbar}-2}{(e^{\hbar t}-1)(e^{-\hbar t}-1)} = -\frac{d}{ds}\bigg|_{s=0}\left(\frac{\Lambda}{x}\right)^s.$$

We have

$$\begin{aligned}
&\gamma_\hbar\left(x+\frac{\hbar}{2};\Lambda\right)-\gamma_\hbar\left(x-\frac{\hbar}{2};\Lambda\right)\\
&= \frac{d}{ds}\bigg|_{s=0}\frac{\Lambda^s}{\Gamma(s)}\int_0^\infty \frac{dt}{t}t^s e^{-tx}\frac{e^{-t\hbar/2}-e^{t\hbar/2}}{(e^{\hbar t}-1)(e^{-\hbar t}-1)}\\
&= \frac{d}{ds}\bigg|_{s=0}\frac{\Lambda^s}{\Gamma(s)}\int_0^\infty \frac{dt}{t}t^s \frac{e^{-t(x+\hbar/2)}}{1-e^{-\hbar t}} = \frac{d}{ds}\bigg|_{s=0}\left(\frac{\Lambda}{\hbar}\right)^s\zeta\left(s,\frac{x}{\hbar}+\frac{1}{2}\right)\\
&= \log\left(\frac{\Lambda}{\hbar}\right)\zeta\left(0,\frac{x}{\hbar}+\frac{1}{2}\right)+\zeta'\left(0,\frac{x}{\hbar}+\frac{1}{2}\right) = \log\left(\frac{1}{\sqrt{2\pi}}\left(\frac{\hbar}{\Lambda}\right)^{x/\hbar}\Gamma\left(\frac{x}{\hbar}+\frac{1}{2}\right)\right),
\end{aligned}$$

where $\zeta(s,a)$ is the Hurwitz zeta function. And in the final equality, we have used the Lerch formula (see [**42**, §XV]).

We have

$$\gamma_\hbar(x;\Lambda)+\gamma_\hbar(-x;\Lambda) = 2\gamma_{\sqrt{-1}\hbar}(\sqrt{-1}x;\Lambda).$$

This can be seen from the expansion (E.1).

E.2. Two-parameter version. Let us introduce a generalization of $\gamma_\hbar(x;\Lambda)$:

$$\gamma_{\varepsilon_1,\varepsilon_2}(x;\Lambda) = \frac{d}{ds}\bigg|_{s=0}\frac{\Lambda^s}{\Gamma(s)}\int_0^\infty \frac{dt}{t}t^s\frac{e^{-tx}}{(e^{\varepsilon_1 t}-1)(e^{\varepsilon_2 t}-1)}.$$

This is formally equal to

$$\sum_{m,n\ge 0}\log\left(\frac{x-m\varepsilon_1-n\varepsilon_2}{\Lambda}\right).$$

The difference equation is

$$\gamma_{\varepsilon_1,\varepsilon_2}(x-\varepsilon_1;\Lambda)+\gamma_{\varepsilon_1,\varepsilon_2}(x-\varepsilon_2;\Lambda)-\gamma_{\varepsilon_1,\varepsilon_2}(x;\Lambda)-\gamma_{\varepsilon_1,\varepsilon_2}(x-\varepsilon_1-\varepsilon_2;\Lambda) = \log\left(\frac{x}{\Lambda}\right).$$

Let k be an integer. We have

$$\begin{aligned}
&\gamma_{\varepsilon_1,\varepsilon_2-\varepsilon_1}(x+\varepsilon_1 k;\Lambda)+\gamma_{\varepsilon_1-\varepsilon_2,\varepsilon_2}(x+\varepsilon_2 k;\Lambda)\\
&= \frac{d}{ds}\bigg|_{s=0}\frac{\Lambda^s}{\Gamma(s)}\int_0^\infty \frac{dt}{t}t^s e^{-tx}\left\{\frac{e^{-t\varepsilon_1 k}}{(e^{\varepsilon_1 t}-1)(e^{(\varepsilon_2-\varepsilon_1)t}-1)}+\frac{e^{-t\varepsilon_2 k}}{(e^{(\varepsilon_1-\varepsilon_2)t}-1)(e^{\varepsilon_2 t}-1)}\right\}\\
&= \frac{d}{ds}\bigg|_{s=0}\frac{\Lambda^s}{\Gamma(s)}\int_0^\infty \frac{dt}{t}t^s e^{-tx}\frac{(1-e^{-\varepsilon_2 t})e^{-k\varepsilon_1 t}-(1-e^{-\varepsilon_1 t})e^{-k\varepsilon_2 t}}{(e^{\varepsilon_1 t}-1)(e^{\varepsilon_2 t}-1)(e^{-\varepsilon_1 t}-e^{-\varepsilon_2 t})}.
\end{aligned}$$

We claim that this is equal to

$$\text{(E.2)}\qquad \gamma_{\varepsilon_1,\varepsilon_2}(x;\Lambda)+\log s^{-k}(\varepsilon_1,\varepsilon_2,x)-\frac{k(k-1)}{2}\log\Lambda,$$

where $s^{-k}(\varepsilon_1,\varepsilon_2,x)$ is given by (4.3). If $k=0$ or 1, this is obvious. Suppose that $k\ge 2$. Then the above is equal to

$$\sum_{\substack{l,m\ge 0\\ l+m=k-1}}\gamma_{\varepsilon_1,\varepsilon_2}(x+l\varepsilon_1+m\varepsilon_2;\Lambda)-\sum_{\substack{l,m\ge 1\\ l+m=k}}\gamma_{\varepsilon_1,\varepsilon_2}(x+l\varepsilon_1+m\varepsilon_2;\Lambda).$$

On the other hand, (E.2) is equal to

$$\gamma_{\varepsilon_1,\varepsilon_2}(x;\Lambda) + \sum_{\substack{l,m\geq 0\\ l+m\leq k-2}} \log\left(\frac{x+(l+1)\varepsilon_1+(m+1)\varepsilon_2}{\Lambda}\right).$$

$$= \left(\sum_{l=0,m=0} + \sum_{\substack{l\geq 0,m\geq 1\\ l+m\leq k-1}} + \sum_{\substack{l\geq 1,m\geq 0\\ l+m\leq k-1}} - \sum_{\substack{l\geq 1,m\geq 1\\ l+m\leq k}} - \sum_{\substack{l\geq 0,m\geq 0\\ l+m\leq k-2}}\right)\gamma_{\varepsilon_1,\varepsilon_2}(x+l\varepsilon_1+m\varepsilon_2;\Lambda)$$

$$= \left(\sum_{l=0,m=0} + \sum_{\substack{l\geq 0,m\geq 0\\ l+m=k-1}} - \sum_{\substack{l\geq 0,m=0\\ l+m\leq k-1}} - \sum_{\substack{l\geq 1,m\geq 1\\ l+m=k}} + \sum_{\substack{l\geq 1,m=0\\ l+m\leq k-1}}\right)\gamma_{\varepsilon_1,\varepsilon_2}(x+l\varepsilon_1+m\varepsilon_2;\Lambda)$$

by the difference equation. Thus we get the assertion when $k \geq 2$.

Similarly we have

$$\gamma_{\varepsilon_1,\varepsilon_2-\varepsilon_1}(x+\varepsilon_1 k;\Lambda) + \gamma_{\varepsilon_1-\varepsilon_2,\varepsilon_2}(x+\varepsilon_2 k;\Lambda) = \gamma_{\varepsilon_1,\varepsilon_2}(x;\Lambda) + \sum_{\substack{l,m\geq 0\\ l+m\leq -k-1}} \log\left(\frac{x-l\varepsilon_1-m\varepsilon_2}{\Lambda}\right)$$

for $k \leq -1$. This is nothing but the assertion.

E.3. Expansion. Let us define c_n ($n = 0, 1, 2, \dots$) by

$$\frac{1}{(e^{\varepsilon_1 t}-1)(e^{\varepsilon_2 t}-1)} = \sum_{n=0}^{\infty} \frac{c_n}{n!} t^{2-n}.$$

We have

$$c_0 = \frac{1}{\varepsilon_1\varepsilon_2}, \qquad c_1 = -\frac{\varepsilon_1+\varepsilon_2}{2\varepsilon_1\varepsilon_2}, \qquad c_2 = \frac{\varepsilon_1^2+\varepsilon_2^2+3\varepsilon_1\varepsilon_2}{6\varepsilon_1\varepsilon_2}, \qquad \dots.$$

Then

$$\text{(E.3)}\quad \gamma_{\varepsilon_1,\varepsilon_2}(x;\Lambda) = \left.\frac{d}{ds}\right|_{s=0} \frac{\Lambda^s}{\Gamma(s)} \sum_{n=0}^{\infty} c_n \int_0^{\infty} \frac{dt}{t} t^{n+s-2} e^{-tx}$$
$$= \left.\frac{d}{ds}\right|_{s=0} \left(\frac{\Lambda}{x}\right)^s \sum_{n=0}^{\infty} c_n x^{2-n} \frac{\Gamma(n+s-2)}{\Gamma(s)}$$
$$= \frac{1}{\varepsilon_1\varepsilon_2}\left\{-\frac{1}{2}x^2\log\left(\frac{x}{\Lambda}\right) + \frac{3}{4}x^2\right\} + \frac{\varepsilon_1+\varepsilon_2}{2\varepsilon_1\varepsilon_2}\left\{-x\log\left(\frac{x}{\Lambda}\right) + x\right\}$$
$$- \frac{\varepsilon_1^2+\varepsilon_2^2+3\varepsilon_1\varepsilon_2}{12\varepsilon_1\varepsilon_2}\log\left(\frac{x}{\Lambda}\right) + \sum_{n=3}^{\infty} \frac{c_n x^{2-n}}{n(n-1)(n-2)}.$$

In particular, we have

$$\text{(E.4)}\quad \gamma_{\varepsilon_1,\varepsilon_2}(x;\Lambda e^u) = \gamma_{\varepsilon_1,\varepsilon_2}(x;\Lambda) + u\left\{\frac{x^2}{2\varepsilon_1\varepsilon_2} + \frac{x(\varepsilon_1+\varepsilon_2)}{2\varepsilon_1\varepsilon_2} + \frac{\varepsilon_1^2+\varepsilon_2^2+3\varepsilon_1\varepsilon_2}{12\varepsilon_1\varepsilon_2}\right\},$$

$$\begin{aligned}\text{(E.5)}\quad &\gamma_{\varepsilon_1,\varepsilon_2}(x;\Lambda)+\gamma_{\varepsilon_1,\varepsilon_2}(-x;\Lambda)\\ &=\frac{2}{\varepsilon_1\varepsilon_2}\left\{-\frac{1}{2}x^2\log\left(\frac{\sqrt{-1}x}{\Lambda}\right)+\frac{3}{4}x^2\right\}+\frac{\varepsilon_1+\varepsilon_2}{2\varepsilon_1\varepsilon_2}\pi\sqrt{-1}x\\ &\quad-\frac{\varepsilon_1^2+\varepsilon_2^2+3\varepsilon_1\varepsilon_2}{6\varepsilon_1\varepsilon_2}\log\left(\frac{\sqrt{-1}x}{\Lambda}\right)+\sum_{g=2}^{\infty}\frac{2c_{2g}x^{2-2g}}{2g(2g-1)(2g-2)}.\end{aligned}$$

Appendix F. A proof of Theorem 3.3

As we mentioned before, we prove Theorem 3.3 also for arbitrary projective surface X. Let us fix the notation. Let H be an ample line bundle over X. For $c_1 \in H^2(X,\mathbb{Z})$, $n \in \mathbb{Q}$, let $M(r,c_1,n)$ be the moduli space of H-stable sheaves E on X with $c_1(E)=c_1$, $c_2(E)-(r-1)/(2r)c_1(E)^2 \equiv \Delta(E) = n$. We assume $\gcd(r,\langle c_1,H\rangle)=1$. Thus $M(r,c_1,n)$ consists of μ-stable sheaves for all n.

Let $p\colon \widehat{X}\to X$ be the blow-up of X at a point $P\in X$ and C the exceptional divisor on $\widehat{X}$. Let $\widehat{M}(r,c_1+kC,n)$ be the moduli space of $(H-\varepsilon C)$-stable sheaves E on $\widehat{X}$ with $c_1(E)=p^*c_1+kC$, $c_2(E)-(r-1)/(2r)c_1(E)^2\equiv\Delta(E)=n$, where c_1, n is as above, $k\in\mathbb{Z}$, and $\varepsilon>0$ is sufficiently small.

For a S-flat coherent sheaf $\mathcal{E}$ on $S\times X$, we set $\mathcal{E}_s := \mathcal{E}_{|\{s\}\times X}$, $s\in S$. For a morphism $f\colon S'\to S$, we denote $(f\times 1)^*\mathcal{E}$ by $f^\#\mathcal{E}$, where 1 is the identity map of X. For $\xi\in H^*(X,\mathbb{Q})$, $\mathrm{Quot}^{\xi}_{\mathcal{E}/S\times X/S}\to S$ denotes the relative quot-scheme:

$$\mathrm{Quot}^{\xi}_{\mathcal{E}/S\times X/S} := \{\mathcal{E}_s\to T \mid s\in S, \mathrm{ch}(T)=\xi\}.$$

$M(r,c_1,n)_{\mathrm{loc}}$ denotes the open subscheme of $M(r,c_1,n)$ consisting of μ-stable vector bundles.

F.1. Uhlenbeck compactification of $M(r,c_1,n)_{\mathrm{loc}}$.

THEOREM F.1 (J. Li). *We set $M_0(r,c_1,n)=\bigsqcup_m M(r,c_1,n-m)_{\mathrm{loc}}\times S^mX$. Then*

(1) *$M_0(r,c_1,n)$ has a scheme structure which is projective.*
(2) *There is a projective morphism $\pi\colon M(r,c_1,n)_{\mathrm{red}}\to M_0(r,c_1,n)$ sending $E\in M(r,c_1,n)_{\mathrm{red}}$ to $\left(E^{\vee\vee},\mathrm{Supp}(E^{\vee\vee}/E)\right)\in M_0(r,c_1,n)$.*

REMARK F.2. Li defined a scheme structure on $M_0(r,c_1,n)$ so that π factors through $M(r,c_1,n)$.

Let us briefly recall Li's construction of $M_0(r,c_1,n)$ [**43, 44**]. We assume that $H' := hH$ is very ample and $r\mid h$. Let L be a line bundle on $C'\in|H'|$ such that $\mathrm{ch}(L)=(0,H,(K_X,H')/2-(c_1,H')/r)$. Then $\chi(L\otimes E)=0$ for a locally free sheaf $E\in M(r,c_1,n)$. Let $\mathcal{L}_n$ be the determinant line bundle on $M(r,c_1,n)$ defined by L. $\mathcal{L}_n$ does not depend on the choice of C and L. Let $D\in|dH'|$ be a smooth curve which does not contain P. Let $M_D(r,\langle c_1,D\rangle)$ be the moduli space of semi-stable vector bundles on D of rank r and degree $\langle c_1,D\rangle$. By sending $E\in M(r,c_1,n)$ to $E|_D$ if $E|_D$ is semi-stable, we get a rational map $M(r,c_1,n)\cdots\to M_D(r,\langle c_1,D\rangle)$. Then $\mathcal{L}_n^{\otimes b}$ is the pull-back of the determinant line bundle $\mathcal{L}_D$ on $M_D(r,\langle c_1,D\rangle)$. Let σ_D^b be a section of $\mathcal{L}_D^{\otimes b}$. Then σ_D^b can be also extended to a section of $\mathcal{L}_n^{\otimes bd}$. More precisely, we take a principal $\mathrm{GL}_N(\mathbb{C})$-bundle $\psi\colon\mathcal{P}\to M(r,c_1,n)$ associated to $V:=\phi_*\left(\mathcal{E}_{|M(r,c_1,n)\times D}(l)\right)$, $l\gg 0$, where N is the rank of V. Then we have a family of quotients $\mathcal{O}_D^{\oplus N}\to E|_D(l)$ on $\mathcal{P}$. Hence we have a $\mathrm{GL}_N(\mathbb{C})$-equivariant morphism $\mathcal{P} \to Q_D := \mathrm{Quot}_{\mathcal{O}_n^{\oplus N}/D}$. Since $M_D(r, \langle c_1, D\rangle) = Q_D /\!/ \mathrm{GL}_N(\mathbb{C})$ (GIT quotient), we have a map $H^0(M_D(r, \langle c_1, D\rangle), \mathcal{L}_D^{\otimes b}) \to H^0(\mathcal{P}, \psi^*(\mathcal{L}_n^{\otimes bd}))^{\mathrm{GL}_N(\mathbb{C})} = H^0(M(r, c_1, n), \mathcal{L}_n^{\otimes kd})$.

We denote this section by σ_n^{bd}. We note that σ_n^{bd} is well-defined up to the multiplication by elements of $H^0(M(r, c_1, n), \mathcal{O}^\times_{M(r,c_1,n)})$. Let W_n^a be the vector subspace of $H^0(M(r, c_1, n)_{\mathrm{red}}, \mathcal{L}_n^{\otimes m})$ generated by sections $c\sigma_n^{bd}$, $a = bd$, where $c \in H^0(M(r, c_1, n)_{\mathrm{red}}, \mathcal{O}^\times_{M(r,c_1,n)_{\mathrm{red}}})$ and $D \in |dH'|$ runs all smooth members with $P \notin D$. For a sufficiently large a, W_n^a is base point free. Let $\pi\colon M(r, c_1, n)_{\mathrm{red}} \to \mathbb{P}(W_n^a)$ be the map associated to $|W_n^a|$. We denote the image by $M_0(r, c_1, n)$.

REMARK F.3. Let $M(r, c_1, n)_{\mathrm{red}} = \coprod_i M_{n,i}$ be the decomposition to connected components. Since $H^0(Y, \mathcal{O}_Y) \cong \mathbb{C}$ for a connected complete reduced scheme, we have a decomposition

$$W_n^a = \bigoplus_i W^a_{n|M_{n,i}}, \tag{F.1}$$

where $W^a_{n|M_{n,i}}$ is the restriction of W_n^a to $M_{n,i}$.

REMARK F.4. In [**44**], Li also considered curves D passing through P. Since $\dim(E^{\vee\vee}/E)_P$ is determined by $E^{\vee\vee}$ and $\dim(E^{\vee\vee}/E)_Q$, $Q \in X \setminus \{P\}$, set-theoretically $M_0(r, c_1, n)$ coincides with the Li's space. In particular $M_0(r, c_1, n)$ is homeomorphic to the Uhlenbeck's compactification.

REMARK F.5. By the following fact, $M_0(r, c_1, n)$ does not depends on the choice of $m \gg 0$:

Let

$$A = A_0 \to A_1 \to A_2 \to \cdots$$

be a sequence of finitely generated rings over a field such that A_i are reduced, $f_i\colon \mathrm{Spec}(A_i) \to \mathrm{Spec}(A)$ are finite and bijective for all i. Then there is an integer m such that $A_m = A_{m+1} = \cdots$: Since A_i are reduced, the total quotient ring $Q(A_i)$ of A_i are products of fields. Since f_i are bijective, $Q(A_i) = Q(A)$. Hence $A_i \to A_{i+1}$ is injective for all i. Since A_i are finite over A, A_i are subring of the normalization $\tilde{A}$ of A in $Q(A)$. Since $\tilde{A}$ is a finite A-module, it is Noetherian A-module. Hence we get our claim.

F.2. A correspondence between $M(r, c_1, n)$ and $M(r, c_1, n+m)$. Let $\mathcal{E}$ be the universal family on $M(r, c_1, n) \times X$. We set

$$\begin{aligned} Q &:= \mathrm{Quot}^{(0,0,m)}_{\mathcal{E}/M(r,c_1,n)\times X/M(r,c_1,n)} \\ &= \{\mathcal{E}_x \to T \mid x \in M(r, c_1, n), \dim T = m\} \\ &= \{F \subset \mathcal{E}_x \mid x \in M(r, c_1, n), \dim \mathcal{E}_x/F = m\}. \end{aligned}$$

Let $q\colon Q \to M(r, c_1, n)$ be the projection. Since F is also μ-stable, we have a morphism $\alpha\colon Q \to M(r, c_1, n+m)$ sending $(F \subset \mathcal{E}_x) \in Q$ to $F \in M(r, c_1, n+m)$.

LEMMA F.6. *$q \times \alpha\colon Q \to M(r, c_1, n) \times M(r, c_1, n+m)$ is a closed immersion. Hence we have an identification*

$$Q \cong \{(E, F) \mid (E, F) \in M(r, c_1, n) \times M(r, c_1, n+m), F \subset E\}.$$

PROOF. Obviously $q\times\alpha$ is injective. Since Q is projective over $M(r,c_1,n)$, $q\times\alpha$ is a proper morphism, and hence a finite morphism. As $M(r,c_1,n)$-schemes, the homomorphism of the relative tangent spaces is $\psi\colon \operatorname{Hom}(F,E/F)\to \operatorname{Ext}^1(F,F)$. Since $\mathbb{C}=\operatorname{Hom}(F,F)\cong\operatorname{Hom}(F,E)$, ψ is injective. □

Since Q is projective over $M(r,c_1,n)$, α is proper over $M(r,c_1,n+m)$. Let $Q\to S^mX$ be the morphism sending $(\mathcal{E}_x\to T)\in Q$ to $\sum_{R\in X}(\dim T_R)[R]\in S^mX$. Let $\dot{Q}$ be the fiber at $m[P]$. Then

$$\alpha(\dot{Q})=\{E\in M(r,c_1,n+m)\mid \dim(E_P^{\vee\vee}/E_P)\geq m\}.$$

REMARK F.7. If $m=1$, then

$$\dot{Q}=\{\varphi\colon \mathcal{E}_x\to\mathbb{C}_P\mid x\in M(r,c_1,n),\ \varphi\text{ is surjective}\}.$$

Thus we can define a scheme structure on $\dot{Q}$ as that of the moduli space of stable pairs in the sense of [**35**]. Then the obstruction for infinitesimal liftings belong to the kernel of

$$\operatorname{Ext}^2(E,E\to\mathbb{C}_P)\to\operatorname{Ext}^2(E,E)\xrightarrow{\operatorname{tr}_2}H^2(X,\mathcal{O}_X).$$

Hence if $(K_X,H)\leq 0$, then $\dot{Q}$ is smooth.

The smoothness of the corresponding correspondence for framed sheaves on $\mathbb{C}^2$ was proved in [**67**, 5.7] by using the ADHM description.

PROPOSITION F.8. *There is a closed immersion $M_0(r,c_1,n)\to M_0(r,c_1,n+1)$ sending $(E,\sum_i n_i[R_i])$ to $(E,\sum_i n_i[R_i]+[P])$. In particular, we have a sequence of closed immersions*

$$M_0(r,c_1,n)\subset M_0(r,c_1,n+1)\subset M_0(r,c_1,n+2)\subset\cdots.$$

PROOF. We use the correspondence with $m=1$:

(F.2)
$$\begin{array}{ccc} & \dot{Q} & \\ {}^{q}\swarrow & & \searrow^{\alpha} \\ M(r,c_1,n) & & M(r,c_1,n+1). \end{array}$$

Let $\mathcal{G}$ be the universal subsheaf and $\mathcal{T}$ the universal quotient on $\dot{Q}\times X$:

$$0\to\mathcal{G}\to q^{\#}\mathcal{E}\to\mathcal{T}\to 0.$$

We take a sufficiently large a and take a decomposition $a=bd$. We take a smooth divisor $D\in|dH'|$ in §F.1. Since D does not contain P, we get an isomorphism $\mathcal{G}_{|\dot{Q}_{\mathrm{red}}\times D}\to q^*\mathcal{E}_{|\dot{Q}_{\mathrm{red}}\times D}$. Hence we get an isomorphism $\alpha^*\mathcal{L}_{n+1}^{\otimes bd}\to q^*\mathcal{L}_n^{\otimes bd}$ such that σ_{n+1}^{bd} corresponds to $q^*\sigma_n^{bd}$. From now on, we fix an isomorphism $\alpha^*\mathcal{L}_{n+1}^{\otimes a}\to q^*\mathcal{L}_n^{\otimes a}$. We take the decomposition $M(r,c_1,n)_{\mathrm{red}}=\coprod_i M_{n,i}$ to connected components as in Remark F.3. We set $Q_i:=q^{-1}(M_i)$. Since every fiber of q is a projective space, Q_i are also connected components of $\dot{Q}_{\mathrm{red}}$. Then we have

(*)
$$(\alpha^*W_{n+1}^a)_{|Q_i}=(q^*W_n^a)_{|Q_i}.$$

For each $M_{n+1,j}$, we set $I_j:=\{i\mid\alpha(Q_i)\subset M_{n+1,j}\}$. Since $\alpha^{-1}(M_{n+1,j})=\coprod_{i\in I_j}Q_i$, $\alpha^*(W_{n+1}^a|M_{n+1,j})\to(\alpha^*W_{n+1}^a)_{|Q_i}$ is an isomorphism for all $i\in I_j$. Hence we have an injective homomorphism $\alpha^*(W^a_{n+1|M_{n+1,j}}) \to \bigoplus_{i\in I_j}(q^*W^a_n)_{|Q_i}$ by using $(*)$. Hence we get

$$\begin{array}{ccccc} H^0(M(r,c_1,n+1)_{\mathrm{red}}, \mathcal{L}^{\otimes a}_{n+1}) & \longrightarrow & H^0(\dot{Q}_{\mathrm{red}}, \alpha^*\mathcal{L}^{\otimes a}_{n+1}) & \cong & H^0(\dot{Q}_{\mathrm{red}}, q^*\mathcal{L}^{\otimes a}_n) \\ \cup & & \cup & & \cup \\ W^a_{n+1} & \longrightarrow & \alpha^* W^a_{n+1} & \hookrightarrow & q^*W^a_n. \end{array}$$

Since q is surjective, we may identify $q^*W^a_n$ with W^a_n. Hence we have the following commutative diagram:

$$\begin{array}{ccccc} M(r,c_1,n)_{\mathrm{red}} & \xleftarrow{q} & \dot{Q}_{\mathrm{red}} & \xrightarrow{\alpha} & M(r,c_1,n+1)_{\mathrm{red}} \\ \downarrow & & \downarrow & & \downarrow \\ \mathbb{P}(W^a_n) & \cdots\xrightarrow{\varpi} & \mathbb{P}(\alpha^*W^a_{n+1}) & \hookrightarrow & \mathbb{P}(W^a_{n+1}), \end{array}$$

where vertical arrows are morphisms associated to the linear systems and ϖ is the linear projection. We denote the image of Q_i to $M_0(r,c_1,n)$ by N_i. Then each N_i is a connected component of $M_0(r,c_1,n)$. By the definition of N_i, N_i is a subset of $\mathbb{P}\big((q^*W^a_n)_{|Q_i}\big)$ $(\hookrightarrow \mathbb{P}(W^a_n))$. By $(*)$, the composition map $\mathbb{P}\big((q^*W^a_n)_{|Q_i}\big) \hookrightarrow \mathbb{P}(W^a_n) \cdots\xrightarrow{\varpi} \mathbb{P}(\alpha^*W^a_{n+1})$ is a closed immersion whose image is $\mathbb{P}((\alpha^*W^a_{n+1})_{|Q_i})$, and hence we get a closed immersion $N_i \hookrightarrow M_0(r,c_1,n+1)$. By the following lemma, $\bigcup_i N_i \to M_0(r,c_1,n+1)$ is a closed immersion. □

LEMMA F.9. *$\pi\big(\alpha(Q_i)\big) \cap \pi\big(\alpha(Q_j)\big) = \varnothing$ for $i \neq j$.*

PROOF. If $\pi\big(\alpha(Q_i)\big) \cap \pi\big(\alpha(Q_j)\big) \neq \varnothing$, then there are elements E_1, $E_2 \in M(r,c_1,n)$ such that $E_1^{\vee\vee} = E_2^{\vee\vee}$, $E_1 \in M_i$ and $E_2 \in M_j$. We set $F := E_1^{\vee\vee}$ and $\dim F/E_1 = t$. Then we get a morphism $\mathrm{Quot}^{(0,0,t)}_{F/X/\mathbb{C}} \to M(r,c_1,n)$. Since E_1 and E_2 belong to the image of $\mathrm{Quot}^{(0,0,t)}_{F/X/\mathbb{C}}$ and $\mathrm{Quot}^{(0,0,t)}_{F/X/\mathbb{C}}$ is connected, we get a contradiction. Thus $\pi\big(\alpha(Q_i)\big) \cap \pi\big(\alpha(Q_j)\big) = \varnothing$. □

F.3. Construction of $\widehat{\pi}$.

LEMMA F.10. *There is an integer l_0 such that for any $E \in \widehat{M}(r,\hat{c}_1,\hat{n})$ and $l \geq l_0$,*

(1) *$R^1p_*\big(E(-lC)\big) = 0$ and*
(2) *there is a canonical inclusion $E(-lC) \hookrightarrow p^*\big(p_*(E(-lC))^{\vee\vee}\big)$.*

PROOF. Since $\widehat{M}(r,\hat{c}_1,\hat{n})$ is bounded and $\mathcal{O}_X(-C)$ is relatively ample, by using Proposition F.23, we see that there is an integer l_0 such that $R^1p_*\big(E(-lC)\big) = 0$ and $p^*p_*\big(E(-lC)\big) \to E(-lC)$ is surjective for $l \geq l_0$. Since E is torsion free, we have

$$p^*p_*\big(E(-lC)\big)/(\text{torsion}) \cong E(-lC).$$

On the other hand, there is an inclusion $p_*\big(E(-lC)\big) \hookrightarrow p_*\big(E(-lC)\big)^{\vee\vee}$. Hence we get a desired inclusion $E(-lC) \hookrightarrow p^*\big(p_*\big(E(-lC)\big)^{\vee\vee}\big)$. □

For a sufficiently large l_0, we have a morphism

$$\begin{aligned} \beta\colon \widehat{M}(r,\hat{c}_1,\hat{n}) &\to M(r,c_1,n') \\ E &\mapsto p_*(E(-l_0C)) \end{aligned} \tag{F.3}$$

where $n' = \Delta(p_*(E(-l_0C)))$. We take an integer l_1 such that $l_1 \equiv -\langle \hat{c}_1, [C]\rangle \bmod r$ and $0 \le l_1 < r$. We set $n := \hat{n} - (r - l_1)l_1/2r$. Then $\operatorname{im}(\pi \circ \beta)$ is contained in $M_0(r, c_1, n)$. Hence we get

THEOREM F.11. *We have a morphism* $\widehat{\pi}\colon \widehat{M}(r, \hat{c}_1, \hat{n})_{\mathrm{red}} \to M_0(r, c_1, n)$ *sending* $E \in \widehat{M}(r, \hat{c}_1, \hat{n})_{\mathrm{red}}$ *to* $(p_*(E)^{\vee\vee}, \operatorname{Supp}(p_*(E)^{\vee\vee}/p_*(E)) + \operatorname{Supp}(R^1p_*(E))) \in M_0(r, c_1, n)$.

F.4. Stratification of $M(r, c_1, n)$ and $\widehat{M}(r, \hat{c}_1, \hat{n})$. Let $i\colon X\backslash\{P\} \to X$ be the immersion. For $E \in M(r, c_1, n)$, $i_*(E_{|X\backslash\{P\}})$ is a stable sheaf on X which is free on a neighborhood of P and $\Delta(i_*(E_{|X\backslash\{P\}})) \le \Delta(E)$. Thus $i_*(E_{|X\backslash\{P\}})_P \cong E_P^{\vee\vee}$. We set

$$M(r, c_1, n)_m := \{E \in M(r, c_1, n) \mid \Delta(i_*(E_{|X\backslash\{P\}})) = n - m\}.$$

For $E \in \widehat{M}(r, \hat{c}_1, \hat{n})$, $i_*(E_{|X\backslash C})$ is also a stable sheaf on X which is free on a neighborhood of P and $\Delta(i_*(E_{|\widehat{X}\backslash C})) \le \Delta(E)$, where we identified $\widehat{X} \setminus C$ with $X \setminus \{P\}$. Thus $i_*(E_{|X\backslash\{P\}})_P \cong (p_*E)_P^{\vee\vee}$. We set

$$\widehat{M}(r, \hat{c}_1, \hat{n})_m := \{E \in \widehat{M}(r, \hat{c}_1, \hat{n}) \mid \Delta(i_*(E_{|\widehat{X}\backslash C})) = \hat{n} - m\}.$$

We remark that m is not an integer in general.

THEOREM F.12. *Assume that* $\gcd(r, \langle c_1, H\rangle) = 1$.

(1) $M(r, c_1, n)$ *has a finite partition* $M(r, c_1, n) = \bigsqcup_m M(r, c_1, n)_m$ *into locally closed subsets.* $M(r, c_1, n)_m \to M(r, c_1, n - m)_0$ $(E \mapsto i_*(E_{|X\backslash\{P\}}))$ *is a locally trivial fibration and its fiber depends only on* m.

(2) $\widehat{M}(r, \hat{c}_1, \hat{n})$ *has a finite partition* $\widehat{M}(r, \hat{c}_1, \hat{n}) = \bigsqcup_m \widehat{M}(r, \hat{c}_1, \hat{n})_m$ *into locally closed subsets.* $\widehat{M}(r, \hat{c}_1, \hat{n})_m \to M(r, c_1, \hat{n} - m)_0$ $(E \mapsto i_*(E_{|\widehat{X}\backslash C}))$ *is a locally trivial fibration and its fiber depends only on* $\langle \hat{c}_1, [C]\rangle$, m.

F.4.1. *Stratification of* $M(r, c_1, n)$. Since

$$M(r, c_1, n)_m = \{E \in M(r, c_1, n) \mid \dim(E_P^{\vee\vee}/E_P) = m\},$$

$M(r, c_1, n)_m$ is a locally closed subset of $M(r, c_1, n)$. Let $Q \subset M(r, c_1, n) \times M(r, c_1, n + m)$ be the scheme in §F.2. We set

$$Q_0 := q^{-1}(M(r, c_1, n)_0), \qquad \dot{Q}_0 := Q_0 \cap \dot{Q}.$$

Then $\alpha\colon \dot{Q}_0 \to M(r, c_1, n + m)_m$ is a proper morphism.

LEMMA F.13. $\dot{Q}_0 \cong M(r, c_1, n + m)_m$. *In particular we have a morphism* $M(r, c_1, n + m)_m \to M(r, c_1, n)_0$.

In Lemma F.15, we shall prove a similar claim. Since the proof of Lemma F.13 is simpler than that of Lemma F.15, we do not give a proof here.

By Lemma F.13, we have a stratification $M(r, c_1, n) = \bigcup_m M(r, c_1, n)_m$ with morphisms $M(r, c_1, n)_m \to M(r, c_1, n - m)_0$. We set

$$\mathfrak{Q}(m) := \operatorname{Quot}^{(0,0,m)}_{(\mathbb{C}[x,y]/(x,y)^d)^{\oplus r}/\mathbb{C}^2/\mathbb{C}}, \quad d \gg 0. \tag{F.4}$$

We shall show that $M(r, c_1, n)_m \to M(r, c_1, n - m)_0$ is a locally trivial fibration with a fiber $\mathfrak{Q}(m)$. Let $\mathfrak{m}$ be the defining ideal of P. For a sufficiently large d, $\mathcal{E} \otimes \mathcal{O}_X/\mathfrak{m}^d$ is a locally free $\mathcal{O}_{M(r,c_1,n-m)_0}$-module. For each point $x \in M(r, c_1, n-m)_0$, there is a neighborhood U of x such that $\mathcal{E}_{|U\times X} \otimes \mathcal{O}_X/\mathfrak{m}^d \cong (\mathcal{O}_U^{\oplus r}) \otimes_{\mathbb{C}} \mathcal{O}_X/\mathfrak{m}^d$.

Then $q^{-1}(U) = U \times \operatorname{Quot}^{(0,0,m)}_{(\mathcal{O}_X/\mathfrak{m}^d)^{\oplus r}/X/\mathbb{C}}$. Since $\mathcal{O}_X/\mathfrak{m}^d \cong \mathbb{C}[x,y]/(x,y)^d$, we get $\operatorname{Quot}^{(0,0,m)}_{(\mathcal{O}_X/\mathfrak{m}^d)^{\oplus r}/X/\mathbb{C}} \simeq \mathfrak{Q}(m)$. Thus Theorem F.12(1) holds.

F.4.2. *Stratification of* $\widehat{M}(r, \hat{c}_1, \hat{n})$. From now on, replacing E by $E(-l_0C)$, we assume that $l_0 = 0$.

By Lemma F.10 and Proposition F.23, there is a number n' and a morphism

$$\beta\colon \widehat{M}(r, \hat{c}_1, \hat{n}) \to M(r, c_1, n')$$

which is defined by sending $E \in \widehat{M}(r, \hat{c}_1, \hat{n})$ to $p_*(E) \in M(r, c_1, n')$. Then

$$\widehat{M}(r, \hat{c}_1, \hat{n})_m = \beta^{-1}(M(r, c_1, n')_{m-\hat{n}+n'}),$$

which is a locally closed subset of $\widehat{M}(r, \hat{c}_1, \hat{n})$. Thus we get a stratification of $\widehat{M}(r, \hat{c}_1, \hat{n}) = \bigcup_m \widehat{M}(r, \hat{c}_1, \hat{n})_m$ into locally closed subsets. If $\hat{c}_1 \equiv p^*c_1 \bmod r$, then we have an isomorphism $M(r, c_1, \hat{n})_0 \cong \widehat{M}(r, \hat{c}_1, \hat{n})_0$. We want to construct a morphism $\widehat{M}(r, \hat{c}_1, \hat{n})_m \to M(r, c_1, \hat{n} - m)_0$. We first introduce schemes which are similar to $Q, \dot{Q}, \dots$. Let $\widetilde{\mathcal{E}}$ be the universal family on $\widehat{M}(r, p^*c_1, \hat{n} - m) \times \widehat{X}$. We set

$$\widehat{Q} := \operatorname{Quot}^{(0,lC,m+l^2/(2r))}_{\widetilde{\mathcal{E}}/\widehat{M}(r,p^*c_1,\hat{n}-m)\times\widehat{X}/\widehat{M}(r,p^*c_1,\hat{n}-m)}.$$

Let $\hat{q}\colon \widehat{Q} \to \widehat{M}(r, p^*c_1, \hat{n} - m)$ be the projection. Let $\mathcal{F}$ be the universal quotient and $\widehat{\mathcal{E}}$ the universal subsheaf:

$$0 \to \widehat{\mathcal{E}} \to \hat{q}^{\#}(\widetilde{\mathcal{E}}) \to \mathcal{F} \to 0.$$

Since

$$\Delta(\widetilde{\mathcal{E}}_{\hat{q}(x)}) - \Delta(\widehat{\mathcal{E}}_x) = -\left(\operatorname{ch}_2(\widetilde{\mathcal{E}}_{\hat{q}(x)}) - \operatorname{ch}_2(\widehat{\mathcal{E}}_x)\right) - \frac{l^2(C^2)}{2r} = -\operatorname{ch}_2(\mathcal{F}_x) + \frac{l^2}{2r},$$

$\widehat{\mathcal{E}}$ is a flat family of stable sheaves which belongs to $\widehat{M}(r, \hat{c}_1, \hat{n})$, where $\hat{c}_1 = p^*c_1 - lC$. Hence we have a morphism

$$\widehat{\alpha}\colon \widehat{Q} \to \widehat{M}(r, \hat{c}_1, \hat{n})$$
$$x \mapsto \widehat{\mathcal{E}}_x.$$

LEMMA F.14. *$\hat{q} \times \widehat{\alpha}\colon \widehat{Q} \to \widehat{M}(r, p^*c_1, \hat{n}-m) \times \widehat{M}(r, \hat{c}_1, \hat{n})$ is a closed immersion. Thus we have an identification*

$$\widehat{Q} \cong \{(E, F) \mid E \in \widehat{M}(r, p^*c_1, \hat{n} - m), F \in \widehat{M}(r, \hat{c}_1, \hat{n}), F \subset E\}$$

PROOF. Obviously $\hat{q} \times \widehat{\alpha}$ is proper and injective. Since

$$\operatorname{Hom}(F, E) \cong \operatorname{Hom}(p_*(F)^{\vee\vee}, p_*(E)^{\vee\vee}) \cong \mathbb{C},$$

$\operatorname{Hom}(F, E/F) \to \operatorname{Hom}(F, F)$ is injective. Hence $\hat{q} \times \widehat{\alpha}$ is a closed immersion. □

We set

$$\dot{\widehat{Q}} := \operatorname{Quot}^{(0,lC,m+l^2/(2r))}_{\widetilde{\mathcal{E}}'/\widehat{M}(r,p^*c_1,\hat{n}-m)\times dC/\widehat{M}(r,p^*c_1,\hat{n}-m)},$$

where $\widetilde{\mathcal{E}}' = \widetilde{\mathcal{E}}_{|\widehat{M}(r,p^*c_1,\hat{n}-m)\times dC}$. $\dot{\widehat{Q}}$ is a closed subscheme of $\widehat{Q}$. We consider

$$\widehat{Q}_0 := \hat{q}^{-1}(\widehat{M}(r, p^*c_1, \hat{n} - m)_0), \qquad \dot{\widehat{Q}}_0 := \widehat{Q}_0 \cap \dot{\widehat{Q}}.$$

If $d \gg 0$, then

$$\dot{Q}_0 = \operatorname{Quot}^{(0,0,m)}_{\mathcal{E}\otimes(\mathcal{O}_X/\mathfrak{m}^d)/M(r,c_1,n)\times X_d/M(r,c_1,n)},$$

where $\mathfrak{m}$ is the defining ideal of P and $X_d := \operatorname{Spec}(\mathcal{O}_X/\mathfrak{m}^d)$.

Since $R^1p_*(\widetilde{\mathcal{E}}_{\hat{q}(x)}) \to R^1p_*(\mathcal{F}_x)$, $x \in \widehat{Q}_0$ is surjective and $\widetilde{\mathcal{E}}_{\hat{q}(x)} \cong p^*(G)$, $G \in M(r, c_1, \hat{n}-m)_0$, Proposition F.23 implies that $(1 \times p)_*(\mathcal{F})$ is flat over $\widehat{Q}_0$. By our assumption, $R^1p_*(\widehat{\mathcal{E}}_x) = 0$, $x \in \widehat{Q}_0$ and hence we have a flat family of quotients $\mathcal{E} \to (1 \times p)_*\mathcal{F}$. Thus we get a morphism $\gamma\colon \widehat{Q}_0 \to Q_0 \subset M(r, c_1, \hat{n}-m)_0 \times M(r, c_1, n')$ such that $\alpha \circ \gamma = \beta \circ \widehat{\alpha}$. Then we have $\widehat{\alpha}(\dot{\widehat{Q}}_0) \subset \widehat{M}(r, \hat{c}_1, \hat{n})_m$. If $d \gg 0$, then $\widehat{\alpha}(\dot{\widehat{Q}}_0) = \widehat{M}(r, \hat{c}_1, \hat{n})_m$. Indeed, assume that $\gamma(x)$, $x \in \widehat{Q}_0$ belongs to $\dot{Q}$. Then there is an integer d which only depends on m such that $p_*(\mathcal{F}_x)$ is an $\mathcal{O}_X/\mathfrak{m}^d$-module. Since $p^*p_*(\mathcal{F}_x) \to \mathcal{F}_x$ is surjective, $\mathcal{F}_x$ is an $\mathcal{O}_{dC}$-module. Hence $\widehat{\alpha}(\dot{\widehat{Q}}_0) = \widehat{M}(r, \hat{c}_1, \hat{n})_m$. We fix a sufficiently large d.

LEMMA F.15. *$\widehat{\alpha}_{|\dot{\widehat{Q}}_0} : \dot{\widehat{Q}}_0 \to \widehat{M}(r, \hat{c}_1, \hat{n})_m$ is an isomorphism.*

PROOF. We shall prove that $\widehat{\alpha}_{|\dot{\widehat{Q}}_0} : \dot{\widehat{Q}}_0 \to \widehat{M}(r, \hat{c}_1, \hat{n}) \setminus \bigcup_{j>m} \widehat{M}(r, \hat{c}_1, \hat{n})_j$ is a closed immersion. We first prove that $\widehat{\alpha}_{|\dot{\widehat{Q}}_0}$ is proper and injective: We note that $\gamma_{|\dot{\widehat{Q}}_0} : \dot{\widehat{Q}}_0 \to \dot{Q}_0$ satisfies $\alpha \circ \gamma_{|\dot{\widehat{Q}}_0} = \beta \circ \widehat{\alpha}_{|\dot{\widehat{Q}}_0}$ and $\gamma_{|\dot{\widehat{Q}}_0}(\dot{\widehat{Q}}_0) = \alpha^{-1}(M(r, c_1, n')_{m+n'-\hat{n}})$. Note that $q \circ \gamma_{|\dot{\widehat{Q}}_0}$ is equal to $\hat{q}_{|\dot{\widehat{Q}}_0}$ via the isomorphism $\widehat{M}(r, p^*c_1, \hat{n}-m)_0 \cong M(r, c_1, \hat{n}-m)_0$. Since the latter is proper over $\widehat{M}(r, p^*c_1, \hat{n}-m)_0$, $\gamma_{|\dot{\widehat{Q}}_0}$ is proper over $\alpha^{-1}(M(r, c_1, n')_{m+n'-\hat{n}})$. Hence $\beta \circ \widehat{\alpha}_{|\dot{\widehat{Q}}_0} = \alpha \circ \gamma_{|\dot{\widehat{Q}}_0}$ is also proper, which implies that $\widehat{\alpha}_{|\dot{\widehat{Q}}_0}$ is proper. Assume that $\widehat{\mathcal{E}}_x \cong \widehat{\mathcal{E}}_{x'}$. Then $\mathcal{E}_{\hat{q}(x)} \cong p_*(\widehat{\mathcal{E}}_x) \cong p_*(\widehat{\mathcal{E}}_{x'}) \cong \mathcal{E}_{\hat{q}(x')}$. Hence $\hat{q}(x) = \hat{q}(x')$ and we get a commutative diagram:

$$\begin{array}{ccc}
\widehat{\mathcal{E}}_x & \longrightarrow & p^*(\mathcal{E}_{\hat{q}(x)}) \\
\downarrow{\scriptstyle \xi} & & \downarrow{\scriptstyle \xi'} \\
\widehat{\mathcal{E}}_{x'} & \longrightarrow & p^*(\mathcal{E}_{\hat{q}(x)})
\end{array}$$

where ξ and ξ' are isomorphisms. Since $\mathcal{E}_{\hat{q}(x)}$ is simple, ξ' is multiplication by a constant. Hence we conclude that $x = x'$. Thus $\widehat{\alpha}_{|\dot{\widehat{Q}}_0}$ is injective.

Let T_x be the Zariski tangent space of $\widehat{Q}$ at $x \in \dot{\widehat{Q}}_0$ and $\eta \in T_x$ a tangent vector. We also denote the corresponding $\mathbb{C}[t]/(t^2)$-valued point of $\dot{\widehat{Q}}_0$ by η. Then we have an exact and commutative diagram.

$$\begin{array}{ccccccccc}
 & & 0 & & 0 & & 0 & & \\
 & & \downarrow & & \downarrow & & \downarrow & & \\
0 & \longrightarrow & \widehat{\mathcal{E}}_x & \longrightarrow & p^*\mathcal{E}_{\hat{q}(x)} & \longrightarrow & \mathcal{F}_x & \longrightarrow & 0 \\
 & & \downarrow & & \downarrow & & \downarrow & & \\
0 & \longrightarrow & \widehat{\mathcal{E}}_\eta & \longrightarrow & p^*\mathcal{E}_{\hat{q}(\eta)} & \longrightarrow & \mathcal{F}_\eta & \longrightarrow & 0 \\
 & & \downarrow & & \downarrow & & \downarrow & & \\
0 & \longrightarrow & \widehat{\mathcal{E}}_x & \xrightarrow{g} & p^*\mathcal{E}_{\hat{q}(x)} & \xrightarrow{h} & \mathcal{F}_x & \longrightarrow & 0 \\
 & & \downarrow & & \downarrow & & \downarrow & & \\
 & & 0 & & 0 & & 0 & &
\end{array}$$

Then the tangent map of $\widehat{\alpha}_{|\dot{\widehat{Q}}_0}$ $\kappa\colon T_x \to \operatorname{Ext}^1(\widehat{\mathcal{E}}_x, \widehat{\mathcal{E}}_x)$ fits in the following exact and commutative diagram:

$$\begin{array}{ccccc} & & & & \operatorname{Ext}^1(\mathcal{F}_x, p^*\mathcal{E}_{\hat{q}(x)}) \\ & & & & \downarrow \\ \operatorname{Hom}(\widehat{\mathcal{E}}_x, \mathcal{F}_x) & \longrightarrow & T_x & \xrightarrow{\hat{q}_*} & \operatorname{Ext}^1(p^*\mathcal{E}_{\hat{q}(x)}, p^*\mathcal{E}_{\hat{q}(x)}) \\ \Big\| & & \Big\downarrow \kappa & & \Big\downarrow -\bar{g} \\ \operatorname{Hom}(\widehat{\mathcal{E}}_x, \mathcal{F}_x) & \xrightarrow{\delta} & \operatorname{Ext}^1(\widehat{\mathcal{E}}_x, \widehat{\mathcal{E}}_x) & \longrightarrow & \operatorname{Ext}^1(\widehat{\mathcal{E}}_x, p^*\mathcal{E}_{\hat{q}(x)}), \end{array}$$

where $\bar{g}\colon \operatorname{Ext}^1(p^*\mathcal{E}_{\hat{q}(x)}, p^*\mathcal{E}_{\hat{q}(x)}) \to \operatorname{Ext}^1(\widehat{\mathcal{E}}_x, p^*\mathcal{E}_{\hat{q}(x)})$ is the natural map induced by g. In order to prove the injectivity of κ, we shall show that

(1) δ is injective,

(2) $\operatorname{Ext}^1(\mathcal{F}_x, p^*\mathcal{E}_{\hat{q}(x)}) = 0$.

Proof of (1): Since

$$\ker\delta = \operatorname{coker}(\operatorname{Hom}(\widehat{\mathcal{E}}_x, \widehat{\mathcal{E}}_x) \to \operatorname{Hom}(\widehat{\mathcal{E}}_x, p^*\mathcal{E}_{\hat{q}(x)}))$$

and $\operatorname{Hom}(\widehat{\mathcal{E}}_x, \widehat{\mathcal{E}}_x) = \mathbb{C}\,\mathrm{id}$, it is enough to show that $\operatorname{Hom}(\widehat{\mathcal{E}}_x, p^*\mathcal{E}_{\hat{q}(x)}) = \mathbb{C}g$. We note that

$$\begin{aligned} \operatorname{Hom}(\widehat{\mathcal{E}}_x, p^*\mathcal{E}_{\hat{q}(x)}) &= H^0(X, p_*\mathcal{H}om_{\mathcal{O}_{\widehat{X}}}(\widehat{\mathcal{E}}_x, p^*\mathcal{E}_{\hat{q}(x)})) \\ &\hookrightarrow H^0(X, p_*\mathcal{H}om_{\mathcal{O}_{\widehat{X}}}(\widehat{\mathcal{E}}_x, p^*\mathcal{E}_{\hat{q}(x)})^{\vee\vee}) = \operatorname{Hom}(\mathcal{E}_{\hat{q}(x)}, \mathcal{E}_{\hat{q}(x)}). \end{aligned}$$

Since $\mathcal{E}_{\hat{q}(x)}$ is simple, we get $\operatorname{Hom}(\widehat{\mathcal{E}}_x, p^*\mathcal{E}_{\hat{q}(x)}) = \mathbb{C}g$. We next prove (2). By Serre duality, it is sufficient to prove that

$$\operatorname{Ext}^1(p^*\mathcal{E}_{\hat{q}(x)}, \mathcal{F}_x(C)) = H^1(\widehat{X}, p^*\mathcal{E}_{\hat{q}(x)}^{\vee} \otimes \mathcal{F}_x(C)) = 0.$$

Since $p_*(\mathcal{F}_x(C))$ is supported on P and $\mathcal{E}_{\hat{q}(x)}$ is locally free at P, we have

$$H^1(\widehat{X}, p^*\mathcal{E}_{\hat{q}(x)}^{\vee} \otimes \mathcal{F}_x(C)) = H^0(X, R^1p_*(\mathcal{F}_x(C)) \otimes \mathcal{E}_{\hat{q}(x)}^{\vee}).$$

Since

$$R^1p_*(p^*\mathcal{E}_{\hat{q}(x)}^{\vee}(C)) = R^1p_*(\mathcal{O}_{\widehat{X}}(C)) \otimes \mathcal{E}_{\hat{q}(x)}^{\vee} = 0$$

and h is surjective, $R^1p_*(\mathcal{F}_x(C)) \otimes \mathcal{E}_{\hat{q}(x)}^{\vee} = 0$. Therefore we conclude that κ is injective. Since $\widehat{\alpha}$ is one to one and finite, $\widehat{\alpha}$ is a closed immersion. □

We next prove that $\hat{q}\colon \dot{\widehat{Q}}_0 \to M(r, c_1, n' - m)_0$ is locally trivial. We set $k := -\langle \hat{c}_1, [C]\rangle$ and

$$\widehat{\mathfrak{Q}}(k, m) := \operatorname{Quot}^{(0, lC, m + l^2/2r)}_{\mathcal{O}_{dC}^{\oplus r}/dC/\mathbb{C}}, \qquad \text{(F.5)}$$

where $d \gg l \gg 0$, $l \equiv -k \mod r$. For each point $x \in M(r, c_1, \hat{n} - m)_0$, there is a neighborhood U of x such that $\mathcal{E}_{|U\times X} \otimes \mathcal{O}_X/\mathfrak{m}^d \cong (\mathcal{O}_U^{\oplus r}) \otimes_{\mathbb{C}} \mathcal{O}_X/\mathfrak{m}^d$. Then we get a trivialization of $(1 \times p)^*\mathcal{E}_{|U\times X}$: $(1\times p)^*\mathcal{E}_{|U\times X} \otimes \mathcal{O}_{dC} \cong (\mathcal{O}_U^{\oplus r}) \otimes_{\mathbb{C}} \mathcal{O}_{dC}$. Hence $\hat{q}^{-1}(U) = U \times \widehat{\mathfrak{Q}}(k, m)$. Since $\mathcal{O}_X/\mathfrak{m}^d \cong \mathbb{C}[x,y]/(x,y)^d$, $\widehat{\mathfrak{Q}}(k,m)$ does not depend on the choice of (X, P). Therefore we complete the proof of Theorem F.12. □

F.4.3. *Fibers of π and $\widehat{\pi}$.* For a partition $P = ({k_1}^{n_1}, {k_2}^{n_2}, \dots, {k_s}^{n_s})$, $n = k_1 n_1 + \dots + k_s n_s$, $1 \le k_1 < k_2 < \dots < k_s$ of n, we set $S^P X := \prod_i S^{n_i} X$. We have a morphism $\phi\colon S^P X \to S^{\sum_i n_i} X$ by the correspondence $\prod_i [Z_i] \mapsto \sum_i [Z_i]$, where $[Z_i] \in S^{n_i} X$. We set $(S^P X)_* := \phi^{-1}(S^{\sum_i n_i} X \setminus (\text{big diagonal}))$. We regard $(S^P X)_*$ as a locally closed subset of $S^n X$ by the correspondence $\prod_i [Z_i] \mapsto \sum_i k_i [Z_i]$.

We set $(S^P_m X)_* := \{m[P] + [Z] \in S^{m+n} X \mid [Z] \in \left(S^P(X \setminus \{P\})\right)_*\}$. Then we have the following theorem whose proof is left to the reader as an exercise.

THEOREM F.16. (1) *$\pi\colon M(r, c_1, n) \to M_0(r, c_1, n)$ is locally trivial along the locally closed subset $M(r, c_1, n-m)_{\mathrm{loc}} \times (S^P X)_*$ ($\subset M(r, c_1, n-m)_{\mathrm{loc}} \times S^m X$) such that its fiber does not depend on X and $M(r, c_1, n)$, where P is a partition of m.*

(2) *$\widehat{\pi}\colon \widehat{M}(r, \hat{c}_1, \hat{n}) \to M_0(r, c_1, n)$ is locally trivial along the locally closed subset $M(r, c_1, n-m)_{\mathrm{loc}} \times (S^P_{m_1} X)_*$ ($\subset M(r, c_1, n-m)_{\mathrm{loc}} \times S^m X$) such that its fiber does not depend on X and $\widehat{M}(r, \hat{c}_1, \hat{n})$, where P is a partition of $m - m_1$.*

F.5. Stratification of $M(r,n)$ and $\widehat{M}(r,k,n)$. Let us turn to framed moduli spaces. As in §F.3, we have a morphism

$$\begin{aligned} \beta\colon \widehat{M}(r, k, \hat{n}) &\to M(r, n') \\ (E, \Phi) &\mapsto \left(p_*(E(-l_0 C)), \Phi\right) \end{aligned}$$

where $n' = \Delta\left(p_*(E(-l_0 C))\right)$. Hence we have a morphism $\pi \circ \beta\colon \widehat{M}(r, k, \hat{n}) \to M_0(r, n')$. We take an integer l_1 such that $l_1 \equiv k \bmod r$ and $0 \le l_1 < r$. We set $n := \hat{n} - (r - l_1) l_1 / 2r$. By the ADHM description, we have a sequence of closed immersions

$$M_0(r, n) \subset M_0(r, n+1) \subset M_0(r, n+2) \subset \cdots.$$

As in §F.3, we see that $\mathrm{im}(\pi \circ \beta) \subset M_0(r, n)$. Hence we have a desired morphism $\widehat{\pi}\colon \widehat{M}(r, k, \hat{n}) \to M_0(r, n)$. We want to show that $\widehat{\pi}$ is projective. By the construction of π, π is a projective morphism. Thus it is sufficient to prove that β is projective. For this purpose, we pick up some results in [**35**].

Let D be a smooth curve on a smooth surface X. Let (E, Ψ) be a pair of coherent sheaf E on X with $\operatorname{rank} E > 0$ and $\Psi\colon E \to \mathcal{O}_D^{\oplus r}$ a homomorphism. We take a polarization H on X.

DEFINITION F.17. A pair (E, Ψ) is semi-stable with respect to H and a polynomial $\delta(x) = \delta_1 x + \delta_2$, $\delta_1 > 0$ if the following two conditions are satisfied:

(1) $\operatorname{rank} E \chi\left(G(n)\right) \le \operatorname{rank} G\left(\chi\left(E(n)\right) - \delta(n)\right)$, $n \gg 0$ for all nontrivial submodules $G \subset \ker \Psi$,

(2) $\operatorname{rank} E\left(\chi\left(G(n)\right) - \delta(n)\right) \le \operatorname{rank} G\left(\chi\left(E(n)\right) - \delta(n)\right)$, $n \gg 0$ for all nontrivial submodules $G \subset E$.

(E, Ψ) is μ-semi-stable with respect to H and $\delta(x) = \delta_1 x + \delta_2$, $\delta_1 > 0$ if the following two conditions are satisfied:

(1) $\operatorname{rank} E \langle c_1(G), H \rangle \le \operatorname{rank} G(\langle c_1(E), H \rangle - \delta_1)$ for all nontrivial submodules $G \subset \ker \Psi$,

(2) $\operatorname{rank} E(\langle c_1(G), H \rangle - \delta_1) \le \operatorname{rank} G(\langle c_1(E), H \rangle - \delta_1)$ for all nontrivial submodules $G \subset E$ with $\operatorname{rank} G < \operatorname{rank} E$.

If the inequalities are strict, then (E, Ψ) is called stable (resp. μ-stable).

In general E is not torsion-free, even if (E, Ψ) is semi-stable.

DEFINITION F.18. A pair (E, Ψ) is torsion-free, if E is torsion-free.

In [35], Huybrechts and Lehn constructed the moduli space of semi-stable pairs, which is a projective scheme. Assume that $D = H$ and $\delta_1 \ll 1$. Let (E, Φ) be a framed sheaf, that is, E is a torsion free sheaf and $\Phi\colon E \to \mathcal{O}_D^{\oplus r}$ is a surjection. Then (E, Φ) is μ-stable. Hence we can regard the moduli space of semi-stable pairs as a compactification of the moduli space of framed sheaves. We can easily prove the following.

LEMMA F.19. *Assume that* $\delta_1 \ll 1$.

(1) *For a semi-stable pair* (E, Ψ), E *is a torsion-free sheaf.*
(2) *For all subsheaf* G *of* E,

$$\frac{\langle c_1(G), H\rangle}{\operatorname{rank} G} - \frac{\langle c_1(E), H\rangle}{\operatorname{rank} E} \neq -\frac{\delta_1}{\operatorname{rank} E}, -\frac{\delta_1}{\operatorname{rank} E} + \frac{\delta_1}{\operatorname{rank} G}.$$

Thus all torsion-free μ-semi-stable pair (E, Ψ) are μ-stable.

Let us consider the blowup $p\colon \widehat{\mathbb{P}}^2 \to \mathbb{P}^2$ and a pair (E, Ψ) on $\widehat{\mathbb{P}}^2$. The following claim is a consequence of Lemma F.19.

LEMMA F.20. *Assume that* $\delta_1 \ll 1$. *If we choose a sufficiently small* $\varepsilon > 0$ *which depends on* $\operatorname{ch}(E)$ *and* δ_1, *then* (E, Ψ) *is semi-stable with respect to* $H - \varepsilon C$ *and* δ *if and only if* $\big(p_*(E(-l_0 C)), \Psi\big)$ *is semi-stable with respect to* H *and* δ. *In particular, the moduli space of semi-stable pairs on* $\widehat{\mathbb{P}}^2$ *contains the moduli space of framed sheaves on* $\widehat{\mathbb{P}}^2$.

Thanks to this lemma, β extends to a morphism between moduli spaces of semi-stable pairs on $\widehat{\mathbb{P}}^2$ and $\mathbb{P}^2$. As these moduli spaces are projective, the extension of β is projective, and β is also projective. Thus we complete the proof of Theorem 3.3. □

We shall construct a similar stratification of $M(r,n)$ and $\widehat{M}(r,k,n)$ as in §F.4. In the same way as in §F.4, we define $Q, M(r,n)_m, \widehat{M}(r,k,n)_m, \dots$. By using Lemma F.19, we also see that α and $\widehat{\alpha}$ are proper, which implies that the arguments in §F.4 work for moduli spaces of framed sheaves. Thus we have the following theorem.

THEOREM F.21. *Let* $\mathfrak{Q}(n)$ *and* $\widehat{\mathfrak{Q}}(k,n)$ *be schemes in* (F.4) *and* (F.5).

(1) $M(r,n)$ *has a finite partition* $M(r,n) = \bigsqcup_m M(r,n)_m$ *into locally closed subsets.* $M(r,n)_m \to M(r,n-m)_0$ $((E,\Phi) \mapsto (i_*(E_{|\mathbb{P}^2\setminus\{P\}}), \Phi))$ *is a locally trivial fibration with a fiber* $\mathfrak{Q}(m)$.
(2) $\widehat{M}(r,k,\hat{n})$ *has a finite partition* $\widehat{M}(r,k,\hat{n}) = \bigsqcup_m \widehat{M}(r,k,\hat{n})_m$ *into locally closed subsets.* $\widehat{M}(r,k,\hat{n})_m \to M(r,\hat{n}-m)_0$ $((E,\Phi) \mapsto (i_*(E_{|\widehat{\mathbb{P}}^2\setminus C}), \Phi))$ *is a locally trivial fibration with a fiber* $\widehat{\mathfrak{Q}}(k,m)$.

Combining Theorem F.12 with Theorem F.21, we get the following.

COROLLARY F.22. *In the Grothendieck group of* $\mathbb{C}$*-varieties* $K_0(V_{\mathbb{C}})$,

$$\sum_n [M(r,c_1,n)]q^n = \Big(\sum_n [M(r,c_1,n)_0]q^n\Big)\Big(\sum_n [\mathfrak{Q}(n)]q^n\Big)$$

$$\sum_n [\widehat{M}(r,p^*c_1 + kC,n)]q^n = \Big(\sum_n [M(r,c_1,n)_0]q^n\Big)\Big(\sum_n [\widehat{\mathfrak{Q}}(k,n)]q^n\Big).$$

Similar relations hold for $M(r,n)$ *and* $\widehat{M}(r,k,n)$*. In particular,*

$$\frac{\sum_n[\widehat{M}(r,p^*c_1+kC,n)]q^n}{\sum_n[M(r,c_1,n)]q^n} = \frac{\sum_n[\widehat{M}(r,k,n)]q^n}{\sum_n[M(r,n)]q^n}.$$

Thus the ratio does not depend on the choice of (X,P)*.*

F.6. Base change theorem.

PROPOSITION F.23. *Let* X *be a* T*-scheme and* Y *a* T*-scheme flat over* T*. Let* $\pi\colon X \to Y$ *be a proper morphism over* T*. Let* F *be a coherent sheaf on* X *which is flat over* T*. For a point* $t \in T$*, if* $R^p\pi_*(F\otimes k(t)) = 0$ *for* $p \neq i$*, then there is an open neighborhood* U *of* t *such that* $R^i\pi_*(F_{|U\times_T Y})$ *is a* U*-flat coherent sheaf on* $U\times_T Y$ *and* $R^i\pi_*(F_{|U\times_T Y})\otimes k(t) \to R^i\pi_*(F\otimes k(t))$ *is an isomorphism for all* $t\in U$*.*

This proposition follows from the following lemma and a standard argument (see [**62**, §1]).

LEMMA F.24. *Keep the notations as above. Assume that* Y *and* T *are affine. We set* $Y := \operatorname{Spec}(B)$ *and* $T := \operatorname{Spec}(A)$*. Then, there is a complex* $K^\bullet$ *of finite* B*-modules which is flat over* A *and a functorial isomorphism* $H^i(X, F\otimes_A M) \cong H^i(K^\bullet \otimes_A M)$ *for all* i *and* A*-module* M*.*

PROOF. Let $C^\bullet$ be a Cech complex of F. By [**64**, 5, Lemma 1], we can choose a complex $K^\bullet$ consisting of finite B-modules K^p such that $K^p \neq 0$ only if $0 \le p \le n$, $n \in \mathbb{Z}$, K^p is free B-module if $1 \le p \le n$ and there is a quasi isomorphism $K^\bullet \to C^\bullet$. Since B is flat over A, K^p, $1\le p\le n$ are flat over A. Let $L^\bullet$ be the mapping cylinder. Then we see that $H^p(L^\bullet) = 0$, i.e., we have an exact sequence

$$0 \to K^0 \to K^1\oplus C^0 \to K^2 \oplus C^1 \to \cdots \to K^n \oplus C^{n-1} \to K^{n+1}\oplus C^n \to 0.$$

Since $K^p \oplus C^{p-1}$, $1\le p\le n+1$ are flat over A, K^0 is also flat over A. Then by [**64**, 5, Lemma 2], we get our claim. □

References

1. M. Aganagic, A. Klemm, M. Mariño, and C. Vafa, *The topological vertex*, hep-th/0305132.
2. P. S. Aspinwall and D. R. Morrison, *Topological field theory and rational curves*, Comm. Math. Phys. **151** (1993), no. 2, 245–262.
3. M. F. Atiyah and R. Bott, *The Yang-Mills equations over Riemann surfaces*, Philos. Trans. Roy. Soc. London Ser. A **308** (1982), no. 1505, 524–615.
4. ______, *The moment map and equivariant cohomology*, Topology **23** (1984), no. 1, 1–28
5. M. F. Atiyah and L. Jeffrey, *Topological Lagrangians and cohomology*, J. Geom. Phys. **7** (1990), no. 1, 119–136.
6. B. C. Berndt, *Ramanujan's notebooks. Part* III, Springer-Verlag, New York, 1991.
7. A. Białynicki-Birula, *Some theorems on actions of algebraic groups*, Ann. of Math. (2) **98** (1973), 480–497.
8. C. De Concini, G. Lusztig, and C. Procesi, *Homology of the zero-set of a nilpotent vector field on a flag manifold*, J. Amer. Math. Soc. **1** (1988), no. 1, 15–34.
9. V. I. Danilov and A. G. Kovanskiĭ, *Newton polyhedra and an algorithm for computing Hodge–Deligne numbers*, Izv. Akad. Nauk SSSR Ser. Mat. 50 (1986), no. 5, 925–945 (Russian); English transl., Math. USSR-Izv. **29** (1987), no. 2, 279–298.
10. E. D'Hoker, I. M. Krichever, and D. H. Phong, *The renormalization group equation in* $N=2$ *supersymmetric gauge theories*, Nuclear Phys. B **494** (1997), no. 1-2, 89–104.
11. ______, *Seiberg–Witten theory, symplectic forms, and Hamiltonian theory of solitons*, hep-th/0212313.

12. R. Y. Donagi, *Seiberg–Witten integrable systems*, Surveys in Differential Geometry: Integrable Systems, Surv. Differ. Geom., vol. 4, Int. Press, Boston, MA, 1998, pp. 83–129.
13. S. K. Donaldson, *Instantons and geometric invariant theory*, Comm. Math. Phys. **93** (1984), no. 4, 453–460.
14. S. K. Donaldson and P. B. Kronheimer, *The geometry of four-manifold*, Oxford Math. Monogr., Oxford Univ. Press, New York, 1990.
15. N. Dorey, T. H. Hollowood, V. V. Khoze, and M. P. Matis, *The calculus of many instantons*, Phys. Rep. **371** (2002), no. 4-5, 231–459.
16. J. D. Edelstein, M. Gómez-Reino, M. Mariño, *Blowup formulae in Donaldson–Witten theory and integrable hierarchies*, Adv. Theor. Math. Phys. **4** (2000), no. 3, 503–543.
17. J. D. Edelstein, M. Mariño, and J. Mas, *Whitham hierarchies, instanton corrections and soft supersymmetry breaking in $N = 2$ SU(N) super Yang–Mills theory*, Nuclear Phys. B **541** (1999), no. 3, 671–697.
18. D. Edidin and W. Graham, *Equivariant intersection theory*, Invent. Math. **131** (1998), no. 3, 595–634.
19. T. Eguchi and H. Kanno, *Topological strings and Nekrasov's formulas*, hep-th/0310235.
20. T. Eguchi and S.-K. Yang, *Prepotential of $N = 2$ supersymmetric gauge theories and solition equations*, Modern Phys. Lett. A **11** (1996), no. 2, 131–138.
21. C. Faber and R. Pandharipande, *Hodge integrals and Gromov-Witten theory*, Invent. Math. **139** (2000), no. 1, 173–199.
22. H. M. Farkas and I. Kra, *Riemann surfaces*, 2nd edition, Grad. Texts in Math., vol. 71, Springer-Verlag, New York, 1992.
23. J. D. Fay, *Theta functions on Riemann surfaces*, Lecture Notes in Math., vol. 352, Springer-Verlag, Berlin–New York, 1973.
24. R. Fintushel and R. J. Stern, *The blowup formula for Donaldson invariants*, Ann. of Math. (2) **143** (1996), no. 3, 529–546.
25. D. S. Freed, *Special Kähler manifolds*, Comm. Math. Phys. **203** (1999), no. 1, 31–52.
26. R. Friedman and J. W. Morgan, *Smooth four-manifolds and complex surfaces*, Ergeb. Math. Grenzgeb. (3), vol. 27, Springer-Verlag, Berlin, 1994.
27. W. Fulton, *Young tableaux*, London Math. Soc. Stud. Texts, vol. 35, Cambridge Univ. Press, Cambridge, 1997.
28. R. Gopakumar and C. Vafa, *On the gauge theory/geometry correspondence*, Adv. Theor. Math. Phys. **3** (1999), no. 5, 1415–1443.
29. A. Gorsky, A. Marshakov, A. Mironov, and A. Morozov, *RG equations from Whitham hierarchy*, Nuclear Phys. B **527** (1998), no. 3, 690–716; hep-th/9802007.
30. L. Göttsche, *Modular forms and Donaldson invariants for 4-manifolds with $b_+ = 1$*, J. Amer. Math. Soc. **9** (1996), no. 3, 827–843.
31. ———, *On the motive of the Hilbert scheme of points on a surface*, Math. Res. Lett. **8** (2001), no. 5-6, 613–627.
32. T. Graber and R. Pandharipande, *Localization of virtual classes*, Invent. Math. **135** (1999), no. 2, 487–518.
33. P. Griffiths and J. Harris, *Principles of algebraic geometry*, Pure Appl. Math. Wiley-Intersci., New York, 1978.
34. M. Haiman, *Vanishing theorems and character formulas for the Hilbert scheme of points in the plane*, Invent. Math. **149** (2002), no. 2, 371–407.
35. D. Huybrechts and M. Lehn, *Stable pairs on curves and surfaces*, J. Algebraic Geom. **4** (1995), no. 1, 67–104.
36. ———, *The geometry of moduli spaces of sheaves*, Aspects Math., vol. E31, Friedr. Vieweg & Sohn, Braunschweig, 1997.
37. A. Iqbal and A.K. Kashani-Poor, *Instanton counting and Chern–Simons theory*, hep-th/0212279.
38. ———, *SU(N) geometries and topological string amplitudes*, hep-th/0306032.
39. S. Katz, A. Klemm, and C. Vafa, *Geometric engineering of quantum field theories*, Nuclear Phys. B **497** (1997), no. 1-2, 173–195.
40. P. B. Kronheimer and T. S. Mrowka, *Recurrence relations and asymptotics for four-manifold invariants*, Bull. Amer. Math. Soc. (N.S.) **30** (1994), no. 2, 215–221.
41. ———, *Embedded surfaces and the structure of Donaldson's polynomial invariants*, J. Differential Geom. **41** (1995), no. 3, 573–734.

42. S. Lang, *Complex analysis*, 4th edition, Grad. Texts in Math., vol. 103, Springer-Verlag, New York, 1999.
43. J. Li, *Algebraic geometric interpretation of Donaldson's polynomial invariants*, J. Differential Geom. **37** (1993), no. 2, 417–466.
44. ———, *Compactification of moduli of vector bundles over algebraic surfaces*, Collection of Papers on Geometry, Analysis and Mathematical Physics, World Sci. Publishing, River Edge, NJ, 1997, pp. 98–113.
45. W.-P. Li and Z. Qin, *On blowup formulae for the S-duality conjecture of Vafa and Witten. II. The universal functions*, Math. Res. Lett. **5** (1998), no. 4, 439–453.
46. ———, *Vertex operator algebras and the blowup formula for the S-duality conjecture of Vafa and Witten*, Math. Res. Lett. **5** (1998), no. 6, 791–798.
47. ———, *On blowup formulae for the S-duality conjecture of Vafa and Witten*, Invent. Math. **136** (1999), no. 2, 451–482.
48. W.-P. Li, Z. Qin, and W. Wang, *Hilbert schemes, integrable hierarchies, and Gromov-Witten theory*, math.AG/0302211.
49. A. Losev, A. Marshakov, and N. Nekrasov, *Small instantons, little strings and free fermions*, hep-th/0302191.
50. A. Losev, N. Nekrasov and S. Shatashvili, *Issues in topological gauge theory*, Nuclear Phys. B **534** (1998), no. 3, 549–611.
51. ———, *Testing Seiberg-Witten solution*, L. D. Faddeev's Seminar on Mathematical Physics, Amer. Math. Soc. Transl. Ser. 2, vol. 201, Amer. Math. Soc., Providence, RI, 2000, pp. 123–135.
52. G. Lusztig, *Cuspidal local systems and graded Hecke algebras*. I, Inst. Hautes Études Sci. Publ. Math. **67** (1988), 145–202.
53. ———, *On quiver varieties*, Adv. Math. **136** (1998), no. 1, 141–182.
54. Yu. I. Manin, *Generating functions in algebraic geometry and sums over trees*, The Moduli Space of Curves (Texel Island, 1994), Birkhäuser, Boston, MA, 1995, pp. 401–417.
55. M. Mariño, *The uses of Whitham hierarchies*, Progr. Theoret. Phys. Suppl. **135** (1999), 29–52.
56. M. Mariño and G. Moore, *The Donaldson-Witten function for gauge groups of rank larger than one*, Comm. Math. Phys. **199** (1998), no. 1, 25–69.
57. A. Marshakov, *Seiberg-Witten theory and integrable systems*, World Sci. Publishing, River Edge, NJ, 1999.
58. M. Matone, *Instantons and recursion relations in $N = 2$ SUSY gauge theory*, Phys. Lett. B **357** (1995), no. 3, 342–348.
59. G. Moore and E. Witten *Integration over the u-plane in Donaldson theory*, Adv. Theor. Math. Phys. **1** (1997), no. 2, 298–387.
60. J. W. Morgan, *Comparison of the Donaldson polynomial invariants with their algebro-geometric analogues*, Topology **32** (1993), no. 3, 449–488.
61. H. R. Morton and S. G. Lukac, *The Homfly polynomial of the decorated Hopf link*, J. Knot Theory Ramifications **12** (2003), no. 3, 395–416.
62. S. Mukai, *Fourier functor and its application to the moduli of bundles on an abelian variety*, Algebraic Geometry (Sendai, 1985), Adv. Stud. Pure Math., vol. 10, North-Holland, Amsterdam, 1987, pp. 515–550.
63. D. Mumford, *Tata lectures on theta*. I, Progr. Math., vol. 28, Birkhäuser, Boston, MA, 1983; II. *Jacobian theta functions and differential equations*, vol. 43, 1984.
64. ———, *Abelian varieties*, Tata Inst. Fund. Res. Stud. Math., vol. 5, Oxford Univ. Press, London, 1970.
65. H. Nakajima, *Instantons on ALE spaces, quiver varieties, and Kac-Moody algebras*, Duke Math. J. **76** (1994), no. 2, 365–416.
66. ———, *Resolutions of moduli spaces of ideal instantons on $\mathbb{R}^4$*, Topology, Geometry and Field Theory, World Sci. Publishing, River Edge, NJ, 1994, pp. 129–136.
67. ———, *Quiver varieties and Kac-Moody algebras*, Duke Math. J. **91** (1998), no. 3, 515–560.
68. ———, *Lectures on Hilbert schemes of points on surfaces*, Univ. Lecture Ser., vol. 18, Amer. Math. Soc., Providence, RI, 1999.
69. H. Nakajima and K. Yoshioka, *Instanton counting on blowup*. I, math.AG/0306198.
70. N. A. Nekrasov, *Seiberg-Witten prepotential from instanton counting*, Proceedings of the International Congress of Mathematicians, Vol. III (Beijing, 2002), Higher Ed. Press, Beijing, 2002, pp. 477–495.

71. N. A. Nekrasov and A. Okounkov, *Seiberg–Witten prepotential and random partitions*, hep-th/0306238.
72. A. Okounkov and R. Pandharipande, *Gromov–Witten theory, Hurwitz numbers, and completed cycles*, math.AG/0204305.
73. ———, *Hodge integrals and invariants of the unknot*, math.AG/0307209.
74. N. Seiberg and E. Witten, *Electric-magnetic duality, monopole condensation, and confinement in $N = 2$ supersymmetric Yang-Mills theory* Nuclear Phys. B **426** (1994), no. 1, 19–52; Erratum, Nuclear Phys. B **430** (1994), no. 2, 485–486.
75. P. Slodowy, *Four lectures on simple groups and singularities*, Commun. Math. Inst. Rijksuniv. Utrecht, vol. 11, Mathematical Institute, Rijksuniversiteit Utrecht, Utrecht, 1980.
76. J. Sonnenschein, S. Theisen and S. Yankielowicz, *On the relation between the holomorphic prepotential and the quantum moduli in SUSY gauge theories*, Phys.Lett. B **367** (1996), no. 1-4, 145–150.
77. C. Voisin, *A mathematical proof of a formula of Aspinwall and Morrison*, Compositio Math. **104** (1996), no. 2, 135–151.
78. E. Witten, *Topological quantum field theory*, Comm. Math. Phys. **117** (1988), no. 3, 353–386.
79. ———, *Monopoles and four-manifolds*, Math. Res. Lett. **1** (1994), no. 6, 769–796.
80. K. Yoshioka, *Chamber structure of polarizations and the moduli of stable sheaves on a ruled surface*, Internat. J. Math. **7** (1996), no. 3, 411–431.
81. J. Zhou, *Localizations on moduli spaces and free field realizations of Feynman rules*, math.AG/0310283.

Department of Mathematics, Kyoto University, Kyoto 606-8502, Japan
E-mail address: nakajima@math.kyoto-u.ac.jp

Department of Mathematics, Faculty of Science, Kobe University, Kobe 657-8501, Japan
E-mail address: yoshioka@math.kobe-u.ac.jp

Centre de Recherches Mathématiques
CRM Proceedings and Lecture Notes
Volume **38**, 2004

Hyper-Kähler Nahm Transforms

Claudio Bartocci and Marcos Jardim

Abstract. Given two hyper-Kähler manifolds M and N and a quaternionic instanton on their product, a hyper-Kähler Nahm transform can be defined, which maps quaternionic instantons on M to quaternionic instantons on N. This construction includes the case of Nahm transform for periodic instantons on $\mathbb{R}^4$, the Fourier–Mukai transform for instantons on K3 surfaces, as well as the Nahm transform for ALE instantons.

1. Introduction

The Nahm transform is an instance of "geometrical Fourier transform," originally introduced by W. Nahm [**18**] as an extension of the ADHM method to construct time-invariant instantons (alias *monopoles*) on $\mathbb{R}^4$. This idea was subsequently applied, by P. Braam and P. van Baal [**8**], to study instantons on flat 4-tori. Nowadays, it is clearly understood that these constructions are special examples of a much wider framework, which provides an unified description of correspondences between solutions of the anti-self-duality equations that are invariant under dual subgroups of translations of $\mathbb{R}^4$ (a detailed exposition can be found in [**14**]).

An alternative approach to the Nahm transform is provided by the so-called Fourier–Mukai transform in algebraic geometry. In this case one obtains correspondences between moduli spaces of stable sheaves over abelian or K3 surfaces [**3**].

Yet another generalization of the Nahm transform can be defined for instantons over asymptotically locally Euclidean (ALE) 4-manifolds [**5**]. ALE spaces are diffeomorphic to minimal resolutions of $\mathbb{C}^2/\Gamma$, where Γ is a finite subgroup of SU(2), and are endowed with complete hyper-Kähler Riemannian metric. The moduli spaces of instantons over ALE spaces were thoroughly studied by Nakajima [**19**] and by Kronheimer and Nakajima [**16**]. Contrary to the previous examples, the Nahm transform on ALE spaces may fail to be invertible.

It turns out that all these various species of Nahm transforms can be adequately described in the setting of hyper-Kähler geometry. Actually, for hyper-Kähler manifolds of any dimension, it is possible to generalize the notion of instanton by defining the so-called quaternionic instantons [**17, 20**]. For these objects, one can prove a generalized Ward correspondence, which relates quaternionic instantons on M to some holomorphic vector bundle on the twistor space of M. Now, given two

2000 *Mathematics Subject Classification.* Primary 53C07; Secondary 56R10, 14D21.
This is the final form of the paper.

hyper-Kähler manifolds M and N and a quaternionic instanton Q on $M \times N$, a hyper-Kähler Nahm transform is obtained roughly according to the following recipe: take a quaternionic instanton on M pull it back to $M \times N$, twist it by Q, and push it down to N. The existence of the hyper-Kähler Ward correspondence ensures (under mild hypotheses) that the result of these operations is a quaternionic instanton on N.

In this expository paper we intend to review, from the unifying hyper-Kähler viewpoint, a number of results on Nahm transforms. Section 2 is devoted to the description of the Ward correspondence for quaternionic instantons. In Sections 3 and 4 we present the hyper-Kähler Nahm transform, discussing a few fundamental examples which point to an interesting new conjecture.

2. Hyper-Kähler Ward correspondence

We shall briefly describe the correspondence between quaternionic instantons on a hyper-Kähler manifold and certain holomorphic bundles over its twistor space (for a more general treatment valid for quaternionic Kähler manifolds see [**4, 17**]).

Let (M,g) be a (possibly noncompact) $4n$-dimensional hyper-Kähler manifold (i.e., the holonomy group H of the Riemannian metric g is contained in $\mathrm{Sp}(n)$). We denote by $\{I_k\}_{k=1,2,3}$ the triple of complex structures and by $\{\omega_k\}_{k=1,2,3}$ the triple of Kähler structures. We recall that the complex nondegenerate 2-form $\Omega = \omega_2 + i\omega_3$ is holomorphic with respect to I_1. The essential tool in the study of hyper-Kähler geometry is the twistor space, which encodes the information about the $\mathbb{CP}^1$ family of complex structures on the hyper-Kähler manifold. The twistor space Z of M is the sphere bundle associated with the rank 3 vector bundle $P \times_H \mathfrak{sp}(1)$, where P is the holonomy bundle of M; we denote by $p : Z \to M$ and by $q : Z \to \mathbb{CP}^1$ the natural projections. On Z there is a complex structure (different from the product structure), and an anti-holomorphic involution τ, preserving the fibers of p, the so-called twistor lines. The differential of τ acts in the following way:

$$d\tau(v,z) = \big(\bar{v}, d\imath(z)\big) \quad \forall (v,z) \in T_u Z = T_{p(u)}M \oplus T_{q(u)}\mathbb{CP}^1,$$

where $\imath \colon \mathbb{CP}^1 \to \mathbb{CP}^1$ is the antipodal map.

For any point $x \in M$, the endomorphism $\Phi = \sum_{k=1}^3 I_k \otimes I_k$ of $\Lambda^2 T_x^* M$ satisfies the identity $\Phi^2 = 2\Phi + 3\,\mathrm{Id}$, so it has eigenvalues 3 and -1. The eigenspace $\mathfrak{E}_x$ corresponding to the eigenvalue 3 is isomorphic to $\mathfrak{sp}(n)$ and can be characterized as

$$\mathfrak{E}_x = \bigcup_{u \in p^{-1}(x)} \Lambda_u^{1,1} T_x^* M,$$

where $\Lambda_u^{1,1} T_x^* M$ is the space of the 2-forms of type $(1,1)$ with respect to the complex structure associated to the point $q(u) \in \mathbb{CP}^1$.

A *quaternionic instanton* on M is a pair (E, ∇), where E is a complex smooth vector bundle on M and ∇ a connection whose curvature at any point $x \in M$ takes values in $\mathfrak{E}_x \otimes \mathrm{End}\, E_x$. When $\dim M = 4$, this definition corresponds to the usual one, since anti-self-dual connections are precisely the 2-forms which are of type $(1,1)$ with respect to all complex structures compatible with the hyper-Kähler metric [**11**].

Let (E, ∇) be a quaternionic instanton on M such that the connection ∇ is compatible with a given Hermitian metric h on E. It is easy to check that the pull-back connection $p^*\nabla$ induces a complex structure on the pull-back bundle

$F = p^*E$ endowed with the pull-back Hermitian metric p^*h. Moreover, we can define a *positive real form* $\sigma : F \to F^*$, i.e., an antilinear anti-holomorphic bundle isomorphism covering the involution $\tau\colon Z \to Z$. We set

$$\sigma(v)(w) = p^*h\big(v, \bar{\tau}(w)\big),$$

where $\bar{\tau}\colon F \to F$ is the lifting of τ to F. The following generalized Ward correspondence holds [**4, 20**].

THEOREM 1. *Gauge equivalence classes of rank k Hermitian quaternionic instantons on the hyper-Kähler manifold M are in one-to-one correspondence with isomorphism classes of rank k holomorphic bundles on the twistor space Z of M, which are holomorphically trivial along the fibers of $p\colon Z \to M$ and carry a positive real form.*

REMARK. When M is compact, irreducible quaternionic $\mathrm{SU}(n)$-instantons correspond to holomorphic bundles which are μ-stable of degree zero with respect to any Kähler structure induced on M by the hyper-Kähler structure [**21**]. In particular, a line bundle fails to be a quaternionic instantons if its first Chern class is not orthogonal to the cohomology classes of the Kähler forms $\{\omega_k\}_{k=1,2,3}$.

3. The Hyper-Kähler Nahm transform

Let us now suppose we are given the following set of data: two hyper-Kähler manifolds M, N and an Hermitian quaternionic instanton (Q, ∇_Q) on the product $M \times N$. Let $(Q_\xi, \nabla_{|Q_\xi})$ denote the restriction of (Q, ∇_Q) to $M \times \{\xi\}$, for any $\xi \in N$. If (E, ∇) is a vector bundle with connection on M, we can define the family of connections

$$\nabla_\xi = \nabla \otimes \mathbf{1}_{Q_\xi} + \mathbf{1}_E \otimes \nabla_{|Q_\xi}$$

on the bundle family $E \otimes Q_\xi$, for any $\xi \in N$.

The twistor space $Z_{M\times N}$ of $M \times N$ is the fiber product along $\mathbb{CP}^1$ of the twistor spaces Z_M and Z_N of M and N. So, we have the commutative diagram:

$$\begin{array}{ccccc} Z_M & \xleftarrow{\ \rho_M\ } & Z_{M\times N} & \xrightarrow{\ \rho_N\ } & Z_N \\ {\scriptstyle p_M}\downarrow & & \downarrow{\scriptstyle q} & & \downarrow{\scriptstyle p_N} \\ M & \xleftarrow[\ \pi_M\ ]{} & M \times N & \xrightarrow[\ \pi_N\ ]{} & N \end{array} \tag{3.1}$$

where the horizontal arrows are holomorphic maps while the vertical arrows are only smooth maps. By our assumptions, the pull-back of (Q, ∇_Q) to $Z_{M\times N}$ is a holomorphic bundle, which is holomorphically trivial along the fibers of $p\colon Z \to M$ and carries a positive real form.

Any hyper-Kähler manifold carries a standard spin structure; the spinor bundles $S^\pm$ are trivial and the projectivization of S^+ coincides with the twistor space. By coupling the connection ∇_ξ with the Dirac operators on $S^\pm$, we get the family of Dirac operators:

$$D^+_\xi : L^2_{k+1}(E \otimes Q_\xi \otimes S^+) \to L^2_k(E \otimes Q_\xi \otimes S^-)$$

and

$$D^-_\xi : L^2_k(E \otimes Q_\xi \otimes S^-) \to L^2_{k-1}(E \otimes Q_\xi \otimes S^+),$$

parametrized by the points of N. If M is noncompact, one has to be more careful, and consider suitable weighted Sobolev space completions of the spaces of C^∞

sections in order to guarantee that the Dirac operators $D_\xi^\pm$ are Fredholm, but we do not want to enter in these details here (for an example, see Section 4.3 below).

THEOREM 2. *Let (E,∇) be an Hermitian quaternionic instanton on M and assume that the index $\widehat{E} = \operatorname{Ind}(D^\pm)$ of the family of Dirac operators $D_\xi^\pm$ is a smooth vector bundle on N. Then $\widehat{E} \to N$ admits an Hermitian quaternionic instanton $\widehat{\nabla}$.*

PROOF. The index bundle $\widehat{E}$ is endowed with the index connection $\widehat{\nabla}$, obtained by projecting the trivial connection on $L_k^2(E \otimes Q_\xi \otimes S^\pm)$ down to $\ker D_\xi^\pm$. It is easy to check that the pull-back of $(\operatorname{Ind}(D^\pm), \widehat{\nabla})$ to Z_N is a holomorphic bundle holomorphically trivial along the fibers of $p\colon Z \to M$ and carrying a positive real form. □

Alternatively, when M and N are compact, we can use the machinery of higher direct images of coherent sheaves to reformulate the index construction above. Let us fix a complex structure I_z on M; we denote by M_z the resulting complex manifold. If (E,∇) is quaternionic instanton, we denote by $\mathcal{E}_z$ its associated sheaf of holomorphic sections with respect to the given complex structure. We say that (E,∇) satisfies the odd (resp. even) IT condition if

$$H^k(M, \mathcal{E}_z \otimes \mathcal{Q}_\xi) = 0 \quad \text{for } k \text{ even (resp. odd) and all } \xi \in N.$$

The definition is well-posed, since the cohomology groups $H^k(M, \mathcal{E}_z \otimes \mathcal{Q}_\xi)$ do not depend on the choice of the complex structure [**21**]. Consequently, we shall drop the subscript z in what follows.

By general results (see [**2**]), if (E,∇) satisfies the odd IT condition, then one has the identification of holomorphic vector bundles

$$-\operatorname{Ind} D^+ = \bigoplus_k R^k\pi_{N*}(\pi_M^*\mathcal{E} \otimes \mathcal{Q})$$

(when the even IT condition is satisfied, one takes D^- instead). Let $\widetilde{\mathcal{E}}$ and $\widetilde{\mathcal{Q}}$ denote the sheaves of holomorphic sections of the pull-back bundles p_M^*E and q^*Q, respectively. It can be proved (being careful, since the vertical maps of the diagram (3.1) are not holomorphic maps) that the pull-back holomorphic bundle $p_N^*(-\operatorname{Ind} D^+)$ coincides with $\bigoplus_k R^k\rho_{N*}(\rho_M^*\widetilde{\mathcal{E}} \otimes \widetilde{\mathcal{Q}})$. Summing up, we get the following result.

THEOREM-DEFINITION 3. *Let M, N be two compact hyper-Kähler manifolds, and let (Q,∇_Q) be an Hermitian quaternionic instanton over the product $M \times N$. Assume that (E,∇) is an Hermitian quaternionic instanton on M satisfying the odd IT condition. Then, $\bigoplus_k R^k\pi_{N*}(\pi_M^*\mathcal{E}\otimes\mathcal{Q})$ is a quaternionic instanton on M, which is called the hyper-Kähler Nahm transform of (E,∇) and is denoted by $(\widehat{E},\widehat{\nabla})$.*

REMARK. Given the Hermitian quaternionic instanton (Q,∇_Q), the correspondence between quaternionic instantons on M and quaternionic instantons on N is not in general invertible. In particular, the hyper-Kähler Nahm transform of an irreducible quaternionic instanton may fail to be irreducible.

4. Hyper-Kähler Nahm transforms for instantons

Let M be a 4-dimensional hyper-Kähler manifold (not necessarily compact) and let N be a connected component of the moduli of (irreducible) instantons on N (with some "framing" at infinity if M is noncompact); i.e., each point $\xi \in N$ can be regarded as an irreducible anti-self-dual connection ∇_ξ on a fixed vector bundle $E \to M$. Then N is also a hyper-Kähler manifold.

Recall that one can define a *universal bundle with connection* over the product $M \times N$ in the following way [**1, 13**]. Let $\mathcal{A}$ denote the set of all connections on E, and let $\mathcal{G}$ denote the group of gauge transformations (i.e., bundle automorphisms). Again, if M is noncompact, one must use the right analytical framework to define $\mathcal{A}$ and $\mathcal{G}$, see Section 4.3 below. Moreover, let G denote the structure group of E, so that E can be associated with a principal G_E-bundle P over M by means of some representation $\rho\colon G \to \mathbb{C}^n$, where $n = \operatorname{rank} E$. The gauge group $\mathcal{G}$ acts on $E \times \mathcal{A}$ by $g(p, A) = (g(p), g(A))$; This action has no fixed points, and it yields a principal $\mathcal{G}$-bundle $E \times \mathcal{A} \to \mathcal{Q}$, where $\mathcal{Q} = E \times \mathcal{A}/\mathcal{G}$.

The structure group G also acts on $E \times \mathcal{A}$, and since this action commutes with the one by $\mathcal{G}$, G acts on $\mathcal{Q}$. Moreover, the G-action on $\mathcal{Q}^* = E \times \mathcal{A}^*/\mathcal{G}$ has no fixed points, where $\mathcal{A}^*$ denotes the set of irreducible connections on F. We end up with a principal G bundle $\mathcal{Q}^* \to M \times (\mathcal{A}^*/\mathcal{G})$, and we denote by $\widetilde{\mathcal{P}}$ the associated vector bundle $\mathcal{Q}^* \times_\rho \mathbb{C}^n$. Since N is a submanifold of $\mathcal{A}^*/\mathcal{G}$, we define the *Poincaré bundle* $\mathcal{P} \to M \times N$ as the restriction of $\widetilde{\mathcal{P}}$.

The principal G-bundle $\mathcal{Q}^*$ has a natural connection $\widetilde{\omega}$, constructed as follows. The space $E \times \mathcal{A}^*$ has a Riemannian metric which is equivariant under $G \times \mathcal{G}$, so that it descends to a G-equivariant metric on $\mathcal{Q}^*$. The orthogonal complements to the orbits of G yields the connection $\widetilde{\omega}$. Passing to the associated vector bundle $\widetilde{\mathcal{P}}$ and restricting it to $M \times N$ gives a connection ω on the *Poincaré bundle* $\mathcal{P}$. The pair $(\mathcal{P}, \omega)$ is universal in the sense that $(\mathcal{P}, \omega)|_{M \times \{\xi\}} \simeq (E, \nabla_\xi)$.

THEOREM 4. *If M is 4-dimensional hyper-Kähler manifold and N is a connected component of the moduli of instantons on N, then the universal connection ω on the Poincaré bundle $\mathcal{P}$ is an Hermitian quaternionic instanton.*

PROOF. For any complex structure on M compatible with its hyper-Kähler metric, the curvature of the universal connection is of type $(1,1)$ with respect to induced complex structure on the $M \times N$ [**13**]. □

As usual, the Weitzenböck formula can be used to guarantee that $\ker D_\xi^+$ is trivial for all $\xi \in N$, so that the index bundle is indeed a smooth vector bundle. Hence, in view of Theorem 2, we may define a hyper-Kähler Nahm transform that maps instantons on M into quaternionic instantons on N. In particular, if N is also 4-dimensional, we get a Nahm transform taking instantons on M into instantons on N.

It is worth noting that if N is noncompact, then one must also do some extra work to determine whether the Nahm transformed quaternionic instanton has finite energy.

Furthermore, if M can also be regarded as a moduli space of (quaternionic) instantons on N, one can define a Nahm transform in the reverse direction, taking (quaternionic) instantons on N into instantons on M. It is then natural to ask whether such transforms are the inverse of one another. Based on the known examples of the hyper-Kähler Nahm transform (see below) we pose the following general conjecture.

CONJECTURE. *Let M and N be two connected 4-dimensional hyper-Kähler manifolds. If N is a component of the moduli of instantons on M and M is a component of the moduli of instantons on N, then:*

- *M is diffeomorphic to N;*
- *The hyper-Kähler Nahm transform is invertible.*

Whenever M and N are compact and algebraic with respect to at least one complex structure, then the second claim of the previous conjecture is actually true, in view of Theorem-Definition 3 and of [**9**, Theorem 1.1].

Let us now analyze some interesting applications of Theorem 4, mentioning a few examples in which the conjecture is expected to be true.

4.1. Algebraic tori. The simplest example of hyper-Kähler Nahm transform is for M being an algebraic torus of complex dimension $2k$ and N being its dual, regarded as a moduli space of flat connections on M; $\mathcal{P}$ is the usual Poincaré bundle. Then the hyper-Kähler Nahm transform is just the usual Fourier–Mukai transform.

In particular, for M being a flat 4-torus, the hyper-Kähler Nahm transform coincides with the usual Nahm transform for instantons on T^4 [**8, 11**], so that the conjecture is true in this case ($M = T^4$ and $N = (T^4)^*$).

4.2. K3 surfaces. The first example of a *nonflat* hyper-Kähler Nahm transform was described in [**3**]. Let M be an algebraic K3 surface, which meets the following requirements:

(1) M admits a Kähler form ω whose cohomology class H satisfies $H^2 = 2$;
(2) M admits a holomorphic line bundle L whose Chern class $\ell = c_1(L)$ is such that $\ell \cdot H = 0$ and $\ell^2 = -12$;
(3) If D is the divisor of a nodal curve on M, one has $D \cdot H > 2$.

We say that M is a reflexive K3 surface.

Now let N be the moduli space of instantons of rank 2 with determinant line bundle L (so that $c_1 = \ell$) and $c_2 = -1$ over M; it can be shown that N is isomorphic to M as a complex algebraic variety [**3**]. Moreover, any instanton E on M, whose dual is not a point of N, satisfies the odd IT condition. Thus, since both M and N are hyper-Kähler manifolds, we get the following result [**3**].

THEOREM 5. *Let (E, ∇) be an irreducible instanton on M such that at least one of the following conditions is satisfied:* $\operatorname{rank} E \neq 2$, $\det E \neq L^*$, $c_2(E) \neq 1$. *Then, the Nahm transform of (E, ∇) is an irreducible instanton on N having the same degree. Moreover, the correspondence is invertible.*

This proves the conjecture for M being reflexive K3 surface and N being the moduli space of instantons on M described above.

It should be pointed out that, also in this case, the hyper-Kähler Nahm transform coincides with Fourier–Mukai transform of coherent sheaves on M.

4.3. ALE spaces. Let Γ be a finite subgroup of $\mathrm{SU}(2)$. Let M be the minimal resolution of the quotient $\mathbb{C}^2/\Gamma$; it can be proved that M carries a hyper-Kähler metric g, which is *asymptotically locally Euclidean* (ALE) in the following sense. Some open neighborhood V of infinity in M has a finite covering $\widetilde{V}$ diffeomorphic to $\mathbb{R}^4 \setminus \overline{B(0,R)}$, for some $R > 0$, and in the induced coordinates x_i the metric g is required to satisfy the relation

$$g_{ij}(x) = \delta_{ij} + a_{ij} \quad \text{with } |\partial^p a_{ij}(x)| = O(|x|^{(-4-p)}),\ p \geq 0.$$

The moduli spaces of instantons over ALE spaces have been studied in detail by Nakajima [**19**], and by Kronheimer and Nakajima [**16**]. Let $E \to M$ be a complex (smooth) vector bundle of rank n and trivial determinant (i.e., an $\mathrm{SU}(n)$ vector bundle). In order to define a suitable notion of connections that are "framed at infinity," we fix a group homomorphism $\rho\colon \Gamma \to \mathrm{SU}(n)$; this homomorphism will be identified with a flat $\mathrm{SU}(n)$ connection over S^3/Γ. By taking coordinates x_i on $V \subset X$ as before, we can extend the function $r(p) = |x(p)|$ to a positive r function on all of M. Given a connection A_0, a weighted Sobolev norm $\|\cdot\|_{l,2,\delta}$ on the space of k-forms $\Omega^k(E)$ is defined as follows:

$$\|\alpha\|_{l,2,\delta} = \sum_{j=0}^{l} \|r^{j-(\delta+2)} \nabla^{(j)}_{A_0} \alpha\|_{L^2} \tag{4.1}$$

for an integer $l \geq 0$ and $\delta \in \mathbb{R}$. We denote the completion of the space $\Omega^k(E)$ in this norm by $W^{l,2}_{\delta}(E \otimes \Lambda^k T^* X)$.

Now fix $l > 2$. We say that a connection A on E is *asymptotic* to ρ in $W^{l,2}_{-2}$ if there is a gauge such that $A = A_0 + \alpha$, where the restriction of (E, A_0) to

$$\{t\} \times S^3/\Gamma \subset (R, \infty) \times S^3/\Gamma \simeq V$$

is the flat bundle with connection ρ, for all $t > R$ and

$$\|\alpha\|_{l,2,-2} < \infty.$$

We denote by $\mathcal{A}^l_X(\rho)$ the space of such connections.

The space of anti-self-dual connections on E asymptotic to ρ and having topological charge k is described as follows:

$$\mathcal{A}^l_{X,\mathrm{asd}}(E, \rho, k) = \left\{ A \in \mathcal{A}^l_X(\rho) \mid A \text{is anti-self-dual and } \frac{1}{8\pi^2} \int_X \|F_A\|^2 = k \right\}.$$

The corresponding moduli space is given by the quotient

$$\mathcal{M}_M(E, \rho, k) = \mathcal{A}^l_{X,\mathrm{asd}}(E, \rho, k)/\mathcal{G}^{l+1}_0$$

where $\mathcal{G}^{l+1}_0$ is the gauge group of automorphisms of E converging to the identity, that is:

$$\mathcal{G}^{l+1}_0 = \left\{ s \in W^{l+1,2}_{-1}(\mathrm{End}(E)) \;\middle|\; \|s - \mathbf{1}_E\|_{l+1,2,-1} < \infty \right\}.$$

The following fundamental result is due to Nakajima [**20**]:

THEOREM 6. *Each nonempty, noncompact 4-dimensional component of the moduli space $\mathcal{M}_M(E, \rho, k)$ is a complete hyper-Kähler ALE space.*

In other words, every such component of the moduli space $\mathcal{M}_M(E, \rho, k)$ is diffeomorphic to a minimal resolution of $\mathbb{C}^2/\widehat{\Gamma}$ for some discrete subgroup $\widehat{\Gamma} \subset \mathrm{SU}(2)$, which might be, in general, distinct from Γ. However, a complete classification (due to Nakajima) of all possible subgroups $\widehat{\Gamma}$ can be achieved [**6**]. Roughly speaking, Nakajima has shown that the Dynkin diagram associated with $\widehat{\Gamma}$ has to be a subgraph of the Dynkin diagram associated with Γ. As a consequence of this classification, one gets two important results (the latter was already proved in [**16**] by different methods).

THEOREM 7. *Let M and N be two ALE spaces. If N is diffeomorphic to a 4-dimensional component of the moduli space of instantons on M and the same holds for M with respect to N, then M is diffeomorphic to N.*

In other words, the first part of the conjecture is true.

THEOREM 8. *Let E be a rank 2 complex vector bundle over an ALE space M. Let $\iota\colon \Gamma \hookrightarrow \mathrm{SU}(2)$ be the inclusion map, with $|\Gamma|$ denoting the order of Γ. Then $\widehat{M} = \mathcal{M}_M(E, \iota, (|\Gamma| - 1)/|\Gamma|)$ is isomorphic to M as a hyper-Kähler manifold.*

Let us now consider the hyper-Kähler manifolds M, $N = \widehat{M}$ and an instanton (F, ∇) over M. The following facts are true:

(1) The universal Atiyah–Singer bundle (Q, ∇_Q) equipped with the universal connection is a quaternionic instanton by Theorem 4 and [**12**];
(2) The index $\operatorname{Ind} D^-$ of the family of Dirac operators parametrized by $\widehat{M}$ (cf. Section 3) is a finite-dimensional smooth, vector bundle on $\widehat{M}$, that we shall denote by $\widehat{F}$ [**16**];
(3) The transformed instanton $\widehat{\nabla}$ has finite energy [**5, 6**].

Therefore, hyper-Kähler Nahm transform takes instantons on M into instantons on $\widehat{M}$, and vice versa. Using the equivalent Fourier–Mukai formulation, one can expect to show that the transform is invertible, thus proving also the second part of the conjecture [**5, 6**].

4.4. Further perspectives. Let us now consider the case $M = \mathbb{T}^2 \times \mathbb{R}^2$; let (x, y) denote flat coordinates in $\mathbb{T}^2$, and (r, θ) denote polar coordinates in $\mathbb{R}^2$. As it was shown in [**7**], the moduli space $\mathcal{M}_{(k,\xi,\mu,\alpha)}$ of SU(2) instantons A on M which are asymptotic to

$$A_0 = d + i\begin{pmatrix} a_0 & 0 \\ 0 & -a_0 \end{pmatrix},$$

with

$$a_0 = \lambda_1 dx + \lambda_2 dy + (\mu_1 \cos\theta - \mu_2 \sin\theta)\frac{dx}{r} + (\mu_1 \sin\theta + \mu_2 \cos\theta)\frac{dy}{r} + \alpha d\theta$$

is a smooth hyper-Kähler manifold of real dimension $8k - 4$, where

$$k = \frac{1}{8\pi^2}\int_M |F_A|^2$$

is the instanton charge. In particular, setting $k = 1$, one also shows that $N = \mathcal{M}_{(1,\xi,\mu,\alpha)} \simeq \mathbb{T}^2 \times \mathbb{R}^2$ with the flat metric, whenever it is nonempty.

Thus there is a hyper-Kähler Nahm transform taking instantons on M into instantons on N; it would be interesting to determine whether the Nahm transformed instanton satisfy $|F_A| \sim O(r^{-2})$ (which is equivalent to the asymptotic condition above, see [**7**]).

Moreover, M can also be regarded as a moduli space of instantons on N, so there is a hyper-Kähler Nahm transform taking instantons on N into instantons on M, giving some evidence for the conjecture being true when $M = \mathbb{T}^2 \times \mathbb{R}^2$.

Finally, it would also be interesting to consider the hyper-Kähler Nahm transform for instantons over $S^1 \times \mathbb{R}^3$ (alias *calorons*) and $M = \mathbb{T}^3 \times \mathbb{R}$. Even more interesting would be to consider different gravitational instantons (other than ALE spaces), like asymptotically locally flat (ALF) 4-manifolds, or the ALG gravitational instantons constructed by Cherkis and Kapustin in [**10**]. The moduli spaces of instantons over these spaces, however, are much less understood.

References

1. M Atiyah and I. Singer, *Dirac operators coupled to vector potentials*, Proc. Nat. Acad. Sci. USA **81** (1984), no. 8, 2597–2600.
2. C. Bartocci, U. Bruzzo, and D. Hernández Ruipérez, *Fourier–Mukai transform and index theory*, Manuscripta Math. **85** (1994), no. 2, 141–163.
3. ———, *A Fourier–Mukai transform for stable bundles on K3 surfaces*, J. Reine. Angew. Math. **486** (1997), 1–16.
4. ———, *A hyper-Kähler Fourier transform*, Differential Geom. Appl. **8** (1998), no. 3, 239–249.
5. C. Bartocci and M. Jardim, *A Nahm transform for instantons over ALE spaces*, Clifford Algebras (R. Ablamowicz, ed.), Progr. Math. Phys., vol. 34, Birkhäuser, Boston, MA, 2004, pp. 155–166.
6. ———, *Nahm–Mukai transform for ALE instantons* (to appear).
7. O. Biquard and M. Jardim, *Asymptotic behaviour and the moduli space of doubly-periodic instantons*, J. Eur. Math. Soc. (JEMS) **3** (2001), no. 4, 335–375.
8. P. Braam and P. van Baal, *Nahm's transformations for instantons*, Comm. Math. Phys. **122** (1989), no. 2, 267–280.
9. T. Bridgeland, *Equivalences of triangulated categories and Fourier–Mukai transforms*, Bull. London Math. Soc. **31** (1999), no. 1, 25–34.
10. S. A. Cherkis and A. Kapustin, *Hyper-Kähler metrics from periodic monopoles*, Phys. Rev. D (3) **65** (2002), no. 8, 084015.
11. S. K. Donaldson and P. B. Kronheimer, *The geometry of four-manifolds*, Oxford Math. Monogr., Oxford University Press, New York, 1990.
12. T. Gocho and H. Nakajima, *Einstein–Hermitian connections on hyper-Kähler quotients*, J. Math. Soc. Japan **44** (1992), no. 1, 43–51.
13. M. Itoh, *Yang–Mills connections and the index bundles*, Tsukuba J. Math. **13** (1989), no. 2, 423–441.
14. M. Jardim, *A survey on Nahm transform*, J. Geom. Phys. (to appear); math.DG/0309305.
15. P. B. Kronheimer, *A Torelli-type theorem for gravitational instantons*, J. Differential Geom. **29** (1989), no. 3, 685–697.
16. P. B. Kronheimer and H. Nakajima, *Yang–Mills instantons on ALE gravitational instantons*, Math. Ann. **288** (1990), no. 2, 263–307.
17. M. Mamone Capria and S. M. Salamon, *Yang–Mills fields on quaternionic spaces*, Nonlinearity **1** (1988), no. 4, 517–530.
18. W. Nahm, *Self-dual monopoles and calorons*, Group Theoretical Methods in Physics (Trieste, 1993), Lecture Notes in Phys., vol. 201, Springer, Berlin, 1984, pp. 189–200.
19. H. Nakajima, *Moduli spaces of anti-self-dual connections on* ALE *gravitational instantons*, Invent. Math. **102** (1990), no. 2, 267–303.
20. T. Nitta, *Vector bundles over quaternionic Kähler manifolds*, Tohoku Math. J. (2) **40** (1988), no. 3, 425–440.
21. M. Verbitsky, *Hyperholomorphic bundles over a hyper-Kähler manifold*, J. Algebraic Geom. **5** (1966), no. 4, 633–669.

Dipartimento di Matematica, Universitá degli Studi di Genova, Via Dodecaneso, 35, 16146 Genova, Italy

E-mail address: bartocci@dima.unige.it

Department of Mathematics & Statistics, University of Massachusetts at Amherst, Amherst, MA 01003-9305, USA

E-mail address: jardim@math.umass.edu

Centre de Recherches Mathématiques
CRM Proceedings and Lecture Notes
Volume **38**, 2004

Instanton Counting via Affine Lie Algebras. I Equivariant J-Functions of (Affine) Flag Manifolds and Whittaker Vectors

Alexander Braverman

ABSTRACT. Let $\mathfrak{g}$ be a simple complex Lie algebra, G—the corresponding simply connected group; let also $\mathfrak{g}_{\text{aff}}$ be the corresponding untwisted affine Lie algebra. For a parabolic subgroup $P \subset G$ we introduce a generating function $\mathcal{Z}^{\text{aff}}_{G,P}$ which roughly speaking counts framed G-bundles on $\mathbb{P}^2$ endowed with a P-structure on the horizontal line (the formal definition uses the corresponding Uhlenbeck type compactifications studied in [**3**]). In the case $P = G$ the function $\mathcal{Z}^{\text{aff}}_{G,P}$ coincides with Nekrasov's partition function introduced in [**23**] and studied thoroughly in [**22, 24**] for $G = \mathrm{SL}(n)$. In the "opposite case" when P is a Borel subgroup of G we show that $\mathcal{Z}^{\text{aff}}_{G,P}$ is equal (roughly speaking) to the Whittaker matrix coefficient in the universal Verma module for the Lie algebra $\check{\mathfrak{g}}_{\text{aff}}$—the Langlands dual Lie algebra of $\mathfrak{g}_{\text{aff}}$. This clarifies somewhat the connection between certain asymptotics of $\mathcal{Z}^{\text{aff}}_{G,P}$ (studied in loc. cit. for $P = G$) and the classical affine Toda system. We also explain why the above result gives rise to a calculation of (not yet rigorously defined) equivariant quantum cohomology ring of the affine flag manifold associated with G. In particular, we reprove the results of [**13, 18**] about quantum cohomology of ordinary flag manifolds using methods which are totally different from loc. cit.

We shall show in a subsequent publication how this allows one to connect certain asymptotic of the function $\mathcal{Z}^{\text{aff}}_{G,P}$ with *the Seiberg–Witten prepotential* (cf. [**2**]), thus proving the main conjecture of [**23**] for an arbitrary gauge group G (for $G = \mathrm{SL}(n)$ it has been proved in [**22, 24**] by other methods).

1. Introduction

1.1. The partition function. This paper has grown out of a (still unsuccessful) attempt to understand the following object. Let K be a simply connected connected compact Lie group and let d be a nonnegative integer. Denote by $\mathcal{M}^d_K$ the moduli space of (framed) K-instantons on $\mathbb{R}^4$ of second Chern class $-d$. This space can naturally be embedded into a larger *Uhlenbeck space* $\mathcal{U}^d_K$. Both spaces admit a natural action of the group K (by changing the framing at ∞) and the torus $(S^1)^2$ acting on $\mathbb{R}^4$ after choosing an identification $\mathbb{R}^4 \simeq \mathbb{C}^2$. Moreover the

---

2000 *Mathematics Subject Classification.* Primary 14D21; Secondary: 17B80.

This research has been partially supported by the NSF.

This is the final form of the paper.

maximal torus of $K \times (S^1)^2$ has unique fixed point on $\mathcal{U}_K^d$. Thus we may consider (cf. [**22, 23**] and Section 2 for precise definitions) the *equivariant integral*

$$\int_{\mathcal{U}_K^d} 1^d$$

of the unit $K \times (S^1)^2$-equivariant cohomology class (which we denote by 1^d) over $\mathcal{U}_K^d$; the integral takes values in the field $\mathcal{K}$ which is the field of fractions of the algebra $\mathcal{A} = H^*_{K\times(S^1)^2}(\mathrm{pt})$.[1] Note that $\mathcal{A}$ is canonically isomorphic to the algebra of polynomial functions on $\mathfrak{k} \times \mathbb{R}^2$ (here $\mathfrak{k}$ denotes the Lie algebra of K) which are invariant with respect to the adjoint action of K on $\mathfrak{k}$. Thus each $\int_{\mathcal{U}_K^d} 1^d$ may naturally be regarded as a rational function of $a \in \mathfrak{k}$ and $(\varepsilon_1, \varepsilon_2) \in \mathbb{R}^2$.

Consider now the generating function

$$\mathcal{Z} = \sum_{d=0}^{\infty} Q^d \int_{\mathcal{U}_K^d} 1^d.$$

It can (and should) be thought of as a function of the variables $\mathfrak{q}$ and a, ε_1, ε_2 as before. In [**23**] it was conjectured that the first term of the asymptotic in the limit $\lim_{\varepsilon_1,\varepsilon_2\to 0} \ln \mathcal{Z}$ is closely related to *Seiberg–Witten prepotential* of K. For $K = \mathrm{SU}(n)$ this conjecture has been proved in [**22, 24**]. Also in [**23**] an explicit combinatorial expression for $\mathcal{Z}$ has been found.

1.2. Algebraic version. In this paper we want to generalize the definition of the function $\mathcal{Z}$ in several directions and compute it in some of these new cases. First of all, it will be convenient for us to make the whole situation completely algebraic.

Namely, let G be a complex semi-simple algebraic group whose maximal compact subgroup is isomorphic to K. We shall denote by $\mathfrak{g}$ its Lie algebra. Let also $\mathbf{S} = \mathbb{P}^2$ and denote by $\mathbf{D}_\infty \subset \mathbf{S}$ the "straight line at ∞"; thus $\mathbf{S}\backslash\mathbf{D}_\infty = \mathbb{A}^2$. It is well-known that $\mathcal{M}_K^d$ is isomorphic to the moduli space $\mathrm{Bun}_G^d(\mathbf{S}, \mathbf{D}_\infty)$ of principal G-bundles on $\mathbf{S}$ endowed with a trivialization on $\mathbf{D}_\infty$ of second Chern class $-d$. When it does not lead to a confusion we shall write Bun_G instead of $\mathrm{Bun}_G(\mathbf{S}, \mathbf{D}_\infty)$. The algebraic analog of $\mathcal{U}_K^d$ has been constructed in [**3**]; we denote this algebraic variety by $\mathcal{U}_G^d$. This variety is endowed with a natural action on $G \times (\mathbb{C}^*)^2$.

1.3. Parabolic generalization of the partition function. Let $\mathbf{C} \subset \mathbf{S}$ denote the "horizontal line." Choose a parabolic subgroup $P \subset G$. Let $\mathrm{Bun}_{G,P}$ denote the moduli space of the following objects:

(1) A principal G-bundle $\mathcal{F}_G$ on $\mathbf{S}$;
(2) A trivialization of $\mathcal{F}_G$ on $\mathbf{D}_\infty \subset \mathbf{S}$;
(3) A reduction of $\mathcal{F}_G$ to P on $\mathbf{C}$ compatible with the trivialization of $\mathcal{F}_G$ on $D_\infty \cap \mathbf{C}$.

Let us describe the connected components of $\mathrm{Bun}_{G,P}$. Let M be the Levi group of P. Denote by $\check{M}$ the *Langlands dual* group of M and let $Z(\check{M})$ be its center. We denote by $\Lambda_{G,P}$ the lattice of characters of $Z(\check{M})$. Let also $\Lambda_{G,P}^{\mathrm{aff}} = \Lambda_{G,P} \times \mathbb{Z}$ be the lattice of characters of $Z(\check{M}) \times \mathbb{C}^*$. Note that $\Lambda_{G,G}^{\mathrm{aff}} = \mathbb{Z}$.

[1] In this paper we always consider cohomology with complex coefficients

The lattice $\Lambda_{G,P}^{\text{aff}}$ contains a canonical semi-group $\Lambda_{G,P}^{\text{aff,pos}}$ of positive elements (cf. [3] and Section 3). It is not difficult to see that the connected components of $\text{Bun}_{G,P}$ are parameterized by the elements of $\Lambda_{G,P}^{\text{aff,pos}}$:

$$\text{Bun}_{G,P} = \bigcup_{\theta_{\text{aff}} \in \Lambda_{G,P}^{\text{aff,pos}}} \text{Bun}_{G,P}^{\theta_{\text{aff}}}.$$

Typically, for $\theta_{\text{aff}} \in \Lambda_{G,P}^{\text{aff}}$ we shall write $\theta_{\text{aff}} = (\theta, d)$ where $\theta \in \Lambda_{G,P}$ and $d \in \mathbb{Z}$.

Each $\text{Bun}_{G,P}^{\theta_{\text{aff}}}$ is naturally acted on by $P \times (\mathbb{C}^*)^2$; by embedding M into P we get an action of $M \times (\mathbb{C}^*)^2$ on $\text{Bun}_{G,P}^{\theta_{\text{aff}}}$. In [3] we define for each $\theta_{\text{aff}} \in \Lambda_{G,P}^{\text{aff,pos}}$ a certain Uhlenbeck scheme $\mathcal{U}_{G,P}^{\theta_{\text{aff}}}$ which contains $\text{Bun}_{G,P}^{\theta_{\text{aff}}}$ as a dense open subset. The scheme $\mathcal{U}_{G,P}^{\theta_{\text{aff}}}$ still admits an action of $M \times (\mathbb{C}^*)^2$.

We want to do some equivariant intersection theory on the spaces $\mathcal{U}_{G,P}^{\theta_{\text{aff}}}$. For this let us denote by $\mathcal{A}_{M\times(\mathbb{C}^*)^2}$ the algebra $H^*_{M\times(\mathbb{C}^*)^2}(\text{pt}, \mathbb{C})$. Of course this is just the algebra of M-invariant polynomials on $\mathfrak{m} \times \mathbb{C}^2$. Let also $\mathcal{K}_{M\times(\mathbb{C}^*)^2}$ be its field of fractions. We can think about elements of $\mathcal{K}_{M\times(\mathbb{C}^*)^2}$ as rational functions on $\mathfrak{m} \times \mathbb{C}^2$ which are invariant with respect to the adjoint action.

Let $T \subset M$ be a maximal torus. Then one can show that $(\mathcal{U}_{G,P}^{\theta_{\text{aff}}})^{T\times(\mathbb{C}^*)^2}$ consists of one point. This guarantees that we may consider the integral $\int_{\mathcal{U}_{G,P}^{\theta_{\text{aff}}}} 1_{G,P}^{\theta_{\text{aff}}}$ where $1_{G,P}^{\theta_{\text{aff}}}$ denotes the unit class in $H^*_{M\times(\mathbb{C}^*)^2}(\mathcal{U}_{G,P}^{\theta_{\text{aff}}}, \mathbb{C})$. The result can be thought of as a rational function on $\mathfrak{m} \times \mathbb{C}^2$ which is invariant with respect to the adjoint action of M. Define

$$\mathcal{Z}_{G,P}^{\text{aff}} = \sum_{\theta \in \Lambda_{G,P}^{\text{aff}}} \mathfrak{q}_{\text{aff}}^{\theta_{\text{aff}}} \int_{\mathcal{U}_{G,P}^{\theta_{\text{aff}}}} 1_{G,P}^{\theta_{\text{aff}}}. \tag{1.1}$$

One should think of $\mathcal{Z}_{G,P}^{\text{aff}}$ as a formal power series in $\mathfrak{q}_{\text{aff}} \in Z(\check{M}) \times \mathbb{C}^*$ with values in the space of ad-invariant rational functions on $\mathfrak{m} \times \mathbb{C}^2$. Typically, we shall write $\mathfrak{q}_{\text{aff}} = (\mathfrak{q}, Q)$ where $\mathfrak{q} \in Z(\check{M})$ and $Q \in \mathbb{C}^*$. Also we shall denote an element of $\mathfrak{m} \times \mathbb{C}^2$ by $(a, \varepsilon_1, \varepsilon_2)$ or (sometimes it will be more convenient) by $(a, \hbar, \varepsilon)$ (note that for general P (unlike in the case $P = G$) the function $\mathcal{Z}_{G,P}^{\text{aff}}$ is not symmetric with respect to switching ε_1 and ε_2).

1.4. Interpretation via maps and the "finite-dimensional" analog. Choose now another smooth projective curve $\mathbf{X}$ of genus 0 and with two marked points $0_{\mathbf{X}}$, $\infty_{\mathbf{X}}$. Choose also a coordinate x on $\mathbf{X}$ such that $x(0_{\mathbf{X}}) = 0$ and $x(\infty_{\mathbf{X}}) = 0$. Let us denote by $\mathcal{G}_{G,P,\mathbf{X}}$ the scheme classifying triples $(\mathcal{F}_G, \beta, \gamma)$, where

(1) $\mathcal{F}_G$ is a principal G-bundle on $\mathbf{X}$;
(2) β is a trivialization of $\mathcal{F}_G$ on the formal neighborhood of $\infty_{\mathbf{X}}$;
(3) γ is a reduction to P of the fiber of $\mathcal{F}_G$ at $0_{\mathbf{X}}$.

We shall usually omit $\mathbf{X}$ from the notations. We shall also write $\mathcal{G}_G^{\text{aff}}$ for $\mathcal{G}_{G,G}^{\text{aff}}$.

Let $e_{G,P}^{\text{aff}} \in \mathcal{G}_{G,P}^{\text{aff}}$ denote the point corresponding to the trivial $\mathcal{F}_G$ with the natural β and γ. It is explained in [3] that the variety $\text{Bun}_{G,P}$ is canonically isomorphic to the scheme classifying *based maps* from $(\mathbf{C}, \infty_{\mathbf{C}})$ to $(\mathcal{G}_{G,P}^{\text{aff}}, e_{G,P}^{\text{aff}})$ (i.e., maps from $\mathbf{C}$ to $\mathcal{G}_{G,P}$ sending $\infty_{\mathbf{C}}$ to $e_{G,P}^{\text{aff}}$).

The scheme $\mathcal{G}^{\mathrm{aff}}_{G,P}$ may (and should) be thought of as a partial flag variety for $\mathfrak{g}_{\mathrm{aff}}$. Thus we may consider the following "finite-dimensional" analog of the above problem: for G and P as above let $\mathcal{G}_{G,P}=G/P$. Let also $e_{G,P}\in\mathcal{G}_{G,P}$ denote the image of $e\in G$. Clearly $e_{G,P}$ is stable under the action of P on $\mathcal{G}_{G,P}$. Let ${}^{\mathrm{b}}\mathcal{M}_{G,P}$ denote the moduli space of based maps from $(\mathbf{C},\infty_{\mathbf{C}})$ to $(\mathcal{G}_{G,P},e_{G,P})$, i.e. the moduli space of maps $\mathbf{C}\to\mathcal{G}_{G,P}$ which send $\infty_{\mathbf{C}}$ to $e_{G,P}$. This space is acted on by the group $M\times\mathbb{C}^*$. Also ${}^{\mathrm{b}}\mathcal{M}_{G,P}$ is a union of connected components ${}^{\mathrm{b}}\mathcal{M}^{\theta}_{G,P}$ where θ lies in a certain sub-semi-group $\Lambda^{\mathrm{pos}}_{G,P}$ of $\Lambda_{G,P}$. For each θ as above one can also consider the space of *based quasi-maps* (or *Zastava space* in the terminology of [**4, 8, 11**]) which we denote by ${}^{\mathrm{b}}\mathcal{QM}^{\theta}_{G,P}$. This is an affine algebraic variety containing ${}^{\mathrm{b}}\mathcal{M}^{\theta}_{G,P}$ as a dense open subset. We also denote by $\mathcal{M}^{\theta}_{G,P},\mathcal{QM}^{\theta}_{G,P}$ the corresponding spaces of all (unbased) maps and quasi-maps.

We now introduce the "finite-dimensional" analog of the partition function (1.1). As before let

$$\mathcal{A}_{M\times\mathbb{C}^*}=H^*_{M\times\mathbb{C}^*}(\mathrm{pt},\mathbb{C})$$

and denote by $\mathcal{K}_{M\times\mathbb{C}^*}$ its field of fractions. Let also $1^{\theta}_{G,P}$ denote the unit class in the $M\times\mathbb{C}^*$-equivariant cohomology of ${}^{\mathrm{b}}\mathcal{QM}^{\theta}_{G,P}$. Then we define

$$\mathcal{Z}_{G,P}=\sum_{\theta\in\Lambda^{\theta}_{G,P}}\mathfrak{q}^{\theta}\int_{{}^{\mathrm{b}}\mathcal{QM}^{\theta}_{G,P}}1^{\theta}_{G,P}. \tag{1.2}$$

This is a formal series in $q\in Z(\check{M})$ with values in the field $\mathcal{K}_{M\times\mathbb{C}^*}$ of M-invariant rational functions on $\mathfrak{m}\times\mathbb{C}$.

In fact the function $\mathcal{Z}_{G,P}$ is a familiar object in Gromov–Witten theory: we shall see later (Theorem 1) that up to a simple factor $\mathcal{Z}_{G,P}$ is the so called *equivariant J-function* of $\mathcal{G}_{G,P}$ (cf. Section 1.6 for the details).

1.5. The Borel case. We believe that it should be possible to express the function $\mathcal{Z}_{G,P}$ (resp. the function $\mathcal{Z}^{\mathrm{aff}}_{G,P}$) in terms of representation theory of the Lie algebra $\check{\mathfrak{g}}$ (resp. $\check{\mathfrak{g}}_{\mathrm{aff}}$)—by the definition this is a Lie algebra whose root system is dual to that of $\mathfrak{g}$ (resp. to that of $\mathfrak{g}_{\mathrm{aff}}$). One of the motivations for this comes from the main results of [**4**] and [**3**] where such a description is found for the intersection cohomology of the varieties ${}^{\mathrm{b}}\mathcal{QM}^{\theta}_{G,P}$ and $\mathcal{U}^{\theta_{\mathrm{aff}}}_{G,P}$ (in this paper we are going to adopt an intersection cohomology approach to the partition functions $\mathcal{Z}_{G,P}$ and $\mathcal{Z}^{\mathrm{aff}}_{G,P}$; this is explained carefully in Section 2). One of the main results of this paper gives such a calculation of the functions $\mathcal{Z}_{G,B}$ and $\mathcal{Z}^{\mathrm{aff}}_{G,B}$ where $B\subset G$ is a Borel subgroup of G. Roughly speaking we show that $\mathcal{Z}_{G,B}$ (resp. $\mathcal{Z}^{\mathrm{aff}}_{G,B}$) is equal to the *Whittaker matrix coefficient* of the Verma module over $\check{\mathfrak{g}}$ (resp. over $\check{\mathfrak{g}}_{\mathrm{aff}}$) whose lowest weight given by $a/\hbar+\rho$ (resp. $(a,\varepsilon_1)/\varepsilon_2+\rho_{\mathrm{aff}}$ where a, $\hbar$, ε_1 and ε_2 are as in the previous subsection (here we regard (a,ε_1) as a weight for the dual affine algebra $\check{\mathfrak{g}}_{\mathrm{aff}}$; this is explained carefully in Section 3). The precise formulation of this result is given by Theorem 2 and Corollary 3.4.

As a corollary we get that the function $\mathfrak{q}^{a/\hbar}\mathcal{Z}_{G,B}$ is an eigenfunction of the *quantum Toda Hamiltonians* associated with $\check{\mathfrak{g}}$ with eigenvalues determined (in the natural way) by a (we refer the reader to [**7**] for the definition of (affine) Toda integrable system and its relation with Whittaker functions). In the affine case one can also show that $\mathfrak{q}^{a/\hbar}\mathcal{Z}^{\mathrm{aff}}_{G,B}$ is an eigenfunction of a certain differential operator which has order 2 ("nonstationary analog" of the affine quadratic Toda Hamiltonian). We shall show how this allows to compute the asymptotics of *all* the functions $\mathcal{Z}^{\mathrm{aff}}_{G,\Gamma}$ when $\varepsilon_1, \varepsilon_2 \to 0$ in another publication (cf. [**2**]).

Let us explain the idea of the proof of the above statement (we shall do it in the "finite-dimensional" (i.e., nonaffine) case; in the affine case the proof is similar). First for a scheme Y let $\mathrm{IH}^*(Y)$ denote the intersection cohomology of Y with complex coefficients (similarly if a group L acts on Y we denote by $\mathrm{IH}^*_L(Y)$ the corresponding L-equivariant intersection cohomology). Let now

$$\mathrm{IH}^\theta_{G,P} = \mathrm{IH}^*_{M\times(\mathbb{C}^*)^2}({}^{\mathrm{b}}\mathcal{QM}^\theta_{G,P}) \otimes_{\mathcal{A}_{M\times\mathbb{C}^*}} \mathcal{K}_{M\times\mathbb{C}^*}.$$

(it is clear that $\mathrm{IH}^*_{M\times(\mathbb{C}^*)^2}({}^{\mathrm{b}}\mathcal{QM}^\theta_{G,P})$ has a natural $\mathcal{A}_{M\times\mathbb{C}^*}$-module structure).

Let also

$$\mathrm{IH}_{G,P} = \bigoplus_{\theta\in\Lambda^{\mathrm{pos}}_{G,P}} \mathrm{IH}^\theta_{G,P}.$$

Each $\mathrm{IH}^\theta_{G,P}$ is a finite-dimensional vector space over $\mathcal{K}_{M\times\mathbb{C}^*}$ endowed with a (non-degenerate) Poincaré pairing $\langle\cdot,\cdot\rangle^\theta_{G,P}$ taking values in $\mathcal{K}_{M\times\mathbb{C}^*}$.

Let us now specialize to the case $P = B$. In Section 5 we show that the Lie algebra $\check{\mathfrak{g}}$ acts naturally on $\mathrm{IH}_{G,B}$. Moreover, this action has the following properties. First of all, let us denote by $\langle\cdot,\cdot\rangle_{G,B}$ the direct sum of the pairings $(-1)^{\langle\theta,\check{\rho}\rangle}\langle\cdot,\cdot\rangle^\theta_{G,B}$.[2]

Recall that the Lie algebra $\check{\mathfrak{g}}$ has its triangular decomposition $\check{\mathfrak{g}} = \check{\mathfrak{n}}_+ \oplus \check{\mathfrak{h}} \oplus \check{\mathfrak{n}}_-$. Let $\kappa\colon \check{\mathfrak{g}} \to \check{\mathfrak{g}}$ denote the Cartan anti-involution which interchanges $\check{\mathfrak{n}}_+$ and $\mathfrak{n}_-$ and acts as identity on $\check{\mathfrak{h}}$. For each $\lambda \in \mathfrak{h} = (\check{\mathfrak{h}})^*$ we denote by $M(\lambda)$ the corresponding Verma module with lowest weight λ; this is a module generated by a vector v_λ with (the only) relations

$$t(v_\lambda) = \lambda(t)v_\lambda \quad \text{for } t \in \check{\mathfrak{h}} \text{ and } \quad n(v_\lambda) = 0 \quad \text{for } n \in \check{\mathfrak{n}}_-.$$

Then:

(1) $\mathrm{IH}_{G,B}$ (with the above action) becomes isomorphic to $M(\lambda)$.
(2) $\mathrm{IH}^\theta_{G,B} \subset \mathrm{IH}_{G,B}$ is the $a/\hbar + \rho + \theta$-weight space of $\mathrm{IH}_{G,B}$.
(3) For each $g \in \check{\mathfrak{g}}$ and $v, w \in \mathrm{IH}_{G,B}$ we have

$$\langle g(v), w\rangle_{G,B} = \langle v, \kappa(g)w\rangle_{G,B}.$$

(4) The vector $\sum_\theta 1^\theta_{G,B}$ (lying is some completion of $\mathrm{IH}_{G,B}$) is a Whittaker vector (i.e., a $\mathfrak{n}_-$-eigenvector) for the above action.

It is easy to see that the assertions (1)–(4) imply the above representation-theoretic description of $\mathcal{Z}_{G,B}$ (for this one has to adopt an "intersection cohomology" point of view at the above equivariant integrals; this is explained in Section 2).

In fact, certain parts of the above statement have been known before: for example in [**8**] the authors compute the dimensions of the spaces $\mathrm{IH}^\theta_{G,B}$ and the answer agrees with (1) and (2) above (in fact that these dimensions are known even if we replace ${}^{\mathrm{b}}\mathcal{QM}^\theta_{G,B}$ by any ${}^{\mathrm{b}}\mathcal{QM}^\theta_{G,P}$—cf. [**4**]). Similarly the computation of dimensions of $\mathrm{IH}^{\theta_{\mathrm{aff}}}_{G,B}$ (defined analogously to $\mathrm{IH}^\theta_{G,B}$) follows from [**3**]. Also the construction of the $\check{\mathfrak{g}}$-action on $\mathrm{IH}_{G,B}$ is very close to the construction in Section 4 of [**8**].

---

[2] Here $\check{\rho}$ denotes the half-sum of the positive roots of $\mathfrak{g}$.

1.6. Connection with J-functions and quantum cohomology of flag manifolds. Let us explain how the above result is connected with the known results on quantum cohomology of flag manifolds.

First, let us recall the general set-up (the main reference is [**12**]). Let X be any smooth projective variety over $\mathbb{C}$. Assume that X is homogeneous, i.e. that the tangent sheaf of X is generated by its global sections. Then for any $\beta \in H_2(X,\mathbb{Z})$ one may consider the moduli space $\overline{\mathcal{M}}_{0,n}(X,\beta)$ of stable maps from a curve Σ of genus zero with n marked points to X of degree β. For every β as above let ev_β denote the evaluation map map at the marked point from $\overline{\mathcal{M}}_{0,1}(X,\beta)$ to X.

The space $\overline{\mathcal{M}}_{0,1}(X,\beta)$ admits canonical line bundle $\mathcal{L}_\beta$ whose fiber over every stable map $C \to X$ is equal to the cotangent space to C at the marked point. We denote by c_β its first Chern class.

In the above set-up one defines the J-function J^X of X in the following way:

$$J^X(t,\hbar) = e^{t/\hbar} \sum_{\beta \in H_2(X,\mathbb{Z})} e^{\langle t,\beta\rangle} (\mathrm{ev}_\beta)_* \left(\frac{1}{\hbar(c_\beta+\hbar)}\right)^3 \tag{1.3}$$

where $t \in H^2(X,\mathbb{C})$. The function J^X takes values in $H^*(X,\mathbb{C}) \otimes \mathbb{C}((\hbar^{-1}))$ which we may think of as a localization of a completion of of $H_{\mathbb{C}^*}(X,\mathbb{C})$ ($\mathbb{C}^*$ acts trivially on X) over $\mathbb{C}[\hbar] = H^*_{\mathbb{C}^*}(\mathrm{pt})$. It is explained in [**12**] that by looking at the differential equations satisfied by the function J^X one may compute explicitly the small quantum cohomology ring of X. Putting $q = e^t$ one may (formally) write

$$J^X = \mathfrak{q}^{1/\hbar} \sum_{\beta \in H_2(X,\mathbb{Z})} \mathfrak{q}^\beta (\mathrm{ev}_\beta)_* \left(\frac{1}{\hbar(c_\beta+\hbar)}\right).$$

In the case when X is acted on by a reductive group G we may consider the equivariant analog of J^X which we shall denote by J^X_G. In this case the function J^X_G takes values in $H^*_{G\times\mathbb{C}^*}(X,\mathbb{C}) \otimes_{\mathbb{C}[\hbar]} \mathbb{C}((\hbar^{-1}))$. By looking at the differential equations satisfied by J^X_G one may compute the equivariant small quantum cohomology ring of X.

Assume now that $X = \mathcal{G}_{G,P}$. Note that in this case we have the natural identification

$$H^*_{G\times\mathbb{C}^*}(\mathcal{G}_{G,P},\mathbb{C}) = H^*_{M\times\mathbb{C}^*}(\mathrm{pt})$$

Note also that $H_2(\mathcal{G}_{G,P},\mathbb{Z}) = \Lambda_{G,P}$ and $H^2(\mathcal{G}_{G,P}) = \mathrm{Lie}\, Z(\check{M})$.

Taking all these identifications into account we now claim the following

THEOREM 1. *For any G and P as above we have*

$$J_G^{\mathcal{G}_{G,P}} = \mathfrak{q}^{a/\hbar} \mathcal{Z}_{G,P}.$$

Theorem 1 is proved in Section 6.

Note that in view of the results announced in the previous section it follows that $J^G_{\mathcal{G}_{G,B}}$ can be interpreted via Whittaker matrix coefficients of Verma modules and in particular it is an eigenfunction of the Toda Hamiltonians for $\check{\mathfrak{g}}$. The latter result is due to B. Kim (cf. [**18**]); in fact (the "nonaffine" part of) this paper may be viewed as a conceptual explanation of Kim's result.

Theorem 1 also allows us to think about $\mathfrak{q}^{a/\hbar}\mathcal{Z}^{\mathrm{aff}}_{G,P}$ as the equivariant (with respect to the loop group) J-function of $\mathcal{G}^{\mathrm{aff}}_{G,P}$. The theory of Gromov–Witten

[3] For general X the definition of push-forward is somewhat tricky—it uses the so-called virtual fundamental class. However for X of the form $\mathcal{G}_{G,P}$ this difficulty is not present.

invariants of $\mathcal{G}^{\text{aff}}_{G,P}$ is not yet constructed. However we see from the above that we have a natural candidate for the notion of J-function of $\mathcal{G}^{\text{aff}}_{G,P}$. In particular, the main result of this paper gives a representation-theoretic interpretation of this J-function. This may be viewed as an explanation of the results of [**15, 19**] (where the authors compute the quantum cohomology of $\mathcal{G}_{G,P}$ making some assumptions about its existence and properties).

1.7. Organization of the paper. This paper is organized as follows. In Section 2 we fix the notation and recall the basic facts about equivariant integration. In Section 3 we give the precise formulation of the main result of this paper discussed above in Section 1.5 (Theorem 2). Section 4 is devoted to an explicit proof of the main result for $G = \mathrm{SL}(2)$. In Section 5 we prove Theorem 2. Finally, Section 6 is devoted to the proof of Theorem 1.

Acknowledgments. We are grateful to N. Nekrasov for his patient explanation of the contents of [**23**] and [**24**] and numerous discussions on the subjects. We also thank T. Coates, P. Etingof, M. Finkelberg, D. Gaitsgory, D. Kazhdan, H. Nakajima and A. Okounkov for interesting conversations and help at various stages of this work.

2. Equivariant integration

2.1. Let L be a reductive algebraic group over $\mathbb{C}$. Set $\mathcal{A}_L = H^*_L(\mathrm{pt}, \mathbb{C})$. It is well known that $\mathcal{A}_L$ is canonically isomorphic to the algebra of invariant polynomials on the Lie algebra $\mathfrak{l}$ of L. Let also $\mathcal{K}_L$ denote the field of fractions of $\mathcal{A}_L$. One can identify $\mathcal{K}_L$ with the field of invariant rational functions on $\mathfrak{l}$. We shall say that an element $f \in \mathcal{K}_L$ is homogeneous of degree $d \in \mathbb{Z}$ if f is a homogeneous rational function of degree d on $\mathfrak{l}$.

2.2. Equivariant (co)homology. Let Y be a scheme of finite type over $\mathbb{C}$ endowed with an action of L. We denote by ω_Y the dualizing complex of Y (in the category of constructible sheaves). We will be interested in the equivariant cohomology $H^*_L(Y, \omega_Y)$ (it is sometimes called the *equivariant Borel-Moore homology* of Y). For any closed L-invariant subscheme $Z \subset Y$ we have the natural map $H^*_L(Z, \omega_Z) \to H^*_L(Y, \omega_Y)$.

We may also consider the ordinary equivariant homology $H_{*,L}(Y, \mathbb{C})$ of Y which is (by definition) equal to $H^*_{c,L}(Y, \omega_Y)$.[4] In particular we have the natural ("forgetting the supports") morphism $H_{*,L}(Y, \mathbb{C}) \to H^*_L(Y, \omega_Y)$. Also since there is a canonical morphism $\mathbb{C}_Y \to \omega_Y[-2\dim Y]$ (here $\mathbb{C}_Y$ denotes the corresponding constant sheaf on Y) we get a canonical map $H^*_L(Y, \mathbb{C}) \to H^{*-2\dim Y}_L(Y, \omega_Y)$ where $H^*_L(Y, \mathbb{C})$ denotes the usual equivariant cohomology of Y.

In addition we have the natural pairing

$$H_{*,L}(Y, \mathbb{C}) \otimes_{\mathcal{A}_L} H^*_L(Y, \mathbb{C}) \to \mathcal{A}_L \tag{2.1}$$

(it is clear that all the (co)homologies in question are modules over $\mathcal{A}_L$). In particular, we have the integration functional

$$\int_Y : H_{*,L}(Y, \mathbb{C}) \to \mathcal{A}_L$$

[4]Here and in what follows the subscript "c" denotes the appropriate cohomology with compact support

which is defined as the pairing with $1 \in H^*_L(Y, \mathbb{C})$. For any closed L-invariant subscheme Z of Y the diagram

$$\begin{array}{ccc} H_{*,L}(Z, \omega_Z) & \longrightarrow & H_{*,L}(Y, \omega_Y) \\ {\scriptstyle \int_Z}\downarrow & & \downarrow{\scriptstyle \int_Y} \\ \mathcal{A}_L & \xrightarrow{\text{id}} & \mathcal{A}_L \end{array}$$

is commutative.

2.3. Integration on Borel–Moore homology. We now want to explain that in certain situations one can extend the above integration functional to the Borel–Moore homology. Assume first that L is a torus and set $Z = Y^L$. This is a closed subscheme of Y. It is well-known that in this case the natural morphism $H^*_L(Z, \omega_Z) = H^*(Z, \omega_Z) \otimes \mathcal{A}_L \to H^*_L(Y, \omega_Y)$ becomes an isomorphism after tensoring with $\mathcal{K}_L$.

Assume now (until the end of this section) that Z is proper. Then $H^*_{c,L}(Z, \omega_Z) \simeq H^*_L(Z, \omega_Z)$ we get the map $H^*_L(Y, \omega_Y) \otimes_{\mathcal{A}_L} \mathcal{K}_L \to H_{*,L}(Y, \mathbb{C}) \otimes_{\mathcal{A}_L} \mathcal{K}_L$ as the composition

$$\begin{aligned} (2.2) \qquad H^*_L(Y, \omega_Y) \otimes_{\mathcal{A}_L} \mathcal{K}_L &\simeq H^*_L(Z, \omega_Z) \otimes_{\mathcal{A}_L} \mathcal{K}_L \\ &\simeq H^*_{c,L}(Z, \omega_Z) \otimes_{\mathcal{A}_L} \mathcal{K}_L \to H_{*,L}(Y, \mathbb{C}) \otimes_{\mathcal{A}_L} \mathcal{K}_L. \end{aligned}$$

Setting $H^*_L(Y, \omega_Y) \otimes_{\mathcal{A}_L} \mathcal{K}_L = H^*_{\mathcal{K}_L}(Y, \omega_Y)$ and $H^*_L(Y, \mathbb{C}) \otimes_{\mathcal{A}_L} \mathcal{K}_L = H^*_{\mathcal{K}_L}(Y, \mathbb{C})$ we see that from (2.1) and (2.2) we get the natural pairing

$$(2.3) \qquad H^*_{\mathcal{K}_L}(Y, \omega_Y) \otimes_{\mathcal{K}_L} H^*_{\mathcal{K}_L}(Y, \mathbb{C}) \to \mathcal{K}_L.$$

In particular, by pairing with the unit class we get the integration functional

$$\int_Y : H^*_{\mathcal{K}_L}(Y, \omega_Y) \to \mathcal{K}_L.$$

Since we have the obvious map $H^*_L(Y, \omega_Y) \to H^*_{\mathcal{K}_L}(Y, \omega_Y)$ the functional $\int_Y$ makes sense on $H^*_L(Y, \omega_Y)$ too (but of course it still takes values in $\mathcal{K}_L$). This functional is "homogeneous" in the following sense. Assume that we are given some $\eta \in H^i_L(Y, \omega_Y)$. If i is odd then $\int_Y \eta = 0$. If i is even then $\int_Y \eta$ considered as a rational function on $\mathfrak{l}$ is homogeneous of degree $i/2$.

It is easy to see that all of the above constructions make sense for an arbitrary reductive L provided that the scheme Y^T is proper where $T \subset L$ is a maximal torus.

2.4. Integration on ordinary cohomology and the integral of the unit class. By composing $\int_Y$ with the map $H^*_L(Y, \mathbb{C}) \to H^{*-2\dim Y}(Y, \omega_Y)$ we may define the integration functional on $H^*_L(Y, \mathbb{C})$ (or $H^*_{\mathcal{K}_L}(Y, \mathbb{C})$). Abusing slightly the notation we shall denote this functional by the same symbol. This functional also vanishes on the odd part of $H^*_L(Y, \mathbb{C})$ and it sends any class $\psi \in H^{2d}_L(Y, \mathbb{C})$ to an element of $\mathcal{K}_L$ which (considered as a rational function on $\mathfrak{l}$) is homogeneous of degree $d - \dim Y$. In particular, we may consider the integral

$$\int_Y 1_Y$$

of the unit class $1_Y \in H^*_L(Y, \mathbb{C})$. This is a rational function on $\mathfrak{l}$ of degree $-\dim Y$.

Here is the basic example of the above situation. Assume that Y is smooth. Let $\mathfrak{t}$ denote the Lie algebra of T. Let also W_L denote the Weyl group of L. Then $\mathcal{A}_L = \mathcal{A}_T^{W_L}$ and $\mathcal{K}_L = \mathcal{K}_T^{W_L}$. For every $y \in Y^T$ let $d_y \in \mathcal{A}_T$ denote the determinant of the $\mathfrak{t}$-action on the cotangent space T_y^*Y to Y at y. We have $d_y \neq 0$ if and only if y is an isolated fixed point. Assume that Y^T is finite. Then it is well known that

$$\int_Y 1_Y = \sum_{y\in Y^T} \frac{1}{d_y} \in \mathcal{K}_T^{W_L} = \mathcal{K}_L. \tag{2.4}$$

More generally, assume that both Y and Y^T are smooth. Let $c \in H^*(Y^T,\mathbb{C})\otimes \mathcal{A}_T$ denote the equivariant Euler class of the conormal bundle to Y^T in Y. Then it is easy to see that c is invertible in $H^*(Y^T,\mathbb{C})\otimes\mathcal{K}_T$ and we have

$$\int_Y 1_Y = \int_{Y^T}\frac{1}{c}.$$

In general, when Y is singular (and, say, Y^T is finite) the integral $\int_Y 1_Y$ can also be written as a sum over all $y\in Y^T$ of certain "local contributions" taking values in $\mathcal{K}_T$. However, in the general case there is no good formula for these local contributions. One of the purposes of this paper is to compute $\int_Y 1_Y$ in certain special cases.

The following lemma follows easily from the definitions and will be used several times in this paper:

LEMMA 2.5. *Let $f\colon Y_1 \to Y_2$ be a proper birational map of L-varieties. Assume also that $\int_{Y_2} 1_{Y_2}$ is well-defined. Then $\int_{Y_1} 1_{Y_1}$ is also well-defined and we have*

$$\int_{Y_1} 1_{Y_1} = \int_{Y_2} 1_{Y_2}.$$

2.6. The intersection cohomology approach. We now want to give yet another definition of the above integration based on intersection cohomology. In fact, in order to do this we will need to make an additional assumption about the action of L on Y (which will be satisfied for the spaces ${}^{\mathrm{b}}\mathcal{QM}^{\theta}_{G,P}$ and $\mathcal{U}^{\theta_{\mathrm{aff}}}_{G,P}$ discussed in the introduction).

Let Y and L be as before and let also IC_Y denote the intersection cohomology complex of Y. Set

$$\mathrm{IH}^*_L(Y) = H^*_L(\mathrm{IC}_Y); \qquad \mathrm{IH}^*_{c,L} = H^*_{c,L}(\mathrm{IC}_Y). \tag{2.5}$$

Both spaces $\mathrm{IH}^*_L(Y)$ and $\mathrm{IH}^*_{c,L}(Y)$ are graded $\mathcal{A}_L$-modules. We have the natural equivariant Poincaré pairing

$$\langle\cdot,\cdot\rangle_Y\colon \mathrm{IH}^*_L(Y)\otimes_{\mathcal{A}_L}\mathrm{IH}^*_{c,L}(Y)\to\mathcal{A}_L.$$

This pairing is homogeneous of degree $-\dim Y$. Also we have the natural (forgetting the supports) morphism

$$\mathrm{IH}^*_{c,L}(Y)\to\mathrm{IH}^*_L(Y)$$

of graded $\mathcal{A}_L$-modules.

Set $\mathrm{IH}_{\mathcal{K}_L}(Y) = \mathrm{IH}_L(Y)\otimes_{\mathcal{A}_L}\mathcal{K}_L$ and $\mathrm{IH}_{c,\mathcal{K}_L}(Y) = \mathrm{IH}_{c,L}(Y)\otimes_{\mathcal{A}_L}\mathcal{K}_L$. Abusing slightly the notation we shall denote the corresponding pairing between $\mathrm{IH}_{c,\mathcal{K}_L}(Y)$ and $\mathrm{IH}_{\mathcal{K}_L}(Y)$ (which now takes values in $\mathcal{K}_L$) also by $\langle\cdot,\cdot\rangle_Y$.

We have the natural morphisms $\mathbb{C}_Y[\dim Y]\to\mathrm{IC}_Y$ and $\mathrm{IC}_Y\to\omega_Y[-\dim Y]$ which give rise to the maps

$$H_L^{*+\dim Y}(Y,\mathbb{C})\to\mathrm{IH}^*_L(Y) \quad\text{and}\quad \mathrm{IH}^*_{c,L}(Y)\to H_{c,L}^{*-\dim Y}(Y,\omega_Y) = H_{*-\dim Y}(Y,\mathbb{C})$$

which are compatible with the above pairings in the obvious sense. In particular we may consider the image of the unit cohomology class 1_Y in $\mathrm{IH}^*_L(Y)$ which we shall denote by the symbol.

Assume now that the map $\mathrm{IH}_{c,\mathcal{K}_L} \to \mathrm{IH}_{\mathcal{K}_L}$ is an isomorphism. Then we may consider the Poincaré pairing as a paring

$$\langle \cdot, \cdot \rangle_Y \colon \mathrm{IH}_{\mathcal{K}_L}(Y) \otimes \mathrm{IH}_{\mathcal{K}_L}(Y) \to \mathcal{K}_L.$$

Let $1_Y \in \mathrm{IH}_{\mathcal{K}_L}(Y)$ denote the unit cohomology class. For every $\omega \in \mathrm{IH}_{\mathcal{K}_L}$ we set

$$\int_Y \omega = \langle \omega, 1_Y \rangle_Y \in \mathcal{K}_L.$$

In particular one may consider the integral $\int_Y 1$.

Here is a condition which guarantees that the map $\mathrm{IH}_{c,\mathcal{K}_L} \to \mathrm{IH}_{\mathcal{K}_L}$ is an isomorphism.

LEMMA 2.7. *Let Y be as above. Assume that*

(a) *Y^T is proper;*
(b) *There exists an embedding $\mathbb{C}^* \subset T$ whose action on Y is contracting (this means that the action morphism $\mathbb{C}^* \times Y \to Y$ extends to a morphism $\mathbb{C} \times Y \to Y$).*

Then the natural map $\mathrm{IH}_{c,\mathcal{K}_L} \to \mathrm{IH}_{\mathcal{K}_L}$ is an isomorphism.

In particular, under the assumptions of the lemma one may consider the integral $\int_Y 1 \in \mathcal{K}_L$. It is easy to see that this definition of $\int_Y 1$ coincides with the one discussed above.

3. The main result

The purpose of this section is to formulate more carefully the results announced in the introduction.

In this section we are going to work over an arbitrary field k of characteristic 0 (later k will become the field $\mathcal{K}_T$ considered above).

Let G, $\mathfrak{g}$ be as in the introduction—i.e., $\mathfrak{g}$ is a simple Lie algebra and G is the corresponding simply connected algebraic group. The Lie algebra $\mathfrak{g}$ has its triangular decomposition

$$\mathfrak{g} = \mathfrak{n}_+ \oplus \mathfrak{h} \oplus \mathfrak{n}_-.$$

Let $T \subset G$ be the corresponding Cartan subgroup. We denote by Λ_G the lattice of cocharacters of T. The lattice Λ_G is spanned by the coroots $\alpha \colon \mathbb{G}_m \to T$ and we shall refer to Λ_G as the coroot lattice. In fact, we let $\mathcal{I}_G$ denote the set of vertices of the Dynkin diagram of G then Λ_G is spanned by the simple coroots α_{i} for $\mathrm{i} \in \mathcal{I}_G$. We also let $\Lambda_G^{\mathrm{pos}} \subset \Lambda_G$ denote the sub-semigroup whose elements are linear combinations of positive coroots with nonnegative coefficients (equivalently, linear combinations of α_{i}'s with nonnegative coefficients).

Let $P \subset G$ be a parabolic subgroup with Levi component M. We denote by $\Lambda_{G,P}$ the lattice of cocharacters of the group $M/[M,M]$. We have the natural map $\Lambda_G \to \Lambda_{G,P}$ and we let $\Lambda_{G,P}^{\mathrm{pos}}$ denote the positive span of the images of α_{i}'s in $\Lambda_{G,P}$ for $\mathrm{i} \in \mathcal{I}_G \backslash \mathcal{I}_M$.

3.1. Verma modules and Whittaker vectors. The finite dimensional case. Let $\check{\mathfrak{g}}$ denote the Langlands dual algebra of $\mathfrak{g}$. We have the triangular decomposition

$$\check{\mathfrak{g}} = \check{\mathfrak{n}}_+ \oplus \check{\mathfrak{h}} \oplus \check{\mathfrak{n}}_-.$$

Note that the set of vertices of the Dynkin diagram of $\check{\mathfrak{g}}$ is equal to $\mathcal{I}_G$. We denote by $e_i \in \check{\mathfrak{n}}_+$ and $f_i \in \check{\mathfrak{n}}_-$ (for $i \in \mathcal{I}_G$) the corresponding Chevalley generators of $\check{\mathfrak{g}}$. We denote by $\kappa\colon \check{\mathfrak{g}} \to \check{\mathfrak{g}}$ the Cartan anti-involution sending e_i to f_i and acting on $\check{\mathfrak{h}}$ as identity.

Recall that the lattice Λ_G defined above may be considered as the coroot lattice of $\check{\mathfrak{g}}$. Also we may think of $\mathfrak{h}$ as the space of all (not necessarily integral) weights of $\check{\mathfrak{g}}$. We let $\rho \in \mathfrak{h}$ denote the half-sum of the positive roots of $\check{\mathfrak{g}}$. Also we denote by $\check{\rho}$ the half-sum of the positive roots of $\mathfrak{g}$.

Choose now any $\lambda \in \mathfrak{h}$. We denote by $M(\lambda)$ the $\check{\mathfrak{g}}$-module generated by a vector v_λ such that:

(1) $\check{\mathfrak{n}}_-$ annihilates v_λ;
(2) For any $t \in \check{\mathfrak{h}}$ one has $t(v_\lambda) = \lambda(t)v_\lambda$.
(3) v_λ is a free generator of $M(\lambda)$ over $U(\check{\mathfrak{n}}_+)$.

We shall call $M(\lambda)$ the *Verma module with lowest weight* λ. We also denote by $\delta M(\lambda)$ the corresponding dual Verma module; by the definition it consists of all $\check{\mathfrak{h}}$-finite linear functionals $m\colon M(\lambda) \to k$ with the $\check{\mathfrak{g}}$-action defined by

$$x(m)(v) = m\big(\kappa(x)(v)\big)$$

for $x \in \check{\mathfrak{g}}$, $v \in M(\lambda)$, $m \in \delta M(\lambda)$.

We have the unique map $\eta_\lambda\colon M(\lambda) \to \delta M(\lambda)$ of $\check{\mathfrak{g}}$-modules satisfying $\eta_\lambda(v_\lambda)(v_\lambda) = 1$. One can interpret this map as a pairing

$$\langle\cdot,\cdot\rangle\colon M(\lambda) \otimes M(\lambda) \to k$$

satisfying

$$\langle x(v), w\rangle = \langle v, \kappa(x)(w)\rangle$$

for each $x \in \check{\mathfrak{g}}$ and $v,\ w \in M(\lambda)$. It is well-known that for generic λ the map η_λ is an isomorphism. In fact for this it is enough to require, for example, that $\langle\lambda, \alpha^\vee\rangle \notin \mathbb{Z}$ for all coroots $\alpha^\vee$ of $\check{\mathfrak{g}}$.

Let $\widehat{M}(\lambda)$ denote the completion of $M(\lambda)$ consisting of all sums

$$\sum m_\mu \quad \text{where } m_\mu \text{ im } M(\lambda)_\mu$$

such that for all $\theta \in \Lambda_G^{\mathrm{pos}}$ the set

$$\{\mu \mid m_\mu \neq 0 \text{ and } \mu \notin \lambda + \theta + \Lambda_G^{\mathrm{pos}}\}$$

is finite. It is clear that the Lie algebra $\check{\mathfrak{g}}$ still acts on $\widehat{M}(\lambda)$.

It is well-known that for any λ for which the map η_λ is an isomorphism and for every $\hbar \in k$ such that $\hbar \neq 0$ there exists unique vector $w(\lambda) \in \widehat{M}(\lambda), w(\lambda) = \sum w(\lambda)_\mu$ (where $w(\lambda)_\mu$ has weight μ) such that

(1) $w(\lambda)_\lambda = v_\lambda$;
(2) For any $i \in \mathcal{I}_G$ we have $f_i(w) = (1/\hbar)w$.

We shall refer to w as the *Whittaker vector* in $M(\lambda)$ (of course it depends on the choice of $\hbar$; however later this choice will become canonical).

We also set

$$w'(\lambda) = \sum_{\mu} (-1)^{\langle \lambda-\mu,\check{\rho}\rangle} w(\lambda)_{\mu}.$$

Clearly $w'(\lambda)$ satisfies the equation $f_{\mathrm{i}}\big(w'(\lambda)\big) = (-1/\hbar)w'(\lambda)$ for all $\mathrm{i} \in \mathcal{I}_G$; $w'(\lambda)$ is uniquely characterized by this property if in addition one requires that the λ-weight part of w' is equal to v_λ.

3.2. The affine case. Let us denote by $\mathfrak{g}_{\mathrm{aff}}$ the full untwisted affine Lie algebra corresponding to $\mathfrak{g}$ (cf. [**16**]). As a vector space the Lie algebra $\mathfrak{g}_{\mathrm{aff}}$ can be written as

$$\mathfrak{g}_{\mathrm{aff}} = \mathfrak{g}[x, x^{-1}] \oplus \mathbb{C}K \oplus \mathbb{C}d$$

where K is a central element of $\mathfrak{g}_{\mathrm{aff}}$ and $\mathrm{ad}(d)$ acts on $\mathfrak{g}[x,x^{-1}]$ as xd/dx. The Cartan subalgebra $\mathfrak{h}_{\mathrm{aff}}$ of $\mathfrak{g}_{\mathrm{aff}}$ is equal to $\mathfrak{h} \oplus \mathbb{C}K \oplus \mathbb{C}d$. We denote by $\mathcal{I}_G^{\mathrm{aff}}$ the set of vertices of the Dynkin diagram of $\mathfrak{g}_{\mathrm{aff}}$. Also we denote by $\check{\mathfrak{g}}^{\mathrm{aff}}$ the Langlands dual Lie algebra of $\mathfrak{g}_{\mathrm{aff}}$. By the definition this is an affine Lie algebra whose Cartan matrix is the transposed of that of $\mathfrak{g}_{\mathrm{aff}}$; note that $\check{\mathfrak{g}}_{\mathrm{aff}}$ might be a twisted affine Lie algebra. For each $\mathrm{i} \in \mathcal{I}_G^{\mathrm{aff}}$ we denote by $\alpha_{\mathrm{i}}^{\mathrm{aff}}$ the corresponding simple root of the algebra $\check{\mathfrak{g}}_{\mathrm{aff}}$.

The Lie algebras $\mathfrak{g}_{\mathrm{aff}}$ and $\check{\mathfrak{g}}_{\mathrm{aff}}$ admit triangular decompositions

$$\mathfrak{g}_{\mathrm{aff}} = \mathfrak{n}_{+,\mathrm{aff}} \oplus \mathfrak{h}_{\mathrm{aff}} \oplus \mathfrak{n}_{-,\mathrm{aff}}$$

and

$$\check{\mathfrak{g}}_{\mathrm{aff}} = \check{\mathfrak{n}}_{+,\mathrm{aff}} \oplus \check{\mathfrak{h}}_{\mathrm{aff}} \oplus \check{\mathfrak{n}}_{-,\mathrm{aff}}.$$

Here $\mathfrak{h}_{\mathrm{aff}} = \mathfrak{h} \oplus \mathbb{C}d \oplus \mathbb{C}K$ and $\check{\mathfrak{h}}_{\mathrm{aff}} = \check{\mathfrak{h}} \oplus \mathbb{C}\check{d} \oplus \mathbb{C}\check{K}$. We have the natural perfect paring between $\mathfrak{h}_{\mathrm{aff}}$ and $\check{\mathfrak{h}}_{\mathrm{aff}}$.

Let $\mathfrak{h}'_{\mathrm{aff}} = \mathfrak{h} \oplus \mathbb{C}K$ and $\mathfrak{h}'' = \mathfrak{h}_{\mathrm{aff}} \oplus \mathbb{C}d$. Similarly we define $\check{\mathfrak{h}}'_{\mathrm{aff}}$ and $\check{\mathfrak{h}}''_{\mathrm{aff}}$. It is well-known that the above perfect pairing identifies $\mathfrak{h}'_{\mathrm{aff}}$ with the dual space of $\check{\mathfrak{h}}''_{\mathrm{aff}}$; it also identifies $\mathfrak{h}''_{\mathrm{aff}}$ with the dual space of $\check{\mathfrak{h}}'_{\mathrm{aff}}$.

We also choose an affine analogue $\rho_{\mathrm{aff}} \in \mathfrak{h}''_{\mathrm{aff}}$ of ρ (cf. [**16**] for a detailed discussion).

Set $\check{\mathfrak{g}}'_{\mathrm{aff}}$ to be the subalgebra of $\check{\mathfrak{g}}_{\mathrm{aff}}$ consisting of elements whose projection to $\mathbb{C}\check{d}$ is equal to 0. We have

$$\check{\mathfrak{g}}' = \check{\mathfrak{n}}_{+,\mathrm{aff}} \oplus \check{\mathfrak{h}}'_{\mathrm{aff}} \oplus \mathfrak{n}_{-,\mathrm{aff}}.$$

For any $\lambda_{\mathrm{aff}} = (\lambda, \varepsilon) \in \mathfrak{h}''_{\mathrm{aff}} = (\check{\mathfrak{h}}'_{\mathrm{aff}})^*$ we let $M(\lambda_{\mathrm{aff}})$ denote the corresponding Verma module with lowest weight λ_{aff} over $\check{\mathfrak{g}}'_{\mathrm{aff}}$. In other words this module is freely generated by a vector $v_{\lambda_{\mathrm{aff}}}$ subject to the conditions:

(1) $t(v_{\lambda_{\mathrm{aff}}}) = \lambda_{\mathrm{aff}}(t)v_{\lambda_{\mathrm{aff}}}$ for any $t \in \check{\mathfrak{h}}'_{\mathrm{aff}}$.
(2) $v_{\lambda_{\mathrm{aff}}}$ is annihilated by $\mathfrak{n}_{+,\mathrm{aff}}$.

It is well-known that for generic ε (more precisely for all but one (critical) value of ε) the $\check{\mathfrak{g}}'_{\mathrm{aff}}$-module structure on $M(\lambda_{\mathrm{aff}})$ extends uniquely to an action of the whole $\check{\mathfrak{g}}_{\mathrm{aff}}$ such that d annihilates $v_{\lambda_{\mathrm{aff}}}$. Thus for such ε we may regard $M(\lambda_{\mathrm{aff}})$ as a $\check{\mathfrak{g}}_{\mathrm{aff}}$-module.

The Lie algebra $\check{\mathfrak{g}}_{\mathrm{aff}}$ carries canonical anti-involution κ_{aff} (interchanging $\check{\mathfrak{n}}_{+,\mathrm{aff}}$ and $\check{\mathfrak{n}}_{-,\mathrm{aff}}$ and acting as identity on $\check{\mathfrak{h}}_{\mathrm{aff}}$). Similarly to the affine case there is a unique pairing

$$\langle\cdot,\cdot\rangle \colon M(\lambda_{\mathrm{aff}}) \otimes M(\lambda_{\mathrm{aff}}) \to \mathbb{C}$$

satisfying

$$\langle x(v), w\rangle = \langle v, \kappa_{\mathrm{aff}}(x)(w)\rangle$$

for each $x \in \check{\mathfrak{g}}$ and $v, w \in M(\lambda_{\mathrm{aff}})$ and such that

$$\langle v_\lambda, v_\lambda\rangle = 1.$$

Of course all the above constructions make sense when the base field $\mathbb{C}$ is replaced by any field k of characteristic 0 (later k will become certain field of rational functions).

As before one may define a completion $\widehat{M}(\lambda_{\mathrm{aff}})$ of $M(\lambda_{\mathrm{aff}})$. Choose $\hbar \in k^*$. Then for generic λ_{aff} this completion will contain unique vector $w(\lambda_{\mathrm{aff}}) = \sum w(\lambda_{\mathrm{aff}})_\mu$ (here $\mu \in \mathfrak{h}''_{\mathrm{aff}}$) subject to the conditions

(1) $w(\lambda_{\mathrm{aff}})_{\lambda_{\mathrm{aff}}} = v_{\lambda_{\mathrm{aff}}}$;
(2) For any $\mathrm{i} \in \mathcal{I}_G^{\mathrm{aff}}$ we have $f_{\mathrm{i}}(w(\lambda_{\mathrm{aff}})) = (1/\hbar)w(\lambda_{\mathrm{aff}})$.

Again we shall call $w(\lambda_{\mathrm{aff}})$ the *Whittaker vector* of $M(\lambda_{\mathrm{aff}})$. Similarly on can define the vector $w'(\lambda_{\mathrm{aff}})$.

3.3. The main result. Set now

$$\mathrm{IH}_{G,P} = \mathrm{IH}_{\mathcal{K}_{M\times\mathbb{C}^*}}(\mathcal{QM}^\theta_{G,P}) \quad \text{and} \quad \mathrm{IH}^{\theta_{\mathrm{aff}}}_{G,P} = \mathrm{IH}_{\mathcal{K}_{M\times(\mathbb{C}^*)^2}}(\mathcal{U}^{\theta_{\mathrm{aff}}}_{G,P}).$$

Also we denote

$$\mathrm{IH}_{G,P} = \bigoplus_{\theta\in\Lambda^{\mathrm{pos}}_{G,P}} \mathrm{IH}^\theta_{G,P} \quad \text{and} \quad \mathrm{IH}^{\mathrm{aff}}_{G,P} = \bigoplus_{\theta_{\mathrm{aff}}\in\Lambda^{\mathrm{aff,pos}}_{G,P}} \mathrm{IH}^{\theta_{\mathrm{aff}}}_{G,P}.$$

We shall further restrict ourselves to the case $P = B$ (where B as before denotes a Borel subgroup of G). In this case $\mathrm{IH}_{G,B}$ is a vector space over the field $\mathcal{K}_{T\times\mathbb{C}^*}$ which we may identify with the field of rational functions on $\mathfrak{h}\times\mathbb{C}$. We shall denote a typical element in this product by $(a,\hbar)$. Similarly $\mathrm{IH}^{\mathrm{aff}}_{G,B}$ is a vector space over the field $\mathcal{K}_{T\times(\mathbb{C}^*)^2}$ which we may identify with the field of rational functions on $\mathfrak{h}\times\mathbb{C}^2$. We shall typically denote an element of the latter product as $(a,\varepsilon,\hbar)$. Note that we may think of (a,ε) as an element of $\mathfrak{h}'_{\mathrm{aff}}$. Note also that we have a canonical element $\hbar$ in both $\mathcal{K}_{T\times\mathbb{C}^*}$ and $\mathcal{K}_{T\times(\mathbb{C}^*)^2}$.

We denote by $\langle\cdot,\cdot\rangle^\theta_{G,B}$ (resp. by $\langle\cdot,\cdot\rangle^{\theta_{\mathrm{aff}}}_{G,B}$) the Poincaré pairing on $\mathrm{IH}^\theta_{G,B}$ (resp. on $\mathrm{IH}^{\theta_{\mathrm{aff}}}_{G,B}$). As in the introduction we let $\langle\cdot,\cdot\rangle_{G,B}$ be the direct sum of the pairings $(-1)^{\langle\theta,\check{\rho}\rangle}\langle\cdot,\cdot\rangle^\theta_{G,B}$. Similarly we let $\langle\cdot,\cdot\rangle^{\mathrm{aff}}_{G,B}$ denote the direct sum of the pairings $(-1)^{\langle\theta_{\mathrm{aff}},\check{\rho}_{\mathrm{aff}}\rangle}\langle\cdot,\cdot\rangle^{\theta_{\mathrm{aff}}}_{G,B}$.

THEOREM 2. *There exists a natural action of $\check{\mathfrak{g}}$ (resp. of $\check{\mathfrak{g}}_{\mathrm{aff}}$) on $\mathrm{IH}_{G,B}$ (resp. on $\mathrm{IH}^{\mathrm{aff}}_{G,B}$) such that:*

(1) *$\mathrm{IH}^\theta_{G,B}$ has weight $a/\hbar+\rho+\theta$ with respect to the action of $\check{\mathfrak{g}}$. Similarly, $\mathrm{IH}^{\theta_{\mathrm{aff}},\mathrm{aff}}_{G,B}$ has weight $(a,\varepsilon)/\hbar+\rho_{\mathrm{aff}}+\theta_{\mathrm{aff}}$ with respect to the action of $\check{\mathfrak{g}}_{\mathrm{aff}}$.*[5]
(2) *The pairing $\langle\cdot,\cdot\rangle_{G,B}$ is $(\check{\mathfrak{g}},\kappa)$-invariant. Similarly, the pairing $\langle\cdot,\cdot\rangle^{\mathrm{aff}}_{G,B}$ is $(\check{\mathfrak{g}}_{\mathrm{aff}},\kappa_{\mathrm{aff}})$-invariant.*
(3) *The $\check{\mathfrak{g}}$-module $\mathrm{IH}_{G,B}$ is isomorphic to $M(a/\hbar+\rho)$ as a $\check{\mathfrak{g}}$-module. Similarly, $\mathrm{IH}^{\mathrm{aff}}_{G,B}$ is isomorphic to $M_{\mathrm{aff}}((a,\varepsilon)/\hbar+\rho_{\mathrm{aff}})$.*

[5]Here we regard (a,ε) as an element of $\mathfrak{h}_{\mathrm{aff}}$ by means of the embedding $\mathfrak{h}'_{\mathrm{aff}} \subset \mathfrak{h}_{\mathrm{aff}}$

(4) *Under the above identifications the vector* $\sum 1_{\mathcal{QM}^\theta_{G,B}}$ (*resp.* $\sum 1_{\mathcal{U}^{\theta_{\mathrm{aff}}}_{G,B}}$) *is the Whittaker vector* $w(a/\hbar+\rho)$ (*resp.* $w((a,\varepsilon)/\hbar+\rho_{\mathrm{aff}})$) *in* $\widehat{M}(a/\hbar+\rho)$ (*resp. in* $\widehat{M}((a,\varepsilon)/\hbar+\rho_{\mathrm{aff}})$).

In what follows we let a_{aff} denote the pair (a,ε) (in the affine case).

COROLLARY 3.4. *We have*

$$\mathcal{Z}_{G,B}=\mathfrak{q}^{-a/\hbar-\rho}\Big\langle w\Big(\frac{a}{\hbar}+\rho\Big),\mathfrak{q}\cdot w'\Big(\frac{a}{\hbar}+\rho\Big)\Big\rangle^{6}$$

Similarly,

$$\mathcal{Z}^{\mathrm{aff}}_{G,B}=\mathfrak{q}^{-a/\hbar-\rho}\Big\langle w\Big(\frac{a_{\mathrm{aff}}}{\hbar}+\rho_{\mathrm{aff}}\Big),\mathfrak{q}_{\mathrm{aff}}\cdot w'\Big(\frac{a_{\mathrm{aff}}}{\hbar}+\rho_{\mathrm{aff}}\Big)\Big\rangle.$$

3.5. The quadratic Toda Hamiltonian. Fix now an invariant quadratic form $(\cdot,\cdot)$ on $\mathfrak{g}$ so that long roots on $\check{\mathfrak{g}}$ have squared norm 2. Denote by Δ the corresponding Laplacian on $\mathfrak{h}$. As before we set $\mathfrak{q}=e^t$ where $t\in\check{\mathfrak{h}}$ and $\mathfrak{q}_{\mathrm{aff}}=e^{t_{\mathrm{aff}}}$. Thus we may write $t_{\mathrm{aff}}=(t,\ln Q)$.

COROLLARY 3.6. (1) *The function* $e^{a(t)/\hbar}\mathcal{Z}_{G,B}$ *is an eigenfunction of the quantum Toda Hamiltonians with the "eigenvalue" defined by* a (*cf.* **[7]** *and references therein for the definition*). *In particular, the function* $\mathcal{Z}_{G,B}$ *satisfies the equation*

$$\Big(\hbar^2\Delta-2\sum_{\mathrm{i}\in\mathcal{I}_G}e^{\alpha_{\mathrm{i}}(t)}\Big)(e^{a(t)/\hbar}\mathcal{Z}_{G,B})=(a,a)e^{a(t)/\hbar}\mathcal{Z}_{G,B}$$

(2) *The function* $\mathcal{Z}^{\mathrm{aff}}_{G,B}$ *satisfies*

$$\Big(\varepsilon Q\frac{\partial}{\partial Q}+\hbar^2\Delta-2\sum_{\mathrm{i}\in\mathcal{I}^{\mathrm{aff}}_G}e^{\alpha^{\mathrm{aff}}_{\mathrm{i}}(t_{\mathrm{aff}})}\Big)e^{a(t)/\hbar}\mathcal{Z}^{\mathrm{aff}}_{G,B}=(a,a)e^{a(t)/\hbar}\mathcal{Z}_{G,B}. \tag{3.1}$$

The operator in the left hand side of (3.1) may be called a *nonstationary analogue* of the quadratic affine Toda Hamiltonian (the latter will be obtained from it by removing the $\varepsilon Q\partial/\partial Q$-term).

4. Example. $G=\mathrm{SL}(2)$

In this section we would like to prove Corollary 3.4 explicitly for $G=\mathrm{SL}(2)$ (in the nonaffine case). So, we shall assume that $G=\mathrm{SL}(2)$ until the end of this section.

4.1. The space $^{\mathrm{b}}\mathcal{M}^d_{G,B}$. Since $G=\mathrm{SL}(2)$ we may identify Λ_G with $\mathbb{Z}$ and we shall typically denote an element of Λ_G by d (instead of θ). Also in this case $\mathcal{G}_{G,B}=\mathbb{P}^1$ (we choose this identification in such a way that $e_{G,B}$ corresponds to $\infty\in\mathbb{P}^1$) and thus $^{\mathrm{b}}\mathcal{M}^d_{G,B}$ is just the space of maps $f\colon\mathbb{P}^1\to\mathbb{P}^1$ of degree d which send ∞ to ∞. This is the same as the set of pairs (g,h) of polynomials in one variable such that:

(1) g and h have no common roots

[6] Of course the torus $\check{T}$ does not act on $M(\frac{a}{\hbar})$ (only its Lie algebra acts). Thus in order to make sense of this formula we must (as before) write $q=e^t$ where $t\in\check{\mathfrak{h}}$ and thus the equality becomes an equality of formal Fourier series in t.

(2) $g(z) = z^d +$ lower order terms
(3) $\deg h < d$

4.2. The space ${}^{\mathrm{b}}\mathcal{QM}^d_{G,B}$ and the integral of 1. The "zastava" space ${}^{\mathrm{b}}\mathcal{QM}^d_{G,B}$ consists of all pairs g, h as above but with condition 1 removed (in particular h is allowed to be identically 0). Thus ${}^{\mathrm{b}}\mathcal{QM}^d_{G,B}$ is isomorphic to the affine space $\mathbb{A}^{2d}$. It inherits the natural action of $T \times \mathbb{C}^*$ with unique fixed point corresponding to $g = z^d$, $h = 0$.

Now for each d we set

$$A_d := \int_{{}^{\mathrm{b}}\mathcal{QM}^d_{G,B}} 1 \in \mathcal{K}_{T\times\mathbb{C}^*}$$

(the integral is taken in the $T \times \mathbb{C}^*$-equivariant cohomology). Note that $T = \mathbb{C}^*$ in our case; thus we shall identify $\mathcal{K}_{T\times\mathbb{C}^*}$ with the field of rational functions in the variables $a, \hbar$. Since $\mathcal{M}^d$ is smooth and we have just one fixed point it is easy to compute this integral explicitly by using (2.4) and we get

$$A_d = \frac{1}{d!\hbar^d \prod_{i=1}^d (a + i\hbar)}.$$

4.3. Verma modules and Whittaker vectors. Let us denote by e, h, f the standard basis for $\mathrm{sl}(2) \simeq \check{\mathfrak{g}}$. For every scalar λ we denote by $M(\lambda)$ (as before) the Verma module over sl(2) with *lowest weight* λ. Explicitly this means that $M(\lambda+1)$ has a basis $\{m_d\}_{d\geq 0}$ in which the action is written in the following way:

$$h(m_d) = (\lambda + 2d + 1)m_d; \quad e(m_d) = m_{d+1}; \quad f(m_d) = -d(\lambda + d)m_{d-1}$$

(thus f acts nilpotently and $\mathbb{C}[e]$ acts freely).

The anti-involution $\kappa\colon \mathrm{sl}(2) \to \mathrm{sl}(2)$ is written in the following way:

$$\kappa(e) = f; \quad \kappa(f) = e \quad \kappa(h) = h.$$

Recall that $\langle\cdot,\cdot\rangle$ denotes the unique $(\mathrm{sl}(2), \kappa)$-invariant form on $M(\lambda + 1)$ satisfying $\langle m_0, m_0\rangle = 1$.

Explicitly this form is written in the following way:

$$\langle m_d, m_{d'}\rangle = \delta_{d,d'}(-1)^d d! \prod_{i=1} (\lambda + i).$$

We now consider the case $\lambda = a/\hbar$. In this case define

$$w_d = \frac{v_d}{d! \prod_{i=1}^d (a + i\hbar)}.$$

Thus

$$f(w_d) = -\frac{1}{\hbar} w_{d-1}.$$

Set also $w = \sum (-1)^d w_d$, $w' = \sum w_d$. Thus $f(w) = (1/\hbar)w$ and $f(w') = (-1/\hbar)w'$, i.e., $w = w(a/\hbar + 1)$ and $w' = w'(a/\hbar + 1)$.

On the other hand we have

$$\langle w_d, w_d\rangle = \frac{\langle v_d, v_d\rangle}{\left(d! \prod_{i=1}^d (a + i\hbar)\right)^2} = \frac{(-1)^d d! \prod_{i=1}^d (a/\hbar + i)}{\left(d! \prod_{i=1}^d (a + i\hbar)\right)^2} \tag{4.1}$$

$$= \frac{(-1)^d}{d!\hbar^d \prod_{i=1}^d (a + i\hbar)} = (-1)^d A_d.$$

This is equivalent to Corollary 3.4 in this case.

4.4. The geometric interpretation of the action. To complete the picture for $G = \mathrm{SL}(2)$ let us explain the geometric origin of the operators (e, h, f) on $\mathrm{IH}_{G,B}$. Since e is adjoint to f with respect to $\langle\cdot,\cdot\rangle$ and $[e,f] = h$ it follows that it is enough to give a construction of f.

For each $d \geq 0$ let $Y_d \subset {}^{\mathrm{b}}\mathcal{QM}^d_{G,B} \times {}^{\mathrm{b}}\mathcal{QM}^{d+1}_{G,B}$ consisting of pairs $((g_1, h_1), (g_2, h_2))$ subject to the following condition: there exists $z_0 \in \mathbb{C}$ so that $g_2(z) = (z - z_0)g_1(z)$ and $h_2(z) = (z - z_0)h_1(z)$.

We denote by pr_1 the projection from Y_d to ${}^{\mathrm{b}}\mathcal{QM}^d_{G,B}$ and by pr_2 the corresponding projection to ${}^{\mathrm{b}}\mathcal{QM}^{d+1}_{G,B}$.

The variety Y_d viewed as a correspondence from ${}^{\mathrm{b}}\mathcal{QM}^{d+1}_{G,B}$ to ${}^{\mathrm{b}}\mathcal{QM}^d_{G,B}$ defines a $\mathcal{K}_{T\times\mathbb{C}^*}$-linear map $H^*_{\mathcal{K}_{T\times\mathbb{C}^*}}({}^{\mathrm{b}}\mathcal{QM}^{d+1}_{G,B}) \to H^*_{\mathcal{K}_{T\times\mathbb{C}^*}}({}^{\mathrm{b}}\mathcal{QM}^d_{G,B})$. Since we are ignoring the grading and since in this case the varieties ${}^{\mathrm{b}}\mathcal{QM}^d_{G,B}$ are smooth we have $H^*_{\mathcal{K}_{T\times\mathbb{C}^*}}({}^{\mathrm{b}}\mathcal{QM}^d_{G,B}) = \mathrm{IH}^*_{\mathcal{K}_{T\times\mathbb{C}^*}}({}^{\mathrm{b}}\mathcal{QM}^d_{G,B})$ and thus we let $f\colon \mathrm{IH}^*_{\mathcal{K}_{T\times\mathbb{C}^*}}({}^{\mathrm{b}}\mathcal{QM}^{d+1}_{G,B}) \to \mathrm{IH}^*_{\mathcal{K}_{T\times\mathbb{C}^*}}({}^{\mathrm{b}}\mathcal{QM}^d_{G,B})$ be the operator defined by Y_d as a correspondence. Since Y_d is clearly isomorphic to ${}^{\mathrm{b}}\mathcal{QM}^d_{G,B} \times \mathbb{A}^1$ to see that $f(1^{d+1}_{G,B}) = 1^d_{G,B}/\hbar$.

We now define $h\colon \mathrm{IH}^*_{\mathcal{K}_{T\times\mathbb{C}^*}}({}^{\mathrm{b}}\mathcal{QM}^d_{G,B}) \to \mathrm{IH}^*_{\mathcal{K}_{T\times\mathbb{C}^*}}({}^{\mathrm{b}}\mathcal{QM}^d_{G,B})$ to be the operator of multiplication by $\frac{a}{\hbar}+2d+1$ and $e\colon \mathrm{IH}^*_{\mathcal{K}_{T\times\mathbb{C}^*}}{}^{\mathrm{b}}(\mathcal{QM}^d_{G,B}) \to \mathrm{IH}^*_{\mathcal{K}_{T\times\mathbb{C}^*}}({}^{\mathrm{b}}\mathcal{QM}^d_{G,B})$ to be the conjugate of f with respect to the pairing $\langle\cdot,\cdot\rangle_{G,B}$.

We now have to check that the operators e, h, f defined in the above way satisfy the relations of the Lie algebra $\mathrm{sl}(2)$. The relations $[h,e] = 2e$ and $[h,f] = -2f$ are obvious. Thus we have to check only that $[e,f] = h$. This is equivalent to (4.1) (since $A_d = (-1)^d \langle 1^d_{G,B}, 1^d_{G,B}\rangle_{G,B}$).

5. Construction of the action

In this section we sketch the proof of Theorem 2 assuming that the reader is familiar with the contents of [**8**].

5.1. Plan of the proof. First we are going to discuss the finite case.
Recall that $\check{\mathfrak{g}}$ has the triangular decomposition

$$\check{\mathfrak{g}} = \check{\mathfrak{n}}_+ \oplus \check{\mathfrak{h}} \oplus \check{\mathfrak{n}}_-.$$

The construction of the $\check{\mathfrak{g}}$-action on $\mathrm{IH}_{G,B}$ will consist of the following 3 steps:

Step 1. Construct an action of $\check{\mathfrak{n}}_+$ on $\mathrm{IH}_{G,B}$.

Step 2. Use the pairing $\langle\cdot,\cdot\rangle_{G,B}\colon \mathrm{IH}_{G,B} \otimes_{\mathcal{K}_{T\times\mathbb{C}^*}} \mathrm{IH}_{G,B} \to \mathcal{K}_{T\times\mathbb{C}^*}$ to construct the action of $\check{\mathfrak{n}}_-$.

Step 3. Check that the two actions of $\check{\mathfrak{n}}_+$ and of $\check{\mathfrak{n}}_-$ generate an action of $\check{\mathfrak{g}}$ and that the resulting $\check{\mathfrak{g}}$-module is isomorphic to $M(\frac{a}{\hbar} + \rho)$.

Step 4. Check that for every Chevalley generator f_{i} corresponding to some $\mathrm{i} \in \mathcal{I}_G$ and every $\theta \in \Lambda^{\mathrm{pos}}_G$ we have

$$f_{\mathrm{i}}(1^\theta_{G,B}) = \frac{1}{\hbar} 1^{\theta-\alpha_{\mathrm{i}}}_{G,B}. \tag{5.1}$$

In fact Steps 1–3 come from almost a word-by-word repetition of the contents of Section 4 of [**8**] and Step 4 is equivalent to the computations of the previous section.

5.2. Step 1. Recall the definition of the *local* Ext-*algebra* from [**8**]. For every $\theta \in \Lambda_G^{\mathrm{pos}}$ we denote (following [**8**]) by $\mathbb{A}^\theta$ the corresponding partially symmetrized power of $\mathbb{A}^1$ (i.e., elements of $\mathbb{A}^\theta$ are formal sums $\sum \theta_i x_i$ with $\theta_i \in \Lambda_G^{\mathrm{pos}}$, $x_i \in \mathbb{A}^1$ and $\sum \theta_i = \theta$). In [**8**] the authors construct a closed embedding $\mathbb{A}^\theta \hookrightarrow {}^{\mathrm{b}}\mathcal{QM}^\theta_{G,B}$ whose image is denoted by $\partial^\theta({}^{\mathrm{b}}\mathcal{QM}^\theta_{G,B})$ (in fact this is the closed stratum of a certain canonical stratification of ${}^{\mathrm{b}}\mathcal{QM}^\theta_{G,B}$).

Now we set (cf. [**8**, Section 2])

$$\mathcal{A}_\theta = \mathrm{Ext}^*_{{}^{\mathrm{b}}\mathcal{QM}^\theta_{G,B}}(\mathrm{IC}_{\partial^\theta({}^{\mathrm{b}}\mathcal{QM}^\theta_{G,B})}, \mathrm{IC}_{{}^{\mathrm{b}}\mathcal{QM}^\theta_{G,B}})$$

and define

$$\mathcal{A} = \bigoplus_{\theta \in \Lambda_G^{\mathrm{pos}}} \mathcal{A}_\theta.$$

In [**8**] the authors define in a geometric way a Hopf algebra structure on $\mathcal{A}$ and show that it is isomorphic (as a Hopf algebra) to $U(\check{\mathfrak{n}}_+)$.

In Section 4 of [**8**] the authors define an action of $\mathcal{A}$ on $\oplus_{\theta\in\Lambda_G^{\mathrm{pos}}} \mathrm{IH}^*(\mathcal{QM}^\theta_{G,B})$. A word-by-word repetition of that construction defines an action of $\mathcal{A}$ on $\mathrm{IH}_{G,B}$. Thus we get an action of the Lie algebra $\check{\mathfrak{n}}_+$ on $\mathrm{IH}_{G,B}$.

5.3. Steps 2, 3 and 4. We already have the action of the operators $f_\mathfrak{i}$ on $\mathrm{IH}_{G,B}$. We define the action of every $t \in \check{\mathfrak{h}}$ on $\mathrm{IH}_{G,B}$ by requiring that t acts on $\mathrm{IH}^\theta_{G,B}$ by means of multiplication by $\langle t, a/\hbar + \rho + \theta\rangle$ on $\mathrm{IH}^\theta_{G,B}$. Also we define $e_\mathfrak{i}$ to be the adjoint operator to $f_\mathfrak{i}$ with respect to $\langle\cdot,\cdot\rangle_{G,B}$. We need to check the following relations:

$$[e_\mathfrak{i}, f_\mathfrak{i}] = h_\mathfrak{i}; \quad [e_\mathfrak{i}, f_\mathfrak{j}] = 0 \quad \text{for } \mathfrak{i} \neq \mathfrak{j}.$$

The first relation is proved exactly as the Proposition in Section 4.9 of [**8**] and the second one is proved exactly as Proposition 4.8 of [**8**].

In order to check that $\mathrm{IH}_{G,B}$ is isomorphic to $M(a/\hbar + \rho)$ as a $\check{\mathfrak{g}}$-module we may argue (for example) as follows. It follows from the main theorem of Section 3 of [**8**] (cf. [**8**, Section 3.1]) that $\mathrm{IH}_{G,B}$ and $M(a/\hbar + \rho)$ have the same character. Also, it is clear that $\mathrm{IH}_{G,B}$ lies in the category $\mathcal{O}$ for $\check{\mathfrak{g}}$ (in other words, $\check{\mathfrak{h}}$ acts on it in a semi-simple way with finite-dimensional weight spaces and $\mathfrak{n}_-$ acts locally nilpotently). However, it is easy to see that $M(a/\hbar + \rho)$ is the only module in this category with such character.

Finally, we need to check (5.1). However, using the same argument as in 4.9 of [**8**] we can reduce it to the case $G = \mathrm{SL}(2)$ and thus it becomes equivalent to the calculation of the previous section.

5.4. The affine case. Let us explain how to extend the above argument to the affine case. Here we need to use the results of [**3**] instead of [**8**]. First of all the algebra $\mathcal{A}$ is defined exactly as above using the varieties $\mathcal{U}^{\theta_{\mathrm{aff}}}_{G,B}$ instead of ${}^{\mathrm{b}}\mathcal{QM}^\theta_{G,B}$ (one has to use the stratification of $\mathcal{U}^{\theta_{\mathrm{aff}}}_{G,B}$ defined in Section 10 of [**3**]). The definition of the Hopf algebra structure is a word-by-word repetition of the construction in Section 2 of [**8**]. The proof of the isomorphism $\mathcal{A} \simeq U(\check{\mathfrak{n}}_{+,\mathrm{aff}})$ is the same as in Section 4 of [**8**] except that one has to use the results of Section 16 of [**3**] for counting the dimension of $\mathcal{A}_\theta$ instead of Section 3 of [**8**]. The rest is proved exactly as before (except that in order to prove the isomorphism $\mathrm{IH}^{\mathrm{aff}}_{G,B} \simeq M(a_{\mathrm{aff}}/\hbar + \rho_{\mathrm{aff}})$ one has to use again Section 16 of [**3**] instead of Section 3 of [**8**] while computing the character of $\mathrm{IH}^{\mathrm{aff}}_{G,B}$).

6. Proof of Theorem 1

The contents of this section are mostly due to A. Givental. We include it for the sake of completeness.

6.1. Some reformulations. In this section we prove Theorem 1. The arguments here are very close to those in [**14**]. Fix G, P as in the formulation of the theorem. We need to prove that for each β we have

$$\int_{{}^{\mathrm{b}}\mathcal{QM}^{\beta}_{G,P}} 1^{\beta}_{G,P} = (\mathrm{ev}_{\beta})_* \left(\frac{1}{\hbar(c_\beta + \hbar)}\right)$$

under the conventions explained before Theorem 1. Let ${}^{\mathrm{b}}\overline{\mathcal{M}}^{\beta}_{0,1}(G,P)$ denote the space of based stable maps from a genus 0 curve with one marked point to $\mathcal{G}_{G,P}$; in other words ${}^{\mathrm{b}}\overline{\mathcal{M}}^{\beta}_{0,1}(G,P)$ is a closed subscheme in $\overline{\mathcal{M}}^{\beta}_{0,1}(G,P) := \overline{\mathcal{M}}^{\beta}_{0,1}(\mathcal{G}_{G,P})$ corresponding to maps sending the marked point to $e_{G,P}$.

It is now clear that the restriction map gives rise to an isomorphism

$$H^*_G(\overline{\mathcal{M}}^{\beta}_{0,1}(G,P),\mathbb{C}) \simeq H^*_M({}^{\mathrm{b}}\overline{\mathcal{M}}^{\beta}_{0,1}(G,P),\mathbb{C}).$$

Moreover, the following diagram commutes:

$$\begin{array}{ccc} H^*_G(\overline{\mathcal{M}}^{\beta}_{0,1}(G,P),\mathbb{C}) & \longrightarrow & H^*_M({}^{\mathrm{b}}\overline{\mathcal{M}}^{\beta}_{0,1}(G,P),\mathbb{C}) \\ \Big\downarrow {\scriptstyle (\mathrm{ev}_\beta)_*} & & \Big\downarrow {\scriptstyle \int_{{}^{\mathrm{b}}\overline{\mathcal{M}}^{\beta}_{0,1}(G,P)}} \\ H^*_G(\mathcal{G}_{G,P},\mathbb{C}) & \longrightarrow & H^*_M(\mathrm{pt},\mathbb{C}), \end{array}$$

where the last map is just the restriction to $e_{G,P}$.

In particular we see that under the above identifications we have

$$(\mathrm{ev}_{\beta})_* \left(\frac{1}{\hbar(c_\beta + \hbar)}\right) = \int_{{}^{\mathrm{b}}\overline{\mathcal{M}}(G,P)} \frac{1}{\hbar(c_\beta|_{{}^{\mathrm{b}}\overline{\mathcal{M}}(G,P)} + \hbar)}.$$

Our next goal will be to rewrite the last integral as the integral of 1 over certain space.

6.2. Graph spaces. For every $\theta \in \Lambda^{\mathrm{pos}}_{G,P}$ let us introduce the *graph* spaces

$$\mathrm{Graph}^{\beta}_{G,P} := \overline{\mathcal{M}}^{\beta,1}_{0,0}(\mathcal{G}_{G,P} \times \mathbb{P}^1).$$

A point of $\mathrm{Graph}^{\beta}_{G,P}$ is a stable map $C \to \mathcal{G}_{G,P} \times \mathbb{P}^1$ which has degree θ over the first multiple and degree one over the second one. In the case when C is nonsingular the second projection gives an isomorphism between C and $\mathbb{P}^1$ and thus our map is actually equal to the graph of some map $\mathbb{P}^1 \to \mathcal{G}_{G,P}$ of degree β. Thus $\mathcal{M}^{\beta}_{G,P}$ is an open subset of $\mathrm{Graph}^{\beta}_{G,P}$.

We denote by ${}^{\mathrm{b}}\,\mathrm{Graph}^{\beta}_{G,P}$ the corresponding "based" version of the graph space; it consists of all stable maps $C \to \mathcal{G}_{G,P}$ such that

(i) over some neighborhood of $\infty \in \mathbb{P}^1$ the projection $C \to \mathbb{P}^1$ is an isomorphism. In this case we denote by ∞_C the (unique) preimage of ∞ in C.

(ii) The image of ∞_C in $\mathcal{G}_{G,P} \times \mathbb{P}^1$ is $(e_{G,P}, \infty)$.

Note that ${}^{\mathrm{b}}\,\mathrm{Graph}^{\beta}_{G,P}$ is not a closed subset of $\mathrm{Graph}^{\beta}_{G,P}$ (it is only locally closed).

6.3. The $\mathbb{C}^*$-action. The group $\mathbb{C}^*$ acts naturally on $\mathrm{Graph}^{\beta}_{G,P}$ (via its action on $\mathbb{P}^1$). The fixed points of this action are discussed in [**14**]; it is shown in loc. cit. that we have

$$(\mathrm{Graph}^{\beta}_{G,P})^{\mathbb{C}^*} = \bigcup_{\beta_1+\beta_2=\beta} \overline{\mathcal{M}}^{\beta_1}_{0,1}(G,P) \times \overline{\mathcal{M}}^{\beta_2}_{0,1}(G,P)$$

This isomorphism is defined as follows. Assume that we have a stable map $f\colon C \to \mathcal{G}_{G,P}\times\mathbb{P}^1$ of degree $(\beta,1)$. Then there exists unique component C' of C which maps isomorphically to the second multiple. Assume now that f is $\mathbb{C}^*$-invariant. Then the composition $p_1\circ f|_{C'}$ must be constant (here p_1 is the projection $\mathcal{G}_{G,P}\times\mathbb{P}^1 \to \mathcal{G}_{G,P}$). Let C'' be the union of all the components of C which are different of C'. f. Then the composition map $C'' \to \mathcal{G}_{G,P}\times\mathbb{P}^1 \to \mathbb{P}^1$ sends every connected component of C'' either to 0 or to ∞. We set $C''_0 = (p_2\circ f|_{C''})^{-1}(0)$ and $C''_\infty = (p_2\circ f|_{C''})^{-1}(\infty)$. We also set $c_0 = C''_0\cap C'$ and $c_\infty = C''_\infty\cap C'$. Thus one can write $C'' = C''_0\cup C''_\infty$ and the restriction of $p_1\circ f$ to either (C''_0, c_0) or (C''_∞, c_∞) is a stable map. In particular, if $(C,f)\in {}^{\mathrm{b}}\mathrm{Graph}^{\beta}_{G,P}$ then $C'' = C''_0$ (since the unique point of C' lying over $\infty\in\mathbb{P}^1$ must be a nonsingular point of C) and the map $(C,f)\mapsto (C'', c_0, p_1\circ f|_{C''})$ defines an isomorphism

$$({}^{\mathrm{b}}\mathrm{Graph}^{\beta}_{G,P})^{\mathbb{C}^*} \simeq {}^{\mathrm{b}}\overline{\mathcal{M}}^{\beta}_{0,1}(G,P)$$

Moreover, it is easy to see (cf. [**14**], Proof of Theorem 2 or [**12**]) that the conormal bundle to ${}^{\mathrm{b}}\overline{\mathcal{M}}^{\beta}_{0,1}(G,P)$ in ${}^{\mathrm{b}}\,\mathrm{Graph}^{\beta}_{G,P}$ is of rank 2 and its $\mathbb{C}^*$-equivariant first Chern class is $h(c_\beta+\hbar)$. Therefore we have

$$\int_{{}^{\mathrm{b}}\,\mathrm{Graph}^{\beta}_{G,P}} 1 = \int_{{}^{\mathrm{b}}\overline{\mathcal{M}}^{\beta}_{0,1}(G,P)} \frac{1}{\hbar(c_\beta+\hbar)} \tag{6.1}$$

6.4. End of the proof. It follows from (6.1) that in order to prove Theorem 1 it is enough to show that

$$\int_{{}^{\mathrm{b}}\,\mathrm{Graph}^{\beta}_{G,P}} 1 = \int_{{}^{\mathrm{b}}\mathcal{QM}^{\beta}_{G,P}} 1. \tag{6.2}$$

For this it is enough to show the existence of a proper birational M-equivariant map ${}^{\mathrm{b}}\,\mathrm{Graph}^{\beta}_{G,P} \to {}^{\mathrm{b}}\mathcal{QM}^{\beta}_{G,P}$. This is proved in [**8**].

References

1. A. Bertram, I. Ciocan-Fontanine, and B. Kim, *Two proofs of a conjecture of Hori and Vafa* (preprint).
2. A. Braverman and P. Etingof, *Instanton counting via affine Lie algebras.* II (in preparation).
3. A. Braverman, M. Finkelberg, and D. Gaitsgory, *Uhlenbeck spaces via affine Lie algebras* (preprint).
4. A. Braverman, M. Finkelberg, D. Gaitsgory, and I. Mirković, *Intersection cohomology of Drinfeld's compactifications*, Selecta Math. (N.S.) **8** (2002), no. 3, 381–418.
5. A. Braverman and D. Gaitsgory, *Geometric Eisenstein series*, Invent. Math. **150** (2002), no. 2, 287–384.
6. S. K. Donaldson and P. Kronheimer, *The geometry of four-manifolds*, Oxford Math. Monogr., Oxford Univ. Press, New York, 1990.
7. P. Etingof, *Whittaker functions on quantum groups and q-deformed Toda operators*, Differential Topology, Infinite-Dimensional Lie Algebras, and Applications, Amer. Math. Soc. Transl. Ser. 2, vol. 194, Amer. Math. Soc., Providence, RI, 1999, pp. 9–25.

8. B. Feigin, M. Finkelberg, A. Kuznetsov, and I. Mirković, *Semi-infinite flags. II. Local and global intersection cohomology of quasimaps' spaces*, Differential Topology, Infinite-Dimensional Lie Algebras, and Applications, Amer. Math. Soc. Transl. Ser. 2, vol. 194, Amer. Math. Soc., Providence, RI, 1999, pp. 113–148.
9. M. Finkelberg, D. Gaitsgory, and A. Kuznetsov, *Uhlenbeck spaces for* $\mathbb{A}^2$ *and affine Lie algebra* $\widehat{\mathfrak{sl}}_n$, Publ. Res. Inst. Math. Sci. **39** (2003), no. 4, 721–766.
10. M. Finkelberg, A. Kuznetsov, N. Markarian, and I. Mirković, *A note on a symplectic structure on the space of G-monopoles*, Comm. Math. Phys. **201** (1999), no. 2, 411–421.
11. M. Finkelberg and I. Mirković, *Semi-infinite flags. I. Case of global curve* $\mathbf{P}^1$, Differential Topology, Infinite-Dimensional Lie Algebras, and Applications, Amer. Math. Soc. Transl. Ser. 2, vol. 194, Amer. Math. Soc., Providence, RI, 1999, pp. 81–112.
12. A. Givental, *Equivariant Gromov–Witten invariants*, Internat. Math. Res. Notices (1996), no. 13, 613–663.
13. A. Givental and B. Kim, *Quantum cohomology of flag manifolds and Toda lattices*, Comm. Math. Phys. **168** (1995), no. 3, 609–641.
14. A. Givental and Y.-P. Lee, *Quantum K-theory on flag manifolds, finite-difference Toda lattices and quantum groups*, Invent. Math. **151** (2003), no. 1, 193–219.
15. M. A. Guest and T. Otofuji, *Quantum cohomology and the periodic Toda lattice*, Comm. Math. Phys. **217** (2001), no. 3, 475–487.
16. V. Kac, *Infinite-dimensional Lie algebras*, 3rd edition, Cambridge Univ. Press, Cambridge, 1990.
17. M. Kashiwara, *The flag manifold of Kac–Moody Lie algebra*, Algebraic Analysis, Geometry, and Number Theory (Baltimore, MD, 1988), The Johns Hopkins Univ. Press, Baltimore, MD, 1989, pp. 161–190.
18. B. Kim, *Quantum cohomology of flag manifolds G/B and quantum Toda lattices*, Ann. of Math. (2) **149** (1999), no. 1, 129–148.
19. A.-L. Mare, *Quantum cohomology of the infinite-dimensional generalized flag manifolds*, math.DG/0105133.
20. J. W. Morgan, *Comparison of the Donaldson polynomial invariants with their algebro-geometric analogues*, Topology **32** (1993), no. 3, 449–488.
21. H. Nakajima, *Lectures on Hilbert schemes of points on surfaces*, Univ. Lecture Ser., vol. 18, Amer. Math. Soc., Providence, RI, 1999.
22. H. Nakajima and K. Yoshioka, *Instanton counting on blow-ups. I* (preprint).
23. N. A. Nekrasov, *Seiberg–Witten prepotential from instanton counting*, Proceedings of the International Congress of Mathematicians, Vol. III (Beijing, 2002), Higher Ed. Press, Beijing, 2002, pp. 477–495.
24. N. Nekrasov and A. Okoun'kov, *Seiberg–Witten theory and random partitions* (preprint).
25. M. Thaddeus (private communication).

Department of Mathematics, Harvard University, 1 Oxford Street, Cambridge, MA 02138, USA

E-mail address: braval@math.harvard.edu

Centre de Recherches Mathématiques
CRM Proceedings and Lecture Notes
Volume **38**, 2004

The Gysin Map Is Compatible With Mixed Hodge Structures

Mark Andrea A. de Cataldo and Luca Migliorini

This paper is dedicated to the memory of Meeyoung Kim.

1. Introduction

The language of Bivariant Theory was developed in [**6**] and it turns out to be extremely useful in Intersection Theory and in Riemann–Roch-type questions (cf. [**5**]). The Chow-theoretic version associates a graded group $A^*(X \xrightarrow{f} Y)$ with a morphism $f\colon X \to Y$ of algebraic schemes.

There are a product structure, a proper push-forward and a pull-back.

In the case of the structural map $X \to \operatorname{Spec} k$, we find the Chow groups $A^*(X \to \operatorname{Spec} k) \simeq A_{-*}(X)$, and in the case of the identity map the groups $A^*(X \xrightarrow{\mathrm{Id}} X)$ are called the Chow cohomology groups.

When X is nonsingular these latter groups agree with the Chow groups:

$$A^*(X \xrightarrow{\mathrm{Id}} X) \simeq A_{\dim X - *}(X)$$

and the bivariant product structure agrees with the usual intersection product.

For varieties defined over the field of complex numbers, the topological counterpart, i.e., Topological Bivariant Theory, admits the characterization

$$H^i(X \xrightarrow{f} Y) \simeq \operatorname{Hom}_{D^b(Y)}(Rf_!\mathbb{Q}_X, \mathbb{Q}_Y[i]),$$

from which it follows immediately that

$$H^*(X \to \{\text{point}\}) \simeq \operatorname{Hom}(H_c^*(X), \mathbb{Q}) = H_*^{\mathrm{BM}}(X),$$

the Borel–Moore homology groups, and

$$H^i(X \xrightarrow{\mathrm{Id}} X) = \operatorname{Hom}_{D^b(Y)}(\mathbb{Q}_X, \mathbb{Q}_X[i]) = H^i(X, \mathbb{Q}),$$

the cohomology groups.

2000 *Mathematics Subject Classification.* Primary 14, 32.

The first author is partially supported by N.S.F. Grant DMS 0202321 and NSA Grant MDA904-02-1-0100.

The second author is partially supported by MIUR project Proprietà Geometriche delle Varietà Reali e Complesse and by GNSAGA.

This is the final form of the paper.

It is quite natural to expect that the topological bivariant groups can be endowed with natural mixed Hodge structures (MHS) and that the maps arising in the context of the bivariant formalism should be compatible with MHS.

It seems that this question can be satisfactorily dealt-with only after having developed the Hodge theory of maps, and in particular the Hodge-theoretic version of the Decomposition Theorem (cf. [**2**]).

In this note we shall be less ambitious and we shall deal with a very special yet useful case which does not seem to be available in the literature in the generality which is needed in some applications. See, for example, those contained in [**1**], where morphisms between cohomology groups associated with correspondences between global finite group quotients of smooth, not necessarily complete, varieties are stated to be maps of MHS.

Consider a codimension d regular embedding $h\colon Y \to X$. There is a refined Gysin homomorphism $[h] \in H^{2d}(Y \xrightarrow{h} X) \simeq H^{2d}(X, X \setminus Y)$, cf. [**5**, 19.2], which gives, for any map $X' \to X$, the so-called refined Gysin maps $h^!\colon H_*^{\mathrm{BM}}(X') \to H_{*-2d}^{\mathrm{BM}}(Y \times_X X')$.

Theorem 4.1 states the compatibility of the Gysin map $h^!$ with the MHS involved and it is proved using the definition of the Gysin map via specialization to the normal cone [**5, 9**].

We work with algebraic varieties and algebraic schemes defined over the field of complex numbers. We use cohomology, etc., with rational coefficients.

2. Remarks on mixed Hodge structures

Given a morphism $f\colon Y \to X$ of algebraic schemes, the relative cohomology groups $H^*(X \bmod Y)$ are given a MHS which is functorial in the map f, in the sense that given a commutative diagram of maps of algebraic schemes

$$\begin{CD} Y' @>{i_1}>> Y \\ @V{f'}VV @VV{f}V \\ X' @>{i_2}>> X, \end{CD}$$

the natural morphism $H^*(X \bmod Y) \to H^*(X' \bmod Y')$ is a morphism of Mixed Hodge structures (cf. [**3**, Example 8.3.8]; see also [**4**, §2]). The relative cohomology groups are the cohomology groups of the simplicial scheme $C(f)$.

The diagram above gives rise to a map of simplicial schemes $C(f') \to C(f)$, hence the morphism of MHS.

Given an algebraic scheme U, one has, for every open immersion $U \to U'$ into a proper algebraic scheme, $H_c^l(U) \simeq H^l(U', U' \setminus U)$.

It follows that cohomology with compact supports admits a natural MHS which is functorial for open immersions and proper maps.

Dually, since $H_l^{\mathrm{BM}}(U) \simeq H_c^l(U)^\vee$, the same is true for Borel–Moore homology.

3. The deformation to the normal cone

For more details on what follows, see [**5, 9**].

Let $h\colon Y \to X$ be a closed embedding of algebraic schemes, and $\mathcal{I}_Y \subseteq \mathcal{O}_X$ be the corresponding sheaf of ideals.

The *normal cone to Y in X* is defined to be the algebraic scheme

$$C_Y X := \mathrm{Spec}_{\mathcal{O}_Y} \bigoplus_{n\geq 0} \mathcal{I}_Y^n / \mathcal{I}_Y^{n+1}$$

There are natural maps $Y \to C_Y X \to Y$, where the second one is the (affine) cone-bundle projection to Y and the first one is the closed embedding of the zero-section.

If the embedding of Y in X is regular of codimension d, then $C_Y X$ is naturally identified with the rank d normal bundle $N_{Y,X}$ of Y in X.

The exceptional divisor of the blowing-up $Bl_Y X \to X$ of X along Y is the projectivization $P(C_Y X)$.

Let us recall how the deformation to the normal cone and the canonical maps which are associated with it are defined.

Let $M = \mathrm{Bl}_{Y\times\{0\}} X \times \mathbb{A}^1$ be the blowing-up of $X \times \mathbb{A}^1$ along $Y \times \{0\}$.

The exceptional divisor can be naturally identified with $P(C_Y X \oplus \mathcal{O}_Y)$.

The resulting map $\tilde{p}\colon M \to \mathbb{A}^1$ is flat.

We have that:

- the blowing-up $\mathrm{Bl}_Y X$ is naturally embedded in $\tilde{p}^{-1}(0) \subseteq M$;
- $\tilde{p}^{-1}(0) = P(C_Y X \oplus \mathcal{O}_Y) \cup \mathrm{Bl}_Y X$;
- $P(C_Y X \oplus \mathcal{O}_Y) \cap \mathrm{Bl}_Y X$ can be viewed either as the exceptional divisor of the blowing-up $\mathrm{Bl}_Y X$, or as the divisor at infinity of $P(C_Y X \oplus \mathcal{O}_Y)$;
- $C_Y X$ embeds in the exceptional divisor $P(C_Y X \oplus \mathcal{O}_Y)$ as the complement of the divisor at infinity;
- Y embeds in $C_Y X \subseteq P(C_Y X \oplus \mathcal{O}_Y)$ as the zero section;
- $Y \times \mathbb{A}^1$ embeds in M, compatibly with the projection to $\mathbb{A}^1$; in particular, over $\mathbb{A}^1 \setminus \{0\}$, it is just the product embedding into $X \times \mathbb{A}^1 \setminus \{0\}$; $Y \times \{0\}$ is embedded in $p^{-1}(0)$ as the zero-section of $C_Y X \subseteq P(C_Y X \oplus \mathcal{O}_Y)$.

Let

$$M \setminus Bl_Y X =: M' \xrightarrow{p} \mathbb{A}^1$$

be the natural flat map. The map p is not proper, even if X is complete.

We say that the embedding $Y \subseteq X$ (i.e., the situation at $p^{-1}(t \neq 0)$) deforms to the embedding $Y \subseteq C_Y X$ (i.e., the situation at $p^{-1}(0)$).

This construction is called the *deformation to the normal cone.*

There are specialization maps [**9**]:

$$h^?\colon H_l^{\mathrm{BM}}(X) \to H_l^{\mathrm{BM}}(C_Y X), \qquad h^{?\vee}\colon H_c^l(C_Y X) \to H_c^l(X),$$

whose construction is recalled in the proof of Theorem 4.1.

In the case of a regular embedding the normal cone is in fact the normal bundle and the flat pull-back gives an isomorphism $H_l^{\mathrm{BM}}(Y) \simeq H_{l+2d}^{\mathrm{BM}}(C_Y X)$.

The Gysin map $h^!$ is, by definition, the composition $H_l^{\mathrm{BM}}(X) \to H_l^{\mathrm{BM}}(C_Y X) \to H_{l-2d}^{\mathrm{BM}}(Y)$.

4. The main result

THEOREM 4.1. *Let $h\colon Y \to X$ and $C_Y X$ be as above. The natural maps*

$$h^?\colon H_l^{\mathrm{BM}}(X) \to H_l^{\mathrm{BM}}(C_Y X), \qquad h^{?\vee}\colon H_c^l(C_Y X) \to H_c^l(X)$$

are maps of MHS of type $(0,0)$.

If h is regular of pure codimension d, then the Gysin map and its dual

$$h^{!}\colon H_l^{\mathrm{BM}}(X) \to H_{l-2d}^{\mathrm{BM}}(Y), \qquad h^{!\vee}: H_c^l(Y) \to H_c^{l+2d}(X)$$

are maps of MHS of type $(-d,-d)$ and (d,d), respectively.

PROOF. We first consider a commutative diagram of maps of algebraic schemes

$$\begin{array}{ccccc} U & \xrightarrow{\ j\ } & \mathcal{X} & \xleftarrow{\ i\ } & B \\ & {\scriptstyle p}\searrow & \downarrow{\scriptstyle \pi} & \swarrow{\scriptstyle b} & \\ & & S & & \end{array}$$

where j is an open immersion, $B := \mathcal{X} \setminus U$ and π is proper.

We have a distinguished triangle in the bounded derived category $D^b_{cc}(\mathcal{X})$ of the category of constructible complexes of rational vector spaces on $\mathcal{X}$:

$$Rj_! j^! \mathbb{Q}_{\mathcal{X}} (\simeq Rj_! \mathbb{Q}_U) \to \mathbb{Q}_{\mathcal{X}} \to Ri_* \mathbb{Q}_B \xrightarrow{+1},$$

to which we apply $R\pi_* \simeq R\pi_!$ and get

$$Rp_! \mathbb{Q}_U \to R\pi_* \mathbb{Q}_{\mathcal{X}} \to Rb_* \mathbb{Q}_B \xrightarrow{+1}.$$

It follows that

$$\mathbb{H}^l(S, Rp_! \mathbb{Q}_U) \simeq H^l(\mathcal{X}, B), \quad \forall l.$$

Let $s \in S$ be a point. By the remarks in §2, the map of pairs:

$$\big(\pi^{-1}(s), b^{-1}(s)\big) \to (\mathcal{X}, B)$$

induces natural maps $H^l(\mathcal{X}, B) \to H^l\big(\pi^{-1}(s), b^{-1}(s)\big)$ of MHS for every l. Since π is proper, $\pi^{-1}(s)$ is compact and the latter group can be identified with $H^l_c\big(p^{-1}(s)\big)$. This endows these latter groups with MHS.

To summarize, the natural maps

$$\mathbb{H}^l(S, Rp_! \mathbb{Q}_U) \to H^l_c\big(p^{-1}(s)\big)$$

are of MHS, for every $s \in S$ and for every l.

We apply what above to the following situation.

Let $Y' \subseteq X'$ be an algebraic compactification of $Y \subseteq X$, $\mathcal{X} := \mathrm{Bl}_{Y' \times \{0\}}(X' \times \mathbb{A}^1)$, $B := (X' \setminus X) \times (\mathbb{A}^1 \setminus \{0\}) \coprod Bl_{Y'}X' \times \{0\} \coprod C_{Y' \setminus Y}(X' \setminus X)$, $U := \mathcal{X} \setminus B$. Let $\pi\colon \mathcal{X} \to \mathbb{A}^1$ be the natural map and p and b be the induced maps.

Note that $p^{-1}(\{0\}) = C_Y X$.

We have maps of MHS by choosing $s = 0$ and $s = s_0 \neq 0$:

$$H^l_c\big(p^{-1}(0)\big) \xleftarrow{a} \mathbb{H}^l(Rp_! \mathbb{Q}_U) \simeq H^l?(\mathcal{X}, B) \to H_c\big(p^{-1}(s_0)\big).$$

CLAIM. *a is an isomorphism of MHS for every l.*

PROOF. It is enough to prove that a is an isomorphism. Consider the fundamental system of neighborhoods of the point $\{0\} \in \mathbb{A}^1$ given by disks $D_r = \{z : |z| < r\}$. Let $\mathcal{X}_r := \pi^{-1}(D_r)$, $U_r : p^{-1}(D_r)$, $B_r = b^{-1}(D_r)$, $p_r := p_{|U_r}$. Since $\mathcal{X}$ is a product away from $\pi^{-1}(\{0\})$, the homotopy axiom for relative cohomology ensures that the restriction maps $\mathbb{H}^l(\mathbb{A}^1, Rp_! \mathbb{Q}_U) \simeq H^l(\mathcal{X}, B) \to H^l(\mathcal{X}_r, B_r) \simeq \mathbb{H}^l(U_r, Rp_{r!} \mathbb{Q}_{U_r})$ are isomorphisms for every $r > 0$.

The complex $Rp_! \mathbb{Q}_U$ is constructible with respect to the obvious stratification of $\mathbb{A}^1$ and, by [**7**, §1.4], this implies that the restriction map $H^l(\mathcal{X}, B) \simeq \mathbb{H}^l(\mathbb{A}^1, Rp_! \mathbb{Q}_U) \xrightarrow{a} H^l_c\big(p^{-1}(0)\big)$ is an isomorphism. □

The first assertion of the theorem follows.

If the embedding $Y \subseteq X$ is regular of codimension d, then $C_Y X$ is the normal bundle of Y in X and we have an isomorphism of MHS $H_l^{\mathrm{BM}}(Y) \simeq H_{l+2d}^{\mathrm{BM}}(C_Y X)$ of type (d, d) and the Gysin map $h^!$ being, by definition, the composition $H_l^{\mathrm{BM}}(X) \to H_l^{\mathrm{BM}}(C_Y X) \to H_{l-2d}^{\mathrm{BM}}(Y)$, is therefore of MHS of type $(-d, -d)$. □

REMARK 4.2. After simple modifications to the statement and to the proof, Theorem 4.1 holds when h is a local complete intersection morphism.

The following corollary can be used when dealing with correspondences:

COROLLARY 4.3. *Let $X \xleftarrow{p} \Gamma \xrightarrow{q} Y$ be algebraic maps of algebraic varieties. Assume that the graph embedding $h\colon \Gamma \to \Gamma \times X$ is regular and that q is proper. Then the natural map*

$$\Gamma_*\colon H_\bullet^{\mathrm{BM}}(X) \to H_{\bullet+2(\dim\Gamma-\dim X)}^{\mathrm{BM}}(Y)$$

is of MHS of type $(\dim\Gamma - \dim X, \dim\Gamma - \dim X)$.

PROOF. We have $\Gamma_*(x) = q_*\big(h^!([\Gamma] \times x)\big)$ and all operations involved are compatible with MHS. □

REMARK 4.4. The same statement above holds when X is assumed to be smooth or a quotient $X = X'/G$ of a smooth variety by a finite group, in which case one modifies the statement by working with X' and with the distinguished irreducible component of $\Gamma \times_X X'$. We leave this task to the reader.

5. Examples

The paper [**1**] contains several examples of correspondences, stemming from maps between surfaces, from Hilbert schemes of points on surfaces, semi-small resolutions of singularities etc. In what follows, for the reader's convenience, we offer two of the applications of Theorem 4.1 contained in [**1**].

5.1. Small resolutions. Let $f_i\colon X_i \to Y$, $i = 1, 2$, be *small* resolutions of the singularities of an algebraic variety Y. This means that the X_i are nonsingular, the f_i are proper and birational morphisms, and the not necessarily irreducible algebraic schemes $X_i \times_Y X_i$ have exactly one component of maximal dimension $\dim Y$. This component is the unique component dominating the spaces X_i under either projection. See [**1**] for more on this notion and more references. Note that the same will be true for $X_1 \times_Y X_2$.

Let $D_{12} \subseteq X_1 \times_Y X_2$ be the unique irreducible component of maximal dimension $\dim Y$.

Theorem 4.1 and a simple calculation using the calculus of correspondences imply the following

PROPOSITION 5.1. *The maps*

$$D_{ij*}\colon H_\bullet^{\mathrm{BM}}(X_i) \to H_\bullet^{\mathrm{BM}}(X_j)$$

are isomorphisms of MHS with inverse D_{ji*}.

In particular, the virtual Hodge-Deligne numbers of the varieties X_i, $i = 1, 2$, coincide.

PROOF. See [**1**, §3]. □

5.2. Wreath products, rational double points and orbifolds. The combinatorial details of this example are rather lengthy. We omit them in favor of the end result. The interested reader can see [**1**, §7.3].

Let Y' be a smooth algebraic surface and $G \subseteq \mathrm{SL}_2(\mathbb{C})$ be a finite group acting on Y' with only isolated fixed points.

Let $Y := Y'/G$ and $f\colon X \to Y$ be its minimal resolution.

The semi-direct product G_n (called the Wreath product) of G^n with the symmetric group in n letters S_n acts on Y'^n.

The Hilbert scheme of $n-$points $X^{[n]}$ is a semi-small resolution of the singularities of Y'^n/G_n.

There is an explicit collection of correspondences that arises in this situation.

There is the notion of orbifold cohomology groups $H^*(Y'^n/G_n)_{\mathrm{orb}}$ for the pair (Y'^n, G_n). These groups carry a natural MHS.

Theorem 4.1 allows to prove, using the aforementioned collection of correspondences, the following

PROPOSITION 5.2. *There is a canonical isomorphism of MHS*

$$H^*(Y'^n/G_n)_{\mathrm{orb}} \simeq H^*(X^{[n]}).$$

References

1. M. A. de Cataldo and L. Migliorini, *The Chow motive of semismall resolutions*, Math. Res. Lett. (to appear); math-AG/0204067.
2. ———, *The Hodge theory of algebraic maps*, math-AG/0306030.
3. P. Deligne, *Théorie de Hodge.* III, Inst. Hautes Études Sci. Publ. Math. **44** (1974), 5–78.
4. A. H. Durfee, *Mixed Hodge structures on punctured neighborhoods*, Duke Math. J. **50** (1983), no. 4, 1017–1040.
5. W. Fulton, *Intersection theory*, Ergeb. Math. Grenzgeb. (3), vol. 2, Springer-Verlag, Berlin, 1984.
6. W. Fulton, and R. MacPherson, *Categorical framework for the study of singular spaces*, Mem. Amer. Math. Soc. **31** (1981), no. 243.
7. M. Goresky and R. MacPherson, *Intersection homology.* II, Invent. Math. **72** (1983), no. 1, 77–129.
8. U. Jannsen, *Deligne homology, Hodge-D-conjecture, and motives*, Beilinson's Conjectures on Special Values of L-Functions, Perspect. Math., vol. 4, Academic Press, Boston, MA, 1988, pp. 305–372.
9. J.-L. Verdier, *Le théorème de Riemann-Roch pour les variétés algébriques éventuellement singulières (d'après P. Baum, W. Fulton et R. MacPherson)*, Séminaire Bourbaki (1974/1975: Exposés Nos. 453–470), Lecture Notes in Math., vol. 514, Springer, Berlin, 1976, pp. 159–175.

DEPARTMENT OF MATHEMATICS, STATE UNIVERSITY OF NEW YORK AT STONY BROOK, STONY BROOK, NY 11794, USA

E-mail address: mde@math.sunysb.edu

DIPARTIMENTO DI MATEMATICA, UNIVERSITÀ DI BOLOGNA, PIAZZA DI PORTA S. DONATO 5, 40126 BOLOGNA, ITALY

E-mail address: migliori@dm.unibo.it

Centre de Recherches Mathématiques
CRM Proceedings and Lecture Notes
Volume **38**, 2004

Representations of Fundamental Groups of Nonorientable 2-Manifolds

Nan-Kuo Ho and Lisa C. Jeffrey

ABSTRACT. We compute the Riemannian volume of the moduli space of flat connections on a nonorientable 2-manifold, for a natural class of metrics which are hyperbolic at all but finitely many points. We also show that Witten's volume formula for these moduli spaces may be derived using Haar measure, and we use Haar measure to give a new proof of Witten's volume formula for the moduli space of flat connections on a Riemann surface.

1. Introduction

In [**3**], Atiyah and Bott studied the moduli space $\mathcal{M}(\Sigma)$ of gauge equivalence classes of flat connections on an orientable 2-manifold: they exhibited this moduli space as an infinite-dimensional symplectic quotient. In [**13**] Witten studied the moduli space $\mathcal{M}(\Sigma)$ of gauge equivalence classes of flat connections on a 2-manifold Σ which may be either orientable or nonorientable. Witten defined a measure on the moduli space for both the orientable and nonorientable cases. Witten's measure was defined using Reidemeister–Ray–Singer torsion. When the 2-manifold is orientable, Witten showed that his volume agrees with the symplectic volume on the moduli space. However, when the 2-manifold is not orientable the moduli space does not have a symplectic form, so it has been unclear how to interpret this volume using symplectic geometry.

The moduli space of gauge equivalence classes of flat connections on a nonorientable 2-manifold Σ turns out to be a Lagrangian submanifold of the moduli space of gauge equivalence classes of flat connections on the orientable double cover $\widetilde{\Sigma}$: it is the fixed point set of the natural involution on the moduli space associated to $\widetilde{\Sigma}$ which arises from the covering transformation. In this article we compute its Riemannian volume for a natural class of metrics on the 2-manifold.

2000 *Mathematics Subject Classification.* 53D99.

The first author was partially supported by grants from OGS and OGSST.

The second author was partially supported by a grant from NSERC.

This article is based on part of the Ph.D. thesis of the first author, under the supervision of the second author. See also [**11**].

Both authors would like to thank Eckhard Meinrenken and Chris Woodward for useful conversations and advice.

This is the final form of the paper.

2. A hyperbolic metric on the 2-manifold

2.1. Hyperbolic metrics on 2-manifolds. All nonorientable 2-manifolds are of the form $\Sigma \# P$ or $\Sigma \# P \# P$, where Σ is an orientable 2-manifold without boundary (in other words a Riemann surface). Here P is the real projective plane $\mathbb{RP}^2$.

A Riemann surface with one boundary component and genus ≥ 1 always has a hyperbolic metric with constant curvature -1 [**6**]. For this reason we construct a metric on $P - D$ (the Möbius band, which is the complement of a disc D in P) for which the boundary is a geodesic and which is hyperbolic with constant curvature -1 at all but one point. We do this by gluing the edges of an isosceles geodesic hyperbolic triangle with opposite orientation: this process identifies all the vertices of the triangle, and the vertex becomes the point where the metric does not have constant curvature -1 (in fact, the metric is singular at this point). It is not possible to form a metric on $P - D$ which has constant curvature -1 and for which the boundary is a geodesic: this would contradict the Gauss-Bonnet formula. (See for example [**7**, Section 4.5]).

REMARK 1. The reason why this procedure gives a metric on the 2-manifold formed by taking the connected sum of a Riemann surface of genus $\ell > 0$ with P is that a Riemann surface of genus $\ell > 0$ with one or two boundary components has a hyperbolic metric with constant curvature for which the boundary components are geodesics. This metric can be obtained from a pants decomposition of the Riemann surface, where each pair of pants is equipped with a hyperbolic structure.

Hence the connected sum of a Riemann surface with P has a metric which is hyperbolic at all but one point, and the connected sum of a Riemann surface with two copies of P has a metric which is hyperbolic at all but 2 points. To form P with one disc removed, we identify the two edges of the triangle using an orientation reversing map. In Figure 1 this corresponds to identifying γ_1 with γ_2 using an orientation reversing map which maps $x_1 \in \gamma_1$ to $x_0 \in \gamma_2$ and maps $x_0 \in \gamma_1$ to $x_2 \in \gamma_2$. It turns out that the metric on the based moduli space on P with one disc removed is singular at the vertex. In this paper, it is only the one boundary component case that is treated extensively since the case of more than one boundary component is similar.

We introduce a geodesic triangle $\triangle$ in the upper half plane (equipped with a hyperbolic metric): see Figure 1.

Let G be a compact connected Lie group; we consider the moduli space of flat connections on $\triangle$ modulo based gauge transformations (i.e., gauge transformations whose value is the identity on the three vertices).

As shown in Figure 1, the geodesic triangle has one distinguished vertex x_0 (which will serve as the origin for geodesic polar coordinates). The sides γ_1, γ_2, γ_3 are geodesics. We introduce geodesic polar coordinates (ρ, σ) (using the hyperbolic metric on the upper half plane, which is assumed to contain the triangle) with x_0 as the origin: here ρ is the distance from the chosen point x_0 and σ is the polar angle. It is assumed that the geodesic γ_1 is defined by the equation $\sigma = 0$, and γ_2 is defined by $\sigma = \phi$ (where ϕ is the angle at x_0). The curves γ_1 and γ_2 have lengths L_1 and L_2 which are the same here. The geodesic γ_3 is specified by Coxeter's equation

$$\coth \rho = \frac{\cos(\sigma - \sigma_0)}{\ell}, \quad \text{see [8, p. 376]}, \tag{2.1}$$

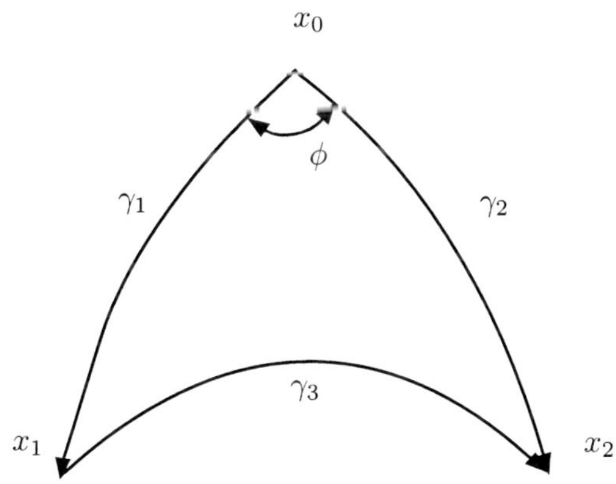

FIGURE 1. Hyperbolic geodesic triangle

where ℓ, σ_0 are constants determined by the geodesic γ_3. Also,

$$\coth\rho = \frac{\cosh\rho}{\sinh\rho} = \frac{\cosh^2(\rho/2)+\sinh^2(\rho/2)}{2\sinh(\rho/2)\cosh(\rho/2)} = \frac{1}{2}\left(\tau + \frac{1}{\tau}\right)$$

for $\tau = \tanh(\rho/2)$. Thus

$$\left(1 - \frac{\cos(\sigma-\sigma_0)}{\ell\tau}\right) d\tau = \frac{-\sin(\sigma-\sigma_0)}{\ell}\, d\sigma.$$

By (2.1), it is clear that since $\rho(\sigma = 0)$ must be equal to $\rho(\sigma = \phi)$ for a hyperbolic geodesic triangle two sides of which will be identified to form $P - D$, we need $\sigma_0 = \phi/2$ in (2.1).

The moduli space of flat connections on $P - D$ modulo gauge transformations which are trivial at one point on the boundary (corresponding to the vertices of the geodesic triangle) is equivalent to the subspace of the moduli space of flat connections on the geodesic triangle (modulo based gauge transformations) which correspond under an orientation reversing map which identifies two sides of the triangle. This is $\mathrm{Hom}(\pi, G)$ where $\pi = \pi_1(P - D)$, in other words

$$\mathrm{Hom}(\pi, G) = \{(u, v) \in G \times G \mid uv^2 = 1\} \cong G.$$

Here v is the holonomy along γ_1 and $-\gamma_2$, while u is the holonomy along γ_3.

On the space of flat connections on $\triangle$, there is a metric

$$\langle a, b\rangle = \int_\triangle \mathrm{Tr}(a \wedge *b), \tag{2.2}$$

where $*$ is the Hodge star operator on differential forms.

We will study the corresponding metric on $G \times G$. Define the metric $\langle\cdot,\cdot\rangle$ at a flat connection A corresponding to a point p in $G \times G$ to be

$$\langle d_A\xi, d_A\eta\rangle =: \int_\triangle \mathrm{Tr}(d_A\xi \wedge *d_A\eta),$$

where $\xi(\cdot)$ and $\eta(\cdot)$ are functions on $\triangle$, so $d_A\xi$ and $d_A\eta$ represent elements of the tangent space to $G \times G$ at p. Here Tr represents the Ad-invariant inner product on the Lie algebra.

In the case when G is abelian, the holonomy of $d\xi$ along γ_1 is

$$\mathrm{Hol}_{\gamma_1} d\xi = \exp \int_{\gamma_1} d\xi = \exp\big(\xi(\rho = L_1, \sigma = 0) - \xi(\rho = 0, \sigma = 0)\big), \tag{2.3}$$

and its holonomy along γ_2 is

$$\mathrm{Hol}_{\gamma_2} d\xi = \exp \int_{\gamma_2} d\xi = \exp\big(\xi(\rho = L_2, \sigma = \phi) - \xi(\rho = 0, \sigma = \phi)\big). \tag{2.4}$$

From now on, we only consider the inner product at $A = 0$. The reason is that the metric $\langle a, b\rangle = \int \mathrm{Tr}(a \wedge *b)$ is ad-invariant (where $*$ denotes the Hodge star operator), and since the gauge action at the infinitesimal level is the adjoint action, this metric is invariant under the action of the gauge group and in particular under the action of the based gauge group. Since any flat connection is gauge equivalent to the trivial connection on a triangle, we need only consider the inner product at $A = 0$.

The space $\triangle$ is contractible, so all flat connections are obtained as the image of one particular flat connection (denoted 0) under a gauge transformation: in other words all infinitesimal flat connections—elements of the tangent space of the space of flat connections modulo based gauge transformations—can be expressed as $d\xi$ for some infinitesimal gauge transformation $\xi\colon \triangle \to \mathfrak{g}$. This means simply that all flat connections are equivalent to the trivial connection modulo a gauge transformation. So we obtain

$$\langle d\xi, d\eta\rangle = \int_\triangle \mathrm{Tr}(d\xi \wedge *d\eta). \tag{2.5}$$

This can be converted to a line integral using

$$\int_\triangle \mathrm{Tr}(d\xi \wedge *d\xi) = \int_{\partial\triangle} \mathrm{Tr}(\xi \wedge *d\xi) - \int_\triangle \mathrm{Tr}(\xi \wedge d * d\xi). \tag{2.6}$$

We assume ξ satisfies

$$d * d\xi = 0. \tag{2.7}$$

This condition represents a transversal to the orbit of the based gauge group; our procedure is analogous to identifying de Rham cohomology classes with harmonic forms (forms α satisfying $d\alpha = 0$ and $d * \alpha = 0$). Note that the space $G \times G$ is isomorphic to the space of flat connections on $\triangle$ modulo based gauge transformations (i.e., gauge transformations which take the value 1 at the vertices). Each equivalence class may be written as $d\xi$ for some ξ (which does not take the value 1 at the vertices, unless one wishes to represent the trivial flat connection).

Suppose that (ρ, σ) are geodesic coordinates on the hyperbolic plane. Recall that the hyperbolic metric in geodesic polar coordinates is

$$ds^2 = d\rho^2 + \sinh^2 \rho d\sigma^2,$$

where ρ is the distance from a chosen point (x_0 in Figure 1) and σ is the polar angle with respect to this point. Let $*$ denote the Hodge star operator on the 2-manifold; the star operator with respect to these polar coordinates is (see [**6**]).

$$*d\rho = \sinh(\rho) d\sigma, \qquad *d\sigma = \frac{-1}{\sinh \rho} d\rho. \tag{2.8}$$

A solution $\xi(\rho, \sigma)$ to $d * d\xi = 0$ is given by

$$\xi(\rho, \sigma) = Y\tau^k(\rho) \cos k\sigma,$$

where $\tau(\rho) = \tanh(\rho/2)$ and Y is a constant element in $\mathfrak{g}$. We prove this as follows. We note that

$$\frac{d\rho}{\sinh(\rho)} = \frac{d\tau}{\tau}$$

so using $u = \ln(\tau)$ the equation (2.8) becomes

$$*du = d\sigma, \qquad *d\sigma = -du. \tag{2.9}$$

We have

$$d * d\xi = (\xi_{uu} + \xi_{\sigma\sigma})\, du \wedge d\sigma \tag{2.10}$$

so (2.7) becomes

$$\xi_{uu} + \xi_{\sigma\sigma} = 0.$$

We introduce a Fourier decomposition

$$\xi(u,\sigma) = \int_{k\in\mathbb{R}} \hat{\xi}(u,k)e^{ik\sigma}\, dk$$

so

$$\xi_{\sigma\sigma} = \int_{k\in\mathbb{R}} \hat{\xi}(u,k)(-k^2)e^{ik\sigma}\, dk$$

and

$$\xi_{uu} = \int_{k\in\mathbb{R}} \frac{\partial^2}{\partial u^2}\hat{\xi}(u,k)e^{ik\sigma}\, dk.$$

Thus

$$\hat{\xi}(u,k) = C_+(k)e^{ku} + C_-(k)e^{-ku} = C_+(k)\tau^k + C_-(k)\tau^{-k}.$$

Imposing the condition that $\hat{\xi}(u,k)$ is finite at $\rho = 0$, we obtain (for $k > 0$)

$$\hat{\xi}(u,k) = C_+(k)\tau^k$$

and (for $k < 0$)

$$\hat{\xi}(u,k) = C_-(k)\tau^{-k}.$$

Let ϕ be the polar angle of the triangle $\triangle$. We impose the constraint that $f(\rho,\sigma)$ attains its extreme values on $\partial\triangle$ (specifically, at $\sigma = 0$ and $\sigma = \phi$), in other words that $\frac{\partial}{\partial\sigma}f(\rho,\sigma) = 0$ at $\sigma = 0$ and $\sigma = \phi$. This leads to

$$\xi(\rho,\sigma) = K\tau^k \cos k\sigma, \tag{2.11}$$

where

$$k = \pi/\phi$$

and K is an arbitrary constant.

We have already described how we form $P - D$, starting from $\triangle$. We let

$$\xi(\rho,\sigma) = Y\tau^k(\rho)\cos k\sigma.$$

The metric is then

$$\langle d\xi, d\xi\rangle = \int_{\gamma_1-\gamma_2+\gamma_3} \mathrm{Tr}(\xi \wedge *d\xi)$$

and this turns out to be

$$\langle Y, Y\rangle h(\phi) \tag{2.12}$$

where $h(\phi)$ is a function which can be calculated explicitly. The volume element of the moduli space using this metric is

$$\big(h(\phi)\big)^{n/2},$$

where $n = \dim(G)$.

3. Volume formulas

Introduce a parameter $t \in T/W$. The moduli space of conjugacy classes of representations of the fundamental group of an orientable surface Σ with one boundary component and genus ℓ is

$$\mathcal{M}(\Sigma, t) = \left\{ (h_1, \dots, h_{2\ell}) \in G^{2\ell} \;\middle|\; \prod_{j=1}^{\ell} [h_j, h_{j+\ell}] = t \right\} \Big/ \operatorname{Stab}(t). \tag{3.1}$$

In algebraic geometry this space is identified with the moduli space of parabolic bundles with weights t.

Every moduli space of parabolic bundles on an orientable 2-manifold is Kähler, so the volume element on such a moduli space is both the symplectic volume and the Riemannian volume. Thus the volume element on the moduli space of parabolic bundles is the Riemannian volume; hence so is the volume we will compute for $M(\Sigma \# P)$ and $M(\Sigma \# P \# P)$.

In [**13**], Witten gave the following formula for the volume of the moduli space $\mathcal{M}(\Sigma_\ell, s)$ of flat connections on a compact orientable surface Σ_ℓ of genus ℓ with one boundary component. If $s \in T/W$ is a conjugacy class, then [**13**, (4.114)] tells us that

$$\operatorname{Vol}(\mathcal{M}(\Sigma_\ell, s)) = C \sum_\alpha \frac{1}{(\dim \alpha)^{2\ell-1}} \chi_\alpha(s) \sqrt{F(s)}, \tag{3.2}$$

where α runs over all isomorphism classes of irreducible representations of G, the representation α has character χ_α and dimension d_α, and the constant C is

$$C = \frac{\sharp Z(G) \operatorname{Vol}(G)^{2\ell-2+1}}{(2\pi)^{\dim \mathcal{M}(\Sigma_\ell, s)} \operatorname{Vol}(T)}. \tag{3.3}$$

This is equivalent to

$$\operatorname{Vol}\big(\mathcal{M}(\Sigma_\ell, s)\big) = v_\ell(s),$$

where

$$v_\ell(s) = C \sum_\alpha \frac{1}{(d_\alpha)^{2\ell-1}} \chi_\alpha(s),$$

since the factor $F(s)$ which results from multiplying two copies of (3.2)) is simply the volume of the conjugacy class through s. Weyl's integral formula tells us that

$$\int_G f(g)\, dg = \int_{T/W} F(s) f(s) \frac{\operatorname{Vol}(G)}{\operatorname{Vol}(T)}\, ds, \tag{3.4}$$

where $F(s)$ is given by

$$v = \frac{\operatorname{Vol}(G)}{\operatorname{Vol}(T)} v_0 F(s). \tag{3.5}$$

Here

$$\dim \mathcal{M}(\Sigma_\ell, s) = (2\ell - 1)|G| - |T|, \tag{3.6}$$

where $|G|$ denotes the dimension of G and $|T|$ denotes the dimension of the maximal torus T. Note also that $F(s)$ is the volume of the conjugacy class through s as defined in [**13**, (4.53)]. In Section 5 we give a new proof of the formula (3.2).

Let us consider the connected sum of a Riemann surface of genus ℓ with $\mathbb{RP}^2$. We have

$$\begin{aligned}(3.7)\qquad \operatorname{Vol}(\mathcal{M}) &= \operatorname{Vol}\big(\mathcal{M}(\Sigma_\ell \,\#\, \mathbb{RP}^2)\big)\\ &= \int_{T/W} \operatorname{Vol}\big(\mathcal{M}(\Sigma_\ell, s)\big) \operatorname{Vol}\big(\mathcal{M}(\mathbb{RP}^2, s)\big)\, ds\\ &= \frac{\operatorname{Vol}(T)}{\operatorname{Vol}(G)} \int_G \sum_\alpha \frac{C\chi_\alpha(g)}{(d_\alpha)^{2\ell-1}} R_*\big(\sqrt{\det(\gamma)}\, dg\big)\end{aligned}$$

(where γ is the metric on $G \times G$ defined by (2.5) and the map $R\colon G \to G$ is defined by $R(g) = g^2$)

$$= \frac{\operatorname{Vol}(T)}{\operatorname{Vol}(G)} \sum_\alpha \int_G \frac{C\chi_\alpha(g^2)}{(d_\alpha)^{2\ell-1}} \sqrt{\det(\gamma)}\, dg.$$

The above is true because $S^2 - D - D$ (in other words, a cylinder) is the double cover of $RP^2 - D$; the space of flat connections on $S^2 - D - D$ (modulo based gauge transformations) is isomorphic to G, so we pull back the integral to G under the covering map. We get

$$(3.8)\qquad \operatorname{Vol}(\mathcal{M}) = \frac{\sharp Z(G) \operatorname{Vol}(G)^{2\ell-1}}{(2\pi)^{(2\ell-2)|G|+|T|}} h(\phi) \sum_\alpha \frac{f_\alpha}{(d_\alpha)^{2\ell-1}}.$$

Here, we have used a classical formula for $\int_G \chi_\alpha(g^2)\, dg$ (quoted in [**13**, (2.70)]). Here, $f_\alpha = 1$ if the representation α has a symmetric bilinear form, $f_\alpha = -1$ if α is equipped with an antisymmetric bilinear form, and $f_\alpha = 0$ otherwise. Also, $h(\phi)$ was described in (2.12) and it turns out that

$$(3.9)\qquad h(\phi) = k\langle Y, Y\rangle \int_{\cos(\phi/2)}^1 \big(x/l - \sqrt{(x/l)^2 - 1}\big)^{2k} \frac{dx}{\sqrt{1 - x^2}}$$

so that $\tau\big(L(\phi)\big)$ is a function of the top angle ϕ. Here $l = \cos(\phi/2)\tanh(L)$; note that the length $L(\phi)$ is determined by hyperbolic trigonometric identities [**6**] once the length of the third side of $\triangle$ is fixed.

Our volume formula differs from Witten's only by the constant multiplicative factor $h(\phi)$, which does not depend on the genus ℓ, but only on the angle ϕ chosen for the triangle.

4. Metric dependence

The torsion volume on the moduli space $\mathcal{M}(\Sigma)$ does not depend on the choice of a metric on Σ. The Riemannian volume on the moduli space of flat connections on a nonorientable surface does depend on the metric (as shown by our formula, and as remarked by Witten [**13**, p. 163]).

For an orientable 2-manifold, the moduli space of gauge equivalence classes of flat connections is Kähler so the Riemannian volume is the same as the symplectic volume, and hence it does not depend on the choice of a metric on the manifold.

5. Another proof of Witten's volume formula for orientable surfaces, using Haar measure

Witten computed the volume of $\mathcal{M}(\Sigma, t)$ (stated in (3.2)): we have introduced the quantity

$$v_\ell(s) = C \sum_\alpha \frac{1}{(d_\alpha)^{2\ell-1}} \chi_\alpha(s) \tag{5.1}$$

and will give an inductive proof (using Haar measure) that

$$v_\ell(s) = \operatorname{Vol} \mathcal{M}(\Sigma, s).$$

The integral is over $G \times G$ using standard Haar measure on G; no further geometrical structure on G is introduced. Here we obtain equation (5.2) below by considering the orientable 2-manifold of genus $\ell + 1$ with one boundary component which is obtained by gluing an orientable 2-manifold of genus ℓ with one boundary component to a torus with two boundary components.

$$\begin{aligned} v_{\ell+1}(s) &= \int_{(g_1,g_2)\in G\times G} v_\ell(s g_1 g_2 g_1^{-1} g_2^{-1})\, dg_1\, dg_2 \\ &= C \sum_\alpha \frac{1}{(d_\alpha)^{(2\ell-1)}} \chi_\alpha(s g_1 g_2 g_1^{-1} g_2^{-1}) dg_1 dg_2 \\ &= C \sum_\alpha \frac{1}{(d_\alpha)^{(2\ell-1)}} \chi_\alpha(s g_1) \chi_\alpha(g_1^{-1}) dg_1 \frac{1}{d_\alpha} \end{aligned} \tag{5.2}$$

using

$$\int_{u\in G} \chi_\alpha(A u B u^{-1})\, du = \frac{\operatorname{Vol}(G)}{d_\alpha} \chi_\alpha(A) \chi_\alpha(B). \tag{5.3}$$

Now we use

$$\int_{g\in G} \chi_\alpha(hg) \chi_\alpha(g^{-1})\, dg = \frac{\operatorname{Vol}(G)}{d_\alpha} \chi_\alpha(h) \tag{5.4}$$

to obtain

$$v_{\ell+1}(s) = C \sum_\alpha \frac{\chi_\alpha(s)}{(d_\alpha)^{2\ell+1}}. \tag{5.5}$$

This completes the inductive proof of (5.1) once we observe that the case $\ell = 0$ is known: by [**13**, (2.43)] we have

$$\sum_\alpha d_\alpha \chi_\alpha(s) = \delta(s-1) \tag{5.6}$$

and the moduli space $\mathcal{M}(\Sigma_0, s)$ is one point if $s = 1$ and empty otherwise, so its volume is indeed equal to $\delta(s-1)$.

References

1. A. Alekseev, A. Malkin, and E. Meinrenken, *Lie group valued moment maps*, J. Differential Geom. **48** (1998), no. 3, 445–495.
2. A. Alekseev, E. Meinrenken, and C. Woodward, *Duistermaat–Heckman measures and moduli spaces of flat bundles over surfaces*, Geom. Funct. Anal. **12** (2002), no. 1, 1–31.
3. M. F. Atiyah and R. Bott, *The Yang–Mills equations over Riemann surfaces*, Philos. Trans. Roy. Soc. London Ser. A **308** (1983), no. 1505, 523–615.
4. ———, *The moment map and equivariant cohomology*, Topology **23** (1984), no. 1, 1–28.

5. T. Bröcker and T. tom Dieck, *Representations of compact Lie groups*, Grad. Texts in Math., vol. 98, Springer-Verlag, New York, 1985.
6. P. Buser, *Geometry and spectra of compact Riemann surfaces*, Progr. Math, vol. 106, Birkhäuser, Boston, MA, 1992.
7. M. P. do Carmo, *Differential geometry of curves and surfaces*, Prentice-Hall, Englewood Cliffs, NJ, 1976.
8. H. S. M. Coxeter, *Introduction to geometry*, Wiley Classics Lib., Wiley, New York, 1989.
9. D. Freed, *Reidemeister torsion, spectral sequences, and Brieskorn spheres*, J. Reine Angew. Math. **429** (1992), 75–89.
10. W. M. Goldman, *The symplectic nature of fundamental groups of surfaces*, Adv. in Math. **54** (1984), no. 2, 200–225.
11. N. Ho and L. Jeffrey, *The volume of the moduli space of flat connections on a nonorientable 2-manifold*, math.SG/0307404.
12. K. Liu, *Heat kernel and moduli space*, Math. Res. Lett. **3** (1996), no. 6, 743–762.
13. E. Witten, *On quantum gauge theories in two dimensions*, Comm. Math. Phys. **141** (1991), no. 1, 153–209
14. ———, *Two-dimensional gauge theories revisited*, J. Geom. Phys. **9** (1992), no. 4, 303–368.

Department of Mathematics, Univ. of Toronto, Toronto, ON M5S 3G3, Canada
Current address: Department of Mathematics, National Cheng Kung University, Taiwan 701, Taiwan
E-mail address: `nankuo@email.ncku.edu.tw`

Department of Mathematics, Univ. of Toronto, Toronto, ON M5S 3G3, Canada
E-mail address: `jeffrey@math.toronto.edu`

Centre de Recherches Mathématiques
CRM Proceedings and Lecture Notes
Volume **38**, 2004

Mukai Flops and Derived Categories. II

Yoshinori Namikawa

Introduction

This paper is a sequel to [**13**]. Let X and Y be birationally equivalent smooth quasi-projective varieties. Then we say that X and Y are K-equivalent if there is a smooth quasi-projective variety Z with proper birational morphisms $f\colon Z \to X$ and $g\colon Z \to Y$ such that $f^*K_X = g^*K_Y$. The following problem is a motivation of this paper.

PROBLEM 1. *Let X and Y be K-equivalent smooth quasi-projective varieties. Then, is there an equivalence of bounded derived categories of coherent sheaves $D(X) \to D(Y)$?*

In some good cases, birational maps $X - - \to Y$ can be decomposed into certain kinds of flops. A flop is a diagram of quasi-projective varieties

$$X \xrightarrow{s} \overline{X} \xleftarrow{s^+} X^+.$$

Here s and s^+ are small, projective, crepant resolutions of the normal variety $\overline{X}$ with $\rho(X/\overline{X}) = \rho(X^+/\overline{X}) = 1$. Moreover, for an s-negative divisor D, its proper transform D' becomes s^+-ample. A special case of Problem 1 is:

PROBLEM 2. *Let*

$$X \xrightarrow{s} \overline{X} \xleftarrow{s^+} X^+$$

be a flop. Then, is there an equivalence of bounded derived categories of coherent sheaves $D(X) \to D(X^+)$?

We already have several affirmative evidences for Problem 2:

EXAMPLE 1 ([**1**]). Let X be a smooth quasi-projective variety of dimension $2h-1$ which contains a subvariety $M \cong \mathbb{P}^{h-1}$ with $N_{M/X} \cong \mathcal{O}(-1)^{\oplus h}$. One can blow up X along M and blow down the exceptional divisor in another direction. In this way, we have a new variety X^+ with a subvariety $M^+ \cong \mathbb{P}^{h-1}$. Let $s\colon X \to \bar{X}$ and $s^+\colon X^+ \to \overline{X}$ be the birational contraction maps of M and M^+ to points respectively. Let $\mu\colon X \times_{\overline{X}} X^+ \to X$ and $\mu^+\colon X \times_{\overline{X}} X^+ \to X^+$ be the projections. Then $\Psi(\bullet) := \mathbb{R}{\mu^+}_*\mathbb{L}\mu^*(\bullet)$ is an equivalence.

2000 *Mathematics Subject Classification.* 14E, 18E30.
This is the final form of the paper.

EXAMPLE 2 ([**10,13**]). Let X be a smooth quasi-projective variety of dimension $2h-2$ which contains a subvariety $M \cong \mathbb{P}^{h-1}$ with $N_{M/X} \cong \Omega^1_{\mathbb{P}^{h-1}}$. Let $s\colon X \to \overline{X}$ be a birational contraction map of M to a point. We have a flop

$$X \xrightarrow{s} \overline{X} \xleftarrow{s^+} X^+$$

along M. Then the functor Ψ defined by the fiber product $X \times_{\overline{X}} X^+$ gives an equivalence $D(X) \to D(X^+)$.

EXAMPLE 3 ([**3**]; cf. also [**4, 10, 16**]). Let

$$X \xrightarrow{s} \overline{X} \xleftarrow{s^+} X^+$$

be a 3-dimensional flop with X and X^+ being smooth quasi-projective 3-folds with trivial canonical line bundles. Then the functor Ψ defined by the fiber product $X \times_{\overline{X}} X^+$ gives an equivalence $D(X) \to D(X^+)$.

In all known examples, the kernels of the Fourier–Mukai transforms are all given by the structure sheaf of the fiber product $X \times_{\overline{X}} X^+$. The purpose of this paper is to show that the functor defined by the fiber product is *not* a correct one in general even if Problem 2 is affirmative. The author believes that it would be an interesting problem to find a correct functor to attain Problem 2.

The flops used in this paper are *stratified Mukai flops* and *stratified Atiyah flops*. These flops naturally appear in the wall-crossing of the moduli spaces of sheaves of K3 surfaces, and Markman [**11**] has given explicit descriptions of these flops. We shall observe that, for a stratified Mukai flop (resp. a stratified Atiyah flop),

(1) the fiber product $X \times_{\overline{X}} X^+$ defines an isomorphism

$$K(X) \cong K(X^+)$$

of Grothendieck groups, but

(2) the functor Ψ is not an equivalence in general.

More precisely, let H be a $\mathbb{C}$-vector space of $\dim h$. For $t \leq h/2$, let $G := G(t, H)$ be the Grassmann variety parameterizing t dimensional vector subspaces of H. We denote by T^*G the cotangent bundle of G. The nilpotent variety $\overline{N}^t(H)$ is defined as

$$\overline{N}^t(H) := \{A \in \mathrm{End}(H);\, A^2 = 0, \mathrm{rank}(A) \leq t\}.$$

Then there is a natural morphism $s\colon T^*G \to \overline{N}^t(H)$ and this gives a resolution of the singular variety $\overline{N}^t(H)$. Let H^* be the dual space of H and put $G^+ := G(t, H^*)$. Now we have the resolution $s^+\colon T^*G^+ \to \overline{N}^t(H^*)$. Since there is a natural identification $\overline{N}^t(H) \cong \overline{N}^t(H^*)$, we have a commutative diagram

$$\begin{array}{ccc} T^*G & \dashrightarrow & T^*G^+ \\ \downarrow & & \downarrow \\ \overline{N}^t(H) & \cong & \overline{N}^t(H^*). \end{array}$$

Moreover, we define 1-parameter deformations $E(H) \to \mathbb{C}^1$ of T^*G and $\overline{N}^t(H) \to \mathbb{C}^1$ of $\overline{N}^t(H)$. The map $s\colon T^*G \to \overline{N}^t(H)$ extends to $\tilde{s}\colon E(H) \to \overline{N}^t(H)$. For the dual space H^* we construct the same, and we have a commutative diagram

$$\begin{array}{ccc} E(H) & \dashrightarrow & E(H^*) \\ \downarrow & & \downarrow \\ \overline{N}^t(H) & \cong & \overline{N}^t(H^*). \end{array}$$

When $t = 1$, these diagrams give Example 2 and Example 1 respectively. We call the first diagram a *stratified Mukai flop*[1] and call the second diagram a *stratified Atiyah flop.*

Let $\Psi_0\colon D(T^*G^+) \to D(T^*G)$ be the functor defined by $T^*G \times_{\overline{N}^t(H)} T^*G^+$. Similarly, let $\Psi\colon D\big(E(H^*)\big) \to D\big(E(H)\big)$ be the functor defined by $E(H) \times_{\overline{N}^t(H)} E(H^*)$. These functor naturally induces homomorphisms of Grothendieck groups. Therefore, we have two commutative diagrams

$$\begin{array}{ccc} D(T^*G^+) & \xrightarrow{\Psi_0} & D(T^*G) \\ \downarrow & & \downarrow \\ K(T^*G^+) & \xrightarrow{\overline{\Psi}_0} & K(T^*G) \end{array} \qquad \begin{array}{ccc} D\big(E(H^*)\big) & \xrightarrow{\Psi} & D\big(E(H)\big) \\ \downarrow & & \downarrow \\ K\big(E(H^*)\big) & \xrightarrow{\overline{\Psi}} & K\big(E(H)\big). \end{array}$$

We shall prove that $\overline{\Psi}$ and $\overline{\Psi}_0$ are isomorphisms (Theorems 2.11 and 2.12). As noted above, when $t = 1$, Ψ_0 and Ψ are both equivalences. But, when $t = 2$ and $h = 4$, none of Ψ_0 or Ψ is an equivalence (Section 4). To show this, we have to describe the birational map $E(H) \dashrightarrow E(H^*)$ as the composite of certain blowing-ups and blowing-downs [**11**]. In Section 3, we shall sketch this when $t = 2$ and $h = 4$. At the moment, when $t \geq 2$, the following question is open:

PROBLEM 3. *For a generalized Mukai flop (resp. a generalized Atiyah flop), does the isomorphism $\overline{\Psi}_0$ (resp. $\overline{\Psi}$) lift to an equivalence of derived categories.*

The above constructions can be also applied to flag varieties. The final section deals with the case of complete flag varieties (cf. [**15**]). In this case, we have an equivalence of the bounded derived categories of coherent sheaves on dual pairs (Theorem 5.12).

Acknowledgment. The author thanks H. Nakajima for a helpful letter concerning the similar question to [**13**] for stratified Mukai flops. He also thanks E. Markman for providing him with the proof of §2.12.1, which enables him to remove an assumption from Theorem 2.12 of the first version.

1. Stratified Mukai flops

1.1. Let H be a $\mathbb{C}$-vector space of dimension h. For a positive integer $t \leq h/2$, let $G(t, H)$ be the Grassmann variety parameterizing t dimensional vector subspaces of H. In the remainder we simply write G for $G(t, H)$. Let τ and q be the universal subbundle and the universal quotient bundle respectively. They fit into the exact sequence

$$0 \to \tau \to H \otimes_{\mathbb{C}} \mathcal{O}_G \to q \to 0.$$

[1]This notation is not exact. In fact, when $t = h/2$, the birational map $T^*G \dashrightarrow T^*G^+$ becomes an isomorphism. But we include here such cases because our functor Ψ_0 is not an trivial one induced by the isomorphism.

1.2. Let T^*G be the cotangent bundle of G and let $\pi\colon T^*G \to G$ be the projection map. The nilpotent variety $\overline{N}^t(H)$ is defined as

$$\overline{N}^t(H) := \{A \in \mathrm{End}(H);\, A^2 = 0, \mathrm{rank}(A) \le t\}.$$

We define a birational morphism

$$s\colon T^*G \to \overline{N}^t(H)$$

as follows. First note that T^*G is identified with the vector bundle $\underline{\mathrm{Hom}}(q, \tau)$ over G. Then a point of T^*G is expressed as a pair (p, ϕ) of a point $p \in G$ and an element $\phi \in \mathrm{Hom}\big(q(p), \tau(p)\big)$. Since there is a surjection $H \to q(p)$ and an injection $\tau(p) \to H$, the element ϕ defines an element of $\mathrm{End}(H)$. We denote this element by the same ϕ. Now we define $s\big((p, \phi)\big) := \phi$. It is easy to check that $\phi \in \overline{N}^t(H)$. We see that, for $A \in \overline{N}^t(H)$,

$$s^{-1}(A) = \{(p, A); \mathrm{Im}(A) \subset \tau(p) \subset \mathrm{Ker}(A)\}.$$

For $i \le t$, put $N^i(H) := \{A \in \mathrm{End}(H); A^2 = 0, \mathrm{rank}(A) = i\}$. Then, for $A \in N^i(H)$,

$$s^{-1}(A) \cong G(t - i, h - 2i).$$

In particular, s is an isomorphism over $N^t(H)$. The map s is called the Springer resolution of $\overline{N}^t(H)$.

1.3. We construct a 1-parameter deformation of the Springer resolution $s\colon T^*G \to \overline{N}^t(H)$. First of all, we shall define a vector bundle $E(H)$ over G and an exact sequence

$$0 \to \underline{\mathrm{Hom}}(q, \tau) \to E(H) \to \mathcal{O}_G \to 0.$$

As in §1.2, $\mathrm{Hom}\big(q(p), \tau(p)\big)$ is embedded in $\mathrm{End}(H)$. For a suitable basis of H, $\mathrm{Hom}\big(q(p), \tau(p)\big)$ is the set of $h \times h$ matrices of the following form:

$$\begin{pmatrix} 0 & * \\ 0 & 0 \end{pmatrix},$$

where the first 0 is the $t \times t$ zero matrix and where $*$ is a $t \times (h - t)$ matrix. Now, let $E(H)(p)$ be the set of $h \times h$ matrices which have the following form:

$$\begin{pmatrix} \alpha I & * \\ 0 & 0 \end{pmatrix},$$

where $\alpha \in \mathbb{C}$ and where I is the $t \times t$ identity matrix. $E(H)(p)$ is characterized as the set of elements $\phi \in \mathrm{End}(H)$ such that $\mathrm{Im}(\phi) \subset \tau(p)$ and $\phi|_{\tau(p)} = \alpha I$; hence it is independent of the choice of the basis of H. By the construction, there is an exact sequence

$$0 \to \mathrm{Hom}\big(q(p), \tau(p)\big) \to E(H)(p) \to \mathbb{C} \to 0.$$

Note that the map $E(H)(p) \to \mathbb{C}$ is defined as $(1/t)$trace; hence it is also independent of the choice of the basis of H. We put $E(H) := \bigcup_{p \in G} E(H)(p)$. Then $E(H)$ becomes a vector bundle over G, and there is an exact sequence of vector bundles

$$0 \to \underline{\mathrm{Hom}}(q, \tau) \to E(H) \to \mathcal{O}_G \to 0.$$

There is a surjective morphism from $E(H)$ to $\mathbb{C}^1$, and its central fiber is T^*G.

1.4. Define

$$\mathcal{N}^t(H) := \{A \in \mathrm{End}(H); A^2 = (\exists\,\mathrm{scalar})A, \mathrm{rank}(A) = t\}.$$

Every elements of $\mathcal{N}^t(H)$ is conjugate to a matrix

$$\begin{pmatrix} \alpha I & * \\ 0 & 0 \end{pmatrix},$$

where αI is a $t \times t$ scalar matrix and where $*$ is a $t \times (h-t)$ matrix. Now let $\overline{\mathcal{N}}^t(H)$ be the Zariski closure of $\mathcal{N}^t(H)$ in $\mathrm{End}(H)$. By taking $(1/t)$trace, we define a morphism $\overline{\mathcal{N}}^t(H) \to \mathbb{C}^1$. Its central fiber is $\overline{N}^t(H)$.

1.5. Each point of $E(H)$ is expressed as a pair (p, ϕ) of $p \in G$ and $\phi \in E(H)(p)$. The extended Springer resolution

$$\tilde{s}\colon E(H) \to \overline{\mathcal{N}}^t(H)$$

is defined as $\tilde{s}\big((p,\phi)\big) := \phi$.

As a consequence, we have a 1-parameter deformation

$$E(H) \to \overline{\mathcal{N}}^t(H) \to \mathbb{C}^1$$

of

$$T^*G \to \overline{N}^t(H) \to 0.$$

1.6. We shall define *dual objects* of those constructed above. Let H^* be the dual space of H. For H^*, we define the similar objects to §1.1–1.4. We write G^+ for $G(t, H^*)$. $s^+\colon T^*G^+ \to \overline{N}^t(H)$ and $\tilde{s}^+\colon E(H^*) \to \overline{\mathcal{N}}^t(H^*)$ are the Springer resolution and the extended Springer resolution, respectively. The relationship between these duals are as follows. There is a canonical isomorphism $\mathrm{End}(H) \cong \mathrm{End}(H^*)$. With respect to dual bases of H and H^*, this isomorphism is given by the transpose $A \to {}^tA$. This isomorphism naturally identifies $\overline{N}^t(H)$ with $\overline{N}^t(H^*)$, and $\overline{\mathcal{N}}^t(H)$ with $\overline{\mathcal{N}}^t(H^*)$, respectively. These identifications induce a birational map between T^*G and T^*G^+, and a birational map between $E(H)$ and $E(H^*)$:

$$\begin{array}{ccc} T^*G & \dashrightarrow & T^*G^+ \\ \downarrow & & \downarrow \\ \overline{N}^t(H) & \cong & \overline{N}^t(H^*), \end{array} \qquad \begin{array}{ccc} E(H) & \dashrightarrow & E(H^*) \\ \downarrow & & \downarrow \\ \overline{\mathcal{N}}^t(H) & \cong & \overline{\mathcal{N}}^t(H^*). \end{array}$$

1.6.1. Example. When $t = 1$ and $h = 2$, the birational map $T^*G \dashrightarrow T^*G^+$ is an isomorphism. T^*G (resp. T^*G^+) is a non-singular surface and its zero section is a (-2)-curve. The nilpotent variety $\overline{N} := \overline{N}^t(H) \cong \overline{N}^t(H^*)$ has an A_1 surface singularity at $0 \in \overline{N}$. The Springer resolution s (or s^+) is nothing but the minimal resolution of $\overline{N}$. On the other hand, $E(H)$ and $E(H^*)$ are both non-singular 3-folds, and their zero sections are $(-1,-1)$-rational curves. The birational map $E(H) \dashrightarrow E(H^*)$ is the *Atiyah flop* along these $(-1,-1)$-curves. $\overline{\mathcal{N}} := \overline{\mathcal{N}}^t(H) \cong \overline{\mathcal{N}}^t(H^*)$ has an ordinary double point at the origin. The extended Springer resolutions $\tilde{s}$ and $\tilde{s}^+$ are mutually different small resolutions of this ordinary double point.

1.6.2. EXAMPLE. When $t = 1$ and $h \geq 3$, T^*G and T^*G^+ are non-singular varieties of dim $2(h-1)$. The zero sections of them are isomorphic to $\mathbb{P}^{h-1}$. The birational map $T^*G \dashrightarrow T^*G^+$ is the *Mukai flop* along these $\mathbb{P}^{h-1}$. On the other hand, $E(H)$ and $E(H^*)$ are non-singular varieties of dim $2h-1$. The zero sections of them are $\mathbb{P}^{h-1}$ whose normal bundles are isomorphic to $\mathcal{O}(-1)^{\oplus h}$. The birational map $E(H) \dashrightarrow E(H^*)$ is the flop treated in [**1**].

2. K-theory

2.1. For an algebraic scheme X, we denote by $K(X)$ the Grothendieck group of X. Namely, let $F(X)$ be the free Abelian group whose basis consists of the set of coherent sheaves on X. Let

$$(E): \quad 0 \to \mathcal{F}' \to \mathcal{F} \to \mathcal{F}'' \to 0$$

be an exact sequence of coherent sheaves on X. The exact sequence (E) defines an element $Q(E) := [\mathcal{F}] - [\mathcal{F}'] - [\mathcal{F}'']$ of $F(X)$. $K(X)$ is the quotient of $F(X)$ by the subgroup generated by all such $Q(E)$.

2.2. Let $G := G(t, H)$ be the same as §1.1. Let $\alpha := (\alpha_1, \ldots, \alpha_n)$ be a sequence of non-negative integers with $\alpha_1 \geq \alpha_2 \cdots \geq \alpha_n$. One can associate a Young diagram with α. Denote by $r(\alpha)$ the number of the rows of this Young diagram, and denote by $c(\alpha)$ the number of the columns of this Young diagram. For such an α and for a vector bundle E of rank n over G, one can define a new vector bundle $\Sigma^\alpha E$ over G (cf. [**8**], [**7**, §8]). The following theorem is well-known.

2.3. THEOREM. *$K(G)$ is a free abelian group generated by $[\Sigma^\alpha \tau]$ with $r(\alpha) \leq t$ and $c(\alpha) \leq h - t$.*

SKETCH OF THE PROOF. By [**8**], the bounded derived category $D(G)$ of coherent sheaves is generated by $[\Sigma^\alpha \tau]$ with $r(\alpha) \leq t$ and $c(\alpha) \leq h - t$. In particular, $K(G)$ is generated by them. On the other hand, the Chern character homomorphism $K(G) \to A(G)_\mathbb{Q}$ induces an isomorphism $K(G)_\mathbb{Q} \cong A(G)_\mathbb{Q}$ (cf. [**6**, Example 15.2.16]). Moreover, $A(G)$ is a free $\mathbb{Z}$-module with the basis $\{\alpha\}$, the Schubert classes for Young diagram α with $r(\alpha) \leq t$ and $c(\alpha) \leq h - t$ (cf. [**6**, 14.7]). □

2.3.1. EXAMPLE. $K(\mathbb{P}^{h-1})$ is generated by $\mathcal{O}$, $\mathcal{O}(-1), \ldots, \mathcal{O}(-h+1)$.

Let $\pi\colon T^*G \to G$ and $\tilde{\pi}\colon E(H) \to G$ be the same as Section 1. Since they are vector bundles over G, the natural maps $\pi^*\colon K(G) \to K(T^*G)$ and $\tilde{\pi}^*\colon K(G) \to K(E(H))$ are both isomorphisms (cf. [**5**, Theorem 5.4.17]). Write τ_{T^*G} for $\pi^*\tau$ and $\tau_{E(H)}$ for $\tilde{\pi}^*\tau$.

2.4. COROLLARY.

(1) *$K(T^*G)$ is a free Abelian group generated by $[\Sigma^\alpha \tau_{T^*G}]$ with $r(\alpha) \leq t$ and $c(\alpha) \leq h - t$.*
(2) *$K(E(H))$ is a free Abelian group generated by $[\Sigma^\alpha \tau_{E(H)}]$ with $r(\alpha) \leq t$ and $c(\alpha) \leq h - t$.*

2.5. Let $\psi\colon E(H) \dashrightarrow E(H^*)$ be the birational map in §1.6. Note that the extended Springer resolution $\tilde{s}\colon E(H) \to \overline{\mathcal{N}}^t(H)$ (resp. $\tilde{s}^+\colon E(H^*) \to \overline{\mathcal{N}}^t(H^*)$) is a small resolution; hence ψ is an isomorphism in codimension 1. Namely, there are Zariski open subsets $U \subset E(H)$ and $U^+ \subset E(H^*)$ such that $U \cong U^+$ and the complement of U in $E(H)$(resp. U^+ in $E(H^*)$) is at least of codimension 2. Let F be a reflexive sheaf on $E(H)$. Since $U \cong U^+$, $F|_U$ is regarded as a sheaf on U^+. Then the direct image $\psi(F) := (j^+)_*(F|_U)$ under the inclusion map $j^+\colon U^+ \to E(H^*)$ is a reflexive sheaf on $E(H^+)$. We call $\psi(F)$ the proper transform of F.

2.6. LEMMA. *$\psi(\tau_{E(H)}) = ({\tau^+}_{E(H^*)})^*$, where $({\tau^+}_{E(H^*)})^*$ is the dual sheaf of ${\tau^+}_{E(H^*)}$.*

PROOF. Put $M := \overline{\mathcal{N}}^t(H) - \mathrm{Sing}(\overline{\mathcal{N}}^t(H))$ and $M^+ := \overline{\mathcal{N}}^t(H^*) - \mathrm{Sing}(\overline{\mathcal{N}}^t(H^*))$. As explained in §1.6, there is an isomorphism, say $\psi_0\colon M \cong M^+$. For $A \in M$, $(\tau_{E(H)})(A) = \mathrm{Im}(A) \subset H$. Since $\psi_0(A) = {}^tA$ (cf.§1.6), $\big((\psi_0)_*\tau_{E(H)}\big)({}^tA) = \mathrm{Im}(A)$. In other words, for $B \in M^+$, $\big((\psi_0)_*\tau_{E(H)}\big)(B) = \mathrm{Im}({}^tB)$. Note that $\tau^+_{E(H^*)}(B) = \mathrm{Im}(B)$, where B is an endomorphism of H^*. Therefore, $({\tau^+}_{E(H^*)})(B)^* = \mathrm{Im}({}^tB)$. This implies that $\big((\psi_0)_*\tau_{E(H)}\big)(B) = ({\tau^+}_{E(H^*)})(B)^*$. Since M and M^+ are naturally embedded as open subsets of $E(H)$ and $E(H^*)$ and their complements are of codimension at least 2, we conclude that $\psi(\tau_{E(H)}) = ({\tau^+}_{E(H^*)})^*$. □

2.7. COROLLARY. *For an α of §2.2, $\psi(\Sigma^\alpha\tau_{E(H)}) = (\Sigma^\alpha\tau^+_{E(H^*)})^*$.*

2.8. Recall that $\overline{N} := \overline{N}^t(H) \cong \overline{N}^t(H^*)$ and $\overline{\mathcal{N}} := \overline{\mathcal{N}}^t(H) \cong \overline{\mathcal{N}}^t(H^*)$. Let $\mu_0\colon T^*G \times_{\overline{N}} T^*G^+ \to T^*G$ and $\mu_0^+\colon T^*G \times_{\overline{N}} T^*G^+ \to T^*G^+$ be the natural projections. Similarly, let $\mu\colon E(H)\times_{\overline{\mathcal{N}}} E(H^*) \to E(H)$ and $\mu^+\colon E(H)\times_{\overline{\mathcal{N}}} E(H^*) \to E(H^+)$ be the natural projections. We define a homomorphism

$$\Psi_0\colon K(T^*G^+) \to K(T^*G)$$

as $\Psi_0(\bullet) := (\mu_0)_* \circ (\mu_0^+)^*(\bullet)$. Similarly, a homomorphism

$$\Psi : K\big(E(H^*)\big) \to K\big(E(H)\big)$$

is defined as $\Psi(\bullet) := \mu_* \circ (\mu^+)^*(\bullet)$.

2.9. PROPOSITION. *For the generators $[\Sigma^\alpha\tau^+_{E(H^*)}]$ of $K\big(E(H^*)\big)$ (cf. Corollary 2.4), $\Psi([\Sigma^\alpha\tau^+_{E(H^*)}]) = [\Sigma^\alpha(\tau_{E(H)})^*]$.*

PROOF. Let F^+ be one of $\Sigma^\alpha\tau^+_{E(H^*)}$'s. By definition,

$$\Psi([F^+]) = \Sigma(-1)^i[R^i\mu_*({\mu^+}^*F^+)].$$

Note that, for $i > 0$, $R^i\mu_*({\mu^+}^*F^+)$ has the support in T^*G. The cokernel of the injection:

$$\mu_*({\mu^+}^*F^+) \to \big(\mu_*({\mu^+}^*F^+)\big)^{**}$$

also has the support in T^*G. Here, $**$ means the double dual. Hence, by the following lemma, $[R^i\mu_*({\mu^+}^*F^+)] = 0$ if $i > 0$, and $[\mu_*({\mu^+}^*F^+)] = \big[\big(\mu_*({\mu^+}^*F^+)\big)^{**}\big]$. By Corollary 2.7, $(\mu_*({\mu^+}^*F^+))^{**} \cong \Sigma^\alpha(\tau_{E(H)})^*$. □

2.10. LEMMA. *Let F be a coherent sheaf on $E(H)$ whose support is contained in T^*G. Then $[F] = 0$ in $K\big(E(H)\big)$.*

PROOF. Let I be the ideal sheaf of T^*G in $E(H)$. For a sufficiently large $n > 0$, $I^n F = 0$. Hence, F can be described as successive extensions of $\mathcal{O}_{T^*G}$-modules. It suffices to show that every coherent $\mathcal{O}_{T^*G}$ module is the zero as an element of $K(E(H))$. Since $K(T^*G)$ is generated by the elements of the form: $[\Sigma^\alpha \tau_{T^*G}]$, we have to prove that $[\Sigma^\alpha \tau_{T^*G}] = 0$ in $K(E(H))$. Note here that $\Sigma^\alpha \tau_{T^*G} = \Sigma^\alpha \tau_{E(H)}|_{T^*G}$. Since $E(H)$ is a 1-parameter deformation of T^*G, there is an exact sequence

$$0 \to \mathcal{O}_{E(H)} \to \mathcal{O}_{E(H)} \to \mathcal{O}_{T^*G} \to 0.$$

By taking the tensor product with $\Sigma^\alpha \tau_{E(H)}$, we have the exact sequence:

$$0 \to \Sigma^\alpha \tau_{E(H)} \to \Sigma^\alpha \tau_{E(H)} \to \Sigma^\alpha \tau_{T^*G} \to 0.$$

This yields that $[\Sigma^\alpha \tau_{T^*G}] = 0$ in $K(E(H))$. □

2.11. THEOREM. *The homomorphism*

$$\Psi\colon K(E(H^*)) \to K(E(H))$$

is an isomorphism.

PROOF. By Proposition 2.9, it is sufficient to prove that $[\Sigma^\alpha(\tau_{E(H)})^*]$ with $r(\alpha) \le t$ and $c(\alpha) \le h-t$ form a basis of $K(E(H))$. We put $\mathcal{O}_{E(H)}(-1) := \tilde{\pi}^*(\bigwedge^t \tau)$, which is a line bundle on $E(H)$. Since $(\tau_{E(H)})^* \cong \bigwedge^{t-1} \tau_{E(H)} \otimes \mathcal{O}_{E(H)}(1)$, one can check that

$$\{\Sigma^\alpha(\tau_{E(H)})^*\} = \{\Sigma^\alpha(\tau_{E(H)})\} \otimes \mathcal{O}_{E(H)}(h-t),$$

where α runs through the Young diagrams with $r(\alpha) \le t$ and $c(\alpha) \le h-t$. The right-hand side is a basis of $K(E(H))$. □

2.12. THEOREM. *The homomorphism*

$$\Psi_0\colon K(T^*G^+) \to K(T^*G)$$

is an isomorphism.

2.12.1. $E(H) \times_{\bar{N}} E(H^+)$ is an integral scheme by [**12**, Corollary 3.15]. For example, when $t = 1$, $E(H) \times_{\bar{N}} E(H^+)$ coincides with the blow-up of $E(H)$ ($E(H^*)$) along the zero section. When $t = 2$ and $h = 4$, $E(H) \times_{\bar{N}} E(H^+)$ is a normal variety with only rational singularities (cf. §3.7). Since $E(H) \times_{\bar{N}} E(H^+)$ is integral, $E(H) \times_{\bar{N}} E(H^+)$ is flat over the base space $\mathbb{C}^1$. We shall use this fact in the proof of Theorem 2.12.

2.12.2. Before proving Theorem 2.12 we review the notion of a *restriction map with supports* of Grothendieck groups (cf. [**5**, p. 246]). Let $f\colon X_0 \to X$ be a closed immersion of schemes with X being a non-singular quasi-projective variety. For a closed subscheme $Z \subset X$, we put $Z_0 := Z \times_X X_0$. We shall define the restriction map with the supports:

$$f^*\colon K(Z) \to K(Z_0).$$

Let $i\colon Z \to X$ be the inclusion map. Let F be a coherent sheaf on Z. Since i_*F has a finite locally free resolution, $\mathrm{Tor}_n^{\mathcal{O}_X}(\mathcal{O}_{X_0}, i_*F)$ are zero except for a finite number of n. Moreover, each $\mathrm{Tor}_n^{\mathcal{O}_X}(\mathcal{O}_{X_0}, i_*F)$ is an $\mathcal{O}_{Z_0}$ module. Now we define $f^*([F]) := \Sigma(-1)^n[\mathrm{Tor}_n^{\mathcal{O}_X}(\mathcal{O}_{X_0}, i_*F)]$. The restriction map f^* depends on the ambient spaces X and X_0.

PROOF OF THEOREM 2.12. Consider the closed immersion

$$f\colon T^*G \times T^*G^+ \to E(H) \times_{\mathbb{C}^1} E(H^*).$$

For $E(H) \times_{\overline{N}} E(H^*) \subset E(H) \times_{\mathbb{C}^1} E(H^*)$, we define the restriction map with supports:

$$f^*\colon K\big(E(H) \times_{\overline{N}} E(H^*)\big) \to K(T^*G \times_{\overline{N}} T^*G^+).$$

On the other hand, one can define the restriction map $i^*\colon K\big(E(H^+)\big) \to K(T^*G^+)$ because $E(H^+)$ is non-singular. First let us check that the following diagram commutes:

$$\begin{array}{ccc} K\big(E(H^*)\big) & \longrightarrow & K(T^*G^+) \\ \big\downarrow{\scriptstyle (\mu^+)^*} & & \big\downarrow{\scriptstyle (\mu_0^+)^*} \\ K\big(E(H) \times_{\overline{N}} E(H^*)\big) & \longrightarrow & K(T^*G \times_{\overline{N}} T^*G^+). \end{array}$$

It is sufficient to check the commutativity for $[F] \in K\big(E(H^*)\big)$ with F a locally free sheaf. We have $(\mu_0^+)^* \circ i^*([F]) = [(\mu_0^+)^* i^* F]$. In turn, $f^* \circ (\mu^+)^*([F]) = f^*([(\mu^+)^*F])$. For simplicity, put $X := E(H) \times_{\mathbb{C}^1} E(H^*)$, $X_0 := T^*G \times T^*G^+$ and $Z := E(H) \times_{\overline{N}} E(H^*)$. By definition, $f^*([(\mu^+)^*F]) = \Sigma(-1)^n[\operatorname{Tor}_n^{\mathcal{O}_X}(\mathcal{O}_{X_0}, (\mu^+)^*F)]$. Note that there is a surjective morphism $X \to \mathbb{C}^1$ and its central fiber is X_0. Since X is flat over $\mathbb{C}^1$, $\operatorname{Tor}_n^{\mathcal{O}_X}(\mathcal{O}_{X_0}, (\mu^+)^*F) = \operatorname{Tor}_n^{\mathcal{O}_{\mathbb{C}^1}}(k(0), (\mu^+)^*F)$. By §2.12.1, $(\mu^+)^*F$ is a flat $\mathcal{O}_{\mathbb{C}^1}$ module. Therefore, $\operatorname{Tor}_n^{\mathcal{O}_{\mathbb{C}^1}}(k(0), (\mu^+)^*F) = 0$ when $n \neq 0$. As a consequence, we have $f^*([(\mu^+)^*F]) = \big[\big((\mu^+)^*F\big)|_{X_0}\big] = [(\mu_0^+)^* i^* F]$ and the diagram commutes. □

Next, by [**5**, Proposition 5.3.15], the following diagram commutes.

$$\begin{array}{ccc} K\big(E(H) \times_{\overline{N}} E(H^*)\big) & \longrightarrow & K(T^*G \times_{\overline{N}} T^*G^+) \\ \big\downarrow{\scriptstyle \mu_*} & & \big\downarrow{\scriptstyle (\mu_0)_*} \\ K\big(E(H)\big) & \longrightarrow & K(T^*G). \end{array}$$

By the two commutative diagrams, we see that homomorphisms Ψ and Ψ_0 are compatible with the pull-backs:

$$\begin{array}{ccc} K\big(E(H^*)\big) & \longrightarrow & K(T^*G^+) \\ \big\downarrow{\scriptstyle \Psi} & & \big\downarrow{\scriptstyle \Psi_0} \\ K\big(E(H)\big) & \longrightarrow & K(T^*G). \end{array}$$

By Corollary 2.4, the horizontal maps are both isomorphisms. Since Ψ is an isomorphism by Theorem 2.12, Ψ_0 is also an isomorphism.

2.12.3. EXAMPLE. Assume that $t = 1$ and $h = 3$. Then $G = \mathbb{P}^2$ and $G^+ = \mathbb{P}^2$. Note that $T^*G \times_{\overline{N}} T^*G^+$ is a normal crossing variety with 2 irreducible components. Let X be the main component, namely, the irreducible component which dominates both T^*G and T^*G^+. Let $p\colon X \to T^*G$ and $p^+\colon X \to T^*G^+$ be the projections. We define a homomorphism

$$\Psi_0'\colon K(T^*G^+) \to K(T^*G)$$

as $\Psi'_0 := p_* \circ (p^+)^*$. In [**13**], we have proved that, this Ψ'_0 is not an equivalence at the level of derived categories. Here we shall show that Ψ'_0 is not an isomorphism even at the level of K-theory.

Put $\mathcal{O}_{T^*G}(k) := \pi^*\mathcal{O}_G(k)$ and $\mathcal{O}_{T^*G^+}(k) := (\pi^+)^*\mathcal{O}_{G^+}(k)$. One can check that

$$\Psi'_0([\mathcal{O}_{T^*G^+}(-1)]) = [\mathcal{O}_{T^*G}(1)], \qquad \Psi'_0([\mathcal{O}_{T^*G^+}]) = [\mathcal{O}_{T^*G^+}],$$

and

$$\Psi'_0([\mathcal{O}_{T^*G^+}(1)]) = [\mathcal{O}_{T^*G}(-1) \otimes I].$$

Here I is the ideal sheaf of the zero section of $\pi\colon T^*G \to G$. The Koszul resolution of I yields the exact sequence

$$0 \to \pi^*\big(\textstyle\bigwedge^2 \Theta_G \otimes \mathcal{O}_G(-1)\big) \to \pi^*\big(\Theta_G \otimes \mathcal{O}_G(-1)\big) \to I \otimes \pi^*\mathcal{O}_G(-1) \to 0.$$

By using the Euler exact sequence, we have

$$\big[\pi^*\big(\Theta_G \otimes \mathcal{O}_G(-1)\big)\big] = 3[\mathcal{O}_{T^*G}] - [\mathcal{O}_{T^*G}(-1)]$$

and

$$\big[\pi^*\big(\textstyle\bigwedge^2 \Theta_G \otimes \mathcal{O}_G(-1)\big)\big] = 3[\mathcal{O}_{T^*G}(1)] - 3[\mathcal{O}_{T^*G}] + [\mathcal{O}_{T^*G}(-1)].$$

Therefore, we have

$$[I \otimes \mathcal{O}_{T^*G}(-1)] = -2[\mathcal{O}_{T^*G}(-1)] + 6[\mathcal{O}_{T^*G}] - 3[\mathcal{O}_{T^*G}(1)].$$

The image of Ψ'_0 is the subgroup of $K(T^*G)$ generated by $[\mathcal{O}_{T^*G}(1)]$, $[\mathcal{O}_{T^*G}]$ and $2[\mathcal{O}_{T^*G}(-1)]$, which does not coincide with $K(T^*G)$. Hence, Ψ'_0 is *not* an isomorphism.

3. Stratified Mukai flop for $G(2,4)$

Markman [**11**] has described a stratified Mukai flop as a sequence of blowing-ups and blowing-downs. Here we shall sketch this when $G = G(2,4)$.

3.1. In the remainder of this section, we assume that H is a 4-dimensional $\mathbb{C}$-vector space, $G = G(2,H)$ and $G^+ = G(2,H^*)$. For $p \in G$, let $\mathrm{Hom}^i\big(q(p),\tau(p)\big) := \{\phi \in \mathrm{Hom}\big(q(p),\tau(p)\big); \mathrm{rank}\,\phi \le i\}$ and put $\underline{\mathrm{Hom}}^i(q,\tau) := \bigcup_{p\in G} \mathrm{Hom}^i\big(q(p),\tau(p)\big)$. By §1.3 we have a sequence of subvarieties of $E(H)$:

$$\underline{\mathrm{Hom}}^0(q,\tau) \subset \underline{\mathrm{Hom}}^1(q,\tau) \subset T^*G \subset E(H).$$

For short, we write $\mathcal{X}$ for $E(H)$, X for T^*G, Z for $\underline{\mathrm{Hom}}^1(q,\tau)$ and M for $\underline{\mathrm{Hom}}^0(q,\tau)$. As in §2.2, $\tilde{\pi}\colon \mathcal{X} \to G$ and $\pi\colon X \to G$ are projection maps. Note that M is the zero section of $\tilde{\pi}\colon \mathcal{X} \to G$. We have $\dim \mathcal{X} = 9$, $\dim X = 8$, $\dim Z = 7$ and $\dim M = 4$. Let $\tilde{s}\colon \mathcal{X} \to \overline{\mathcal{N}}^2(H)$ be the extended Steinberg resolution (cf. §1.5). Let Σ be the singular locus of $\overline{\mathcal{N}}^2(H)$. Every element of Σ is conjugate to a matrix

$$A = \begin{pmatrix} \alpha I & * \\ 0 & 0 \end{pmatrix}$$

whose rank ≤ 1 (cf. §1.4). This implies that $\alpha = 0$; hence $A^2 = 0$. Thus, we have $\Sigma \cong \overline{N}^1(H)$. Note that $\dim \Sigma = 6$. The exceptional locus $\mathrm{Exc}(\tilde{s})$ coincides with Z. There is a fibration $\tilde{s}|_Z\colon Z \to \Sigma$. For $p \in \Sigma \setminus \{0\}$, $\tilde{s}^{-1}(p) \cong \mathbb{P}^1$. Clearly, $\tilde{s}^{-1}(0) = M$.

3.2. Let $\nu_1\colon \mathcal{X}_1 \to \mathcal{X}$ be the blowing up of $\mathcal{X}$ along M.

3.2.1. Let $\mathrm{Exc}(\nu_1)$ be the exceptional locus of ν_1. Then $\mathrm{Exc}(\nu_1) \cong \mathbb{P}(E(H))$, which is a $\mathbb{P}^4$ bundle over M.

3.2.2. Let $\widetilde{Z}$ be the proper transform of Z by ν_1. One can check that Z has 3-dimensional ordinary double points along M. Hence $(\nu_1)|_{\widetilde{Z}} \colon \widetilde{Z} \to Z$ is a resolution of Z. The exceptional locus $\mathrm{Exc}((\nu_1)|_{\widetilde{Z}})$ is a $\mathbb{P}^1 \times \mathbb{P}^1$ bundle over M.

3.2.3. $\widetilde{Z} \cap \mathrm{Exc}(\nu_1)$ is described as follows. Let $p \in M$ ($\cong G$). If we choose a suitable basis of H, then $E(H)(p)$ consists of the matrices of the following form

$$\begin{pmatrix} \alpha & 0 & x & y \\ 0 & \alpha & z & w \\ 0 & 0 & 0 & 0 \\ 0 & 0 & 0 & 0 \end{pmatrix}.$$

We can regard $(\alpha : x : y : z : w)$ as homogeneous coordinates of the projective space $\mathbb{P}(E(H)(p))$. Then

$$\widetilde{Z} \cap \mathbb{P}(E(H)(p)) = \{\alpha = xw - yz = 0\},$$

which is isomorphic to $\mathbb{P}^1 \times \mathbb{P}^1$. In this way, $\widetilde{Z} \cap \mathbb{P}(E(H))$ becomes a $\mathbb{P}^1 \times \mathbb{P}^1$ bundle over M.

3.2.4. Let $T^*G(1, H) \to \overline{N}^1(H)$ be the Springer resolution (cf. §1.2). Let $\widetilde{N}$ be the blowing up of $T^*G(1, H)$ along the zero section. Identify $\overline{N}^1(H)$ with Σ (cf. §3.1). Note that $\widetilde{Z} \to Z$ is the blowing-up along M, and $\widetilde{N} \to \Sigma$ is the blowing-up at $0 \in \Sigma$. Then the composite $\widetilde{Z} \to Z \to \Sigma$ is factorized as

$$\begin{array}{ccc} \widetilde{Z} & \longrightarrow & Z \\ \downarrow & & \downarrow \\ \widetilde{N} & \longrightarrow & \Sigma. \end{array}$$

Moreover, $\widetilde{Z} \to \widetilde{N}$ is a $\mathbb{P}^1$ bundle. The proof of these facts are omitted.

3.3. Let $\nu_2 \colon \widetilde{\mathcal{X}} \to \mathcal{X}_1$ be the blowing up of $\mathcal{X}_1$ along $\widetilde{Z}$. We put $E := \mathrm{Exc}(\nu_2)$. E is a $\mathbb{P}^1$ bundle over $\widetilde{Z}$. Moreover, let F be the proper transform (=total transform) of $\mathrm{Exc}(\nu_1)$. $(\nu_1 \circ \nu_2)|_F \colon F \to M$ is a smooth morphism, whose fibers are blown-up $\mathbb{P}^4$ along $\mathbb{P}^1 \times \mathbb{P}^1$ described in §3.2.3.

3.4. For the dual $G^+ = G(2, H^*)$, we also have the varieties

$$M^+ \subset Z^+ \subset X^+ \subset \mathcal{X}^+.$$

By the same way as §§3.1, 3.2 and 3.3, we have a sequence of blowing-ups:

$$\widetilde{\mathcal{X}}^+ \xrightarrow{\nu_2^+} \mathcal{X}_1^+ \xrightarrow{\nu_1^+} \mathcal{X}^+.$$

The birational map $\mathcal{X} \dashrightarrow \mathcal{X}^+$ in §1.6 induces a birational map $\widetilde{\mathcal{X}} \dashrightarrow \widetilde{\mathcal{X}}^+$. This birational map is actually an isomorphism by [**11**].

3.5. We put $\overline{\mathcal{N}} := \overline{\mathcal{N}}^2(H) \cong \overline{\mathcal{N}}^2(H^*)$. The fiber product $\mathcal{X} \times_{\overline{\mathcal{N}}} \mathcal{X}^+$ birationally dominates $\mathcal{X}$. Now we have a natural birational map over $\mathcal{X}$:

$$\mathcal{X}_1 \dashrightarrow \mathcal{X} \times_{\overline{\mathcal{N}}} \mathcal{X}^+.$$

Composing this with $\mu_2 : \widetilde{\mathcal{X}} \to \mathcal{X}_1$, we have a birational map

$$\widetilde{\mathcal{X}} \dashrightarrow \mathcal{X} \times_{\overline{\mathcal{N}}} \mathcal{X}^+ ,$$

which is, in fact, a morphism because $\widetilde{\mathcal{X}} \dashrightarrow \mathcal{X}^+$ is a morphism. Note that these birational maps are defined over $\overline{\mathcal{N}}$. Over $0 \in \overline{\mathcal{N}}$, the first birational map induces a rational map

$$\gamma\colon\ \mathbb{P}(E(H)) \dashrightarrow M \times M^+ ,$$

and the second birational morphism induces a morphism

$$F \to M \times M^+.$$

We shall describe this last map as the blowing-up of $M \times M^+$ with a suitable center.

3.5.1. Let $\gamma(p)\colon\ \mathbb{P}(E(H)(p)) \dashrightarrow M^+$ be the restriction of γ over $p \in M$. We shall give an explicit description of $\gamma(p)$. By a suitable choice of the basis of H, $E(H)(p)$ is identified with the set of matrices in §3.2.3. Here we introduce the dual basis in H^*, which will be used later. Fix a matrix

$$\begin{pmatrix} \alpha & 0 & x & y \\ 0 & \alpha & z & w \\ 0 & 0 & 0 & 0 \\ 0 & 0 & 0 & 0 \end{pmatrix}$$

in $E(H)(p)$ in such a way that $\alpha \neq 0$. Let $l := \{q(t)\}_{t \in \mathbb{C}}$ be the line of $E(H)(p)$ passing through this matrix and 0. Here

$$q(t) := \begin{pmatrix} \alpha t & 0 & xt & yt \\ 0 & \alpha t & zt & wt \\ 0 & 0 & 0 & 0 \\ 0 & 0 & 0 & 0 \end{pmatrix}$$

is a 1-parameter family of matrices. Let $[l] \in \mathbb{P}(E(H)(p))$ be the point represented by l. The image $\gamma(p)([l]) \in M^+$ of $[l]$ is described as follows. We have the following commutative diagram:

$$\begin{array}{ccc} E(H)\tilde{s} & \dashrightarrow & E(H^*)\tilde{s}^+ \\ \downarrow & & \downarrow \\ \overline{\mathcal{N}}^t(H) & \cong & \overline{\mathcal{N}}^t(H^*). \end{array}$$

For $(p, q(t)) \in E(H)$, $\tilde{s}(p, q(t)) = q(t)$. The isomorphism $\overline{\mathcal{N}}^2(H) \cong \overline{\mathcal{N}}^2(H^*)$ is given by the transposition. Hence this isomorphism sends $q(t)$ to

$${}^t q(t) := \begin{pmatrix} \alpha t & 0 & 0 & 0 \\ 0 & \alpha t & 0 & 0 \\ xt & zt & 0 & 0 \\ yt & wt & 0 & 0 \end{pmatrix}.$$

Assume that $t \neq 0$. Then, since $\alpha \neq 0$, the inverse image $(\tilde{s}^+)^{-1}({}^tq(t))$ is uniquely determined and is given by $(?(t), {}^tq(t)) \in E(H^*)$. Here $?(t) \in G^+$ is the 2-dimensional subspace of H^* generated by two vectors

$$\begin{pmatrix} \alpha t \\ 0 \\ xt \\ yt \end{pmatrix}, \quad \begin{pmatrix} 0 \\ \alpha t \\ zt \\ wt \end{pmatrix}.$$

Now $\lim_{t\to 0}(?(t), {}^tq(t)) = (\gamma(p)([l]), 0)$. Therefore, $\gamma(p)([l]) \in M^+$ $(\cong G^+)$ is the 2-dimensional subspace of H^* generated by two vectors

$$\begin{pmatrix} \alpha \\ 0 \\ x \\ y \end{pmatrix}, \quad \begin{pmatrix} 0 \\ \alpha \\ z \\ w \end{pmatrix}.$$

Let $M^+ \to \mathbb{P}^5$ be the Plücker embedding with respect to the basis of H^* given above. Let $(p_{12} : p_{13} : p_{14} : p_{23} : p_{24} : p_{34})$ be the Plücker coordinates. Note that M^+ is a quadratic hypersurface of $\mathbb{P}^5$ defined by $p_{12}p_{34} - p_{13}p_{24} + p_{14}p_{23} = 0$. Then $\gamma(p)\colon \mathbb{P}(E(H)(p)) \dashrightarrow M^+ \subset \mathbb{P}^5$ is given by

$$\gamma(p)((\alpha : x : y : z : w)) = (\alpha^2 : \alpha z : \alpha w : -\alpha x : -\alpha y : xw - yz).$$

The indeterminancy of $\gamma(p)$ is the subvariety $\{\alpha = xw - yz = 0\}$. This coincides with $\widetilde{Z} \cap \mathbb{P}(E(H)(p))$ (cf. §3.2.3). Let $F(p)$ be the fiber of the morphism $F \to M$ over $p \in M$ (cf. §3.3). Then, $F(p)$ is the blow-up of $\mathbb{P}(E(H)(p))$ along $\widetilde{Z} \cap \mathbb{P}(E(H)(p))$. It is immediately checked that the composite $F(p) \to \mathbb{P}(E(H)(p)) \dashrightarrow M^+$ is a birational morphism. The following are also checked.

3.5.1.1. Let $R(p)$ be the proper transform of $\{\alpha = 0\} \subset \mathbb{P}(E(H)(p))$ by the blowing-up $F(p) \to \mathbb{P}(E(H)(p))$. Then $R(p)$ is isomorphic to $\{\alpha = 0\}$ $(\cong \mathbb{P}^3)$. Moreover, $R(p)$ is contracted to the point $(0 : 0 : 0 : 0 : 0 : 1) \in \mathbb{P}^5$ by the birational morphism $F(p) \to M^+$. Actually, $F(p)$ is the blowing-up of M^+ at $(0 : 0 : 0 : 0 : 0 : 1)$.

3.5.1.2. Let $S(p)$ be the exceptional divisor of the blow-up $F(p) \to \mathbb{P}(E(H)(p))$. Then $S(p)$ is mapped onto the divisor $\{p_{12} = 0\} \cap M^+$ of M^+ by the birational map $F(p) \to M^+$. This divisor has an ordinary double threefold singularity at $(0 : 0 : 0 : 0 : 0 : 1)$. Actually, $S(p)$ is the blow-up of $\{p_{12} = 0\} \cap M^+$ at $(0 : 0 : 0 : 0 : 0 : 1)$. The exceptional locus of this blowing-up is $S(p) \cap R(p)$, which is isomorphic to $\mathbb{P}^1 \times \mathbb{P}^1$.

3.5.2. In general, we have a natural isomorphism $G(t, H) \cong G(h - t, H^*)$, where $\dim H = h$. Now, since $t = 2$ and $h = 4$, there is an isomorphism $\iota\colon G(2, H) \cong G(2, H^*)$. The center $p^+ \in M^+$ of the blowing-up $F(p) \to M^+$ depends on $p \in M$. By §3.5.1.1, we see that $p^+ = \iota(p)$. Let $\Gamma \subset M \times M^+$ be the graph of ι. By §3.5.1 the birational map $F \to M \times M^+$ is the blowing-up of $M \times M^+$ along Γ:

$$F \cong \mathrm{Bl}_\Gamma(M \times M^+).$$

3.6. As in §3.5, for $\widetilde{\mathcal{X}}^+$, there is a birational morphism

$$\widetilde{\mathcal{X}}^+ \to \mathcal{X} \times_{\overline{N}} \mathcal{X}^+$$

as $\overline{\mathcal{N}}$ schemes. Over $0 \in \overline{\mathcal{N}}$, this birational morphism induces a morphism

$$\gamma^+ : F^+ \to M \times M^+.$$

By the same argument as §3.5.2 we see that

$$F^+ \cong \mathrm{Bl}_\Gamma(M \times M^+).$$

3.7. One can check that $\mathcal{X} \times_{\overline{\mathcal{N}}} \mathcal{X}^+$ is a normal variety with rational singularities. In fact, set-theoretically, $\mathcal{X} \times_{\overline{\mathcal{N}}} \mathcal{X}^+$ is obtained from $\widetilde{\mathcal{X}}$ by contracting $\{R(p)\}_{p\in M}$ to $\{p \times p^+ \in M \times M^+\}_{p\in M}$. Therefore, $\mathcal{X} \times_{\overline{\mathcal{N}}} \mathcal{X}^+$ has singularities along Γ in §3.5.2. Moreover, by a direct calculation, we see that these singularities are locally of the following form: $(V, 0) \times (\mathbb{C}^4, 0)$, where

$$V := \left\{ (x, y, z, w, s, t, u, v) \in \mathbb{C}^8; \operatorname{rank} \begin{pmatrix} x & y & z & w \\ -v & t & u & -s \end{pmatrix} \leq 1 \right\}.$$

3.8. Let Σ be the singular locus of $\overline{\mathcal{N}}$ and put $\Sigma^* := \Sigma \setminus \{0\}$. Z (resp. Z^+) has a fibration over Σ (cf. §3.1). Let Z^* (resp. $(Z^+)^*$) be the inverse image of Σ^* by this map. By §3.1, $Z^* \to \Sigma^*$ and $(Z^+)^* \to \Sigma^*$ are both $\mathbb{P}^1$ bundles. Outside M and M^+, the birational map *birational*$\mathcal{X}\mathcal{X}^+$ is a family of Atiyah flops along $Z^* \to \Sigma^*$ and $(Z^+)^* \to \Sigma^*$.

4. Derived categories

Let

$$\Psi : D(E(H^*)) \to D(E(H))$$

be the functor defined by $\Psi(\bullet) := \mathbb{R}\mu_* \circ \mathbb{L}(\mu^+)^*(\bullet)$ (cf. §2.8), where $D(E(H))$ (resp. $D(E(H^*))$) is the bounded derived category of coherent sheaves on $E(H)$ (resp. $E(H^*)$). Let

$$\Psi_0 : D(T^*G^+) \to D(T^*G)$$

be the functor defined by $\Psi_0(\bullet) := \mathbb{R}(\mu_0)_* \circ \mathbb{L}(\mu_0^+)^*(\bullet)$ (cf. §2.8). In this section, we show that these functors are not equivalences when $G = G(2, 4)$. In the remainder, we use the same notation as Section 3.

4.1. LEMMA. *Let* $\mathrm{pr} : F \to M$ *and* $\mathrm{pr}^+ : F \to M^+$ *be the projections in* §3.5. *Then*

$$(\mathrm{pr}^+)^*\mathcal{O}_{M^+}(1) \otimes \mathcal{O}_F(E + 2F) \in \mathrm{pr}^* \operatorname{Pic}(M),$$

where $\mathcal{O}_{M^+}(1)$ *is the tautological line bundle of the Grassmannian* M^+.

PROOF. It is sufficient to prove that, for $p \in M$,

$$(\mathrm{pr}^+)^*\mathcal{O}_{M^+}(1) \otimes \mathcal{O}_{F(p)}(E + 2F) \cong \mathcal{O}_{F(p)}.$$

Here $F(p)$ is the blowing-up of $\mathbb{P}^4$ $(= \mathbb{P}(E(H)(p)))$ along $\widetilde{Z} \cap \mathbb{P}(E(H)(p))$. We call this blowing-up $\nu_2(p)$ (§3.5.1). Note that $\mathrm{Exc}(\nu_2(p)) = E \cap F(p)$. Then, by §§3.5.1.1 and 3.5.1.2,

$$(\mathrm{pr}^+|_{F(p)})^*\mathcal{O}_{M^+}(1) = (\nu_2)^*\mathcal{O}_{\mathbb{P}^4}(2) \otimes \mathcal{O}_{F(p)}(-E) = \mathcal{O}_{F(p)}(-2F - E). \qquad \square$$

4.2. Recall that $X := T^*G$ and $\pi\colon X \to G$ is the projection. Since $X \cong \underline{\mathrm{Hom}}(q, \tau)$, there is a universal homomorphism

$$f_{\mathrm{univ}}\colon \pi^* q \to \pi^* \tau.$$

Then Z is the divisor of X defined by $\bigwedge^2 f_{\mathrm{univ}} = 0$, where $\bigwedge^2 f_{\mathrm{univ}}$ is an element of $\Gamma\big(X, \underline{\mathrm{Hom}}(\bigwedge^2 \pi^* q, \bigwedge^2 \pi^* \tau)\big) = \Gamma\big(X, \pi^* \mathcal{O}_G(-2)\big)$. We have an exact sequence:

$$0 \to \pi^* \mathcal{O}_G(2) \to \mathcal{O}_X \to \mathcal{O}_Z \to 0.$$

By §1.5, X is the central fiber of the morphism $\mathcal{X} := E(H) \to \mathbb{C}^1$. Hence, there is an exact sequence:

$$0 \to \mathcal{O}_{\mathcal{X}} \to \mathcal{O}_{\mathcal{X}} \to \mathcal{O}_X \to 0.$$

4.3. Put $\nu := \nu_1 \circ \nu_2$ and $\nu^+ := (\nu_1^+) \circ (\nu_2^+)$. Then we have a diagram

$$\mathcal{X} \xleftarrow{\nu} \widetilde{\mathcal{X}} \cong \widetilde{\mathcal{X}}^+ \xrightarrow{\nu^+} \mathcal{X}^+.$$

This diagram is defined over the parameter space $\mathbb{C}^1$. The restriction of this to the central fibers becomes

$$X \xleftarrow{\nu_0} \widetilde{X} \cong \widetilde{X}^+ \xrightarrow{\nu_0^+} X^+.$$

By the isomorphism $\widetilde{\mathcal{X}} \cong \widetilde{\mathcal{X}}^+$, we regard ν^+ as a morphism from $\widetilde{\mathcal{X}}$ to $\mathcal{X}^+$. We simply write $\mathcal{O}(k)$ for $\tilde{\pi}^* \mathcal{O}_G(k)$, and $\mathcal{O}^+(k)$ for $(\tilde{\pi}^+)^* \mathcal{O}_{G^+}(k)$. Define a functor Φ as $\mathbb{R}\nu_* \circ \mathbb{L}\nu^{+*}$ and a functor Φ_0 as $\mathbb{R}\nu_{0*} \circ \mathbb{L}\nu_0^{+*}$:

$$\Phi\colon D(\mathcal{X}^+) \to D(\mathcal{X}), \qquad \Phi_0\colon D(X^+) \to D(X).$$

4.4. We first show that Φ is not an equivalence. Since $\overline{\mathcal{N}}^2(H)$ is a Stein space with rational singularities and $\mathcal{X}$ is a resolution of $\overline{\mathcal{N}}^2(H)$(cf. §1.5), we have $\mathrm{Ext}^i\big(\mathcal{O}^+(1), \mathcal{O}^+(1)\big) = 0$ for $i > 0$. Let us compute $\mathrm{Ext}^i\big(\Phi(\mathcal{O}^+(1)), \Phi(\mathcal{O}^+(1))\big)$.

4.5. By Lemma 4.1 and §3.8, we see that the restriction of the line bundle $(\nu^+)^* \mathcal{O}^+(1)(E + 2F)$ to each fiber of ν is a trivial line bundle. This implies that $(\nu^+)^* \mathcal{O}^+(1)(E + 2F)$ is a trivial line bundle around each fiber of ν. Hence, $\nu_*(\nu^+)^* \mathcal{O}^+(1)(E + 2F)$ is a line bundle on $\mathcal{X}$ and its pull-back by ν coincides with $(\nu^+)^* \mathcal{O}^+(1)(E + 2F)$. By Corollary 2.7, $\nu_*(\nu^+)^* \mathcal{O}^+(1)(E + 2F) \cong \mathcal{O}(-1)$. Hence we have $(\nu^+)^* \mathcal{O}^+(1)(E + 2F) \cong \nu^* \mathcal{O}(-1)$. Now the exact sequence

$$0 \to (\nu^+)^* \mathcal{O}^+(1) \to (\nu^+)^* \mathcal{O}^+(1)(E + 2F) \to (\nu^+)^* \mathcal{O}^+(1)|_{E+2F} \to 0$$

is identified with the exact sequence

$$0 \to \nu^* \mathcal{O}(-1)(-E - 2F) \to \nu^* \mathcal{O}(-1) \to \nu^* \mathcal{O}(-1)|_{E+2F} \to 0.$$

Apply ν_* to this last sequence. Since $R^1 \nu_* (\nu^+)^* \mathcal{O}^+(1) = 0$, we have the exact sequence

$$0 \to \mathcal{O}(-1) \otimes \nu_* \mathcal{O}(-E - 2F) \to \mathcal{O}(-1) \to \mathcal{O}(-1) \otimes \nu_* \mathcal{O}_{E+2F} \to 0.$$

Let Z' be the scheme theoretic image of $E + 2F$ by ν. By definition, the ideal sheaf $I_{Z'}$ of Z' is $\nu_* \mathcal{O}(-E - 2F)$. Then, this sequence is obtained from the exact sequence

$$0 \to I_{Z'} \to \mathcal{O}_{\mathcal{X}} \to \mathcal{O}_{Z'} \to 0$$

by taking the tensor product with $\mathcal{O}(-1)$. Now Z coincides with the scheme theoretic image of $E + F$. By the next lemma, we have an exact sequence

$$0 \to \mathcal{O}_M \to \mathcal{O}_{Z'} \to \mathcal{O}_Z \to 0. \tag{4.1}$$

Since $R^i\nu_*(\nu^+)^*\mathcal{O}^+(1) = 0$ for $i > 0$, we have
$$\Phi(\mathcal{O}^+(1)) = \mathcal{O}(-1) \otimes I_{Z'}.$$

4.6. LEMMA. *There is an exact sequence*
$$0 \to \nu_*\mathcal{O}_F(-E-F) \to \mathcal{O}_{Z'} \to \mathcal{O}_Z \to 0,$$
and $\nu_*\mathcal{O}_F(-E-F) \cong \mathcal{O}_M$.

PROOF. The first claim easily follows from the definitions of Z and Z'. For the second claim, we first show that $\nu_*\mathcal{O}_F(-E-F)$ is a line bundle on M. It is enough to prove that, for $p \in M$, $h^0(F(p), \mathcal{O}_{F(p)}(-E-F)) = 1$. By §3.5.1, $F(p)$ is the blowing-up of $\mathbb{P}(E(H)(p))$ along $e(p) := \widetilde{Z} \cap \mathbb{P}(E(H)(p))$. We call this blowing-up $\nu_2(p)$ as in the proof of Lemma 4.1. We shall use the homogeneous coordinates $(\alpha : x : y : z : w)$ of $\mathbb{P}^4 = \mathbb{P}(E(H)(p))$ in §3.5.1. Then $e(p) = \{\alpha = xw - yz = 0\}$. Let $I_{e(p)}$ be the ideal sheaf of $e(p)$ in $\mathbb{P}^4$. Since $\mathcal{O}_{F(p)}(F) = (\nu_2(p))^*\mathcal{O}_{\mathbb{P}^4}(-1)$, we only have to prove that $h^0(\mathbb{P}^4, \mathcal{O}_{\mathbb{P}^4}(1) \otimes I_{e(p)}) = 1$. But this is checked directly. We next show that $\nu_*\mathcal{O}_F(-E-F)$ has a nowhere-vanishing section. Let $R(p)$ be the same as §3.5.1.1. Then $R(p)$ is a non-zero section of $H^0(F(p), \mathcal{O}_{F(p)}(-E-F))$. Now, $R := \{R(p)\}_{p \in M}$ gives a nowhere-vanishing section of $\nu_*\mathcal{O}_F(-E-F)$. □

4.7. By §4.5 we have
$$\operatorname{Ext}^i\Big(\Phi(\mathcal{O}^+(1)), \Phi(\mathcal{O}^+(1))\Big) \cong \operatorname{Ext}^i(I_{Z'}, I_{Z'}).$$
We shall prove that $\operatorname{Ext}^5(I_{Z'}, I_{Z'}) \neq 0$.

4.8. LEMMA. $H^i(\mathcal{X}, I_{Z'}) = 0$ *for* $i > 1$.

PROOF. Let I_Z be the ideal sheaf of Z. Then, by ((4.1), there is an exact sequence
$$0 \to I_{Z'} \to I_Z \to \mathcal{O}_M \to 0.$$
Since $H^i(\mathcal{O}_M) = 0$ for $i > 0$, it is enough to prove that $H^i(I_Z) = 0$ for $i > 0$. We use the diagram in §3.2.4:
$$\begin{array}{ccc} \widetilde{Z} & \longrightarrow & Z \\ \downarrow & & \downarrow \\ \widetilde{N} & \longrightarrow & \Sigma. \end{array}$$
Since Z has rational singularities (cf. §3.2.2) and $\Sigma \cong \overline{N}^1(H)$ also has a rational singularity, we have $H^i(\mathcal{O}_Z) = 0$ for $i > 0$. Then the results follow from the exact sequence
$$0 \to I_Z \to \mathcal{O}_{\mathcal{X}} \to \mathcal{O}_Z \to 0$$
because $H^i(\mathcal{O}_{\mathcal{X}}) = 0$ for $i > 0$. □

4.9. LEMMA.
$$\operatorname{Ext}^i(\mathcal{O}_X, \mathcal{O}_X) = \begin{cases} 0 & (i \neq 1) \\ H^0(X, \mathcal{O}_X) & (i = 1) \end{cases}$$
$$\operatorname{Ext}^i(\mathcal{O}_X, \mathcal{O}(-2)) = \begin{cases} 0 & (i \neq 1, 2) \\ H^0(X, \mathcal{O}(-2)|_X) & (i = 1) \\ H^1(X, \mathcal{O}(-2)|_X) & (i = 2) \end{cases}$$

PROOF. By the exact sequence in §4.2:

$$0 \to \mathcal{O}_{\mathcal{X}} \to \mathcal{O}_{\mathcal{X}} \to \mathcal{O}_X \to 0,$$

we have

$$\underline{\mathrm{Ext}}^i(\mathcal{O}_X, \mathcal{O}_{\mathcal{X}}) = \begin{cases} 0 & (i \neq 0) \\ \mathcal{O}_X & (i = 1) \end{cases}$$

$$\underline{\mathrm{Ext}}^i(\mathcal{O}_X, \mathcal{O}(-2)) = \begin{cases} 0 & (i \neq 1) \\ \mathcal{O}(-2)|_X & (i = 1) \end{cases}$$

Note that $H^j(\mathcal{O}(-2)|_X) = 0$ for $j \geq 2$, and $H^j(\mathcal{O}_X) = 0$ for $j \geq 1$. The results follow from the local to global spectral sequence of Ext. □

4.10. LEMMA.

$$\mathrm{Ext}^i(\mathcal{O}_Z, \mathcal{O}_{\mathcal{X}}) = \begin{cases} 0 & (i \neq 3) \\ H^1(X, \mathcal{O}(-2)|_X) & (i = 3). \end{cases}$$

PROOF. Apply $\mathrm{Ext}(\bullet, \mathcal{O}_{\mathcal{X}})$ to the exact sequence

$$0 \to \mathcal{O}(2)|_X \to \mathcal{O}_X \to \mathcal{O}_Z \to 0.$$

Since $\mathrm{Ext}^2(\mathcal{O}_X, \mathcal{O}_{\mathcal{X}}) = 0$ by Lemma 4.9, we have an exact sequence

$$\mathrm{Ext}^1(\mathcal{O}_X, \mathcal{O}_{\mathcal{X}}) \to \mathrm{Ext}^1(\mathcal{O}(2)|_X, \mathcal{O}_{\mathcal{X}}) \to \mathrm{Ext}^2(\mathcal{O}_Z, \mathcal{O}_{\mathcal{X}}) \to 0.$$

By Lemma 4.9 and §4.2, this sequence is identified with the exact sequence

$$H^0(X, \mathcal{O}_X) \to H^0(X, \mathcal{O}(-2)|_X) \to H^0(Z, \mathcal{O}(-2)|_Z) \to 0.$$

It is easily checked that $H^0(Z, \mathcal{O}(-2)|_Z) = 0$. This implies that $\mathrm{Ext}^2(\mathcal{O}_Z, \mathcal{O}_{\mathcal{X}}) = 0$. Since $\mathrm{Ext}^2(\mathcal{O}_X, \mathcal{O}_{\mathcal{X}}) = \mathrm{Ext}^3(\mathcal{O}_X, \mathcal{O}_{\mathcal{X}}) = 0$, we see that $\mathrm{Ext}^2(\mathcal{O}(2)|_X, \mathcal{O}_{\mathcal{X}}) \cong \mathrm{Ext}^3(\mathcal{O}_Z, \mathcal{O}_{\mathcal{X}})$. By Lemma 4.9, $\mathrm{Ext}^2(\mathcal{O}(2)|_X, \mathcal{O}_{\mathcal{X}}) = H^1(X, \mathcal{O}(-2)|_X)$. □

4.11. LEMMA.

$$\mathrm{Ext}^i(\mathcal{O}_Z, \mathcal{O}_Z) = \begin{cases} 0 & (i \neq 0, 3) \\ H^0(Z, \mathcal{O}_Z) & (i = 0) \\ H^1(Z, \mathcal{O}(-2)|_Z) & (i = 3) \end{cases}$$

PROOF. Since $Z \subset \mathcal{X}$ is locally of complete intersection, $\underline{\mathrm{Ext}}^i(\mathcal{O}_Z, \mathcal{O}_Z) \cong \bigwedge^i N_{Z/\mathcal{X}}$. Since $N_{Z/X} \cong \mathcal{O}(-2)|_Z$ and $N_{X/\mathcal{X}}|_Z \cong \mathcal{O}_Z$, there is an exact sequence

$$0 \to \mathcal{O}(-2)|_Z \to N_{Z/\mathcal{X}} \to \mathcal{O}_Z \to 0.$$

From this sequence we know that

$$\underline{\mathrm{Ext}}^i(\mathcal{O}_Z, \mathcal{O}_Z) = \begin{cases} \mathcal{O}_Z & (i = 0) \\ N_{Z/\mathcal{X}} & (i = 1) \\ \mathcal{O}(-2)|_Z & (i = 2) \\ 0 & (\text{otherwise}). \end{cases}$$

We shall prove that $H^i(Z, \underline{\mathrm{Ext}}^j(\mathcal{O}_Z, \mathcal{O}_Z)) = 0$ except for $(i,j) = (0,0)$, $(1,2)$. For $(i,0)$ with $i > 0$, the cohomology clearly vanishes because $h^i(\mathcal{O}_Z) = 0$ for $i > 0$. For $(0,2)$ and for $(i,2)$ with $i > 1$, one can check that the cohomologies also vanish. Hence we only have to prove that $h^i(Z, N_{Z/\mathcal{X}}) = 0$ for all i. By the exact sequence above, we immediately see that $h^i(Z, N_{Z/\mathcal{X}}) = 0$ for $i > 1$. Now let us consider the commutative diagram in §3.2.4:

$$\begin{array}{ccc} \widetilde{Z} & \xrightarrow{\phi} & Z \\ {\scriptstyle \tilde{p}}\downarrow & & \downarrow{\scriptstyle p} \\ \widetilde{N} & \xrightarrow{\bar{\phi}} & \Sigma. \end{array}$$

For $q \in \Sigma$, let Z_q be the fiber of p over q. If $q \neq 0$, then $Z_q \cong \mathbb{P}^1$ and $N_{Z/\mathcal{X}}|_{Z_q} \cong \mathcal{O}_{\mathbb{P}^1}(-1) \oplus \mathcal{O}_{\mathbb{P}^1}(-1)$. Therefore, $H^0(Z, N_{Z/\mathcal{X}}) = 0$. Let $p_*\mathcal{O}_Z \to R^1p_*(\mathcal{O}(-2)|_Z)$ be the connecting homomorphism induced from the sequence above. The right hand side can be written as

$$\begin{aligned} R^1p_*(\mathcal{O}(-2)|_Z) &= R^1p_*\big(\phi_*\phi^*(\mathcal{O}(-2)|_Z)\big) = R^1(p\circ\phi)_*\big(\phi^*(\mathcal{O}(-2)|_Z)\big) \\ &= R^1(\bar{\phi}\circ\tilde{p})_*\big(\phi^*(\mathcal{O}(-2)|_Z)\big) \subset \bar{\phi}_*R^1\tilde{p}_*\big(\phi^*(\mathcal{O}(-2)|_Z)\big). \end{aligned}$$

Here, in the second equality, we used the fact Z has only rational singularities. The last map is an inclusion, because $\tilde{p}_*\big(\phi^*(\mathcal{O}(-2)|_Z)\big) = 0$ and hence $R^1\bar{\phi}_*\big(\phi^*(\mathcal{O}(-2)|_Z)\big) = 0$. In particular, we see that $R^1p_*(\mathcal{O}(-2)|_Z)$ is a torsion free sheaf of rank 1. Note that $R^1p_*N_{Z/\mathcal{X}}$ is zero outside $0 \in \Sigma$. Hence, the connecting homomorphism is an isomorphism outside 0. Since $p_*\mathcal{O}_Z$ is reflexive and $\dim\Sigma = 6$, we conclude that the connecting homomorphism is an isomorphism. Therefore, $R^1p_*N_{Z/\mathcal{X}} = 0$ and $H^1(Z, N_{Z/\mathcal{X}}) = 0$. The results of the lemma follow from the local to global spectral sequence of Ext. □

4.12. LEMMA.

$$\mathrm{Ext}^i(\mathcal{O}_{Z'}, \mathcal{O}_{\mathcal{X}}) = \begin{cases} 0 & (i \neq 3, 9) \\ H^1(X, \mathcal{O}(-2)|_X) & (i = 3) \\ \mathbb{C} & (i = 9). \end{cases}$$

PROOF. Apply $\mathrm{Ext}(\bullet, \mathcal{O}_{\mathcal{X}})$ to the exact sequence (4.1)

$$0 \to \mathcal{O}_M \to \mathcal{O}_{Z'} \to \mathcal{O}_Z \to 0.$$

We have an exact sequence

$$\to \mathrm{Ext}^i(\mathcal{O}_Z, \mathcal{O}_{\mathcal{X}}) \to \mathrm{Ext}^i(\mathcal{O}_{Z'}, \mathcal{O}_{\mathcal{X}}) \to \mathrm{Ext}^i(\mathcal{O}_M, \mathcal{O}_{\mathcal{X}}) \to .$$

Since M is compact and $\omega_{\mathcal{X}}|_M \cong \mathcal{O}_M$, $\mathrm{Ext}^i(\mathcal{O}_M, \mathcal{O}_{\mathcal{X}})$ is dual to $\mathrm{Ext}^{9-i}(\mathcal{O}_{\mathcal{X}}, \mathcal{O}_M) = H^{9-i}(M, \mathcal{O}_M)$. Hence, we have

$$\mathrm{Ext}^i(\mathcal{O}_M, \mathcal{O}_{\mathcal{X}}) = \begin{cases} 0 & (i \neq 9) \\ \mathbb{C} & (i = 9). \end{cases}$$

The exact sequence above and Lemma 4.10 now gives the result. □

4.13. LEMMA.

$$\mathrm{Ext}^i(\mathcal{O}_M, \mathcal{O}_X) = \begin{cases} 0 & (i \neq 8, 9) \\ \mathbb{C} & (i = 8, 9), \end{cases}$$

$$\mathrm{Ext}^i(\mathcal{O}_M, \mathcal{O}(2)|_X) = 0 \quad (\forall i),$$

and

$$\mathrm{Ext}^i(\mathcal{O}_M,\mathcal{O}_Z)=\begin{cases}0 & (i\neq 8,9)\\ \mathbb{C} & (i=8,9).\end{cases}$$

PROOF. By the exact sequence (cf. §4.2)

$$0\to\mathcal{O}_{\mathcal{X}}\to\mathcal{O}_{\mathcal{X}}\to\mathcal{O}_X\to 0$$

we have an exact sequence

$$\mathrm{Ext}^i(\mathcal{O}_M,\mathcal{O}_{\mathcal{X}})\xrightarrow{t}\mathrm{Ext}^i(\mathcal{O}_M,\mathcal{O}_{\mathcal{X}})\to\mathrm{Ext}^i(\mathcal{O}_M,\mathcal{O}_X)\to\mathrm{Ext}^{i+1}(\mathcal{O}_M,\mathcal{O}_{\mathcal{X}}),$$

where t is the local coordinate of $\mathbb{C}^1$ with $t(0)=0$ (cf. §4.2). Since $t\mathcal{O}_M=0$, the first map in the sequence is zero. On the other hand, since M is compact and $\omega_{\mathcal{X}}|_M\cong\mathcal{O}_M$, $\mathrm{Ext}^i(\mathcal{O}_M,\mathcal{O}_{\mathcal{X}})$ is the dual space of $H^{9-i}(M,\mathcal{O}_M)$. Now the first claim follows from the exact sequence above. Next apply $\mathrm{Ext}(\mathcal{O}_M,\bullet)$ to the exact sequence

$$0\to\mathcal{O}(2)\to\mathcal{O}(2)\to\mathcal{O}(2)|_X\to 0.$$

Note that $\mathrm{Ext}^i(\mathcal{O}_M,\mathcal{O}(2))$ is the dual space of $H^{9-i}(M,\mathcal{O}_M(-2))$ by the Serre duality. Since $H^i(M,\mathcal{O}_M(-2))=0$ for all i, the second claim follows. Finally apply $\mathrm{Ext}(\mathcal{O}_M,\bullet)$ to the exact sequence in §4.2:

$$0\to\mathcal{O}(2)|_X\to\mathcal{O}_X\to\mathcal{O}_Z\to 0.$$

By the first and the second claims, we have the third statements. □

4.14. LEMMA.

$$\mathrm{Ext}^i(\mathcal{O}_{Z'},\mathcal{O}_Z)=\begin{cases}0 & (i\neq 0,3,8,9)\\ H^0(Z,\mathcal{O}_Z) & (i=0)\\ H^1(Z,\mathcal{O}(-2)|_Z) & (i=3)\\ \mathbb{C} & (i=8,9).\end{cases}$$

PROOF. By the exact sequence (4.1)

$$0\to\mathcal{O}_M\to\mathcal{O}_{Z'}\to\mathcal{O}_Z\to 0$$

we have an exact sequence

$$\to\mathrm{Ext}^i(\mathcal{O}_Z,\mathcal{O}_Z)\to\mathrm{Ext}^i(\mathcal{O}_{Z'},\mathcal{O}_Z)\to\mathrm{Ext}^i(\mathcal{O}_M,\mathcal{O}_Z)\to.$$

Here we use Lemmas 4.11 and 4.13. □

4.15. LEMMA. $\mathrm{Ext}^5(\mathcal{O}_M,\mathcal{O}_M)\neq 0$.

PROOF. Since $M\subset\mathcal{X}$ is of locally complete intersection, $\underline{\mathrm{Ext}}^i(\mathcal{O}_M,\mathcal{O}_M)\cong\bigwedge^i N_{M/\mathcal{X}}$. Since $N_{M/X}\cong\Omega^1_M$ and $N_{X/\mathcal{X}}|_M\cong\mathcal{O}_M$, we have an exact sequence

$$0\to\Omega^1_M\to N_{M/\mathcal{X}}\to\mathcal{O}_M\to 0.$$

This sequence, in particular, yields the exact sequence

$$0\to\Omega^3_M\to\textstyle\bigwedge^3 N_{M/\mathcal{X}}\to\Omega^2_M\to 0.$$

The following sequence is exact.

$$H^2(\textstyle\bigwedge^3 N_{M/\mathcal{X}})\to H^2(M,\Omega^2_M)\to H^3(M,\Omega^3_M).$$

Since $M \cong G(2,4)$, $h^2(M, \Omega^2_M) = 2$ and $h^3(M, \Omega^3_M) = 1$. Therefore, we have $H^2(M, \bigwedge^3 N_{M/\mathcal{X}}) \neq 0$. We use the spectral sequence

$$E_2^{i,j} := H^i(\mathcal{X}, \underline{\mathrm{Ext}}^j(\mathcal{O}_M, \mathcal{O}_M)) \implies \mathrm{Ext}^{i+j}(\mathcal{O}_M, \mathcal{O}_M)$$

to compute $\mathrm{Ext}^5(\mathcal{O}_M, \mathcal{O}_M)$. By the argument above, $E_2^{2,3} \neq 0$. Moreover, we can check that $E_\infty^{2,3} = E_2^{2,3}$. In particular, $\mathrm{Ext}^5(\mathcal{O}_M, \mathcal{O}_M) \neq 0$. □

4.16. LEMMA. *$\mathrm{Ext}^5(\mathcal{O}_{Z'}, \mathcal{O}_M) \neq 0$ and $\mathrm{Ext}^5(\mathcal{O}_{Z'}, \mathcal{O}_{Z'}) \neq 0$.*

PROOF. By the exact sequence (4.1)

$$0 \to \mathcal{O}_M \to \mathcal{O}_{Z'} \to \mathcal{O}_Z \to 0$$

we have an exact sequence

$$\mathrm{Ext}^5(\mathcal{O}_Z, \mathcal{O}_M) \to \mathrm{Ext}^5(\mathcal{O}_{Z'}, \mathcal{O}_M) \to \mathrm{Ext}^5(\mathcal{O}_M, \mathcal{O}_M) \to \mathrm{Ext}^6(\mathcal{O}_Z, \mathcal{O}_M).$$

By the Serre duality, $\mathrm{Ext}^i(\mathcal{O}_Z, \mathcal{O}_M) \cong \left(\mathrm{Ext}^{9-i}(\mathcal{O}_M, \mathcal{O}_Z)\right)^*$. For $i = 5$, 6, these vanish by Lemma 4.13. Now the first claim follows from Lemma 4.15. We apply $\mathrm{Ext}(\mathcal{O}_{Z'}, \bullet)$ to the exact sequence (4.1) above. Since $\mathrm{Ext}^i(\mathcal{O}_{Z'}, \mathcal{O}_Z) \neq 0$ for $i = 4, 5$ by Lemma 4.14, we have an isomorphism $\mathrm{Ext}^5(\mathcal{O}_{Z'}, \mathcal{O}_M) \cong \mathrm{Ext}^5(\mathcal{O}_{Z'}, \mathcal{O}_{Z'})$. The second claim follows from the first claim. □

4.17. LEMMA. *$\mathrm{Ext}^5(I_{Z'}, I_{Z'}) \neq 0$.*

PROOF. In the exact sequence

$$\mathrm{Ext}^5(\mathcal{O}_{\mathcal{X}}, I_{Z'}) \to \mathrm{Ext}^5(I_{Z'}, I_{Z'}) \to \mathrm{Ext}^6(\mathcal{O}_{Z'}, I_{Z'}) \to \mathrm{Ext}^6(\mathcal{O}_{\mathcal{X}}, I_{Z'})$$

the first term and the last term both vanish by Lemma 4.8. Hence

$$\mathrm{Ext}^5(I_{Z'}, I_{Z'}) \cong \mathrm{Ext}^6(\mathcal{O}_{Z'}, I_{Z'}).$$

In the exact sequence

$$\mathrm{Ext}^5(\mathcal{O}_{Z'}, \mathcal{O}_{\mathcal{X}}) \to \mathrm{Ext}^5(\mathcal{O}_{Z'}, \mathcal{O}_{Z'}) \to \mathrm{Ext}^6(\mathcal{O}_{Z'}, I_{Z'}) \to \mathrm{Ext}^6(\mathcal{O}_{Z'}, \mathcal{O}_{\mathcal{X}})$$

the first and the last terms both vanish by Lemma 4.12. Hence

$$\mathrm{Ext}^5(\mathcal{O}_{Z'}, \mathcal{O}_{Z'}) \cong \mathrm{Ext}^6(\mathcal{O}_{Z'}, I_{Z'}).$$

By Lemma 4.16,

$$\mathrm{Ext}^5(\mathcal{O}_{Z'}, \mathcal{O}_{Z'}) \neq 0.$$ □

4.18. OBSERVATION. *Φ is not fully faithful.*

PROOF. We have, by Lemma 4.17, $\mathrm{Ext}^5\left(\Phi(\mathcal{O}^+(1), \Phi(\mathcal{O}^+(1))\right) \neq 0$. Since $\mathrm{Ext}^i\left(\mathcal{O}^+(1), \mathcal{O}^+(1)\right) = 0$ for all $i > 0$, this implies that Φ is not fully faithful. □

4.19. LEMMA. *$\Phi = \Psi$ and $\Phi_0 = \Psi_0$.*

PROOF. We put $\widehat{\mathcal{X}} := E(H) \times_{\overline{N}} E(H^*)$ and $\widehat{X} := T^*G \times_{\overline{N}} T^*G^+$. Let $\alpha \colon \widetilde{\mathcal{X}} \to \widehat{\mathcal{X}}$ and $\alpha_0 \colon \widetilde{X} \to \widehat{X}$ be the natural morphisms (cf. §3.7). Note that $\nu = \mu \circ \alpha$ and $\nu_0 = \mu_0 \circ \alpha_0$. Since $\widehat{\mathcal{X}}$ has only rational singularities by §3.7, we can write

$$\mathbb{R}\nu_* \circ \mathbb{L}(\nu^+)^* = \mathbb{R}\mu_* \circ \mathbb{R}\alpha_* \circ \mathbb{L}\alpha^* \circ \mathbb{L}(\mu^+)^* = \mathbb{R}\mu_* \circ \mathbb{L}(\mu^+)^*.$$

Therefore, $\Phi = \Psi$. We next claim that $\mathbb{R}(\alpha_0)_* \mathcal{O}_{\widetilde{X}} = \mathcal{O}_{\widehat{X}}$. Since $R^1\alpha_* \mathcal{O}_{\widetilde{\mathcal{X}}} = 0$, we have the exact sequence

$$0 \to \alpha_* \mathcal{O}_{\widetilde{\mathcal{X}}} \xrightarrow{t} \alpha_* \mathcal{O}_{\widetilde{\mathcal{X}}} \to \alpha_* \mathcal{O}_{\widetilde{X}} \to 0.$$

Since $\alpha_*\mathcal{O}_{\widetilde{\mathcal{X}}} = \mathcal{O}_{\widehat{\mathcal{X}}}$, this sequence is identified with

$$0 \to \mathcal{O}_{\widehat{\mathcal{X}}} \xrightarrow{t} \mathcal{O}_{\widehat{\mathcal{X}}} \to \mathcal{O}_{\widehat{X}} \to 0.$$

Hence, $\alpha_*\mathcal{O}_{\widetilde{X}} \cong \mathcal{O}_{\widehat{X}}$. Therefore, $(\alpha_0) * \mathcal{O}_{\widetilde{X}} = \mathcal{O}_{\widehat{X}}$. For $i > 0$, the exact sequence

$$R^i\alpha_*\mathcal{O}_{\widetilde{\mathcal{X}}} \to R^i(\alpha_0)_*\mathcal{O}_{\widetilde{X}} \to R^{i+1}\alpha_*\mathcal{O}_{\widetilde{\mathcal{X}}}$$

yields $R^i(\alpha_0)_*\mathcal{O}_{\widetilde{X}} = 0$. Our claim is now justified. Then we can write

$$\mathbb{R}(\nu_0)_* \circ \mathbb{L}(\nu_0^+)^* = \mathbb{R}(\mu_0)_* \circ \mathbb{R}(\alpha_0)_* \circ \mathbb{L}(\alpha_0)^* \circ \mathbb{L}(\mu_0^+)^* = \mathbb{R}(\mu_0)_* \circ \mathbb{L}(\mu_0^+)^*.$$

Therefore, $\Phi_0 = \Psi_0$. □

4.20. OBSERVATION. (1) Ψ *is not fully faithful.*
(2) Ψ_0 *is not an equivalence.*

PROOF. (1) This is clear from Observation 4.18 and Lemma 4.19.
(2) (i) The functor $\Phi' := \mathbb{R}\nu^+{}_*(\mathbb{L}\nu^*(\bullet) \otimes \omega_{\widetilde{\mathcal{X}}/\mathcal{X}})$ is the right adjoint of Φ, where $\omega_{\widetilde{\mathcal{X}}/\mathcal{X}} := \omega_{\widetilde{\mathcal{X}}} \otimes \nu^*(\omega_{\mathcal{X}}^{-1})$. On the other hand, the functor $\Phi'' := \mathbb{R}\nu^+{}_*(\mathbb{L}\nu^*(\bullet) \otimes \omega_{\widetilde{\mathcal{X}}/\mathcal{X}^+})$ is the left adjoint of Φ. But, since $\nu^*\omega_{\mathcal{X}} = (\nu^+)^*\omega_{\mathcal{X}^+}$, these functors coincides: $\Phi' = \Phi''$. Now, $\Phi_0' := \mathbb{R}\nu_0^+{}_*(\mathbb{L}(\nu_0)^*(\bullet) \otimes \omega_{\widetilde{X}/X})$ becomes the adjoint of Φ_0.
(ii) By the projection formula, Φ' coincides with the functor

$$\Psi' := \mathbb{R}(\mu^+)_*(\mathbb{L}\mu^*(\bullet) \otimes \mathbb{R}\alpha_*\omega_{\widetilde{\mathcal{X}}/\mathcal{X}}).$$

Similarly, Φ_0' coincides with the functor

$$\Psi_0' := \mathbb{R}\big((\mu_0)^+\big)_*(\mathbb{L}(\mu_0)^*(\bullet) \otimes \mathbb{R}(\alpha_0)_*\omega_{\widetilde{X}/X}\big).$$

By the Grauert–Riemmenschneider vanishing, we have $\mathbb{R}\alpha_*\omega_{\widetilde{\mathcal{X}}/\mathcal{X}} = \alpha_*\omega_{\widetilde{\mathcal{X}}/\mathcal{X}}$. By the exact sequence

$$R^i\alpha_*\omega_{\widetilde{\mathcal{X}}/\mathcal{X}} \xrightarrow{t} R^i\alpha_*\omega_{\widetilde{\mathcal{X}}/\mathcal{X}} \to R^i(\alpha_0)_*\omega_{\widetilde{X}/X} \to R^{i+1}\alpha_*\omega_{\widetilde{\mathcal{X}}/\mathcal{X}}$$

we see that $\mathbb{R}(\alpha_0)_*\omega_{\widetilde{X}/X} = (\alpha_0)_*\omega_{\widetilde{X}/X}$ and $(\alpha_0)_*\omega_{\widetilde{X}/X} \cong \alpha_*\omega_{\widetilde{\mathcal{X}}/\mathcal{X}} \otimes_{\mathcal{O}_{\widehat{\mathcal{X}}}} \mathcal{O}_{\widehat{X}}$. Let $j\colon \widehat{X} \to \widehat{\mathcal{X}}$ be the inclusion map. Since $\alpha_*\omega_{\widetilde{\mathcal{X}}/\mathcal{X}}$ is flat over $\mathbb{C}^1$ by §3.7, $\mathbb{L}j^*\alpha_*\omega_{\widetilde{\mathcal{X}}/\mathcal{X}} = (\alpha_0)_*\omega_{\widetilde{X}/X}$. Therefore,

$$\Psi' = \mathbb{R}(\mu^+)_*(\mathbb{L}\mu^*(\bullet) \otimes \alpha_*\omega_{\widetilde{\mathcal{X}}/\mathcal{X}}), \tag{4.2}$$

$$\Psi_0' = \mathbb{R}\big((\mu_0)^+\big)_*(\mathbb{L}(\mu_0)^*(\bullet) \otimes \mathbb{L}j^*\alpha_*\omega_{\widetilde{\mathcal{X}}/\mathcal{X}}\big). \tag{4.3}$$

Since $\Psi = \Phi$ and $\Psi_0 = \Phi_0$ by Lemma 4.19, Ψ' and Ψ_0' are adjoint of Ψ and Ψ_0 by (i). Let $i\colon X^+ \to \mathcal{X}^+$ be the inclusion map. By (4.2) and (4.3) we can apply [**4**, Lemma 6.1] and [**10**, Lemma 5.6] to conclude that the following diagram commutes

$$\begin{array}{ccc} D(X^+) & \xrightarrow{\Psi_0'\circ\Psi_0} & D(X^+) \\ \Big\downarrow{\scriptstyle \mathbb{R}i_*} & & \Big\downarrow{\scriptstyle \mathbb{R}i_*} \\ D(\mathcal{X}^+) & \xrightarrow{\Psi'\circ\Psi} & D(\mathcal{X}^+). \end{array}$$

(iii) Assume now that Ψ_0 is an equivalence. Then the quasi-inverse of Ψ_0 coincides with the adjoint Ψ'_0. In particular, $\Psi'_0 \circ \Psi_0 \cong \mathrm{id}_{D(X^+)}$. Let $\Omega := \{\mathcal{O}_p\}_{p \in \mathcal{X}^+}$, where $\mathcal{O}_p$ are structure sheaves of the closed points of $\mathcal{X}^+$. Then Ω is the spanning class for $D(\mathcal{X}^+)$ (cf. [**2**, Example 2.2]). It is clear that $\Psi' \circ \Psi(\mathcal{O}_p) = \mathcal{O}_p$ for $p \in \mathcal{X}^+ \setminus X^+$. For $p \in X^+$, since $\mathbb{R}i_*\mathcal{O}_p = \mathcal{O}_p$, we have $\Psi' \circ \Psi(\mathcal{O}_p) = \mathcal{O}_p$ by the commutative diagram because $\Psi'_0 \circ \Psi_0 \cong \mathrm{id}_{D(X^+)}$. Then, by the next lemma, we see that Ψ is fully faithful; but this contradicts (1). □

4.21. LEMMA ([**9**, Lemma 5.4]). *Let $f\colon A \to B$ be a functor with the right adjoint g and the left adjoint h. Let Ω be a spanning class for A. Assume that $\forall \omega \in \Omega$, $\omega \cong g \circ f(\omega)$ and $h \circ f(\omega) \cong \omega$. Then, $g \circ f(a) \cong h \circ f(a) \cong a$ for all $a \in A$. Moreover, f is fully faithful.*

PROOF. For $a \in A$ and $\omega \in \Omega$, we have:

$$\mathrm{Hom}(h \circ f(a), \omega) \cong \mathrm{Hom}\big(f(a), f(\omega)\big) \cong \mathrm{Hom}\big(a, g \circ f(\omega)\big) \cong \mathrm{Hom}(a, \omega),$$
$$\mathrm{Hom}\big(\omega, g \circ f(a)\big) \cong \mathrm{Hom}\big(f(\omega), f(a)\big) \cong \mathrm{Hom}(h \circ f(\omega), a) \cong \mathrm{Hom}(\omega, a).$$

Therefore, $a \cong g \circ f(a)$ and $h \circ f(a) \cong a$. Since

$$\mathrm{Hom}\big(f(a), f(b)\big) \cong \mathrm{Hom}(h \circ f(a), b) \cong \mathrm{Hom}(a, b),$$

f is fully faithful. □

5. Complete flag varieties

5.1. Let H be a $\mathbb{C}$-vector space of $\dim = h$. We denote by F the complete flag variety. Namely,

$$F := \{V_1 \subset V_2 \subset \cdots \subset V_{h-1} \subset H; \dim V_i = i\}.$$

Let

$$\tau_1 \subset \tau_2 \subset \cdots \subset \tau_{h-1} \subset H \otimes_{\mathbb{C}} \mathcal{O}_F$$

be the universal sub-bundles. We put $q_i := H \otimes_{\mathbb{C}} \mathcal{O}_F / \tau_i$ and call them the universal quotient bundles.

5.2. Let $\pi\colon T^*F \to F$ be the cotangent bundle of F. The nilpotent variety $\overline{N}(H)$ is defined as

$$\overline{N}(H) := \{A \in \mathrm{End}(H); A^h = 0\}.$$

A point of T^*F is expressed as a pair (p, ϕ) of $p \in F$ and $\phi \in \mathrm{End}(H)$ such that

$$\phi(H) \subset \tau_{h-1}(p), \phi\big(\tau_{h-1}(p)\big) \subset \tau_{h-2}(p), \ldots, \phi\big(\tau_1(p)\big) = 0.$$

The Springer resolution

$$s\colon T^*F \to \overline{N}(H)$$

is defined as $s\big((p, \phi)\big) := \phi$. The Springer resolution s has the following properties.

5.2.1. $s^{-1}(0) = F$.

5.2.2. Let $\overline{N}(H)_{\text{sing}}$ be the singular locus of $\overline{N}(H)$. Then

$$N(H)_{\text{sing}} := \{A \in \overline{N}(H); \text{rank}(A) \leq h-2\}.$$

Let $\overline{N}(H)^0_{\text{sing}}$ be the open orbit consisting of the matrices conjugate to

$$\begin{pmatrix} 0 & 0 & \cdot & \cdot & \cdot & \cdot \\ 0 & 0 & 1 & 0 & \cdot & \cdot \\ 0 & 0 & 0 & 1 & 0 & \cdot \\ \cdot & \cdot & \cdot & \cdot & 1 & \cdot \\ \cdot & \cdot & \cdot & \cdot & \cdot & 1 \\ 0 & 0 & \cdot & \cdot & \cdot & 0 \end{pmatrix}.$$

For $A \in \overline{N}(H)^0_{\text{sing}}$, $s^{-1}(A)$ is a tree of $\mathbb{P}^1$ with the A_{h-1}-configuration.

5.2.3. For $A \in \overline{N}(H) \setminus \overline{N}(H)_{\text{sing}}$, $s^{-1}(A)$ is one point.

5.3. The complete flag variety F has $h-1$ natural fibrations

$$f_1 \colon F \to F(2,3,\ldots,h-1,H), \quad f_2 \colon F \to F(1,3,\ldots,h-1,H), \quad \ldots,$$
$$f_{h-1} \colon F \to F(1,2,\ldots,h-2,H).$$

Each fibration is a $\mathbb{P}^1$ bundle. Assume that $f_i = \Phi_{|L_i|}$ for a line bundle L_i on F. Then $\pi^* L_i \in \text{Pic}(T^*F)$ defines a birational morphism

$$s_i \colon T^*F \to X_i$$

over $\overline{N}(H)$, where X_i is a normal variety which factorize s as $T^*F \xrightarrow{s_i} X_i \to \overline{N}(H)$. Over $0 \in \overline{N}(H)$, s_i restricts to the fibration

$$f_i \colon F \to F(1,2,\ldots,i-1,i+1,\ldots,h-1,H).$$

Let E_i be the exceptional locus of s_i. Then $E_i \to s_i(E_i)$ is a $\mathbb{P}^1$ bundle. In T^*F, each fiber of this $\mathbb{P}^1$-bundle has the normal bundle $\mathcal{O}_{\mathbb{P}^1}^{\oplus 2\dim F-2} \oplus \mathcal{O}_{\mathbb{P}^1}(-2)$. Note that, in the family of rational curves: $E_i \to s_i(E_i)$, each fiber of f_i deforms to one of $\mathbb{P}^1$ in the tree of §5.2.2.

5.4. We put $\text{sl}(H) := \{A \in \text{End}(H); \text{tr}(A) = 0\}$. For $A \in \text{End}(H)$, let $\phi_A(x)$ be the characteristic polynomial of A:

$$\phi_A(x) := \det(xI - A).$$

Let $\phi_i(A)$ be the coefficient of x^{h-i} in $\phi_A(x)$. If $A \in \text{sl}(H)$, then $\phi_1(A) = 0$. We define the *characteristic map*

$$\text{ch} \colon \text{sl}(H) \to \mathbb{C}^{h-1}$$

as $\text{ch}(A) := \big(\phi_2(A), \ldots, \phi_h(A)\big)$. Note that $\text{ch}^{-1}(0) = \overline{N}(H)$.

5.5. We shall define a simultaneous resolution of $\text{ch} \colon \text{sl}(H) \to \mathbb{C}^{h-1}$ up to a finite cover. First, we shall define a vector bundle $\mathcal{E}(H)$ over F and an exact sequence

$$0 \to T^*F \to \mathcal{E}(H) \xrightarrow{\eta} \mathcal{O}_F^{\oplus h-1} \to 0.$$

Let $T^*F(p)$ be the cotangent space of F at $p \in F$. Then, by §5.2, for a suitable basis of H, $T^*F(p)$ consists of the matrices of the following form

$$\begin{pmatrix} 0 & * & \cdot & \cdot & \cdot \\ 0 & 0 & * & \cdot & \cdot \\ \cdot & \cdot & \cdot & \cdot & \cdot \\ \cdot & \cdot & \cdot & 0 & * \\ 0 & 0 & \cdot & 0 & 0 \end{pmatrix}.$$

Let $\mathcal{E}(H)(p)$ be the vector subspace of $\mathrm{sl}(H)$ consisting of the matrices A of the following form

$$\begin{pmatrix} \alpha_1 & * & \cdot & \cdot & \cdot \\ 0 & \alpha_2 & * & \cdot & \cdot \\ \cdot & \cdot & \cdot & \cdot & \cdot \\ \cdot & \cdot & \cdot & \alpha_{h-1} & * \\ 0 & 0 & \cdot & 0 & \alpha_h \end{pmatrix}.$$

Here $\alpha_1 + \cdots + \alpha_h = 0$. We define a map $\eta(p)\colon \mathcal{E}(H)(p) \to \mathbb{C}^{\oplus h-1}$ as $\eta(p)(A) := (\alpha_1, \dots, \alpha_{h-1})$. Then we have an exact sequence of vector spaces

$$0 \to T^*F(p) \to \mathcal{E}(H)(p) \xrightarrow{\eta(p)} \mathbb{C}^{\oplus h-1} \to 0.$$

We put $\mathcal{E}(H) := \bigcup_{p\in F} \mathcal{E}(H)(p)$. Then $\mathcal{E}(H)$ becomes a vector bundle over F, and we get the desired exact sequence. Each point of $\mathcal{E}(H)$ is expressed as a pair of $p \in F$ and $\phi \in \mathcal{E}(H)(p)$. Now we define

$$\tilde{s}\colon \mathcal{E}(H) \to \mathrm{sl}(H)$$

as $\tilde{s}\big((p, \phi)\big) := \phi$. The birational morphisms $s_i\colon T^*G \to X_i$ extend to the birational morphisms

$$\tilde{s}_i\colon \mathcal{E}(H) \to \mathcal{X}_i$$

and $\tilde{s}$ is factorized as $\mathcal{E}(H) \to \mathcal{X}_i \to \mathrm{sl}(H)$. Let $\mathcal{E}_i$ be the exceptional locus of $\tilde{s}_i$. Then $\mathcal{E}_i \to \tilde{s}_i(\mathcal{E}_i)$ is a $\mathbb{P}^1$ bundle. In $\mathcal{E}(H)$, each fiber of this $\mathbb{P}^1$-bundle has the normal bundle $\mathcal{O}_{\mathbb{P}^1}^{\oplus 2\dim F+h-4} \oplus \mathcal{O}_{\mathbb{P}^1}(-1) \oplus \mathcal{O}_{\mathbb{P}^1}(-1)$. Let $\tilde{\eta}\colon \mathcal{E}(H) \to \mathbb{C}^{h-1}$ be the morphism induced by η. We define a finite Galois cover

$$\varphi\colon \mathbb{C}^{h-1} \to \mathbb{C}^{h-1}$$

as $\varphi(\alpha_1, \dots, \alpha_{h-1}) := \big(\phi_2(A), \dots, \phi_h(A)\big)$ where A is the diagonal matrix

$$\begin{pmatrix} \alpha_1 & 0 & \cdot & \cdot & \cdot \\ 0 & \alpha_2 & 0 & \cdot & \cdot \\ \cdot & \cdot & \cdot & \cdot & \cdot \\ \cdot & \cdot & \cdot & \alpha_{h-1} & 0 \\ 0 & 0 & \cdot & 0 & \alpha_h \end{pmatrix}$$

with $\alpha_h = -\alpha_1 - \cdots - \alpha_{h-1}$. Then we have a commutative diagram

$$\begin{array}{ccc} \mathcal{E}(H) & \xrightarrow{\tilde{s}} & \mathrm{sl}(H) \\ {\scriptstyle\tilde{\eta}}\downarrow & & \downarrow{\scriptstyle\mathrm{ch}} \\ \mathbb{C}^{h-1} & \xrightarrow{\varphi} & \mathbb{C}^{h-1}. \end{array}$$

Let $\mathrm{ch}'\colon \mathrm{sl}(H)\times_{\mathbb{C}^{h-1}}\mathbb{C}^{h-1} \to \mathbb{C}^{h-1}$ be the pull-back of ch by φ. By the commutative diagram, we have a morphism

$$\beta\colon \mathcal{E}(H) \to \mathrm{sl}(H) \times_{\mathbb{C}^{h-1}} \mathbb{C}^{h-1}.$$

By this morphism, $\tilde{\eta}\colon \mathcal{E}(H) \to \mathbb{C}^{h-1}$ becomes a simultaneous resolution of ch$'$.

5.6. For the dual space H^*, we define the complete flag variety F^+. We denote by T^*F^+ the cotangent bundle of F^+. We define a nilpotent variety $\overline{N}(H^*)$ and the Springer resolution $s^+\colon T^*F^+ \to \overline{N}(H^*)$ in the same way as §5.2. Let $\tilde{\eta}^+\colon \mathcal{E}(H^*) \to \mathbb{C}^{h-1}$ be the corresponding objects of §5.5 for H^*. The natural isomorphism $\mathrm{End}(H) \cong \mathrm{End}(H^*)$ induces an isomorphism $\iota\colon \mathrm{sl}(H) \cong \mathrm{sl}(H^*)$. Let $\mathrm{ch}^+\colon \mathrm{sl}(H^*) \to \mathbb{C}^{h-1}$ be the characteristic map. Then ι is compatible with ch and ch^+: $\mathrm{ch}^+ \circ \iota = \mathrm{ch}$. We have an isomorphism

$$\mathrm{sl}(H) \times_{\mathbb{C}^{h-1}} \mathbb{C}^{h-1} \xrightarrow{\iota\times\mathrm{id}} \mathrm{sl}(H^*) \times_{\mathbb{C}^{h-1}} \mathbb{C}^{h-1}.$$

This induces a birational map

$$f\colon \mathcal{E}(H) \dashrightarrow \mathcal{E}(H^*)\,.$$

5.7. Let $L^+ \in \mathrm{Pic}(\mathcal{E}(H))$ be a β^+-ample line bundle, where $\beta^+\colon \mathcal{E}(H^*) \to \mathrm{sl}(H^*) \times_{\mathbb{C}^{h-1}} \mathbb{C}^{h-1}$ is the simultaneous resolution defined in §5.5. Denote by $L \in \mathrm{Pic}(\mathcal{E}(H))$ its proper transform by f. For $\sigma \in \mathrm{Gal}(\varphi) \simeq S_h$, we consider the isomorphism $\mathrm{sl}(H) \times_{\mathbb{C}^{h-1}} \mathbb{C}^{h-1} \xrightarrow{\mathrm{id}\times\sigma} \mathrm{sl}(H)\times_{\mathbb{C}^{h-1}}\mathbb{C}^{h-1}$. This isomorphism induces a birational map $\phi_\sigma\colon\ \mathcal{E}(H) \dashrightarrow \mathcal{E}(H)\,.$

5.8. LEMMA. *For a suitable $\sigma \in \mathrm{Gal}(\varphi)$, the proper transform of L by ϕ_σ becomes β-ample. In particular, for this $\sigma \in \mathrm{Gal}(\varphi)$, the composite $f_\sigma := f \circ \phi_\sigma\colon\ \mathcal{E}(H) \dashrightarrow \mathcal{E}(H^*)$ becomes an isomorphism.*

PROOF. Note that $\mathrm{ch}'^{-1}(0) = \overline{N}(H)$. For $p \in \overline{N}(H)^0_{\mathrm{sing}}$ (cf. §5.2.2), $\beta^{-1}(p)$ is a tree of $\mathbb{P}^1$ with A_{h-1}-configuration. Let $C_1, \dots, C_{h-1}$ be the irreducible components of this tree. By [**14**, §7], we can take $\sigma \in \mathrm{Gal}(\varphi)$ in such a way that the proper transform L_σ of L by ϕ_σ has positive intersections with all C_i. We next observe the central fiber $\beta^{-1}(0) = F$. As in §5.3, F has $h-1$ different $\mathbb{P}^1$ fibrations $f_1, :, f_{h-1}$. Let l_j be a fiber of f_j. The cone $NE(F)$ of effective 1-cycles is a polyhedral cone generated by $h-1$ rays $\mathbb{R}^+[l_j]$. Since l_i deforms to one of C_i's in $\mathcal{E}(H)$, we see that $(L_\sigma.l_j) > 0$ for all j. This implies that L_σ is β-ample. □

5.9. $\mathrm{Gal}(\varphi) = \{\phi_\sigma\}$ contains the Atiyah flops along $\mathcal{E}_i \subset \mathcal{E}(H)$ for $i = 1, \dots, h-1$. They are generators of $\mathrm{Gal}(\varphi)$. We denote them by ϕ_i. One can choose an isomorphism $\mathrm{Gal}(\varphi) \cong S_h$ in such a way that ϕ_i is sent to $(i, i+1)$. For the flop ϕ_i:

$$\mathcal{E}(H) \to \mathcal{X}_i \leftarrow \mathcal{E}(H),$$

let

$$\Psi_i\colon D(\mathcal{E}(H)) \to D(\mathcal{E}(H))$$

be the functor defined by the fiber product $\mathcal{E}(H) \times_{\mathcal{X}_i} \mathcal{E}(H)$.

5.10. PROPOSITION. *Ψ_i is an equivalence.*

PROOF. Let $\Psi_i'\colon D(\mathcal{E}(H)) \to D(\mathcal{E}(H))$ be the adjoint functor of Ψ_i (cf. the proof of Observation 4.20(2)). Let $\Omega := \{\mathcal{O}_p\}_{p\in\mathcal{E}(H)}$. Then Ω is a spanning class for $D(\mathcal{E}(H))$. We show that $\Psi_i \circ \Psi_i'(\omega) \cong \omega$ and $\Psi_i' \circ \Psi_i(\omega) \cong \omega$ for all $\omega \in \Omega$. Then, by Lemma 4.21, we conclude that $\Psi_i \circ \Psi_i' \cong \mathrm{id}$ and $\Psi_i' \circ \Psi_i \cong \mathrm{id}$. The problem being local, we can replace $\mathcal{E}(H)$ by $\mathcal{X} := X \times S$ with X a smooth quasi-projective threefold containing $(-1,-1)$-curve C and with S a smooth quasi-projective variety. Let $\overline{X}$ be the threefold obtained from X by contracting C to a point. Let

$$X \to \overline{X} \leftarrow X^+$$

be the Atiyah flop along $C \subset X$, and let

$$\mathcal{X} \to \overline{X} \times S \leftarrow \mathcal{X}^+$$

be the product of the Atiyah flop with S, where $\mathcal{X}^+ := X^+ \times S$. Let $\Psi\colon D(\mathcal{X}) \to D(\mathcal{X}^+)$ be the functor defined by the fiber product $\mathcal{X} \times_{\overline{X}\times S} \mathcal{X}^+$. Denote by Ψ' its adjoint. Moreover, let $\Psi_0\colon D(X) \to D(X^+)$ be the functor defined by $X \times_{\overline{X}} X^+$. We already know that Ψ_0 is an equivalence (cf. [**10, 13**]). Then, by the same argument as Observation 4.20(2) we see that $\Psi \circ \Psi'(\mathcal{O}_p) \cong \mathcal{O}_p$ for $p \in \mathcal{X}^+$ and $\Psi' \circ \Psi(\mathcal{O}_p) \cong \mathcal{O}_p$ for $p \in \mathcal{X}$. □

5.11. For $\sigma \in \mathrm{Gal}(\varphi)$, the birational automorphism ϕ_σ can be decomposed into a finite sequence of Atiyah flops ϕ_i. By Lemma 5.8, $f = f_\sigma \circ \phi_{\sigma^{-1}}$ for a suitable $\sigma \in \mathrm{Gal}(\varphi)$. The σ corresponds to $(12)(23)\cdots(h-2,h-1)(h-1,h)(h-2,h-1)\cdots(12)$ by the identification $\mathrm{Gal}(\varphi) \cong S_h$ in §5.9. Now $\phi_{\sigma^{-1}}$ is decomposed as

$$\phi_{\sigma^{-1}} = \phi_1 \circ \cdots \circ \phi_{h-2} \circ \phi_{h-1} \circ \phi_{h-2} \circ \cdots \circ \phi_1.$$

5.12. THEOREM. *f induces an equivalence of derived categories*

$$\Psi_f := \Psi_{f_\sigma} \circ \Psi_1 \circ \cdots \circ \Psi_{h-2} \circ \Psi_{h-1} \circ \Psi_{h-2} \circ \cdots \circ \Psi_1\colon D(\mathcal{E}(H)) \to D(\mathcal{E}(H^*)).$$

Moreover, Ψ_f induces an equivalence

$$(\Psi_f)_0\colon D(T^*F) \to D(T^*F^+).$$

PROOF. The first claim is clear from Proposition 5.10. The second claim follows from [**10**, Lemma 5.6, Corollary 5.7]. □

References

1. A. I. Bondal and D. O. Orlov, *Semiorthogonal decompositions for algebraic varieties*, math.AG/9506012.
2. T. Bridgeland, *Equivalences of triangulated categories and Fourier–Mukai transform*, Bull. London Math. Soc. **31** (1999), no. 1, 25–34.
3. ———, *Flops and derived categories*, Invent. Math. **147** (2002), n. 3, 613–632.
4. J.-C. Chen, *Flops and equivalences of derived categories for threefolds with only terminal Gorenstein singularities*, J. Differential Geom. **61** (2002), no. 2, 227–261.
5. M. Chriss and V. Ginzburg, *Representation theory and complex geometry*, Birkhäuser, Boston, MA, 1997.
6. W. Fulton, *Intersection theory*, Ergeb. Math. Grenzgeb. (3), vol. 2, Springer-Verlag, Berlin, 1984.
7. ———, *Young tableaux*, London Math. Soc. Stud. Texts, vol. 35, Cambridge Univ. Press, Cambridge, 1997.
8. M. M. Kapranov, *Derived category of coherent sheaves on Grassmann manifolds*, Izv. Akad. Nauk SSSR Ser. Mat. **48** (1984), no. 1, 192–202 (Russian).
9. Y. Kawamata, *Francia's flip and derived categories*, Algebraic Geometry, de Gruyter, Berlin, 2002, pp. 197–215.
10. ———, *D-equivalence and K-equivalence*, J. Differential Geom. **61** (2002), no. 1, 147–171.

11. E. Markman, *Brill–Noether duality for moduli spaces of sheaves of K3 surfaces*, J. Alg. Geom. **10** (2001), no. 4, 623–694.
12. ———, *On the monodromy of moduli spaces of sheaves on K3 surfaces*. II, math.AG/0305043.
13. Y. Namikawa, *Mukai flops and derived categories*, J. Reine Angew. Math **560** (2003), 65–76.
14. M. Reid, *Minimal models of canonical 3-folds*, Algebraic Varieties and Analytic Varieties (Tokyo, 1981), Adv. Stud. Pure Math., vol. 1, North-Holland, Amsterdam, 1983, pp. 131–180.
15. P. Slodowy, *Simple singularities and simple algebraic groups*, Lecture Notes in Math., vol. 815, Springer, Berlin, 1980.
16. M. Van den Bergh, *Three-dimensional flops and non-commutative rings*, math.AG/0207170.

Department of Mathematics, Graduate School of Science, Kyoto University, Kita-shirakawa Oiwake-cho, Kyoto, 606-8502, Japan
E-mail address: namikawa@kusm.kyoto-u.ac.jp

Centre de Recherches Mathématiques
CRM Proceedings and Lecture Notes
Volume **38**, 2004

Fourier–Mukai Number of a K3 Surface

Shinobu Hosono, Bong H. Lian, Keiji Oguiso, and Shing-Tung Yau

Abstract. We shall give a Counting Formula for the number of Fourier–Mukai partners of a K3 surface and consider three applications.

Introduction

In his papers [**12–14**], Mukai discovered the fundamental importance of manifolds with equivalent bounded derived categories of coherent sheaves. Such manifolds are now known as Fourier–Mukai (FM) partners, which have been the focus of attention in several contexts of mathematics.

The aim of this note is to derive an explicit counting formula for the FM number, i.e., the cardinality of the set $\mathrm{FM}(X)$ of FM partners, for a projective K3 surface X in terms of the Néron–Severi lattice $\mathrm{NS}(X)$ and the Hodge structure $(T(X), \mathbb{C}\omega_X)$ of the transcendental lattice $T(X)$ (Counting Formula 2.3), and to give then three applications (Corollary 2.7, Theorems 3.3 and 4.1).

The Counting Formula 2.3 is inspired by an earlier work of Bridgeland and Maciocia [**2**]. Our proof here is based on the fundamental result of Mukai [**14**] and Orlov [**19**] (see also [**2, 6**] and Theorem 2.2 for summary), and the theory of primitive embeddings of lattices due to Nikulin [**15**]. What we actually need here is a sort of equivariant version of the theory in [**15**]. Namely, we introduce the notion of G-equivalence classes of primitive embeddings of an even nondegenerate lattice T into an even unimodular indefinite lattice Λ, where G is a prescribed subgroup of the orthogonal group $O(T)$ (Definition 1.1), and we find a formula for the cardinality of the set $\mathcal{P}^{G\text{-eq}}(T, \Lambda)$ of G-equivalence classes of the primitive embeddings $\iota\colon T \to \Lambda$ (Theorem 1.4).

We give a precise formula for $|\mathcal{P}^{G\text{-eq}}(T, \Lambda)|$ in §1. Such a natural and useful formula like Theorem 1.4 is clearly not unexpected. However, we are unable to find this general formula anywhere in the literature, and so we give a proof in Appendix A. We remark that in the special case $G = \{\mathrm{id}\}$, the formula is given explicitly in [**10**].

2000 *Mathematics Subject Classification.* 14J28, 11R29, 11A41.

The first and third would like to express their thanks to the Harvard University and the Education Ministry of Japan for financial support.

The second author is supported by NSF grant DMS-0072158.

This is the final form of the paper.

We deduce the Counting Formula 2.3 in §2.

We shall then give three applications.

First, as an immediate consequence of the Counting Formula, we see that a K3 surface X such that $\rho(X) \geq 3$ and $\det \mathrm{NS}(X)$ is square free, has no FM partner other than X itself (Corollary 2.7(3)). This is in sharp contrast to the cases of $\rho = 1, 2$ described in [17] (See also Corollary 2.7(4) and Theorem 3.3). We shall also show how some of known cases are derived from our Counting Formula (Corollary 2.7(1) and (4)).

Next, as an arithmetical application, we recast two famous questions on class numbers, after Gauss, circa 1800, into geometric terms of FM partners of K3 surfaces of Picard number 2 (Theorem 3.3, Questions 1′ and 2′ in §3). Here we recall the two questions on the class number $h(p)$ of the real quadratic field $\mathbb{Q}(\sqrt{p})$ for a prime number $p \equiv 1 \bmod 4$:

QUESTION 1. Are there infinitely many primes p such that $h(p) = 1$?

QUESTION 2. Is there a sequence of primes $p_1, p_2, \ldots,$ such that $h(p_k) \to \infty$?

Our reformulations (Questions 1′ and 2′) of these arithmetical questions might afford us a new geometrical approach to studying them via moduli of stable sheaves on a family of K3 surfaces (cf. Theorem 2.2 and an argument in §4).

Finally, for a more geometrical application, we shall construct explicitly a pair of smooth projective families of K3 surfaces, $f\colon \mathcal{X} \to \Delta$ and $g\colon \mathcal{Y} \to \Delta$, with a sequence $\{t_k\}_{k=1}^{\infty} \subset \Delta$ such that $\lim_{k\to\infty} t_k = 0$ and $\mathcal{X}_{t_k} \simeq \mathcal{Y}_{t_k}$, but $\mathcal{X}_0 \not\simeq \mathcal{Y}_0$ (Theorem 4.1). Here Δ is a unit disk. This explicitly shows the non-Hausdorff nature of the moduli space of unpolarized K3 surfaces. We carry out the construction in §4.

We remark that, aside from the three applications here, the Counting Formula 2.3 has also been applied to a study of Kummer structures on a K3 surface [7].

Acknowledgments. We thank Professors B. Gross and K. Zuo for their interest in this work. We also thank C. Madonna for communications on our previous note [6]. The main part of this collaboration was done during the first and third authors' stay at Harvard University. The third author would like to express his thanks to the organizers for their invitation to the conference.

1. G-equivalence classes of primitive embeddings

The goal of this section is Theorem 1.4. Here we shall need a G-equivariant version of the notion of primitive embeddings [15].

Unless otherwise stated, by a lattice $L := \big(L, (\ ,\ )\big)$, we mean a pair of a free $\mathbb{Z}$-module L of finite rank and its non-degenerate symmetric integral-valued bilinear form $(\ ,\ )\colon L \times L \to \mathbb{Z}$. We set $L^* = \mathrm{Hom}(L, \mathbb{Z})$.

Let $\Lambda = \big(\Lambda, (\ ,\ )\big)$ be an even, unimodular, indefinite lattice. Consider an even lattice T which admits at least one primitive embedding $\iota_0\colon T \hookrightarrow \Lambda$, and a subgroup $G \subset O(T)$. The group G is not necessarily a finite group. Let S be a lattice isomorphic to the orthogonal lattice $\iota_0(T)^{\perp}$ in Λ and choose an isomorphism $\iota_0^S\colon S \simeq \iota_0(T)^{\perp}$. Throughout §1, the objects Λ, T, ι_0, G, S and ι_0^S are fixed.

DEFINITION 1.1. Two primitive embeddings $\iota\colon T \hookrightarrow \Lambda$, $\iota'\colon T \hookrightarrow \Lambda$ are called G-equivalent if there exist $\Phi \in O(\Lambda)$ and $g \in G$ such that the following diagram commutes:

$$\begin{array}{ccc} \Lambda & \overset{\iota}{\hookrightarrow} & \Lambda \\ {\scriptstyle g}\downarrow & & \downarrow{\scriptstyle \Phi} \\ T & \overset{\iota'}{\hookrightarrow} & T. \end{array}$$

Our main interest in this section is the set of all primitive embeddings:

$$\mathcal{P}(T, \Lambda) := \{\iota\colon T \hookrightarrow \Lambda; \text{primitive embedding}\},$$

and the set of G-equivalence classes, i.e., the quotient set

$$\mathcal{P}^{G\text{-eq}}(T, \Lambda) := \mathcal{P}(T, \Lambda)/G\text{-equivalence}.$$

We denote the restriction of the bilinear form $(\ ,\ )$ on Λ to S and T, respectively, by $(\ ,\ )_S$, $(\ ,\ )_T$. These coincide with the original bilinear forms of S and T. (Here and hereafter, we often identify T and S with the primitive sublattices $\iota_0(T)$ and $\iota_0(T)^{\perp} = \iota_0^S(S)$ of Λ *via the fixed isomorphisms* ι_0 and ι_0^S.) Since Λ is unimodular and S and T are primitive in Λ, we have surjective maps $\pi_S\colon \Lambda = \Lambda^* \to S^*$, $\pi_T\colon \Lambda = \Lambda^* \to T^*$ defined by $\pi_S(l) = (l,\ )|_S$, $\pi_T(l) = (l,\ )|_T$, and natural isomorphisms induced by π_S and π_T [**15**, Corollary 1.6.2]):

$$S^*/S \xleftarrow[\bar{\pi}_S]{\sim} \Lambda/(S \oplus T) \xrightarrow[\bar{\pi}_T]{\sim} T^*/T.$$

Here the last isomorphisms are written explicitly as $\bar{\pi}_S(l \bmod S \oplus T) = l_S \bmod S$ and $\bar{\pi}_T(l \bmod S \oplus T) = l_T \bmod T$, where l_S, l_T are defined by the orthogonal decomposition $l = l_S + l_T$ made in $\Lambda \otimes \mathbb{Q} = (S \oplus T) \otimes \mathbb{Q}$. We make the identifications $S \subset S^* \subset S \otimes \mathbb{Q}$ and $T \subset T^* \subset T \otimes \mathbb{Q}$ via the respective nondegenerate forms $(\ ,\ )_S$ and $(\ ,\ )_T$.

As it is defined by [**15**, §1], the discriminant group (A_s, q_s) of the lattice $(L, (\ ,\ )_L)$ is the pair of the abelian group $A_L := L^*/L$ and a natural quadratic form $q_L\colon A_L \to \mathbb{Q}/2\mathbb{Z}$ defined by $q_L(x \bmod L) := (x, x)_{L\otimes\mathbb{Q}} \bmod 2\mathbb{Z}$. Here $(\ ,\ )_{L\otimes\mathbb{Q}}$ is the natural $\mathbb{Q}$-linear extension of the bilinear form $(\ ,\ )_L$ of L.

PROPOSITION 1.2 ([**15**, Corollary 1.6.2]). *Put* $\sigma_0 = \bar{\pi}_S \circ \bar{\pi}_T^{-1}\colon A_T \xrightarrow{\sim} A_S$. *Then* σ_0 *is an isometry*: $\sigma_0\colon (A_T, -q_{A_T}) \cong (A_S, q_{A_S})$.

As it is well-known, primitive embeddings of T into Λ are closely related to the so-called genus of the lattice S. We shall briefly recall this here.

DEFINITION-THEOREM 1.3. Let M be an even lattice. The set of isomorphism classes of the lattices M' satisfying the following equivalent conditions (1) and (2), are called the *genus* of M, and will be denoted by $\mathcal{G}(M)$:

(1) $M' \otimes \mathbb{Z}_p \cong M \otimes \mathbb{Z}_p$ for all primes p, and $M' \otimes \mathbb{R} \cong M \otimes \mathbb{R}$.
(2) $(A_{M'}, q_{M'}) \cong (A_M, q_M)$, and $\operatorname{sgn} M' = \operatorname{sgn} M$ where $\operatorname{sgn} M$ is the signature of the lattice M.

The equivalence of (1) and (2) is due to [**15**, Corollary 1.9.4]. From the condition (2), we have $\det M' = \det M$ and $\operatorname{rank} M' = \operatorname{rank} M$ if $M' \in \mathcal{G}(M)$. It is well-known that the genus $\mathcal{G}(M)$ is a finite set (see, e.g., [**3**, p. 128, Theorem 1.1]).

Consider the genus $\mathcal{G}(S)$ of S and put: $\mathcal{G}(S) = \{S_1, S_2, \dots, S_m\}$ $(S_1 := S)$. By definition, we have an isomorphism $(A_S, q_S) \cong (A_{S_j}, q_{S_j})$ for each $S_j \in \mathcal{G}(S)$. Since

$\sigma_0 : (A_T, -q_T) \cong (A_S, q_S)$ by Proposition 1.2, we can then choose an isomorphism for each $S_j \in \mathcal{G}(S)$: $\varphi_j \colon (A_T, -q_T) \xrightarrow{\sim} (A_{S_j}, q_{S_j})$.

Throughout §1, these isomorphisms φ_j $(1 \le j \le m)$ are fixed. Now consider an arbitrarily primitive embedding $\iota \colon T \hookrightarrow \Lambda$. We have, by Proposition 1.2,

$$(A_{\iota(T)^\perp}, q_{\iota(T)^\perp}) \cong (A_T, -q_T) \cong (A_S, q_S)$$

and also $\operatorname{sgn} \iota(T)^\perp = \operatorname{sgn} \Lambda - \operatorname{sgn} T$. Therefore there exists a unique $j = j(\iota) \in \{1, 2, \dots, m\}$ and an isomorphism $S_j \xrightarrow{\sim} \iota(T)^\perp$. If two primitive embeddings $\iota \colon T \hookrightarrow \Lambda$ and $\iota' \colon T \hookrightarrow \Lambda$ are G-equivalent, there exist elements $\Phi \in O(\Lambda)$ and $g \in G$ such that $\Phi \circ \iota = \iota' \circ g$. Since the restriction $\Phi|_{\iota(T)}$ is then an isometry from $\iota(T)$ to $\iota'(T)$, we have $S_{j(\iota)} \cong \iota(T)^\perp \cong \iota'(T)^\perp \cong S_{j(\iota')}$, which implies that $j(\iota) = j(\iota')$, i.e., $S_{j(\iota)} = S_{j(\iota')}$ in $\mathcal{G}(S)$. Therefore setting

$$\mathcal{P}_j(T, \Lambda) := \{\iota \colon T \hookrightarrow \Lambda \in \mathcal{P}(T, \Lambda) \mid \iota(T)^\perp \cong S_j\},$$
$$\mathcal{P}_j^{G\text{-eq}}(T, \Lambda) := \{[\iota \colon T \hookrightarrow \Lambda] \in \mathcal{P}^{G\text{-eq}}(T, \Lambda) \mid \iota(T)^\perp \cong S_j\},$$

we have the following well-defined disjoint unions:

$$\mathcal{P}(T, \Lambda) = \bigcup_{j=1}^{m} \mathcal{P}_j(T, \Lambda), \qquad \mathcal{P}^{G\text{-eq}}(T, \Lambda) = \bigcup_{j=1}^{m} \mathcal{P}_j^{G\text{-eq}}(T, \Lambda).$$

The following is a generalization of [**10**], which will be proved in Appendix A:

THEOREM 1.4. *For each j, we have $|\mathcal{P}_j^{G\text{-eq}}(T, \Lambda)| = |(S_j) \setminus O(A_{S_j})/G|$. In particular,*

$$|\mathcal{P}^{G\text{-eq}}(T, \Lambda)| = \sum_{j=1}^{m} |O(S_j) \setminus O(A_{S_j})/G|.$$

Here $O(S_j)$ acts on $O(A_{S_j})$ through the natural map $O(S_j) \to O(A_{S_j})$, $h \mapsto \bar{h}$. G acts on $O(A_{S_j})$ through the composite of the natural map $G \subset O(T) \to O(A_T)$ and the adjoint map $\operatorname{ad}(\varphi_j) \colon O(A_T) \to O(A_{S_j})$.

2. FM number of a K3 surface

Let $D(X)$ be the derived category of bounded complexes of coherent sheaves on a smooth projective variety X. A smooth projective variety Y is called a *Fourier-Mukai (FM) partner* of X if there is an equivalence $D(Y) \cong D(X)$ as triangulated categories. We denote by $\operatorname{FM}(X)$ the set of isomorphism classes of FM partners of X.

DEFINITION 2.1. We call the cardinality $|\operatorname{FM}(X)|$ of the set $\operatorname{FM}(X)$ the Fourier–Mukai (FM) number of X.

A FM partner of a K3 surface is again a K3 surface by a result of Mukai [**14**] and Orlov [**19**] (See also [**2**]).

By a K3 surface, we mean a smooth *projective* surface X over $\mathbb{C}$ with $\mathcal{O}_X(K_X) \cong \mathcal{O}_X$ and $h^1(\mathcal{O}_X) = 0$. We denote by $(\, , \,)$ the symmetric bilinear form on $H^2(X, \mathbb{Z})$ given by the cup product. Then $\big(H^2(X, \mathbb{Z}), (\, , \,)\big)$ is an even unimodular lattice of signature $(3, 19)$. This lattice is isomorphic to the K3 lattice $\Lambda_{\mathrm{K3}} := E_8(-1)^{\oplus 2} \oplus U^{\oplus 3}$ where U is the hyperbolic lattice (an even unimodular lattice of signature (1,1)). An isomorphism $\tau \colon H^2(X, \mathbb{Z}) \to \Lambda_{\mathrm{K3}}$ is called a marking and a pair (X, τ) is called a marked K3 surface. We denote by $\operatorname{NS}(X)$ the Néron–Severi lattice of X and by $\rho(X)$ the Picard number, i.e., the rank of $\mathrm{NS}(X)$. The lattice $\mathrm{NS}(X)$ is primitive in $H^2(X,\mathbb{Z})$ and has signature $(1,\rho(X)-1)$. We call the orthogonal lattice $T(X) := \mathrm{NS}(X)^{\perp}$ in $H^2(X,\mathbb{Z})$ the transcendental lattice. $T(X)$ is primitive in $H^2(X,\mathbb{Z})$ and has signature $\big(2, 20-\rho(X)\big)$. We denote by ω_X a nowhere vanishing holomorphic two form on X. Then one has a natural inclusion

$$\mathbb{C}\omega_X \oplus \mathbb{C}\bar{\omega}_X \subset T(X)\otimes\mathbb{C}.$$

This defines a Hodge structure $(T(X),\mathbb{C}\omega_X)$ of weight 2 on $T(X)$. Note that $T(X)$ is the minimal primitive sublattice of $H^2(X,\mathbb{Z})$ such that $T(X)\otimes\mathbb{C}$ contains $\mathbb{C}\omega_X$.

The following is a fundamental theorem due to Mukai and Orlov. This connects three aspects of FM partners of K3 surfaces. Statement (1) is categorical, (2) is arithmetical, and (3) is geometrical.

THEOREM 2.2 ([**14, 19**]). *Let X and Y be K3 surfaces. Then the following statements are equivalent*:

(1) $D(Y)\cong D(X)$;
(2) *There exists a Hodge isometry* $g\colon (T(Y),\mathbb{C}\omega_Y) \xrightarrow{\sim} (T(X),\mathbb{C}\omega_X)$;
(3) *Y is a two-dimensional fine compact moduli space of stable sheaves on X with respect to some polarization on X.*

For the rest of this section, X will be a K3 surface. Our Counting Formula 2.3 follows from specializing Theorem 1.4 to the case

$$\Lambda = H^2(X,\mathbb{Z}),\quad (T,\mathbb{C}\omega) = (T(X),\mathbb{C}\omega_X),\quad S=\mathrm{NS}(X),\quad G = O_{\mathrm{Hodge}}(T(X),\mathbb{C}\omega_X).$$

Note that in this case the group G is a finite cyclic group of order $2I$ such that $\varphi(2I) \mid \operatorname{rank} T(X)$. (See Appendix B for proof.)

The aim of this section is to prove the following theorem:

COUNTING FORMULA 2.3. *Set $\mathcal{G}(NS(X)) = \{S_1 = S, S_2, \dots, S_m\}$ where $m = \big|\mathcal{G}(\mathrm{NS}(X))\big|$. Then,*

$$|\mathrm{FM}(X)| = \sum_{j=1}^{m} |O(S_j)\setminus O(A_{S_j})/O_{\mathrm{Hodge}}(T(X),\mathbb{C}\omega_X)|.$$

Moreover the jth summand here coincides with the number of FM partners $Y\in \mathrm{FM}(X)$ with $\mathrm{NS}(Y)\simeq S_j$.

PROOF. By Theorem 1.4, it suffices to show the following:

THEOREM 2.4. *There is a natural bijective correspondence*

$$\mathcal{P}^{G\text{-eq}}(T(X), H^2(X,\mathbb{Z})) \leftrightarrow \mathrm{FM}(X).$$

Moreover, under this bijection a primitive embedding ι with $\iota(T)^{\perp}\simeq S_j$ corresponds to a FM partner Y of X with $\mathrm{NS}(Y)\simeq S_j$.

Our proof of Theorem 2.4 is based on Theorem 2.2 and the two most fundamental facts about K3 surfaces, namely, the global Torelli theorem and the surjectivity of period map (See, e.g., [**1**, Chapter VIII] and the references therein).

In what follows, for convenience, we identify $H^2(X,\mathbb{Z}) = \Lambda_{K3}$ through a fixed marking τ_X and set $(T,\mathbb{C}\omega) = (T(X),\mathbb{C}\omega_X) \subset H^2(X,\mathbb{Z})$ under the natural inclusion. We proceed in five steps.

STEP 0. *For an arbitrary primitive embedding* $\iota\colon T \hookrightarrow \Lambda_{\mathrm{K3}}$, *there exists a marked K3 surface* (Y, τ_Y) *with the commutative diagram:*

$$\begin{array}{ccc} \tau_Y\colon H^2(Y,\mathbb{Z}) & \xrightarrow{\ \sim\ } & \Lambda_{\mathrm{K3}} \\ \cup & & \uparrow\iota \\ \iota^{-1}\circ\tau_Y|_{T(Y)}\colon (T(Y),\mathbb{C}\omega_Y) & \longrightarrow & (T,\mathbb{C}\omega). \end{array}$$

PROOF. This follows immediately from the surjectivity of the period map and the minimality of T. □

STEP 1. *There is a map* $\tilde{c}\colon \mathcal{P}(T,\Lambda_{\mathrm{K3}}) \to FM(X)$, *i.e., there exists a FM partner* $Y(\iota)$ *for each primitive embedding* $\iota\colon T \hookrightarrow \Lambda_{\mathrm{K3}}$.

PROOF. Let $\iota\colon T \hookrightarrow \Lambda_{\mathrm{K3}}$ be a primitive embedding, then by Step 0, we have a marked K3 surface (Y,τ_Y) with the commutative diagram in Step 0. Then, by Theorem 2.2, we have $Y \in \mathrm{FM}(X)$. Assume there exists another marked K3 surface $(Y',\tau_{Y'})$ with the commutative diagram in Step 0 (Y replaced by Y'). Then

$$\tau_Y^{-1}\circ\tau_{Y'}\colon (H^2(Y',\mathbb{Z}),\mathbb{C}\omega_{Y'}) \to (H^2(Y,\mathbb{Z}),\mathbb{C}\omega_Y)$$

is a Hodge isometry, which means that $Y' \cong Y$ by the Torelli theorem (See, e.g., [**1**]). Therefore there exists a well-defined map $\tilde{c}\colon \mathcal{P}(T,\Lambda_{\mathrm{K3}}) \to \mathrm{FM}(X)$. □

STEP 2. *The map* $\tilde{c}$ *is surjective.*

PROOF. Let $Y \in FM(X)$ and fix a marking $\tau_Y\colon H^2(Y,\mathbb{Z}) \to \Lambda_{\mathrm{K3}}$. Then by Theorem 2.2, we have a Hodge isometry $g\colon (T(Y),\mathbb{C}\omega_Y) \xrightarrow{\sim} (T(X),\mathbb{C}\omega_X)$ and the following commutative diagram in which $T' = \tau_Y(T(Y))$, $\omega' = \tau_Y(\omega_Y)$:

$$\begin{array}{ccccccc} \Lambda_{\mathrm{K3}} & \xleftarrow[\tau_Y]{\sim} & H^2(Y,\mathbb{Z}) & & H^2(X,\mathbb{Z}) & \xrightarrow[\tau_X]{\sim} & \Lambda_{\mathrm{K3}} \\ \cup\,\iota_Y & & \cup & & \cup & & \cup\,\iota_X \\ (T',\mathbb{C}\omega') & \xleftarrow{\sim} & (T(Y),\mathbb{C}\omega_Y) & \xrightarrow[g]{\sim} & (T(X),\mathbb{C}\omega_X) & \xrightarrow{\sim} & (T,\mathbb{C}\omega) \end{array}$$

The map $\iota := \iota_{Y'}\circ\tau_Y|_{T(Y)}\circ g^{-1}\circ\tau_X^{-1}|_T\colon T \hookrightarrow \Lambda_{\mathrm{K3}}$ is then a primitive embedding and $Y = \tilde{c}(\iota)$. □

STEP 3. *If two primitive embeddings* ι, $\iota'\colon (T,\mathbb{C}\omega) \hookrightarrow \Lambda_{\mathrm{K3}}$ *are G-equivalent, then* $\tilde{c}(\iota) = \tilde{c}(\iota')$. *This together with the surjectivity in Step 2 entails that the map* $\tilde{c}$ *descends to a surjective map* $c\colon \mathcal{P}^{G\text{-eq}}(T,\Lambda_{\mathrm{K3}}) \to \mathrm{FM}(X)$.

PROOF. Suppose ι, $\iota'\colon (T,\mathbb{C}\omega) \hookrightarrow \Lambda_{\mathrm{K3}}$ are G-equivalent, and denote $Y = \tilde{c}(\iota)$, $Y' = \tilde{c}(\iota')$. We also fix markings τ_Y, $\tau_{Y'}$ of Y and Y'. Then we have the following commutative diagram:

$$\begin{array}{ccccccc} H^2(Y,\mathbb{Z}) & \xrightarrow[\tau_Y]{\sim} & \Lambda_{\mathrm{K3}} & \xrightarrow[\exists\Phi]{\sim} & \Lambda_{\mathrm{K3}} & \xleftarrow[\tau_{Y'}]{\sim} & H^2(Y',\mathbb{Z}) \\ \cup & & \uparrow\iota & & \uparrow\iota' & & \cup \\ (T(Y),\mathbb{C}\omega_Y) & \xrightarrow{\sim} & (T,\mathbb{C}\omega) & \xrightarrow[\exists g]{\sim} & (T,\mathbb{C}\omega) & \xleftarrow{\sim} & (T(Y'),\mathbb{C}\omega_{Y'}), \end{array}$$

where $\Phi \in O(\Lambda_{\mathrm{K3}})$ and $g \in G = O_{\mathrm{Hodge}}(T,\mathbb{C}\omega)$. From this diagram, we have a Hodge isometry $\tau_{Y'}^{-1}\circ\Phi\circ\tau_Y\colon (H^2(Y,\mathbb{Z}),\mathbb{C}\omega_Y) \to (H^2(Y',\mathbb{Z}),\mathbb{C}\omega_{Y'})$, which implies $Y \cong Y'$ by the global Torelli theorem. □

STEP 4. *The induced map* $c\colon \mathcal{P}^{G\text{-eq}}(T,\Lambda_{\mathrm{K3}}) \to \mathrm{FM}(X)$ *is injective.*

PROOF. Take two G-equivalence classes $[\iota]$, $[\iota'] \in \mathcal{P}^{G\text{-eq}}(T, \Lambda_{\mathrm{K3}})$ and denote $Y = c([\iota]) = \tilde{c}(\iota)$, $Y' = c([\iota']) = \tilde{c}(\iota')$. We fix markings τ_Y, $\tau_{Y'}$ for Y and Y', respectively. To see the injectivity we show that $f\colon Y' \cong Y$ implies $[\iota'] = [\iota]$. Denote the Hodge isometry $f^*\colon (H^2(Y,\mathbb{Z}), \mathbb{C}\omega_Y) \to (H^2(Y',\mathbb{Z}), \mathbb{C}\omega_{Y'})$ associated with the isomorphism $f\colon Y' \cong Y$. Then we have the following commutative diagram:

$$\begin{array}{ccccccc} \Lambda_{\mathrm{K3}} & \xleftarrow[\tau_Y]{\sim} & H^2(Y,\mathbb{Z}) & \xrightarrow[f^*]{\sim} & H^2(Y',\mathbb{Z}) & \xrightarrow[\tau_{Y'}]{\sim} & \Lambda_{\mathrm{K3}} \\ {\scriptstyle\iota}\uparrow & & \cup & & \cup & & \uparrow{\scriptstyle\iota'} \\ (T,\mathbb{C}\omega) & \xleftarrow{\sim} & (T(Y),\mathbb{C}\omega_Y) & \xrightarrow{\sim} & (T(Y'),\mathbb{C}\omega_{Y'}) & \xrightarrow{\sim} & (T,\mathbb{C}\omega). \end{array}$$

From this diagram we obtain;

$$\begin{array}{ccc} \Lambda_{\mathrm{K3}} & \xrightarrow[\exists\Phi]{\sim} & \Lambda_{\mathrm{K3}} \\ {\scriptstyle\iota}\Big\uparrow & & \Big\uparrow{\scriptstyle\iota'} \\ (T,\mathbb{C}\omega) & \xrightarrow[\exists g]{\sim} & (T,\mathbb{C}\omega), \end{array}$$

where $g \in O_{\mathrm{Hodge}}(T, \mathbb{C}\omega)$. We then conclude $[\iota] = [\iota']$ in $\mathcal{P}^{G\text{-eq}}(T, \Lambda_{\mathrm{K3}})$. □

It now follows that c is bijective. The second assertion in the theorem is clear from the construction of c. □

In an actual application of the Counting Formula 2.3, we need to determine the genus of $\mathrm{NS}(X)$. The following theorem due to [**15**, Theorem 1.14.2] is very useful when $\rho(X)$ is large:

THEOREM 2.5. *Let S be an even indefinite nondegenerate lattice. If* $\operatorname{rank} S \geq 2 + \ell(S)$, *then* $\mathcal{G}(S) = \{S\}$ *and the natural map* $O(S) \to O(A_S)$ *is surjective. Here* $\ell(S)$ *is the minimal number of generators of* A_S, *e.g.,* $\ell(A_S) = 1$ *when* A_S *is cyclic.*

COROLLARY 2.6. *If* $\rho(X) \geq \ell(NS(X)) + 2$, *then* $\mathrm{FM}(X) = \{X\}$.

PROOF. By Theorem 2.5, we have $\mathcal{G}(\mathrm{NS}(X)) = \{\mathrm{NS}(X)\}$. Then, by the Counting Formula, we have $|\,\mathrm{FM}(X)| = |O(\mathrm{NS}(X)) \setminus O(A_{\mathrm{NS}(X)})/O_{\mathrm{Hodge}}(T(X), \mathbb{C}\omega_X)|$. Since the natural map $O(\mathrm{NS}(X)) \to O(A_{\mathrm{NS}(X)})$ is surjective again by Theorem 2.5, we have $|\,\mathrm{FM}(X)| = 1$. □

COROLLARY 2.7. (1) (*cf.* [**12**, Proposition 6.2]) *If* $\rho(X) \geq 12$, *then* $\mathrm{FM}(X) = \{X\}$. *In particular,* $\mathrm{FM}(\mathrm{Km}\, A) = \{\mathrm{Km}\, A\}$.

(2) *If* $\rho(X) \geq 3$ *and* $\det \mathrm{NS}(X)$ *is square-free, then* $\mathrm{FM}(X) = \{X\}$.

(3) *If* $X \to \mathbb{P}^1$ *is a Jacobian K3 surface, i.e., an elliptic K3 surface with a section, then* $\mathrm{FM}(X) = \{X\}$.

(4) (*cf.* [**17**, Proposition 1.10]) *Assume that* $\rho(X) = 1$ *and* $\mathrm{NS}(X) = \mathbb{Z}H$ *with* $(H^2) = 2n$. *Then* $|\,\mathrm{FM}(X)| = 2^{\tau(n)-1}$. *Here* $\tau(1) = 1$, *and* $\tau(n)$ *is the number of prime factors of* $n \geq 2$, *e.g.,* $\tau(4) = \tau(2) = 1$, $\tau(6) = 2$.

PROOF. (1) This follows from Corollary 2.6 and the following calculation: $\ell(NS(X)) = \ell(T_X) \leq \operatorname{rank} T(X) \leq 10 \leq \rho(X) - 2$.

(2) Since $\det \mathrm{NS}(X)$ is square-free, it follows that $A_{NS(X)}$ is cyclic. Then $\ell(\mathrm{NS}(X)) = 1$. Since $\rho(X) \geq 3$, the result follows from Corollary 2.6.

(3) The unimodular hyperbolic lattice U is a direct summand of $\mathrm{NS}(X)$, say $\mathrm{NS}(X) = U \oplus N$, by the assumption. Thus, $\ell(NS(X)) = \ell(N) \leq \operatorname{rank} N = \rho(X) - 2$. Therefore, the result follows from Corollary 2.6.

(4) We assume $n \geq 2$. The case $n = 1$ is similar and easier. Since $\mathrm{NS}(X) = \mathbb{Z}H$ with $(H^2) = 2n$, we have $\mathcal{G}(\mathrm{NS}(X)) = \{\mathrm{NS}(X)\}$, $O(\mathrm{NS}(X)) \simeq O_{\mathrm{Hodge}}(T(X), \mathbb{C}\omega_X) \simeq \mathbb{Z}/2$ and $(A_{\mathrm{NS}(X)}, q_{\mathrm{NS}(X)}) \simeq (\mathbb{Z}/2n, q_{2n})$. Here q_N on $\mathbb{Z}/N$ is defined by $q_N(1) = 1/N$. Therefore, $|\mathrm{FM}(X)| = |O(\mathbb{Z}/2n, q_{2n})|/2$ by the Counting Formula. By a straightforward calculation (cf. [**17**]), we have also $|O(\mathbb{Z}/2n, q_{2n})| = 2^{\tau(n)}$. □

3. Class numbers and Fourier–Mukai numbers

In this section, we focus on K3 surfaces of $\rho(X) = 2$ (Theorem 3.3) and recast the two class number problems, Questions 1 and 2 in the introduction, in corresponding geometrical terms (Questions 1′ and 2′).

We begin by recalling the following fact which is well-known to experts (see [**11, 15, 16**]). This proposition shows that every question about the class numbers of real quadratic fields can, in principle, be recast in terms of K3 surfaces.

PROPOSITION 3.1. *For any even hyperbolic lattice with* $\operatorname{rank} S \leq 10$, *there is a projective K3 surface* X *with* $\mathrm{NS}(X) \simeq S$. *In particular, for every rank 2 even hyperbolic lattice* S, *there is a projective K3 surface* X *with* $\mathrm{NS}(X) \simeq S$.

Next we briefly recall some basic background on class numbers from [**3, 22**]. Let D be an odd fundamental discriminant, i.e., a rational integer D such that $D \equiv 1 \bmod 4$ and D square-free. Put $K := \mathbb{Q}(\sqrt{D})$, and let $\mathcal{O}_K$ denote its ring of integers. Throughout this section, this D is fixed. Let $H(D)$ denote the narrow ideal class group of $\mathcal{O}_K$, i.e., the quotient of the group of fractional ideals of $\mathcal{O}_K$, by the subgroup of principal fractional ideals with *positive* norm. A theorem of Hurwitz says that $H(D)$ is finite. The number $h(D) = |H(D)|$ is *the class number* of K.

THEOREM 3.2 ([**3, 22**]). *The following three sets are in natural* 1–1 *correspondence:*

(1) *The class group* $H(D)$.
(2) *The set* $\mathcal{B}(D)$ *of proper, i.e.,* $\mathrm{SL}(2, \mathbb{Z})$*-equivalence classes of integral binary quadratic forms* $ax^2 + bxy + cy^2$ *with discriminant* $b^2 - 4ac = D$.
(3) *The set* $\mathcal{L}(D)$ *of proper, i.e., orientation-preserving, isomorphism classes of even hyperbolic rank 2 lattices* S *with* $\det S = -D$.

Note that every class in $\mathcal{L}(D)$ can be represented by a pair $\left(\mathbb{Z}^2, \left(\begin{smallmatrix} 2a & b \\ b & 2c \end{smallmatrix}\right)\right)$, where the matrix represents the bilinear form with respect to the standard base of $\mathbb{Z}^2$. Two such pairs are properly isomorphic if and only if their matrices are conjugate under $\mathrm{SL}(2, \mathbb{Z})$.

We recall a few more facts about the sets $H(D)$, $\mathcal{B}(D)$, $\mathcal{L}(D)$. For details, see [**22**, §12] and [**3**, Chap. 14 §3]. Consider the exact sequence

$$0 \to \mathcal{J} \to H(D) \xrightarrow{\mathrm{sq}} H(D)^2 \to 0$$

where $\mathrm{sq}(x) = x^2$ and $\mathcal{J}$ is the kernel of sq. It is known that $|\mathcal{J}| = |H(D)/H(D)^2| = 2^{n-1}$, where n is the number of prime factors of D. Given a rank 2 even lattice L with $\det L = -D$, let $\widetilde{\mathcal{G}}(L)$ be the set of *proper* isomorphism classes of lattices in the same genus as L. The set of genera $\{\widetilde{\mathcal{G}}(L) \mid [L] \in \mathcal{L}(D)\}$ is a principal homogeneous space of the group $H(D)/H(D)^2$. Therefore, there are $[L_1], \ldots, [L_{2^{n-1}}] \in \mathcal{L}(D)$, and a decomposition:

$$\mathcal{L}(D) = \coprod_{k=1}^{2^{n-1}} \widetilde{\mathcal{G}}(L_k)$$

where the genera $\widetilde{\mathcal{G}}(L_k)$ all have the same cardinality. Now the subset $\mathcal{J} \subset H(D)$ consists of the ideal classes of the so-called 2-sided ideals. This means that, under the correspondence $H(D) \leftrightarrow \mathcal{L}(D)$, each element in $\mathcal{J}$ corresponds to a class in $\mathcal{L}(D)$ representable as a pair $\left(\mathbb{Z}^2, \left(\begin{smallmatrix} 2a & b \\ b & 2c \end{smallmatrix}\right)\right)$ such that $\left(\mathbb{Z}^2, \left(\begin{smallmatrix} 2a & b \\ b & 2c \end{smallmatrix}\right)\right) \simeq \left(\mathbb{Z}^2, \left(\begin{smallmatrix} 2a & -b \\ -b & 2c \end{smallmatrix}\right)\right)$ under $\mathrm{SL}(2, \mathbb{Z})$. In particular the number of such classes in $\mathcal{L}(D)$ is $|\mathcal{J}| = 2^{n-1}$.

Now we have:

THEOREM 3.3. *If $D = p$ is a prime and X is a K3 surface with $\rho(X) = 2$ and $\det \mathrm{NS}(X) = -p$, then $|\,\mathrm{FM}(X)| = (h(p) + 1)/2$.*

REMARK. For each prime number $p \equiv 1 \bmod 4$, there is a K3 surface X such that $\mathrm{NS}(X) \simeq S_0$ by Proposition 3.1. Here $S_0 = \left(\mathbb{Z}^2, \left(\begin{smallmatrix} 2 & 1 \\ 1 & (1-p)/2 \end{smallmatrix}\right)\right)$. Note that S_0 is an even hyperbolic lattice of rank 2 and of $\det S_0 = -p$.

PROOF. Put $S = \mathrm{NS}(X)$. Since $D = p$, i.e., $n = 1$, it follows that $\widetilde{\mathcal{G}}(S) = \mathcal{L}(D)$, hence $h(p) = |\widetilde{\mathcal{G}}(S)|$. Since $\mathcal{G}(S)$ consists of isomorphism (not necessarily proper) classes of lattices, it follows that $\mathcal{G}(S)$ is the orbit space of the group $\langle\sigma\rangle = \mathrm{GL}(2, \mathbb{Z})/\mathrm{SL}(2, \mathbb{Z})$ ($\simeq \mathrm{C}_2$) acting on $\widetilde{\mathcal{G}}(S)$. Since $|\mathcal{J}| = 1$ because $n = 1$, it follows that there is just one class $[L] \in \widetilde{\mathcal{G}}(S)$ (in fact, $[L] = [S_0]$) which is fixed by σ acting on $\widetilde{\mathcal{G}}(S)$. Since σ is an involution, it follows that $|\mathcal{G}(S)| = (|\widetilde{\mathcal{G}}(S)| + 1)/2$. On the other hand, the group A_S has p elements. Since p is prime, it follows that $A_S \simeq \mathbb{Z}/p$, and that $O(A_S) = \{\pm \mathrm{id}\}$. Since the group $O_{\mathrm{Hodge}}(T(X), \mathbb{Z}\omega_X)$ obviously contains $\pm\mathrm{id}$, it follows that each of the quotients $O(A_{S_i})/O_{\mathrm{Hodge}}(T(X), \mathbb{C}\omega_X)$ in the Counting Formula for $|\,\mathrm{FM}(X)|$ has just 1 element. Therefore $|\,\mathrm{FM}(X)| = |\mathcal{G}(S)|$. Now we are done. $\square$

From a table [**5**], for a prime 1297 ($\equiv 1 \bmod 4$), we have $h(1297) = 11$. Then for any K3 surface X with $\rho(X) = 2$ and $\det \mathrm{NS}(X) = -1297$, we have $|\,\mathrm{FM}(X)| = 6$. In this case the value $|\,\mathrm{FM}(X)|$ is not a power of 2, either. This example

TABLE 1

p	$h(p)$	$\lvert\mathrm{FM}(X)\rvert$
229	3	2
257	3	2
401	5	3
577	7	4
733	3	2
761	3	2
1009	7	4
1093	5	3
1129	9	5
1229	3	2
1297	11	6
1373	3	2
1429	5	3
1489	3	2

may be contrasted to the case $\rho(X) = 1$, where $|\operatorname{FM}(X)|$ is always a power of 2 (Corollary 2.7(4).)

By Theorem 3.3, Gauss' questions can now be stated in the following geometrical form:

QUESTION 1$'$. *Are there infinitely many K3 surfaces X, with each X having $\rho(X) = 2$ and $\operatorname{FM}(X) = \{X\}$, such that the numbers $|\det \operatorname{NS}(X)|$ are distinct primes?*

QUESTION 2$'$. *Is there a sequence of K3 surfaces $X_1, X_2, \dots$, with $\rho(X_k) = 2$ and $|\operatorname{FM}(X_k)| \to \infty$ such that the numbers $|\det \operatorname{NS}(X_k)|$ are primes?*

4. Non-Hausdorff properties of the moduli of K3 surfaces

In this section, we shall show the following:

THEOREM 4.1. *Let Δ be a unit disk in $\mathbb{C}$. Then there is a pair of smooth projective families of K3 surfaces $\mathcal{X} \to \Delta$, $\mathcal{Y} \to \Delta$ and a sequence $\{t_k\} \subset \Delta$ such that*

$$\lim_{k\to\infty} t_k = 0, \qquad \mathcal{X}_{t_k} \simeq \mathcal{Y}_{t_k} \qquad \textit{but} \qquad \mathcal{X}_0 \not\simeq \mathcal{Y}_0.$$

In our construction, we shall make use of the 3rd characterization of FM partners of a K3 surface in Theorem 2.2. We begin by recalling a few facts about the 3-rd characterization, i.e., 2-dimensional fine compact moduli spaces of stable sheaves on a K3 surface X.

Set $\widetilde{\operatorname{NS}}(X) := H^0(X,\mathbb{Z}) \oplus \operatorname{NS}(X) \oplus H^4(X,\mathbb{Z})$. We call $\widetilde{\operatorname{NS}}(X)$ the extended Néron–Severi lattice of X. For $v = (r, H, s) \in \widetilde{\operatorname{NS}}(X)$ and an ample class $A \in \operatorname{NS}(X)$, we denote by $M_A(v)$ (resp. $\overline{M}_A(v)$) the coarse moduli space of stable sheaves (resp. the coarse moduli space of S-equivalence classes of semi-stable sheaves) $\mathcal{F}$ with respect to the polarization A with $\mu(\mathcal{F}) = v$. Here $\mu(\mathcal{F})$ is the so-called Mukai vector of $\mathcal{F}$, which is defined by $\mu(\mathcal{F}) := \operatorname{ch}(\mathcal{F})\sqrt{td_X}$. By [**8, 9**] (see also [**14**]), the space $\overline{M}_A(v)$ is a projective compactification of $M_A(v)$, and $\overline{M}_A(v) = M_A(v)$ if all the semi-stable sheaf $\mathcal{F}$ with $\mu(\mathcal{F}) = v$ is stable, for instance if $(r, s) = 1$.

The following theorem essentially due to Mukai [**14**] is proved in [**19**] in the course of the proof of Theorem 2.2:

THEOREM 4.2. *If $Y \in FM(X)$, then $Y \simeq M_H((r, H, s))$ where H is ample, $r > 0$ and s are integers such that $(r, s) = 1$ and $2rs = (H^2)$, and vice versa.*

In our proof, we need is the following relative version:

LEMMA 4.3. *Let $\pi\colon (\mathcal{X}, \mathcal{H}) \to \mathcal{B}$ be a smooth projective family of K3 surfaces. Let $f\colon \overline{\mathcal{M}} \to \mathcal{B}$ be a relative moduli space of the S-equivalence classes of semi-stable sheaves of π with respect to the polarization $\mathcal{H}$ with Mukai vectors $(r, \mathcal{H}_t, s)$. If $r > 0$, $(r, s) = 1$ and $2rs = (\mathcal{H}_t^2)$, then f is projective and gives a relative FM family of π, i.e., $\overline{\mathcal{M}}_t \in FM(\mathcal{X}_t)$ for all $t \in \mathcal{B}$.*

PROOF. The existence of f and its projectivity over $\mathcal{B}$ are shown by Maruyama [**8, 9**]. By the definition, we have $\overline{\mathcal{M}}_t = \overline{\mathcal{M}}_{\mathcal{H}_t}\big((r, \mathcal{H}_t, s)\big)$. Since $(r, s) = 1$, $2rs = (\mathcal{H}_t^2)$ and $\mathcal{H}_t$ is ample on $\mathcal{X}_t$, it follows that $\overline{\mathcal{M}}_t = \mathcal{M}_{H_t}\big((r, \mathcal{H}_t, s)\big) \in \operatorname{FM}(\mathcal{X}_t)$ by Theorem 4.2. $\square$

PROOF OF THEOREM 4.1. Consider the number

$$m := \max\{\rho(X) \mid X \text{ is a K3 surface with } |\mathrm{FM}(X)| > 2\}.$$

Such a number m exists and satisfies $m \le 11$ by Corollary 2.7(1) and (4). Take one such X and choose then $Y \in FM(X)$ such that $Y \not\simeq X$. Write $Y = M_H(r, H, s)$ as in Theorem 4.2.

Let us consider the Kuranishi family $u \colon \mathcal{U} \to \mathcal{K}$ of X. The base space $\mathcal{K}$ is assumed to be a small polydisk in $H^1(X, T_X) \simeq \mathbb{C}^{20}$. Choosing a marking $\tau \colon R^2u_*\mathbb{Z}_{\mathcal{U}} \simeq \Lambda \times \mathcal{K}$, where Λ is the K3 lattice, we consider the period map

$$\pi \colon \mathcal{K} \to \mathcal{P} = \{[\omega] \in \mathbb{P}(\Lambda \otimes \mathbb{C}) \mid (\omega.\omega) = 0, (\omega.\overline{\omega}) > 0\} \subset \mathbb{P}(\Lambda \otimes \mathbb{C}) \simeq \mathbb{P}^{21}.$$

By the local Torelli theorem (see, e.g., [**1**, Chapter VIII]), $\mathcal{K}$ is isomorphic, via π, to an open neighborhood in $\mathcal{P}$, which we again denote by $\mathcal{P}$. In what follows, we identify $\mathcal{K}$ with $\mathcal{P}$.

Set $\Lambda_0 := \tau_0(\mathrm{NS}(X))$. Let us consider the subset $\mathcal{K}^0$ of $\mathcal{K} = \mathcal{P}$ defined by the linear equations $([\omega], \Lambda_0) = 0$. This $\mathcal{K}^0$ parametrizes K3 surfaces $\mathcal{U}_t$ such that $\Lambda_0 \subset \tau_t(\mathrm{NS}(\mathcal{U}_t))$, or in other words, $\mathcal{K}^0$ is the subspace of $\mathcal{K}$ in which $\mathrm{NS}(X)$ is kept to be algebraic. Since $\operatorname{rank}\Lambda_0 = \rho(X) = m$, we have $\dim \mathcal{K}^0 = 20 - m \ge 1$. Let us take a generic small one-dimensional disk Δ such that $0 \in \Delta \subset \mathcal{K}^0$ and consider the family $\varphi \colon \mathcal{X} \to \Delta$ induced from $u \colon \mathcal{U} \to \mathcal{K}$. By definition, there is an invertible sheaf $\mathcal{H}$ on $\mathcal{X}$ such that $\mathcal{H}_0 = H$. Since ampleness is an open condition, $\mathcal{H}$ is φ-ample. Therefore, the morphism φ is projective. By [**18**, Main Theorem], we see that $\mathrm{NS}(\mathcal{X}_t) \simeq \Lambda_0 \sim \mathrm{NS}(X)$ for generic $t \in \Delta$ and that the set $\mathcal{S} := \{s \in \Delta \mid \rho(\mathcal{X}_s) > m\}$ is everywhere dense in Δ with respect to the Euclidean topology.

Let us consider the family of FM partners $f \colon \mathcal{Y} \to \Delta$ of $\varphi \colon \mathcal{X} \to \Delta$ with Mukai vectors $(r, \mathcal{H}_t, s)$. This exists by Lemma 4.3. By the definition of m, $\mathcal{S}$ and f, one has $\mathcal{Y}_0 \simeq Y \not\simeq X \simeq \mathcal{X}_0$ and $\mathcal{Y}_s \simeq \mathcal{X}_s$ for $s \in \mathcal{S}$. Then, any sequence $\{t_k\} \subset \mathcal{S}$ with $\lim_{k \to \infty} t_k = 0$ satisfies the desired property. □

Appendix A. Proof of Theorem 1.4

In this appendix, for completeness, we shall give a sketch of a proof of Theorem 1.4. Throughout this appendix, we use the same notation as in §1.

Let T and S be the even lattices defined at the beginning of §1 and fix $S_j \in \mathcal{G}(S)$. The goal is to show the following equality:

$$|\mathcal{P}_j^{G\text{-eq}}(T, \Lambda)| = |O(S_j) \setminus O(A_{S_j})/G|,$$

which clearly implies the first equality in Theorem 1.4. This equality is a consequence of Propositions A.7 and A.9 below.

DEFINITION A.1 ([**15**, §1]). An over-lattice $L = (L, (\ ,\ )_L)$ of $S_j \oplus T$ is a pair consisting of a $\mathbb{Z}$-module such that $S_j \oplus T \subset L \subset S_j^* \oplus T^*$ and the bilinear form $(\ ,\ )_L$ given by the restriction of $(\ ,\ )_{S_j^*} \oplus (\ ,\ )_{T^*}$ to L. An over-lattice L is called *integral* if its bilinear form $(\ ,\ )_L$ is $\mathbb{Z}$-valued. An integral over-lattice is a lattice in our sense.

DEFINITION A.2. We denote by $\mathcal{U}_j := \mathcal{U}(S_j \oplus T)$ the set of over-lattices $L = (L, (\ ,\ )_L)$ of $S_j \oplus T$ such that L is integral, unimodular and such that the inclusion $T \subset L$ is primitive. We also denote by $\mathcal{E}_j$ the subset of $\mathcal{U}_j$ which consists of even lattices.

LEMMA A.3 ([**15**, §1]). *Let $L \in \mathcal{U}_j$. Then:*

(1) *The orthogonal lattice $T^{\perp}$ in L coincides with S_j, and the inclusion $S_j \subset L$ is primitive.*
(2) *The natural projections $\bar{\pi}_{L,S_j}$ and $\bar{\pi}_{L,T}$ (see §1 for definition) are group isomorphisms.*
(3) *L is even, i.e., $L \in \mathcal{E}_j$ if and only if $q_{S_j}(\varphi_L(x)) = -q_T(x)$ $(\forall x \in A_T)$, where $\varphi_L := \bar{\pi}_{L,S_j} \circ \bar{\pi}_{L,T}^{-1} \colon A_T \xrightarrow{\sim} A_{S_j}$.*

PROOF. This follows directly from Proposition 1.2. See also [**15**, §1]. □

REMARK. (1) The relation $q_{S_j}(\varphi_L(x)) = -q_T(x)$ $(\forall x \in A_T)$ is nothing but that the isomorphism φ_L is in fact an isometry of the discriminant lattices

$$\varphi_L \colon (A_T, -q_T) \xrightarrow{\sim} (A_{S_j}, q_{S_j}).$$

(2) The isometry $\varphi_L : (A_T, -q_T) \xrightarrow{\sim} (A_{S_j}, q_{S_j})$ associated to $L \in \mathcal{E}_j$ recovers L explicitly as: $L/(S_j \oplus T) = \{\varphi_L(a) \oplus a \mid a \in A_T\} \subset S_j^*/S_j \oplus T^*/T$.

LEMMA A.4. *For an arbitrary isometry $f \colon (A_T, -q_T) \xrightarrow{\sim} (A_{S_j}, q_{S_j})$, we define the over-lattice L^f of $S_j \oplus T$ by: $L^f/(S_j \oplus T) = \{f(a) \oplus a \mid a \in A_T\}$. Then L^f is an element of $\mathcal{E}_j$, and the associated isometry φ_{L^f} $(= \bar{\pi}_{L^f,S_j} \circ \bar{\pi}_{L^f,T}^{-1})$ coincides with f.*

PROOF. This is an old result known as gluing methods, see [**4**, Chap. 4] for example, and also follows immediately from Proposition 1.2. □

By Lemma A.3(3), each $L \in \mathcal{E}_j$ gives rise to an isometry $\varphi_L \colon (A_T, -q_T) \xrightarrow{\sim} (A_{S_j}, q_{S_j})$. Combining this with Proposition 1.2, we obtain the following bijective correspondence such that $L \mapsto \varphi_L$, $f \mapsto L^f$ are inverses of each other:

$$\mathcal{E}_j = \mathcal{E}(S_j \oplus T) \leftrightarrow \{\varphi \colon (A_T, -q_T) \xrightarrow{\sim} (A_{S_j}, q_{S_j})\}.$$

LEMMA A.5. *Fix an isometry $\varphi_j \colon (A_T, -q_T) \xrightarrow{\sim} (A_{S_j}, q_{S_j})$. Then the following map is bijective: $\mu_j \colon O(A_{S_j}) \to \mathcal{E}_j$; $\sigma \mapsto L^{\sigma \circ \varphi_j}$.*

PROOF. The assignment $\sigma \mapsto \sigma \circ \varphi_j$ gives a bijection between $O(A_{S_j})$ and the set $\{\varphi \colon (A_T, -q_T) \xrightarrow{\sim} (A_{S_j}, q_{S_j})\}$. Combining this bijection with Lemma A.3, we obtain the result. □

DEFINITION A.6. (1) Two over-lattices $L, L' \in \mathcal{E}_j$ are called G-equivalent if there exist isometries $\Phi \colon L \xrightarrow{\sim} L'$ and $g \colon T \xrightarrow{\sim} T$ such that $g \in G$ and $\Phi|_T = g$:

$$\begin{array}{rcl} \Phi \colon L & \xrightarrow{\sim} & L' \\ \cup & & \cup \\ g : T & \xrightarrow{\sim} & T. \end{array}$$

(2) We denote the set of G-equivalence classes of the elements of $\mathcal{E}_j$ by $\mathcal{E}_j^{G\text{-eq}}$.

REMARK. By Lemma A.3(1), the orthogonal lattice $T^{\perp}$ in L coincides with S_j for all $L \in \mathcal{U}_j$. Therefore when L and L' are G-equivalent, the isometry $\Phi \colon L \xrightarrow{\sim} L'$ determines $f := \Phi_{S_j} \in O(S_j)$ as well as $g = \Phi|_T \in G$. Conversely any pair $(f, g) \in O(S_j) \times G$ induces an isometry $f \oplus g \colon S_j^* \oplus T^* \xrightarrow{\sim} S_j^* \oplus T^*$. Moreover, this isometry sends an over-lattice $L \subset S_j^* \oplus T^*$ to an over-lattice $L' = (f \oplus g)(L) \subset S_j^* \oplus T^*$. Clearly L and L' are G-equivalent.

PROPOSITION A.7. *The bijective map $\mu_j \colon O(A_{S_j}) \to \mathcal{E}_j$ in Lemma A.5 descends to the bijective map: $\bar{\mu}_j \colon O(S_j) \backslash O(A_{S_j})/G \to \mathcal{E}_j^{G\text{-eq}}$, where the double coset is defined as in Theorem 1.4.*

PROOF. Let $\sigma_1, \sigma_2 \in O(A_{S_j})$. Assume $L^{\sigma_1 \circ \varphi_j}$ and $L^{\sigma_2 \circ \varphi_j}$ are G-equivalent over-lattices of $S_j \oplus T$, where $\varphi_j \colon (A_T, -q_T) \xrightarrow{\sim} (A_{S_j}, q_{S_j})$ is a fixed isometry. By definition of G-equivalence, we have isometries $\Phi \colon L^{\sigma_1 \circ \varphi_j} \xrightarrow{\sim} L^{\sigma_2 \circ \varphi_j}$ and $g : T \xrightarrow{\sim} T$ such that $\Phi|_T = g \in G$. By Lemma A.3(1), $S_j = T^\perp$ in both $L^{\sigma_1 \circ \varphi_j}$ and $L^{\sigma_2 \circ \varphi_j}$, and thus we have $\Phi(S_j \oplus T) = S_j \oplus T$ and $\Phi_{S_j} := \Phi|_{S_j} \in O(S_j)$, $\Phi_T := \Phi|_T = g \in G$. Now we have $L^{\sigma_2 \circ \varphi_j}/(S_j \oplus T) = \{\sigma_2 \circ \varphi_j(a) \oplus a \mid a \in A_T\}$, and

$$\begin{aligned}\Phi(L^{\sigma_1 \circ \varphi_j}/(S_j \oplus T)) &= \{\bar{\Phi}_{S_j} \circ \sigma_1 \circ \varphi_j(a) \oplus \bar{g}(a) \mid a \in A_T\} \\ &= \{\bar{\Phi}_{S_j} \circ \sigma_1 \circ \varphi_j \circ \bar{g}^{-1}(a) \oplus a \mid a \in A_T\}.\end{aligned}$$

Comparing these two equations, we obtain $\sigma_2 = \bar{\Phi}_{S_j} \circ \sigma_1 \circ (\varphi_j \circ \bar{g}^{-1} \circ \varphi_j^{-1})$, which implies $[\sigma_1] = [\sigma_2]$ in $O(S_j) \setminus O(A_{S_j})/G$.

To see that the map μ_j is well-defined on the double coset $O(S_j) \setminus O(A_{S_j})/G$, let us assume $[\sigma_1] = [\sigma_2]$ in the double coset. Then there exist $f \in O(S_j)$ and $g \in G$ such that $\sigma_2 = \bar{f} \circ \sigma_1 \circ (\varphi_j \circ \bar{g} \circ \varphi_j^{-1})$. Now it is clear from the above calculations that for the $\mathbb{Q}$-linear extensions $f \oplus g \colon S_j^* \oplus T^* \to S_j^* \oplus T^*$ we have $(f \oplus g)(L^{\sigma_1 \circ \varphi_j}) = L^{\sigma_2 \circ \varphi_j}$. Then, by the remark after Definition A.6, $L^{\sigma_1 \circ \varphi_j}$ and $L^{\sigma_2 \circ \varphi_j}$ are G-equivalent.

These two observations show that the map μ_j descends to a well-defined injective map $\bar{\mu}_j$. Since μ_j is surjective, so is $\bar{\mu}_j$. □

Let $[\iota] \in \mathcal{P}_j^{G\text{-eq}}(T, \Lambda)$. Then, by definition, $\iota(T)^\perp \cong S_j$. We choose a representative $\iota \colon T \hookrightarrow \Lambda$ of this G-equivalence class. We also fix an isometry $\sigma \colon S_j \xrightarrow{\sim} \iota(T)^\perp$. Now consider the isomorphism $\sigma \oplus \iota \colon S_j \oplus T \xrightarrow{\sim} \iota(T)^\perp \oplus \iota(T)$ and its extension to the dual lattices. Then one has the following commutative diagram:

$$\begin{array}{ccc} \sigma \oplus \iota \colon S_j^* \oplus T^* & \xrightarrow{\sim} & (\iota(T)^\perp)^* \oplus \iota(T)^* \\ \cup & & \cup \\ L(\sigma, \iota) & \xrightarrow{\sim} & \Lambda \\ \cup & & \cup \\ \sigma \oplus \iota \colon S_j \oplus T & \xrightarrow{\sim} & \iota(T)^\perp \oplus \iota(T) \end{array}$$

where we define the over-lattice $L(\sigma, \iota)$ of $S_j \oplus T$ by $L(\sigma, \iota) := (\sigma \oplus \iota)^{-1}(\Lambda)$. Since Λ is even integral unimodular, so is $L(\sigma, \iota)$, i.e., $L(\sigma, \iota) \in \mathcal{E}_j$.

LEMMA A.8. *The over-lattices $L(\sigma, \iota)$ above satisfy the following:*

(1) *If $\sigma \colon S_j \xrightarrow{\sim} \iota(T)^\perp$ and $\sigma' \colon S_j \xrightarrow{\sim} \iota(T)^\perp$, then $[L(\sigma, \iota)] = [L(\sigma', \iota)]$ as an element of $\mathcal{E}_j^{G\text{-eq}}$.*
(2) *If $\iota \colon T \hookrightarrow \Lambda$ and $\iota' \colon T \hookrightarrow \Lambda$ are G-equivalent primitive embeddings $\mathcal{P}_j(T, \Lambda)$, then $[\iota] = [\iota']$ in $\mathcal{P}_j^{G\text{-eq}}(T, \Lambda)$ and $[L(\sigma, \iota)] = [L(\sigma, \iota')]$ in $\mathcal{E}_j^{G\text{-eq}}$.*

PROOF. Both claims are immediate from a diagram similar to the above and the definition of G-equivalence. Details are left for readers. □

PROPOSITION A.9. (1) *The following map is well-defined:*

$$\xi\colon \mathcal{P}_j^{G\text{-eq}}(T,\Lambda) \to \mathcal{E}_j^{G\text{-eq}}, \quad [\iota] \mapsto [L(\sigma,\iota)]$$

(2) *The map* $\xi\colon \mathcal{P}_j^{G\text{-eq}}(T,\Lambda) \to \mathcal{E}_j^{G\text{-eq}}$ *is bijective.*

PROOF. The well-definedness of the map ξ is immediate by Lemma A.8. To show surjectivity, let $[K]$ be in $\mathcal{E}_j^{G\text{-eq}}$. Then we have $S_j \oplus T \subset K \subset S_j^* \oplus T^*$, and $T \hookrightarrow K$ is primitive. Let $\iota_K\colon T \hookrightarrow K$ be an inclusion. Since K is an even unimodular indefinite lattice, there exists an isometry $f\colon K \xrightarrow{\sim} \Lambda$ by the theorem of Milnor (see [**20**]). Then for the embedding $\iota := f \circ \iota_K\colon T \hookrightarrow \Lambda$ and $\sigma := f|_{S_j}\colon S_j \xrightarrow{\sim} \iota(T)^\perp$ we have $K = L(\sigma,\iota)$, that is $[K] = \xi([\iota])$. To show the injectivity of ξ, assume $[L(\sigma,\iota)] = [L(\sigma',\iota')]$ in $\mathcal{E}_j^{G\text{-eq}}$. Then we have:

$$\begin{array}{ccccccc}
(\iota(T)^\perp)^* \oplus \iota(T)^* & \xleftarrow{\sim} & S_j^* \oplus T^* & \xrightarrow{\sim} & S_j^* \oplus T^* & \xrightarrow{\sim} & (\iota'(T)^\perp)^* \oplus \iota'(T)^* \\
\cup & & \cup & & \cup & & \cup \\
\Lambda & \xleftarrow{\sim} & L(\sigma,\iota) & \xrightarrow[\exists\Phi]{\sim} & L(\sigma',\iota') & \xrightarrow{\sim} & \Lambda \\
\cup & & \cup & & \cup & & \cup \\
\iota(T)^\perp \oplus \iota(T) & \xleftarrow[\sigma\oplus\iota]{\sim} & S_j \oplus T & \xrightarrow[\Phi_{S_j}\oplus\Phi_T]{\sim} & S_j \oplus T & \xrightarrow[\sigma'\oplus\iota']{\sim} & \iota'(T)^\perp \oplus \iota'(T),
\end{array}$$

where $\Phi_{S_j} := \Phi|_{S_j} \in O(S_j)$ and $\Phi_T := \Phi|_T \in G$. If we define $\Phi' = (\sigma' \oplus \iota') \circ \Phi \circ (\sigma \oplus \iota)^{-1} : \Lambda \xrightarrow{\sim} \Lambda$, we see that $\Phi' \circ \iota = \iota' \circ \Phi_T$, which means that $[\iota] = [\iota']$ in $\mathcal{P}_j^{G\text{-eq}}(T,\Lambda)$. This completes the proof. $\square$

Appendix B. The group of Hodge isometries

In this appendix, we shall give a proof of the following Proposition applied in previous sections. This is probably well-known to experts. Our proof here is similar to an argument in [**16, 21**].

PROPOSITION B.1. *Let T be a lattice of signature $(2, \operatorname{rank} T - 2)$ and $(T, \mathbb{C}\omega)$ be a weight two Hodge structure on T, i.e. ω is an element of $T \otimes \mathbb{C}$ such that $(\omega,\omega) = 0$ and $(\omega,\overline{\omega}) > 0$. Assume the following minimality condition on T: if $T' \subset T$ is primitive and $\omega \in T' \otimes \mathbb{C}$, then $T' = T$. Then $O_{\mathrm{Hodge}}(T,\mathbb{C}\omega)$ is a cyclic group of even order, say, $2I$ such that $\varphi(2I) \mid \operatorname{rank} T$. Here $\varphi(J) = |(\mathbb{Z}/J)^\times|$ is the Euler function. Moreover, if g is a generator of $O_{\mathrm{Hodge}}(T,\mathbb{C}\omega)$, then $g(\omega) = \zeta_{2I}\omega$, where ζ_{2I} is a primitive $2I$th root of unity.*

We proceed in five steps.

STEP 1. $|O_{\mathrm{Hodge}}(T,\mathbb{C}\omega)| < \infty$.

PROOF. Put $t := \operatorname{rank} T$. Let $P := \langle \operatorname{Re}\omega, \operatorname{Im}\omega \rangle$. Then P is a positive definite 2-dimensional plane by $(\omega,\omega) = 0$, $(\omega,\overline{\omega}) > 0$. Set $N := P^\perp$ in $T \otimes \mathbb{R}$. Since $\operatorname{sgn} T = (2, t-2)$, N is a negative definite. Let $g \in O_{\mathrm{Hodge}}(T,\mathbb{C}\omega)$. By definition of $O_{\mathrm{Hodge}}(T,\mathbb{C}\omega)$, we have $g(P) = P$ and $g(N) = N$. Therefore, via the assignment $g \mapsto (g|P, g|N)$, we have an embedding $O_{\mathrm{Hodge}}(T,\mathbb{C}\omega) \subset O(P) \times O(N)$. This should be continuous. Since P and N are both definite, $O(P) \times O(N)$ is compact. Since $O_{\mathrm{Hodge}}(T,\mathbb{C}\omega) \subset O(T)$ is discrete, this implies the result. $\square$

Let $g \in O_{\mathrm{Hodge}}(T,\mathbb{C}\omega)$. Then $g(\omega) = \alpha(g)\omega$ for some $\alpha(g) \in \mathbb{C}^\times$. The assignment $g \mapsto \alpha(g)$ defines a group homomorphism $\alpha : O_{\mathrm{Hodge}}(T,\mathbb{C}\omega) \to \mathbb{C}^\times$.

STEP 2. *α is injective.*

PROOF. Let $g \in \ker\alpha$. Since g is defined over $\mathbb{Z}$ (i.e., g is represented by a matrix of integral entries with respect to integral basis of T), the invariant set $T^g := \{x \in T \mid g(x) = x\}$ is a primitive sublattice of T such that $T^g \otimes \mathbb{C} = (T \otimes \mathbb{C})^g \ni \omega$. Thus, by the minimality condition, we have $T^g = T$. This means $g = \mathrm{id}_T$. □

STEP 3. *$O_{\mathrm{Hodge}}(T, \mathbb{C}\omega) \simeq \mathbb{Z}/2I$ for some $I \in \mathbb{N}$.*

PROOF. By Steps 1 and 2, $O_{\mathrm{Hodge}}(T, \mathbb{C}\omega)$ is isomorphic to a finite subgroup of the multiplicative group $\mathbb{C}^\times$ of the field $\mathbb{C}$. Such groups are always cyclic. Therefore $O_{\mathrm{Hodge}}(T, \mathbb{C}\omega)$ is a finite cyclic group. Since the involution $-\mathrm{id}_T$ is an element of $O_{\mathrm{Hodge}}(T, \mathbb{C}\omega)$, the order of $O_{\mathrm{Hodge}}(T, \mathbb{C}\omega)$ is also even. □

Set $O_{\mathrm{Hodge}}(T, \mathbb{C}\omega) = \langle g \rangle$. Then $\mathrm{ord}(g) = 2I$.

STEP 4. *All the eigenvalues of g (on $T \otimes \mathbb{C}$) are primitive $2I$th roots of unity.*

PROOF. Since $\mathrm{ord}(g) = 2I$, all the eigenvalues of g are $2I$th roots of unity. Since α is injective, $g(\omega) = \zeta_{2I}\omega$, where ζ_{2I} is a primitive $2I$th root of unity. Let d be a positive integer such that $d \mid 2I$ and $d \neq 2I$. Suppose that g admits a primitive dth root of unity ζ_d as its eigenvalue. Then g^d has an eigenvalue 1. Since g is defined over $\mathbb{Z}$, there is $x \in T - \{0\}$ such that $g^d(x) = x$. Since $g^d(\omega) = \zeta_e\omega$, where $e = 2I/d > 1$ and ζ_e is a primitive eth root of unity, we have

$$(x, \omega) = (g^d(x), g^d(\omega)) = (x, \zeta_e\omega).$$

This implies $(x, \omega) = 0$, because $\zeta_e \neq 1$ by $e > 1$. However, we have then $\omega \in (x)^\perp \otimes \mathbb{C}$, a contradiction to the minimality condition of T. Therefore all the eigenvalues of g are primitive $2I$th roots of unity. □

STEP 5. $\varphi(2I) \mid t = \mathrm{rank}\, T$.

PROOF. Since g is defined over $\mathbb{Q}$ and the primitive $2I$th roots of unity are mutually conjugate over $\mathbb{Q}$, their multiplicities as eigenvalue of g are the same, say m. Combining this with the fact that the number of primitive $2I$th roots of unity is $\varphi(2I)$, we have $m \cdot \varphi(2I) = t$. This implies $\varphi(2I) \mid t$. □

References

1. W. Barth, C. Peters, and A. Van de Ven, *Compact complex surfaces*, Ergeb. Math. Grenzgeb. (3), vol. 4, Springer-Verlag, Berlin, 1984.
2. T. Bridgeland and A. Maciocia, *Complex surfaces with equivalent derived categories*, Math. Z. **236** (2001), no. 4, 677–697.
3. J. W. S. Cassels, *Rational quadratic forms*, London Math. Soc. Monogr., vol. 13, Academic Press, London–New York, 1978.
4. J. H. Conway and N. J A. Sloane, *Sphere packings, lattices and groups*, Grundlehren Math. Wiss., vol. 290, Springer-Verlag, New York, 1988.
5. D. Flath, *Introduction to number theory*, Wiley-Intersci. Publ., Wiley, New York, 1989.
6. S. Hosono, B. Lian, K. Oguiso, and S.-T. Yau, *Autoequivalences of derived category of a K3 surface and monodromy transformations* , J. Algebraic Geom. (to appear); math.AG/0201047.
7. ———, *Kummer structures on a K3 surface- an old question of T. Shioda*, Duke Math. J. **120** (2003), no. 3, 635–647.
8. M. Maruyama *Stable vector bundles on an algebraic surface*, Nagoya Math. J. **58** (1975), 25–68.
9. ———, *Moduli of stable sheaves.* II, J. Math. Kyoto Univ. **18** (1978), no. 3, 557–614.
10. R. Miranda and D. R. Morrison, *The number of embeddings of integral quadratic forms.* I, Proc. Japan Acad. Ser. A Math. Sci. **61** (1985) no. 10, 317–320.

11. D. R. Morrison, *On K3 surfaces with large Picard number*, Invent. Math. **75** (1984), no. 1, 105–121.
12. S. Mukai, *Duality between $D(X)$ and $D(\widehat{X})$ with its application to Picard sheaves*, Nagoya Math. J. **81** (1981), 101–116.
13. ______, *Symplectic structure of the moduli space of sheaves on an abelian or K3 surface*, Invent. Math. **77** (1984), no. 1, 101–116.
14. ______, *On the moduli space of bundles on K3 surfaces.* I, Vector Bundles on Algebraic Varieties (Bombay, 1984), Tata Inst. Fund. Res. Stud. Math., vol. 11, Tata Inst. Fund. Res., Bombay, 1987, pp. 341–413.
15. V. V. Nikulin, *Integral symmetric bilinear forms and some of their geometric applications*, Izv. Akad. Nauk SSSR Ser. Mat. **43** (1979), no. 1, 111–177 (Russian); English transl., Math. USSR-Izv. **14** (1979) no. 1, 103–167.
16. ______, *Finite groups of automorphisms of Kählerian surfaces of type K3*, Trans. Moscow Math. Soc. **38** (1980), no. 2, 71–137.
17. K. Oguiso, *K3 surfaces via almost-primes* , math.AG/0110282, Math. Res. Lett. **9** (2002), no. 1, 47–63.
18. ______, *Local families of K3 surfaces and applications*, J. Algebraic Geom. **12** (2003), no. 3, 405–433.
19. D. Orlov, *Equivalences of derived categories and K3 surfaces*, J. Math. Sci. (New York) **84** (1997), no. 5, 1361–1381.
20. J.-P. Serre, *A course in arithmetic*, Grad. Texts in Math. vol. 7, Springer-Verlag, New York–Heidelberg, 1973.
21. H. Sterk, *Finiteness results for algebraic K3 surfaces*, Math. Z. **189** (1985), no. 4, 507–513.
22. D. Zagier, *Zetafunktionen und quadratische Körper*, Hochschultext, Springer-Verlag, Berlin–New York, 1981.

Graduate School of Mathematical Sciences, University of Tokyo, Komaba Meguro-ku, Tokyo 153-8914, Japan
E-mail address, S. Hosono: `hosono@ms.u-tokyo.ac.jp`
E-mail address, K. Oguiso: `oguiso@ms.u-tokyo.ac.jp`

Department of Mathematics, Brandeis University, Waltham, MA 02154, USA
E-mail address: `lian@brandeis.edu`

Department of Mathematics, Harvard University, Cambridge, MA 02138, USA
E-mail address: `yau@math.harvard.edu`

Centre de Recherches Mathématiques
CRM Proceedings and Lecture Notes
Volume **38**, 2004

Derived Equivalence of Holomorphic Symplectic Manifolds

Justin Sawon

ABSTRACT. We use twisted Fourier–Mukai transforms to study the relation between an abelian fibration on a holomorphic symplectic manifold and its dual fibration. Our reasoning leads to an equivalence between the derived category of coherent sheaves on one space and the derived category of twisted sheaves on the other space.

1. Introduction

Fourier–Mukai transforms were introduced by Mukai [**26**] in the early 80s, and have recently been instrumental in the understanding of several important problems in higher dimensional algebraic geometry (for example, see Bridgeland, King, and Reid [**7**] or Bridgeland [**6**]). The original example of Mukai related the derived categories of an abelian variety and its dual. In this paper we investigate a relative version of this equivalence using the twisted Fourier–Mukai transforms of Căldăraru [**9**]; this leads to (twisted) derived equivalences for holomorphic symplectic manifolds.

The main idea is to begin with a holomorphic symplectic manifold which is fibred by abelian varieties (which we call an *abelian fibration*), construct its dual fibration, and then relate the derived category of coherent sheaves on the original manifold to the derived category of twisted sheaves on the dual fibration. The twisting comes from a gerbe which arises as the natural obstruction to extending the fibrewise equivalence to a family.

The motivation for studying this problem came from a desire to understand the relation between different abelian fibrations. An elliptic K3 surface can be deformed, through elliptic surfaces, to an elliptic K3 surface with a section. For a higher dimensional abelian fibration X we would like to construct a fibration X^0 which is locally isomorphic to X and which admits a section (so that X is a torsor over X^0). We would also like X^0 to be a deformation of X, though the example in Section 5.4 indicates that this may not always be the case. Elliptic K3 surfaces which admit sections are easily classified via their Weierstraß models, and the hope

---

2000 *Mathematics Subject Classification.* Primary 14J60; Secondary 14D06, 18E30, 53C26.
The author was supported in part by NSF Grant #0305865.
This is the final form of the paper.

is that in higher dimensions one may be able to classify (holomorphic symplectic) abelian fibrations which admits sections (see Sawon [**31**]).

Section 2 contains some preliminaries on holomorphic gerbes, twisted sheaves, and derived categories. In Section 3 we review Fourier–Mukai transforms, their twisted version, and some results important for their application. In Section 4 we discuss fibrations by abelian varieties on holomorphic symplectic manifolds. We then describe how one can construct a dual fibration and a derived equivalence coming from a twisted Fourier–Mukai transform. Singular fibres provide a significant technical obstacle: even in the 'nicest' cases our results depend on several as-yet unverified assumptions (these are explained in the text; one involves extending an autoduality result to compactified Jacobians, the other involves extending a Fourier–Mukai result to the twisted case). In the final section we present examples of abelian fibrations and their duals. The author intends to present the first of these examples in complete rigor in a future article [**32**]. The remaining examples are meant to illustrate some interesting behavior that warrants further investigation.

The author would like to thank the organizers of the *Workshop on algebraic structures and moduli spaces* for putting together such an interesting schedule of talks, and for support to attend the workshop and present this work. Tom Bridgeland, Andrei Căldăraru, Eduardo de Sequeira Esteves, Manfred Lehn, Daisuke Matsushita, and Richard Thomas have also been very helpful in answering the author's questions.

2. Preliminaries

2.1. Holomorphic gerbes. In trying to extend certain local sheaves to global objects we will come up against obstructions in the form of holomorphic gerbes. Gerbes are higher-dimensional analogues of line bundles, as we now explain. Let M be a complex manifold with Stein cover $\{U_i\}$. An element in $\mathrm{H}^1(M,\mathcal{O}^*)$ is an equivalence class of holomorphic line bundles on M. A holomorphic gerbe is the geometric object whose equivalence classes are elements of $\mathrm{H}^2(M,\mathcal{O}^*)$.

DEFINITION 2.1. A (holomorphic) gerbe is a collection $\{\mathcal{L}_{ij}\}$ of line bundles on two-fold intersections $U_{ij} := U_i \cap U_j$ such that

(1) $\mathcal{L}_{ii} = \mathcal{O}_{U_i}$,
(2) $\mathcal{L}_{ji} = \mathcal{L}_{ij}^{-1}$,
(3) $\mathcal{L}_{ijk} := \mathcal{L}_{ij} \otimes \mathcal{L}_{jk} \otimes \mathcal{L}_{ki}$ has a given trivialization,
(4) the trivialization of $\mathcal{L}_{ijk} \otimes \mathcal{L}_{jkl}^{-1} \otimes \mathcal{L}_{kli} \otimes \mathcal{L}_{lij}^{-1}$ induced by the given trivializations of $\mathcal{L}_{ijk}$ is the canonical one.

REMARK 2.2. Since we chose our cover to be Stein, each line bundle $\mathcal{L}_{ij}$ is trivializable. If we fix trivializations of each $\mathcal{L}_{ij}$, this gives a trivialization of $\mathcal{L}_{ijk}$, but this may differ by a nonzero holomorphic function

$$\beta_{ijk} \in \Gamma(U_{ijk}, \mathcal{O}^*_{U_{ijk}})$$

from the trivialization given in the definition. The fourth condition then states that $\delta\beta$ is trivial, so that β is a cocycle representing a class in $\mathrm{H}^2(M,\mathcal{O}^*)$.

DEFINITION 2.3. The isomorphism class of a gerbe is the class $[\beta] \in \mathrm{H}^2(M,\mathcal{O}^*)$. We will usually abuse notation by writing simply β for $[\beta]$.

REMARK 2.4. Isomorphism of gerbes can be defined in the expected way, by taking into account different choices of trivialization of $\mathcal{L}_{ij}$ and refinement of the cover (for a more detailed discussion see Hitchin [17]). There is a bijection between the set of all gerbes isomorphic to a given gerbe and the set of all cocycles representing $[\beta]$.

In this article, all our constructions can be made to depend only on the isomorphism class of the gerbe involved, not the gerbe itself. We will often abuse terminology by referring to the isomorphism class simply as a gerbe.

EXAMPLE 2.5. The exponential exact sequence

$$0 \to \mathbb{Z} \to \mathcal{O} \to \mathcal{O}^* \to 0$$

induces the long exact sequence

$$\cdots \to \mathrm{H}^2(M, \mathbb{Z}) \to \mathrm{H}^2(M, \mathcal{O}) \to \mathrm{H}^2(M, \mathcal{O}^*) \to \mathrm{H}^3(M, \mathbb{Z}) \to \cdots$$

On a Calabi–Yau three-fold $\mathrm{H}^2(M, \mathcal{O})$ vanishes and therefore $\mathrm{H}^2(M, \mathcal{O}^*)$ injects into the lattice $\mathrm{H}^3(M, \mathbb{Z})$ and is discrete. We say that each topological gerbe has a unique holomorphic structure.

On a K3 surface $\mathrm{H}^2(M, \mathcal{O}) = \mathbb{C}$ and $\mathrm{H}^3(M, \mathbb{Z})$ vanishes. Therefore $\mathrm{H}^2(M, \mathcal{O}^*)$ is a quotient of $\mathbb{C}$ (possibly non-Hausdorff, depending on the rank of the image of $\mathrm{H}^2(M, \mathbb{Z})$). We say that there are no nontrivial topological gerbes, only nontrivial holomorphic structures.

As explained by Căldăraru [**9**], a gerbe can also be described as a sheaf of Azumaya algebras over M. However, this applies only to gerbes whose class in $\mathrm{H}^2(M, \mathcal{O}^*)$ is torsion, a restriction that we would prefer to avoid.

2.2. Twisted sheaves. Suppose we are given a gerbe, denoted β, on the complex manifold M with corresponding trivializations

$$\beta_{ijk} \in \Gamma(U_{ijk}, \mathcal{O}^*_{U_{ijk}}).$$

There is a natural notion of sheaves twisted by β.

DEFINITION 2.6. A β-twisted sheaf on M, denoted $\mathcal{F}$, is a collection $\{\mathcal{F}_i\}$ of local coherent sheaves (on U_i) and isomorphisms $\phi_{ij} \colon \mathcal{F}_i \to \mathcal{F}_j$ satisfying

(1) $\phi_{ji} = \phi_{ij}^{-1}$,
(2) $\phi_{ki} \circ \phi_{jk} \circ \phi_{ij} = \beta_{ijk}\mathrm{Id}$.

REMARK 2.7. The second condition means that the composition on the left, which is an automorphism of $\mathcal{F}_i$ restricted to U_{ijk}, is given by β_{ijk} times the identity isomorphism.

If there exists a locally free β-twisted sheaf of rank n, then one can show that β must be torsion of order dividing n. In this case, when β is torsion, twisted sheaves can also be describes as sheaves of modules over the corresponding sheaf of Azumaya algebras (see Căldăraru [**9**]).

The space of gerbes forms a group which we write multiplicatively. We can also pull back a gerbe β by a map $f \colon N \to M$ to get a gerbe $f^*\beta$ on N.

LEMMA 2.8 (Căldăraru [**9**]). *The pull-back of a β-twisted sheaf $\mathcal{F}$ on M by $f \colon N \to M$ is a $f^*\beta$-twisted sheaf $f^*\mathcal{F}$ on N. The tensor product of a β_1-twisted sheaf $\mathcal{F}_1$ and a β_2-twisted sheaf $\mathcal{F}_2$ on M is a $\beta_1 \otimes \beta_2$-twisted sheaf $\mathcal{F}_1 \otimes \mathcal{F}_2$ on*

M. In particular, if $\beta_2 = \beta_1^{-1}$, the inverse gerbe, then $\mathcal{F}_1 \otimes \mathcal{F}_2$ is untwisted, i.e., a genuine sheaf on M.

Strictly speaking, the tensor product makes sense for locally free sheaves; for more general sheaves we should pass to the derived category and use derived tensor product, which we shall do in the next subsection.

2.3. Derived categories. The basic idea behind working with derived categories is that an object in an abelian category should be identified with all its resolutions. This ensures that certain functors behave nicely. For example, exactness of sequences of coherent sheaves is not preserved under tensor product or push-forward, but if we restrict to projective or injective sheaves (respectively) then exactness is preserved. In this case, rather than working with the abelian category of coherent sheaves, we should take its derived category so that we can replace sheaves by projective or injective resolutions. See Thomas [**34**] or Gelfand and Manin [**16**, p. 143] for further motivation behind the derived category.

DEFINITION 2.9. Let $\mathcal{C}$ be an abelian category (basically, the set of morphisms between any two objects is an abelian group, and kernels and cokernels exist; see [**34**, Definition 4.1]). Examples include the category $\mathrm{Coh}(M)$ of coherent sheaves on a complex manifold M, or in the presence of a gerbe β, the category $\mathrm{Coh}(M, \beta)$ of β-twisted sheaves.

Let $\mathrm{Kom}(\mathcal{C})$ be the category of chain complexes over $\mathcal{C}$. A quasi-isomorphism is a morphism between two complexes which induces an isomorphism on the level of cohomology.

The derived category $\mathcal{D}(\mathcal{C})$ of $\mathcal{C}$ is the category whose objects are chain complexes over $\mathcal{C}$ and whose morphisms are usual morphisms of chain complexes, plus formal inverses of quasi-isomorphisms. Thus in the derived category, quasi-isomorphic objects become isomorphic. The bounded derived category is given by considering only bounded chain complexes.

In this article we will use the bounded derived categories of $\mathrm{Coh}(M)$ and $\mathrm{Coh}(M, \beta)$ on M, which we denote by $\mathcal{D}^b_{\mathrm{coh}}(M)$ and $\mathcal{D}^b_{\mathrm{coh}}(M, \beta)$ respectively.

The derived category is an additive category, but not abelian as kernels and cokernels do not exist. Instead, given a morphism $B^\bullet \to C^\bullet$ of chain complexes, there exists an "exact triangle"

$$A^\bullet \to B^\bullet \to C^\bullet \to A^\bullet[1]$$

where $A^\bullet[1]$ is the complex $A^\bullet$ shifted one place to the right. It is these exact triangles which are preserved under tensor product and push-forward, and which induce long exact sequences in cohomology. Thus derived categories are examples of triangulated categories, and when we talk about functors between derived categories we shall always mean functors which preserve exact triangles.

The derived category $\mathcal{D}^b_{\mathrm{coh}}(M)$ of a projective variety M has recently emerged as an important tool in studying the birational geometry of M. It contains a great deal of information about M. Indeed Bondal and Orlov [**4**] proved that if M has ample canonical or anti-canonical bundle, it can be recovered up to isomorphism from its derived category. This is no longer the case when the canonical bundle is trivial. In the case of K3 surfaces, Orlov [**29**] showed that two K3 surfaces have equivalent derived categories if and only if there is a Hodge isometry between their transcendental lattices.

In this article we deal with holomorphic symplectic manifolds, higher dimensional analogues of K3 surfaces. Our aim is to find pairs of nonisomorphic manifolds with equivalent derived categories; more accurately, the derived category of one manifold should be equivalent to the twisted derived category of the other (so this is a not quite a direct analogue of the K3 case, as studied by Orlov).

3. Fourier–Mukai transforms

3.1. Untwisted. Fourier-Mukai transforms were first introduced in [**26**] as a way of producing equivalences of derived categories of nonisomorphic varieties. Recently there have been some striking applications of these methods, including Bridgeland, King, and Reid's "derived" McKay correspondence [**7**] and Bridgeland's construction of three-fold flops [**6**].

Let M be a fine moduli space of stable sheaves on some projective variety X. In other words, M parametrizes a complete family of stable sheaves on X and there exists a universal sheaf $\mathcal{U}$ on $X \times M$. Denote by π_X and π_M the projections to X and M respectively.

DEFINITION 3.1 (Mukai [**26**]). We call the functor

$$\Phi^{\mathcal{U}}_{M\to X}\colon \mathcal{D}^b_{\mathrm{coh}}(M) \to \mathcal{D}^b_{\mathrm{coh}}(X)$$

$$\mathcal{E}^\bullet \mapsto \mathbf{R}\pi_{X*}(\mathcal{U} \overset{\mathbf{L}}{\otimes} \pi_M^*\mathcal{E}^\bullet)$$

an integral transform. If it is an equivalence of categories we call it a Fourier–Mukai transform.

REMARK 3.2. Defining this operation on the category of coherent sheaves would be unsatisfactory, as it would not preserve exactness. Instead we use derived categories, and derived functors which preserve exact triangles.

REMARK 3.3. The dual sheaf $\mathcal{U}^\vee$ allows us to define a functor

$$\Phi^{\mathcal{U}^\vee}_{X\to M}\colon \mathcal{D}^b_{\mathrm{coh}}(X) \to \mathcal{D}^b_{\mathrm{coh}}(M),$$

which is the inverse equivalence in the case of a Fourier–Mukai transform. Note that the dimensions of X and M must agree if their derived categories are equivalent.

EXAMPLE 3.4. Let X be an elliptic curve E, and let M the dual elliptic curve $\widehat{E}$, which parametrizes degree zero line bundles on E. The Poincaré line bundle $\mathcal{P}$ is a universal bundle on $E \times \widehat{E}$. In higher dimensions, we could let X be an abelian variety and M its dual. Once again the Poincaré line bundle is a universal bundle. Mukai [**26**] showed these universal bundles induce equivalences of derived categories.

Let $\mathcal{O}_m$ be the skyscraper sheaf supported at $m \in M$. Then $\mathcal{U}_m := \Phi^{\mathcal{U}}_{M\to X}\mathcal{O}_m$ is the sheaf on X which the point $m \in M$ parametrizes. The skyscraper sheaves on M "span" the derived category and in some sense are "orthonormal" with respect to the $\mathrm{Ext}^\bullet$-pairing. If the integral transform $\Phi^{\mathcal{U}}_{M\to X}$ is an equivalence it should preserve these properties. Based on these ideas, the following theorem was proved by Bridgeland, extending work of Mukai, Bondal, and Orlov.

THEOREM 3.5 (Bridgeland [**5**]). *Assume that M has the same dimension as X, and both are smooth varieties. The functor $\Phi^{\mathcal{U}}_{M\to X}$ is an equivalence of categories if and only if the following criteria are satisfied:*

(1) *for all* $m \in M$, $\mathcal{U}_m$ *is a simple sheaf, i.e.,*

$$\mathrm{Hom}(\mathcal{U}_m, \mathcal{U}_m) = \mathbb{C},$$

and $\mathcal{U}_m \otimes \mathcal{K}_X = \mathcal{U}_m$ *where* $\mathcal{K}_X$ *is the canonical bundle of* X,

(2) *for all* $m_1 \neq m_2 \in M$ *and all integers* i,

$$\mathrm{Ext}^i_X(\mathcal{U}_{m_1}, \mathcal{U}_{m_2}) = 0.$$

EXAMPLE 3.6. In the previous example, when X is an abelian variety and M its dual, $\mathcal{U}_m$ is a line bundle and

$$\mathrm{Ext}^i_X(\mathcal{U}_{m_1}, \mathcal{U}_{m_2}) = \mathrm{H}^i(X, \mathcal{U}^\vee_{m_1} \otimes \mathcal{U}_{m_2})$$

vanishes for $m_1 \neq m_2$ by standard results on the cohomology of line bundles on abelian varieties (see Lange and Birkenhake [**19**, Chapter 3]). The first criterion is also easy to verify, and hence the integral transform is indeed an equivalence.

In general the second criterion can be more difficult to verify. However, Bridgeland and Maciocia proved the following theorem, which says it is enough to check the condition on a set of sufficiently large codimension.

THEOREM 3.7 (Bridgeland–Maciocia [**8**]). *Assume that* M *has the same dimension* n *as* X, *and that* X *is a smooth variety. Suppose that the following criteria are satisfied:*

(1) *for all* $m \in M$, $\mathcal{U}_m$ *is a simple sheaf and* $\mathcal{U}_m \otimes \mathcal{K}_X = \mathcal{U}_m$,

(2) *for all* $m_1 \neq m_2 \in M$

$$\mathrm{Hom}_X(\mathcal{U}_{m_1}, \mathcal{U}_{m_2}) = 0,$$

and the closed subscheme

$$\Gamma(\mathcal{U}) := \{(m_1, m_2) \in M \times M \mid \mathrm{Ext}^i_X(\mathcal{U}_{m_1}, \mathcal{U}_{m_2}) \neq 0 \text{ for some } i \in \mathbb{Z}\}$$

of $M \times M$ *has dimension at most* $n+1$.

Then in fact $\Gamma(\mathcal{U})$ *is the diagonal,* M *is also smooth, and* $\Phi^{\mathcal{U}}_{M \to X}$ *is an equivalence of categories.*

REMARK 3.8. If the dimension n is greater than two, it can also be difficult to show directly that M is smooth; but the theorem takes care of that.

3.2. Twisted. In [**9**] Căldăraru generalized the ideas of the previous subsection to twisted sheaves by introducing twisted Fourier-Mukai transforms. Before giving the definition, let us explain how twisted sheaves naturally arise in this context.

Perhaps the clearest motivation comes from considering relative moduli spaces. Suppose that $p_X \colon X \to B$ is a proper fibration and let M be an irreducible component of the relative moduli space of stable sheaves on X. In other words, each point of M corresponds to a stable sheaf supported on a fibre of $X \to B$. Thus M is also fibred over B, and we write $p_M \colon M \to B$.

EXAMPLE 3.9. Let $X \to B$ be an elliptic fibration and let M be its compactified relative Jacobian $\overline{\mathrm{Jac}}(X/B)$ (see D'Souza [**12**], Altman–Iarrobino–Kleiman [**1**], or Rego [**30**]). Then M parametrizes torsion-free rank one degree zero sheaves on the fibres of X. Since these sheaves are rank one, a destabilizing sheaf cannot exist, and hence M is a component of the moduli space of stable sheaves on X (we use Simpson's definition of stability and construction of the moduli space [**33**]).

To construct M in this example, smooth fibres of X are replaced by their dual elliptic curves. Of course elliptic curves are self-dual, but the global structure of M may also differ from X. For instance, M has a canonical section (given by the trivial sheaf on each fibre of X) whereas X may not admit any global section.

Let $t \in B$, and consider the fibres X_t and M_t. One can show that provided X_t is reduced (it need not be irreducible) then M_t is a moduli space of stable sheaves on X_t. This may not be the case if X_t is nonreduced because M_t might include 'fat' sheaves whose scheme-theoretic support is nonreduced. We assume that M_t is a fine moduli space of sheaves on X_t, so that a universal sheaf exists, and moreover that these fibrewise universal sheaves fit together locally to give a local universal sheaf over an open subset in B containing t.

EXAMPLE 3.10. In the previous example, the Poincaré line bundle is a universal sheaf for any smooth fibre of X_t and its dual M_t. One can extend it over a neighborhood B_i of $t \in B$.

Now assume that B is covered by open sets B_i such that there exists a local universal sheaf $\mathcal{U}_i$ over each $X \times M_i$, where $M_i := p_M^{-1}(B_i)$. More precisely, the local universal sheaf is, a priori, really a sheaf over $X_i \times_{B_i} M_i$; but we can extend it (by zero) to $X \times_B M_i$ and then push it forward by the inclusion of the fibre-product in $X \times M_i$. On the intersection $X \times M_{ij}$ of $X \times M_i$ and $X \times M_j$, the restrictions of the two sheaves $\mathcal{U}_i$ and $\mathcal{U}_j$ are universal sheaves for the same moduli problem. Hence they are isomorphic and there exists a line bundle $\mathcal{L}_{ij}$ on M_{ij} such that

$$\mathcal{U}_i|_{X\times M_{ij}} = \pi_M^*\mathcal{L}_{ij} \otimes \mathcal{U}_j|_{X\times M_{ij}}.$$

The fibre of $\mathcal{L}_{ij}$ over a point $m \in M_{ij}$ is identified with the set of morphisms from $\mathcal{U}_i|_{X\times m}$ to $\mathcal{U}_j|_{X\times m}$. These sheaves are both isomorphic to $\mathcal{U}_m$, and thus we have implicitly used the fact that M parametrizes stable, and hence simple sheaves on X; i.e., the fibres of $\mathcal{L}_{ij}$ are isomorphic to $\mathbb{C}$.

THEOREM 3.11 (Căldăraru [**9**]). *In the situation described above, the collection* $\{\mathcal{L}_{ij}\}$ *defines a gerbe on* M. *The class* $\beta \in \mathrm{H}^2(M, \mathcal{O}^*)$ *of this gerbe represents the obstruction to patching the local universal sheaves* $\mathcal{U}_i$ *together*; *i.e., there exists a global universal sheaf on* $X \times M$ *if and only if* $\beta = 0$.

EXAMPLE 3.12. For an elliptic fibration $X \to B$ and its compactified relative Jacobian $M := \overline{\mathrm{Jac}}(X/B)$ one can show that the following are equivalent (provided the singular fibres are well enough behaved):

(1) X is isomorphic to M,
(2) X admits a global section,
(3) there is a global universal sheaf on $X \times M$,
(4) β vanishes.

The first two statements are equivalent because X is a torsor over M. The last two statements are equivalent by the previous theorem. For the remaining equivalence, one shows that a local section of $X \to B$ can be used to locally extend the Poincaré bundle, thereby producing a local universal bundle (see Căldăraru [**9**] for details).

Next observe that the collection $\{\mathcal{U}_i\}$ of local universal sheaves forms a $\pi_M^*\beta$-twisted sheaf on $X \times M$, which we denote simply by $\mathcal{U}$. So in the twisted case, the failure of the existence of a universal sheaf is precisely controlled by β, and moreover we have instead a *twisted universal sheaf* $\mathcal{U}$. We can also construct integral transforms with twisted sheaves.

DEFINITION 3.13. If the functor

$$\Phi^{\mathcal{U}}_{M\to X}\colon \mathcal{D}^b_{\mathrm{coh}}(M,\beta^{-1}) \to \mathcal{D}^b_{\mathrm{coh}}(X)$$

$$\mathcal{E}^\bullet \mapsto \mathbf{R}\pi_{X*}(\mathcal{U} \overset{\mathbf{L}}{\otimes} \pi^*_M\mathcal{E}^\bullet)$$

is an equivalence of categories we call it a twisted Fourier–Mukai transform.

REMARK 3.14. In the definition, $\mathcal{E}^\bullet$ denotes a complex of β^{-1}-twisted sheaves on M. Therefore $\pi^*_M\mathcal{E}^\bullet$ is a complex of $\pi^*_M\beta^{-1}$-twisted sheaves on $X \times M$, and $\mathcal{U} \overset{\mathbf{L}}{\otimes} \pi^*_M\mathcal{E}^\bullet$ is an untwisted sheaf on $X \times M$. The final push-forward takes us to the derived category of X.

Căldăraru [**9**] generalized Bridgeland's criteria for when an integral transform is an equivalence (Theorem 3.5) to the twisted case. Skyscraper sheaves can always be regarded as twisted sheaves, so defining $\mathcal{U}_m := \Phi^{\mathcal{U}}_{M\to X}\mathcal{O}_m$ still makes perfect sense in the twisted case. The criteria are then precisely the same as in the untwisted case, and we will not repeat them here.

EXAMPLE 3.15. If β is nonzero in Example 3.12, then provided the singular fibres are well enough behaved, we can apply the criteria and hence show that

$$\Phi^{\mathcal{U}}_{M\to X}\colon \mathcal{D}^b_{\mathrm{coh}}(M,\beta^{-1}) \to \mathcal{D}^b_{\mathrm{coh}}(X)$$

is an equivalence. Various examples where X is an elliptic Calabi–Yau three-fold are discussed in Căldăraru [**9**]. Dealing with the singular fibres is one of the biggest difficulties, so usually one begins with an example whose singular fibres are fairly mild.

4. Holomorphic symplectic manifolds

4.1. Abelian fibrations.

DEFINITION 4.1. Let X be a compact Kähler manifold. We call X a holomorphic symplectic manifold if it admits a closed nondegenerate holomorphic two-form

$$\sigma \in \mathrm{H}^0(X, \Lambda^2 T^*) = \mathrm{H}^{2,0}(X)$$

(σ is the holomorphic analogue of a symplectic structure). Note that the canonical bundle $\mathcal{K}_X$ is trivialized by $\sigma^{\wedge n}$, where n is half the dimension of X. If X is simply connected and $\mathrm{H}^{2,0}(X) = \mathbb{C}$ is generated by σ then we say X is irreducible.

REMARK 4.2. By Bogomolov's decomposition theorem a holomorphic symplectic manifold has a finite unramified cover which is a product of complex tori and irreducible holomorphic symplectic manifolds. The latter are therefore the *building blocks* of the theory, and we will always mean "irreducible holomorphic symplectic" when we write simply "holomorphic symplectic," unless stated otherwise.

In two dimensions the only (irreducible) holomorphic symplectic manifolds are K3 surfaces. A proper morphism from a K3 surface to a curve is necessarily an elliptic fibration, with base $\mathbb{P}^1$ and generic fibre an elliptic curve. In higher dimensions we have the following theorem of Matsushita.

THEOREM 4.3 (Matsushita [**24,25**]). *Let X be a projective irreducible holomorphic symplectic manifold of dimension $2n$. Let $p_X : X \to B$ be a proper surjective projective morphism, with connected generic fibre, and with projective base B of dimension strictly between zero and $2n$. Then*

(1) *the generic fibre is a (holomorphic) Lagrangian abelian variety of dimension n,*
(2) *the base is Fano with the same Hodge numbers as $\mathbb{P}^n$.*

It follows from the second statement and the classification of surfaces that the base must be $\mathbb{P}^2$ when $n = 2$.

DEFINITION 4.4. If there exists a morphism $p_X : X \to B$ as in the theorem, we call X an abelian fibration. In this case X is also known as an algebraically complete integrable system.

A K3 surface is elliptic if and only if it contains a nontrivial divisor with square zero (with respect to intersection pairing). Similarly, in higher dimensions we expect that X is an abelian fibration if and only if it contains a divisor with some special properties. Presently this is only conjectural (see Sawon [**31**]). Moreover, we don't yet know whether every holomorphic symplectic manifold can be deformed to an abelian fibration.

4.2. Dual fibrations. Now we come to the heart of this article. Given an abelian fibration, our aim is to construct its dual fibration and then relate it, via a twisted Fourier–Mukai transform, to the original fibration. In this subsection we outline the general theory, which is a higher dimensional analogue of the elliptic fibrations discussed in Examples 3.12 and 3.15. In the next section we will present some examples that indicate what kind of behavior we can expect.

Let $p_X : X \to B$ be an abelian fibration of dimension $2n$. We will assume that all the fibres are geometrically integral (i.e., reduced and irreducible).

DEFINITION 4.5. The compactified relative Picard scheme $\overline{\mathrm{Pic}}^0(X/B)$ of X is the moduli space parameterizing torsion-free rank one sheaves of degree zero (vanishing first Chern class) supported on the fibres of X. Its construction, which is due to Altman and Kleiman [**2**], already requires that the fibres of X be geometrically integral.

Let M be the compactified relative Picard scheme $\overline{\mathrm{Pic}}^0(X/B)$ of X. Then M parametrizes stable sheaves: they are pure-dimension with geometrically integral support, so there can be no destabilizing subsheaf (we are using Simpson's definition of stability [**33**]).

A priori we don't know very much about M. It need not be a smooth space. There is a projection $p_M : M \to B$, though we cannot expect this to be flat in general, as we now explain. If X_t is a smooth fibre of X, $t \in B$, then clearly the corresponding fibre M_t of M is simply the dual abelian variety. More generally, if X_t is a singular fibre one can show that M_t parametrizes sheaves on X of the form $\iota_*\mathcal{E}$, where $\iota : X_t \hookrightarrow X$ is the inclusion of the fibre and $\mathcal{E}$ is a stable rank one degree zero sheaf on X_t. In particular, M_t is always isomorphic to a moduli space of stable sheaves on X_t (our singular fibres are geometrically integral, though in fact this appears to be true provided X_t contains no nonreduced components).

This gives us a precise description of the singular fibres of M, but unfortunately moduli spaces of stable sheaves on singular varieties have not been extensively studied. Undesirable behavior cannot automatically be ruled out: for example, the singular fibres of M need not be irreducible and they may contain components of dimension greater than n. One example we can try to understand is the (compactified) Jacobian of a degeneration of a curve.

EXAMPLE 4.6. Let C be a singular curve contained in a surface. Assume that C has at worst double point singularities (for example, nodes, cusps, or tacnodes). Altman, Kleiman, and Iarrobino [**1**] showed that the compactified Jacobian $J := \overline{\mathrm{Jac}}\, C$ of such a curve is geometrically integral; it is a degeneration of an abelian variety. We can then apply Altman and Kleiman's construction to J, and hence obtain the compactified Picard scheme $\overline{\mathrm{Pic}}^0\, J$. A recent theorem of Esteves, Gagné, and Kleiman [**14**] gives an isomorphism of uncompactified Picard schemes

$$\mathrm{Pic}^0\, C \cong \mathrm{Pic}^0\, J.$$

It is expected [**13**] that this isomorphism will extend to the compactifications

$$\overline{\mathrm{Pic}}^0\, C \cong \overline{\mathrm{Pic}}^0\, J.$$

The left-hand side is just J itself, up to isomorphism.

In terms of our abelian fibration X and its dual fibration M, this means that if the singular fibre X_t can be identified as the compactified Jacobian of a mildly singular curve then the corresponding singular fibre M_t, which is $\overline{\mathrm{Pic}}^0(X_t)$, will be isomorphic to X_t. In particular M_t is geometrically integral and has dimension n, so $M \to B$ will be flat. Singular fibres like this will arise in the examples of the next section.

The example shows that in some special cases we can acquire an exact understanding of the singular fibres of M, so we will proceed under the assumption that the singular fibres of M are sufficiently well-behaved. In fact, we would like M to be locally isomorphic to X as a fibration over B. However, this is too much to ask for in general: if the fibres of X are not principally polarized then even a smooth fibre X_t of X need not be isomorphic to the corresponding fibre M_t of M. This problem does not arise with elliptic fibres, which are always principally polarized. The remedy is to define $p_0 \colon X^0 \to B$ to be the compactified relative Picard scheme $\overline{\mathrm{Pic}}^0(M/B)$ of $p_M \colon M \to B$. For a smooth fibre X_t of X, the corresponding fibre X^0_t of X^0 is the double dual of, and hence canonically isomorphic to, X_t. We will assume the same is true for singular fibres (this should be true when X_t is a mild degeneration of a Jacobian, as in the last example).

Thus $p_X \colon X \to B$ and $p_0 \colon X^0 \to B$ are locally isomorphic as fibrations. Since the compactified relative Picard scheme over B always has a canonical section, X is therefore a torsor over X^0.

LEMMA 4.7. *The set of torsors X over X^0 is one-to-one correspondence with the sheaf cohomology $\mathrm{H}^1(B, X^0)$, where by X^0 we really mean the sheaf of local sections of $X^0 \to B$.*

PROOF. Choose an open cover (in the analytic topology) $\{B_i\}$ of B such that X and X^0 are isomorphic fibrations over each B_i, i.e., there exists local isomorphisms

$$f_i \colon X_i := p_X^{-1}(B_i) \to X^0_i := p_0^{-1}(B_i)$$

over B_i. On the overlap $X^0_{ij} := p_0^{-1}(B_{ij})$ the composition $f_j \circ f_i^{-1}$ gives an automorphism α_{ij} of the fibration $X^0_{ij} \to B_{ij}$, which is given by a translation in each fibre. Since we have a basepoint in each fibre (X^0 has a global section), the family of translations α_{ij} is equivalent to a section of $X^0_{ij} \to B_{ij}$. These local sections form a 1-cocyle $\alpha \in \mathrm{H}^1(B, X^0)$. Conversely, choosing a representative of α gives us the families of translations α_{ij} on overlaps needed to glue the local fibrations $X^0_i \to B_i$ together to obtain X. □

LEMMA 4.8. *To each torsor X over X^0 we can assign a gerbe $\beta \in \mathrm{H}^2(M, \mathcal{O}^*)$ on M. This is the same gerbe that arises as the obstruction to the existence of a global universal sheaf on $X \times M$.*

PROOF. We have seen how to construct $\alpha \in \mathrm{H}^1(B, X^0)$ from X. Taking a representative, we obtain local sections α_{ij} of X^0; but each point of X^0 represents a rank one degree zero sheaf (generically a line bundle) supported on a fibre of $M \to B$, so the local section α_{ij} gives a line bundle $\mathcal{L}_{ij}$ on $M_{ij} := p_M^{-1}(B_{ij})$. One can show that this collection of line bundles forms a gerbe $\beta \in \mathrm{H}^2(M, \mathcal{O}^*)$, which does not depend on our choice of representative of α.

To relate this to the gerbe arising as an obstruction in Theorem 3.11, we first consider X^0 as a moduli space on M. On an abelian variety and its dual, the Poincaré line bundle is a universal sheaf. Since both X^0 and M have global sections, these Poincaré line bundles fit together to give a global universal sheaf on $M \times X^0$. Restricting to

$$M \times X^0_i \cong M \times X_i$$

gives a bundle whose support is contained in $M_i \times X_i$. Interchanging the factors, and extending by zero, gives a local universal sheaf on $X \times M_i$. Note that we have implicitly used the relation between the Poincaré bundle of an abelian variety and the Poincaré bundle of its dual (see Lange and Birkenhake [**19**, Exercise 16, p. 45]). Finally, it is not difficult to see that these local universal sheaves no longer patch together to give a global sheaf on $X \times M$; moreover they differ on overlaps precisely by tensoring with $\mathcal{L}_{ij}$. □

REMARK 4.9. The lemma implies that there is a map

$$\mathrm{H}^1(B, X^0) \to \mathrm{H}^2(M, \mathcal{O}^*).$$

Our next theorem says that this is an inclusion, but it will not be an isomorphism in general because the gerbes we obtain from torsors X over X^0 have a somewhat specialized form.

Now we come to our main result, the generalization of Example 3.12 to abelian fibrations.

THEOREM 4.10. *Let $X \to B$ be an abelian fibration, M the compactified relative Picard scheme of X, and X^0 the compactified relative Picard scheme of M. Assume that for all $t \in B$ the fibres X^0_t and X_t are isomorphic. Let $\beta \in \mathrm{H}^2(M, \mathcal{O}^*)$ be the gerbe associated to the torsor X over X^0, as in the previous lemma. Then the following are equivalent:*

(1) *X is isomorphic to X^0,*
(2) *X admits a global section,*
(3) *there is a global universal sheaf on $X \times M$,*

(4) *β vanishes.*

PROOF. The first two statements are equivalent because X is a torsor over X^0. The last two statements are equivalent by the previous lemma and Theorem 3.11. The remaining equivalence is shown using the same method as in the elliptic case: namely one shows that a local section of $X \to B$ can be used to locally extend the Poincaré bundle, thereby producing a local universal bundle. □

Finally we want to generalize Example 3.15, and show that we have an equivalence of derived categories for abelian fibrations. So assume we are in the same situation as in Theorem 4.10, and that β is nonzero. We want to use Căldăraru's generalization of Bridgeland's criteria to show that the twisted universal sheaf $\mathcal{U}$ induces an equivalence

$$\Phi^{\mathcal{U}}_{M\to X} \colon \mathcal{D}^b_{\mathrm{coh}}(M, \beta^{-1}) \to \mathcal{D}^b_{\mathrm{coh}}(X).$$

For all $m \in M$, $\mathcal{U}_m$ is a stable and hence simple sheaf on X. Moreover X is holomorphic symplectic, with trivial canonical bundle, and thus $\mathcal{U}_m \otimes \mathcal{K}_X = \mathcal{U}_m$. The first criterion is therefore satisfied, and it remains to show that if $m_1 \neq m_2 \in M$ then

$$\mathrm{Ext}^i_X(\mathcal{U}_{m_1}, \mathcal{U}_{m_2}) = 0$$

for all integers i. We can consider various cases:

(1) if m_1 and m_2 belong to different fibres of M, then $\mathcal{U}_{m_1}$ and $\mathcal{U}_{m_2}$ are supported on different fibres of X, and the Ext-groups must vanish,
(2) if m_1 and m_2 belong to the same smooth fibre of M, then $\mathcal{U}_{m_1}$ and $\mathcal{U}_{m_2}$ are (push-forwards of) distinct line bundles L_1 and L_2 on the same smooth fibre X_t of X; in this case all cohomology groups $\mathrm{H}^i(X_t, L_1^* \otimes L_2)$ vanish (see Lange and Birkenhake [**19**, Chapter 3]), the normal bundle to the fibre is holomorphically trivial, and the Koszul spectral sequence then shows that the Ext-groups also vanish (as in Bridgeland and Maciocia [**8**, Section 7.2]),
(3) if m_1 and m_2 belong to the same singular fibre of M, it is more difficult to proceed with this calculation.

For elliptic fibrations, a direct approach to the third case is feasible. For abelian fibrations, we instead want to apply Bridgeland and Maciocia's Theorem 3.7. Since the proof of that theorem uses only local arguments, the result should still be valid in the twisted case. We need to bound the dimension of $\Gamma(\mathcal{U})$. Let X have dimension $2n$. Singular fibres occur over a codimension one subset of the base B, and the fibres themselves have dimension n, so the pairs (m_1, m_2) corresponding to case three above are contained in a subset of $M \times M$ of dimension $3n - 1$. The theorem requires this to be at most one larger than the dimension of X, and therefore n can be at most two.

Using these arguments, we can expect to produce equivalences

$$\Phi^{\mathcal{U}}_{M\to X} \colon \mathcal{D}^b_{\mathrm{coh}}(M, \beta^{-1}) \to \mathcal{D}^b_{\mathrm{coh}}(X).$$

for holomorphic symplectic four-folds X fibred by abelian surfaces, under some additional hypotheses. In higher dimensions the same equivalence ought to hold, but there are some additional technical difficulties yet to be overcome. We will discuss some examples of abelian fibrations in the next section.

5. Examples

5.1. The Hilbert scheme of two points on a K3.

DEFINITION 5.1. Let S be a K3 surface. The Hilbert scheme of two points on S is the space

$$S^{[2]} := \mathrm{Blow}_\Delta(S \times S)/\mathbb{Z}_2$$

obtained by blowing up the diagonal in $S \times S$ and quotienting by the involution interchanging the two factors.

Fujiki [**15**] showed that the four-fold $S^{[2]}$ has a holomorphic symplectic structure; this was the first example of a higher dimensional irreducible holomorphic symplectic manifold. We will be interested in certain deformations of $S^{[2]}$ which are abelian fibrations, i.e., fibred over $\mathbb{P}^2$ by abelian surfaces.

EXAMPLE 5.2. Let S be a double cover of the plane $(\mathbb{P}^2)^\vee$ ramified over a generic sextic. The linear system of lines in $(\mathbb{P}^2)^\vee$ is simply the dual plane $\mathbb{P}^2$. Pulling these lines back to S gives a complete linear system of genus two curves on S. Denote the family of these curves by $\mathcal{C} \to \mathbb{P}^2$. One can show that all of these curves are irreducible (they develop at worst two nodes or a cusp), and we can therefore consider the compactified relative Jacobian $Z^2 := \overline{\mathrm{Jac}}(\mathcal{C}/\mathbb{P}^2)$, and more generally the compactified degree d relative Picard scheme $Z^d := \overline{\mathrm{Pic}}^d(\mathcal{C}/\mathbb{P}^2)$. Each space Z^d is clearly a fibration

$$\begin{array}{ccc} \overline{\mathrm{Pic}}^d & \hookrightarrow & Z^d \\ & & \downarrow \\ & & \mathbb{P}^2 \end{array}$$

with generic fibre a smooth abelian surface.

Now given a degree d line bundle (or more generally, a torsion-free rank one sheaf) on a curve C in the linear system $\mathbb{P}^2$, we can take the push-forward by the inclusion $C \hookrightarrow S$. This gives a (torsion) sheaf on S, which moreover is stable. We can therefore regard Z^d as an irreducible component of the moduli space of stable sheaves on S. Mukai [**27**] showed that these moduli spaces are smooth and admit holomorphic symplectic forms, and hence Z^d is an irreducible holomorphic symplectic four-fold and an abelian fibration. Moreover, it is a deformation of the Hilbert scheme $S^{[2]}$ (see Yoshioka [**35**], for instance).

We summarize some further properties of the four-folds Z^d, which hold for all $d \in \mathbb{Z}$ (these statements are mostly straightforward, and will be proved in Sawon [**32**]):

- all fibres of Z^d are irreducible,
- Z^d is locally isomorphic to Z^{d+1} as a fibration over $\mathbb{P}^2$,
- $Z^0 \to \mathbb{P}^2$ admits a global section, given by taking the trivial degree zero line bundle on each curve,
- $Z^1 \to \mathbb{P}^2$ admits a birational two-valued section,
- Z^d is globally isomorphic to Z^{d+2}, the isomorphism given by tensoring each degree d line bundle by the canonical bundle of the genus two curve; thus Z^d admits a global section if d is even, and a birational two-valued section if d is odd,
- Z^1 is not isomorphic to Z^0, as they have different weight-two Hodge structures; thus Z^d does not admit a section if d is odd.

Thus we essentially have just two different spaces, with Z^1 a torsor over Z^0, and Z^1 corresponds to a two-torsion element $\alpha \in \mathrm{H}^1(\mathbb{P}^2, Z^0)$. These spaces were investigated by Markushevich [**22, 23**]; he showed that a holomorphic symplectic four-fold which is also the compactified relative Jacobian of a family of curves must be isomorphic to Z^0. Our approach will be slightly different, as we next want to interpret Z^0 as a moduli space of sheaves on the *four-fold* Z^1, so that we can set up a twisted Fourier–Mukai transform between them.

THEOREM 5.3. *Under a mild assumption (explained in the proof below), Z^0 is the compactified relative Picard scheme $\overline{\mathrm{Pic}}^0(Z^1/\mathbb{P}^2)$ of $Z^1 \to \mathbb{P}^2$.*

PROOF. The fibres of Z^1 are compactified Jacobians of surficial curves with at worst double point singularities. As explained in Example 4.6, we expect that the autoduality result of Esteves, Gagné, and Kleiman [**14**] should extend to compactified Picard schemes, and thus the fibres of $\overline{\mathrm{Pic}}^0(Z^1/\mathbb{P}^2)$ will be isomorphic to the corresponding fibres of Z^1, and of Z^0. Thus $\overline{\mathrm{Pic}}^0(Z^1/\mathbb{P}^2)$ will be a torsor over Z^0; since they both admit global sections they must be isomorphic. $\square$

REMARK 5.4. Thus Z^0 is the dual fibration of Z^1. We also obtain the smoothness of the compactified relative Picard scheme of Z^1. This is difficult to prove directly, as the compactified relative Picard scheme is a moduli space of sheaves on a four-fold, and moduli spaces on higher-dimensional varieties are poorly understood.

There exists a gerbe $\beta \in \mathrm{H}^2(Z^0, \mathcal{O}^*)$ and a twisted universal sheaf $\mathcal{U}$ on $Z^1 \times Z^0$. Since Z^0 and Z^1 are not isomorphic, β is nontrivial by Theorem 4.10.

THEOREM 5.5. *Assuming that the theorem of Bridgeland and Maciocia (Theorem 3.7) extends to the twisted case, we obtain an equivalence of derived categories*

$$\Phi^{\mathcal{U}}_{Z^0 \to Z^1} : \mathcal{D}^b_{\mathrm{coh}}(Z^0, \beta^{-1}) \to \mathcal{D}^b_{\mathrm{coh}}(Z^1).$$

PROOF. This follows from the arguments at the end of Section 4.2. In this example the degeneracy locus is a curve in $\mathbb{P}^2$, so pairs of points (m_1, m_2) in the same singular fibre contribute a five dimensional component to $\Gamma(\mathcal{U}) \subset Z^0 \times Z^0$. Since $\dim Z^1 + 1 = 5$, $\Gamma(\mathcal{U})$ is not too large. $\square$

5.2. The Hilbert scheme of more than two points on a K3. Generalizing the example of the last subsection, the Hilbert scheme $S^{[n]}$ of n points on a K3 surface S is a minimal desingularization of the symmetric product $\mathrm{Sym}^n S$. Beauville [**3**] showed that $S^{[n]}$ is an irreducible holomorphic symplectic manifold of dimension $2n$.

EXAMPLE 5.6. Let S be a genus g K3 surface, $g > 2$, i.e., S contains a smooth genus g curve C. Assume that S is otherwise generic. The curve moves in a g-dimensional linear system; denote this family of curves by $\mathcal{C} \to \mathbb{P}^g$. One can show that all the curves are irreducible, and hence we can consider the compactified degree d relative Picard scheme $Z^d := \overline{\mathrm{Pic}}^d(\mathcal{C}/\mathbb{P}^g)$. These give fibrations

$$\begin{array}{ccc} \overline{\mathrm{Pic}}^d & \hookrightarrow & Z^d \\ & & \downarrow \\ & & \mathbb{P}^g \end{array}$$

whose generic fibres are g-dimensional abelian varieties.

As in the previous subsection, the spaces Z^d can also be interpreted as irreducible components of the Mukai moduli space of stable sheaves on S [**27**]. Thus they are smooth irreducible holomorphic symplectic manifolds. They have dimension $2g$ and are deformation equivalent to the Hilbert scheme $S^{[g]}$ (see [**35**]), which can therefore be deformed to an abelian fibration.

We can generalize some of the statements of the previous subsection:

- all fibres of Z^d are irreducible,
- Z^d is locally isomorphic to Z^{d+1} as a fibration over $\mathbb{P}^g$,
- $Z^0 \to \mathbb{P}^g$ admits a global section, given by taking the trivial degree zero line bundle on each curve,
- Z^d is globally isomorphic to Z^{d+2g-2}, the isomorphism given by tensoring each degree d line bundle by the canonical bundle of the genus g curve; thus Z^d admits a global section if d is a multiple of $2g-2$.

Thus $Z^1, Z^2, \dots, Z^{2g-1}$ are all torsors over Z^0, and there are corresponding elements $\alpha_1, \alpha_2, \dots, \alpha_{2g-1} \in \mathrm{H}^1(\mathbb{P}^g, Z^0)$. These $2g-1$ spaces won't all be different; for example, there are isomorphisms

$$Z^{2g-1} \cong Z^{-1} \cong Z^1$$

where the second isomorphism is given by taking a line bundle of degree -1 to its dual. Nevertheless, it should be possible to show that we obtain at least some spaces which are not isomorphic to Z^0 (for instance, by looking at their weight-two Hodge structures). Let Z^i be such a space.

CONJECTURE 5.1. *The space Z^0 is the compactified relative Picard scheme $\overline{\mathrm{Pic}}^0(Z^i/\mathbb{P}^g)$ of $Z^i \to \mathbb{P}^g$. Hence there is a (nontrivial) gerbe $\beta \in \mathrm{H}^2(Z^0, \mathcal{O}^*)$ and a twisted universal sheaf $\mathcal{U}$ on $Z^i \times Z^0$, which induces an equivalence of derived categories*

$$\Phi^{\mathcal{U}}_{Z^0 \to Z^i} : \mathcal{D}^b_{\mathrm{coh}}(Z^0, \beta^{-1}) \to \mathcal{D}^b_{\mathrm{coh}}(Z^i).$$

REMARK 5.7. We first need to look carefully at the curves in the family $\mathcal{C} \to \mathbb{P}^g$ to see what kind of singularities they can acquire; the first statement should then follow from a compact version of the autoduality result of Esteves, Gagné, and Kleiman [**14**], as in the proof of Theorem 5.3.

To show that we get an equivalence we need the twisted version of Bridgeland and Maciocia's theorem, as in the proof of Theorem 5.5. However, this is not enough since the degeneracy locus is codimension one in $\mathbb{P}^g$, so the dimension of $\Gamma(\mathcal{U})$ is

$$(g-1) + 2\dim\{\text{fibre of } Z^0\} = 3g - 1 > 2g + 1 = \dim Z^i + 1$$

as $g > 2$. To make $\Gamma(\mathcal{U})$ smaller, we need to show directly that

$$\mathrm{Ext}^j_{Z^i}(\mathcal{U}_{m_1}, \mathcal{U}_{m_2}) = 0$$

for all integers j in some of the cases when m_1 and m_2 belong to the same singular fibre of Z^0. This can be done: the generic point m_1 on a singular fibre of Z^0 parametrizes a locally free sheaf on the corresponding singular fibre of Z^i, and then some perseverance with the standard machinery of homological algebra shows that the above Ext-groups vanish (the same can be done when m_2 parametrizes a locally free sheaf). This leaves the case when m_1 and m_2 both parametrize nonlocally free sheaves, so both m_1 and m_2 are codimension one in the fibre of Z^0. This gives

$$\dim \Gamma(\mathcal{U}) = 3g - 3 \leq 2g + 1 = \dim Z^i + 1$$

provided $g \leq 4$. We expect the equivalence should still exist when $g > 4$, but new techniques must be developed to verify this.

5.3. The ten-dimensional example of O'Grady. In the previous subsection we considered a K3 surface containing a genus g curve, which is otherwise generic. The K3 surface S of Section 5.1 (the double cover of the plane) contains a genus five curve E, the pull-back of a conic from the plane; however, it is not otherwise generic as it also contains a genus two curve. So we don't expect the conjecture of the previous subsection to apply. Nevertheless, this example is still of interest as we now explain.

The curve E moves in a five dimensional linear system $\mathbb{P}^5$; we denote the family of curves by $\mathcal{E} \to \mathbb{P}^5$. These curves are certainly not all irreducible: for example, the family includes the pull-backs from the plane of singular conics (pairs of lines). The family also includes nonreduced curves (pull-backs of double lines). Consequently it does not make immediate sense to define the compactified relative Picard schemes of $\mathcal{E} \to \mathbb{P}^5$. However, we can define Y^k to be a component of the Mukai moduli space of stable sheaves on S; its generic element will be a torsion sheaf obtained by pushing forward a degree k line bundle on a smooth curve in the family $\mathcal{E} \to \mathbb{P}^5$.

We now consider two cases:

(1) when $k = 2m + 1$ is odd Y^{2m+1} is smooth and compact, and deformation equivalent to $S^{[5]}$; Y^{2m+1} can therefore be regarded as a smooth compactification of $\mathrm{Pic}^{2m+1}(\mathcal{E}/\mathbb{P}^5)$,

(2) when $k = 2m$ is even Y^{2m} is noncompact; we can compactify to $\overline{Y}^{2m}$ by adding semi-stable sheaves but this introduces singularities.

By Hironaka's theorem $\overline{Y}^{2m}$ can be desingularized, but to find a holomorphic symplectic resolution is more difficult.

Recently O'Grady [**28**] produced a new ten dimensional irreducible holomorphic symplectic manifold by showing that a similar singular moduli space of semi-stable sheaves $\mathcal{M}$ admits a symplectic desingularization $\widetilde{\mathcal{M}}$. In fact the singular spaces $\mathcal{M}$ and $\overline{Y}^6$ are birational, and the structure of their singularities agree, at least locally [**21**]. So we expect that there exists a similar symplectic desingularization $\widetilde{Y}^6$ of $\overline{Y}^6$, which would be birational to O'Grady's example $\widetilde{\mathcal{M}}$ (and hence also deformation equivalent to it, by a result of Huybrechts [**18**]). The space $\widetilde{Y}^6$ could be regarded as a smooth compactification of $\mathrm{Pic}^6(\mathcal{E}/\mathbb{P}^5)$.

Fix now a specific example from the first case: Y^5 for instance. For a smooth genus five curve in the family $\mathcal{E} \to \mathbb{P}^5$, the fibres Pic^5 and Pic^6 of $Y^5 \to \mathbb{P}^5$ and $\widetilde{Y}^6 \to \mathbb{P}^5$ are isomorphic smooth abelian varieties. Denote by $U^5 \subset Y^5$ and $U^6 \subset \widetilde{Y}^6$ the Zariski open subsets consisting of the union of these smooth fibres; then in fact U^5 and U^6 are locally isomorphic fibrations, but globally we expect a twist.

We summarize this in the following diagram:

$$\begin{array}{ccccccc} S^{[5]} & \stackrel{\text{deform}}{\rightsquigarrow} & Y^5 & & \widetilde{Y}^6 & \stackrel{\text{deform}}{\rightsquigarrow} & \widetilde{\mathcal{M}} \\ & & \cup & & \cup & & \\ & & U^5 & \stackrel{\text{twist}}{\rightsquigarrow} & U^6 & & \end{array}$$

The Hilbert scheme $S^{[5]}$ and O'Grady's example $\widetilde{\mathcal{M}}$ are not deformation equivalent as their second Betti numbers differ; therefore Y^5 and $\widetilde{Y}^6$ cannot be deformation equivalent. It is therefore interesting that large subsets of them both (i.e., U^5 and U^6) are so closely related.

The hope of constructing a twisted Fourier-Mukai transform relating Y^5 and $\widetilde{Y}^6$ (and thereby $S^{[5]}$ and $\widetilde{\mathcal{M}}$) seems remote, as their singular fibres are quite different: they could contain different numbers of irreducible components, for instance. Nevertheless, some interesting questions still arise: can one define a dual fibration of Y^5 in a sensible way? It should agree with Y^0 over smooth fibres; how is Y^0 related to Y^6? Cho, Miyaoka, and Shepherd-Barron [**10**] described how to construct an abelian fibration X^0 with a section from an abelian fibration X with only stable fibres, such that X is a torsor over X^0: over smooth fibres X^0 is just $\mathrm{Pic}^0(\mathrm{Pic}^0(X/B))$, which is then completed by adding the same singular fibres as occur in X. In our case it is not clear that Y^5 does not contain nonreduced, and hence unstable, fibres (which might occur over nonreduced curves in the family $\mathcal{E} \to \mathbb{P}^5$).

5.4. The generalized Kummer four-fold. We return to irreducible holomorphic symplectic four-folds. Currently just two examples are known: the Hilbert scheme of two points on a K3 surface, that we saw in Section 5.1, and the generalized Kummer four-fold K_4. Beginning with an abelian surface A, the Hilbert scheme $A^{[3]}$ of three points on A is a minimal resolution of the symmetric product $\mathrm{Sym}^3 A$. It is holomorphic symplectic but not irreducible. However, there is a composition of maps

$$A^{[3]} \to \mathrm{Sym}^3 A \to A$$

(the second map is addition of the three points in A) and all fibres of this composition are isomorphic, irreducible holomorphic symplectic four-folds, which we denote by K_4 (see Beauville [**3**]).

EXAMPLE 5.8. Suppose that A is polarized by a smooth genus four curve C (this is a type $(1,3)$ polarization). The curve C moves in a two dimensional linear system $\mathbb{P}^2$. The induced map $A \to (\mathbb{P}^2)^\vee$ is six-to-one, ramified over a degree eighteen curve (see Lange and Sernesi [**20**]), and the family $\mathcal{C} \to \mathbb{P}^2$ of curves are pull-back of lines. If A is chosen to be otherwise generic, then the curves in this family are all irreducible and we can define the compactified relative Picard schemes $Z^k := \overline{\mathrm{Pic}}^k(\mathcal{C}/\mathbb{P}^2)$, which are six-folds.

For each k, there is a map $\pi_k \colon Z^k \to A$ which is just the Albanese map of Z^k. The kernel Y^k is a smooth four-fold, and it was shown by Debarre [**11**] that Y^k is deformation equivalent to K_4. Restricting π_k to a fibre of Z^k gives a surjective map from a principally polarized four dimensional abelian variety (or a degeneration thereof) to the $(1,3)$ polarized abelian surface A; the kernel of this map is thus a $(1,3)$ polarized abelian surface (or degeneration thereof). It follows that Y^k is an abelian fibration over $\mathbb{P}^2$ whose generic fibre is a $(1,3)$ polarized abelian surface.

Unlike the examples we have already seen, whose generic fibres were principally polarized Jacobians, the fibres of Y^k will not be self-dual. Assuming that the fibres of Y^k are irreducible, we can define the compactified relative Picard scheme $P := \overline{\mathrm{Pic}}^0(Y^k/\mathbb{P}^2)$. Since this is the dual fibration, even around smooth fibres it won't be isomorphic to Y^k, only isogeneous. Numerous questions now arise: is P a deformation of K_4? Is it even smooth or holomorphic symplectic?

If P is a deformation of K_4, then we observe that the exponential long exact sequence gives

$$\mathrm{H}^2(P,\mathcal{O}) \to \mathrm{H}^2(P,\mathcal{O}^*) \to \mathrm{H}^3(P,\mathbb{Z}) \to 0$$

where the first term is $\mathbb{C}$ and the third term is $\mathbb{Z}^{\oplus 8}$. Thus the space of gerbes $\mathrm{H}^2(P,\mathcal{O}^*)$ contains an infinite number of connected components. Can we construct torsors Y_β for all gerbes β? If so, and if β_1 and β_2 belong to different connected components of $\mathrm{H}^2(P,\mathcal{O}^*)$, then it would appear that Y_{β_1} cannot be deformed through abelian fibrations to Y_{β_2}. Whether Y_{β_1} and Y_{β_2} are related through any deformation at all is another interesting question.

References

1. A. B. Altman, A. Iarrobino, and S. L. Kleiman, *Irreducibility of the compactified Jacobian*, Real and Complex Singularities (Oslo, 1976), Sijthoff and Noordhoff, Alphen aan den Rijn, 1977, pp. 1–12.
2. A. B. Altman and S. L. Kleiman, *Compactifying the Picard scheme*, Adv. in Math. **35** (1980), no. 1, 50–112.
3. A. Beauville, *Variétés kählériennes dont le premiére classe de Chern est nulle*, J. Differential Geom. **18** (1983), no. 4, 755–782.
4. A. Bondal and D. Orlov, *Reconstruction of a variety from the derived category and groups of autoequivalences*, Compositio Math. **125** (2001), no. 3, 327–344.
5. T. Bridgeland, *Equivalences of triangulated categories and Fourier-Mukai transforms*, Bull. London Math. Soc. **31** (1999), no. 1, 25–34.
6. ______, *Flops and derived categories*, Invent. Math. **147** (2002), no. 3, 613–632.
7. T. Bridgeland, A. King, and M. Reid, *The McKay correspondence as an equivalence of derived categories*, J. Amer. Math. Soc. **14** (2001), no. 3, 535–554 (electronic).
8. T. Bridgeland and A. Maciocia, *Fourier-Mukai transforms for K3 and elliptic fibrations*, J. Algebraic Geom. **11** (2002), no. 4, 629–657.
9. A. Căldăraru, *Derived categories of twisted sheaves on Calabi–Yau manifolds*, Ph.D. thesis, Cornell University, 2000 (available from `www.math.umass.edu/~andreic/`).
10. K. Cho, Y. Miyaoka, and N. I. Shepherd-Barron, *Characterizations of projective space and applications to complex symplectic manifolds*, Higher Dimensional Birational Geometry (Kyoto, 1997), Adv. Stud. Pure Math., vol. 35, Math. Soc. Japan, Tokyo, 2002, pp. 1–88.
11. O. Debarre, *On the Euler characteristic of generalized Kummer varieties*, Amer. J. Math. **121** (1999), no. 3, 577–586.
12. C. D'Souza, *Compactification of generalized Jacobians*, Proc. Indian Acad. Sci. Sect. A Math. Sci. **88** (1979), no. 5, 419–457.
13. E. Esteves, private communication, 2003.
14. E. Esteves, M. Gagné, and S Kleiman, *Autoduality of the compactified Jacobian*, J. London Math. Soc. (2) **65** (2002), no. 3, 591–610.
15. A. Fujiki, *On primitively symplectic compact Kähler V-manifolds of dimension four*, Classification of Algebraic and Analytic Manifolds (Katata, 1982), Progr. Math., vol. 39, Birkhäuser, Boston, MA, 1983, pp. 71–250.
16. S. I. Gelfand and Yu. I. Manin, *Methods of homological algebra*, Springer-Verlag, Berlin, 1996.
17. N. Hitchin, *Lectures on special Lagrangian submanifolds*, Winter School on Mirror Symmetry, Vector Bundles and Lagrangian Submanifolds (Cambridge, MA, 1999), AMS/IP Stud. Adv. Math., vol. 23, Amer. Math. Soc., Providence, RI, 2001, pp. 151–182.
18. D. Huybrechts, *Compact hyper-Kähler manifolds: basic results*, Invent. Math. **135** (1999), no. 1, 63–113.
19. H. Lange and C. Birkenhake, *Complex abelian varieties*, Grundlehren Math. Wiss., vol. 302, Springer-Verlag, Berlin, 1992.
20. H. Lange and E. Sernesi, *Severi varieties and branch curves of abelian surfaces of type* $(1,3)$, Internat. J. Math. **13** (2002), no. 3, 227–244.
21. M. Lehn, private communication, 2003.

22. D. G. Markushevich, *Completely integrable projective symplectic 4-dimensional varieties*, Izv. Ross. Akad. Nauk Ser. Mat. **59** (1995), no. 1, 157–184 (Russian); English transl., Izv. Math. **59** (1995), no. 1, 159–187.

23. ———, *Lagrangian families of Jacobians of genus 2 curves*, J. Math. Sci. **82** (1996), no. 1, 3268–3284.

24. D. Matsushita, *On fibre space structures of a projective irreducible symplectic manifold*, Topology **38** (1999), no. 1, 79–83; Addendum, **40** (2001), no. 2, 431–432.

25. ———, *Higher direct images of Lagrangian fibrations*, math.AG/0010283.

26. S. Mukai, *Duality between* $D(X)$ *and* $D(\widehat{X})$ *with its application to Picard sheaves*, Nagoya Math. J. **81** (1981), 153–175.

27. ———, *Symplectic structure of the moduli space of simple sheaves on an abelian or K3 surface*, Invent. Math. **77** (1984), no. 1, 101–116.

28. K. G. O'Grady, *Desingularized moduli spaces of sheaves on a K3*, J. Reine Angew. Math. **512** (1999), 49–117.

29. D. O. Orlov, *Equivalences of derived categories and K3 surfaces*, J. Math. Sci. (New York) **84** (1997), no. 5, 1361–1381.

30. C. J. Rego, *The compactified Jacobian*, Ann. Sci. École Norm. Sup. (4) **13** (1980), no. 2, 211–223.

31. J. Sawon, *Abelian fibred holomorphic symplectic manifolds*, Turkish J. Math. **27** (2003), no. 1, 197–230.

32. ———, *Twisted Fourier-Mukai transforms for holomorphic symplectic manifolds* (in preparation).

33. C. T. Simpson, *Moduli of representations of the fundamental group of a smooth projective variety*. I, Inst. Hautes Études Sci. Publ. Math. **79** (1994), 47–129.

34. R. P. Thomas, *Derived categories for the working mathematician*, Winter School on Mirror Symmetry, Vector Bundles and Lagrangian Submanifolds (Cambridge, MA, 1999), AMS/IP Stud. Adv. Math., vol. 23, Amer. Math. Soc., Providence, RI, 2001, pp. 349–361.

35. K. Yoshioka, *Irreducibility of moduli spaces of vector bundles on K3 surfaces*, math.AG/9907001.

Department of Mathematics, State University of New York at Stony Brook, Stony Brook, NY 11794-3651, USA

E-mail address: sawon@math.sunysb.edu

Centre de Recherches Mathématiques
CRM Proceedings and Lecture Notes
Volume **38**, 2004

The Affine Stratification Number and the Moduli Space of Curves

Mike Roth and Ravi Vakil

Abstract. We define the *affine stratification number* $\operatorname{asn} X$ of a scheme X. For X equidimensional, it is the minimal number $\operatorname{asn} X$ such that there is a stratification of X by locally closed affine subschemes of codimension at most $\operatorname{asn} X$. We show that the affine stratification number is well-behaved, and bounds many aspects of the topological complexity of the scheme, such as vanishing of cohomology groups of quasicoherent, constructible, and ℓ-adic sheaves. We explain how to bound $\operatorname{asn} X$ in practice. We give a series of conjectures (the first by E. Looijenga) bounding the affine stratification number of moduli spaces of pointed curves, in which the filtration by number of rational components (which first arose in [**4**]) plays a role. This investigation is based on work and questions of Looijenga.

One relevant example (Example 4.9) turns out to be a proper integral variety with no embeddings in a smooth algebraic space. This one-paragraph construction appears to be simpler and more elementary than the earlier examples, due to Horrocks [**9**] and Nori [**12**].

1. Introduction

The *affine stratification* number of a scheme X bounds the "topological complexity" of a scheme. For example, it bounds the *cohomological dimension* $\operatorname{cd} X$ of X, which is the largest integer n such that $H^n(X, \mathcal{F}) \neq 0$ for some quasicoherent sheaf $\mathcal{F}$ (Proposition 4.12). Similarly, the cohomology of any constructible or ℓ-adic sheaf vanishes in degree greater than $\operatorname{asn} X + \dim X$ (Proposition 4.19). We expect that if the base field is $\mathbb{C}$, then X has the homotopy type of a finite complex of dimension at most $\operatorname{asn} X + \dim X$ (Conjecture 4.21), but have not completed a proof. (Unless otherwise stated, all schemes and stacks are assumed to be separated and of finite type over an arbitrary base field.)

A related, previously studied invariant is the *affine covering number* $\operatorname{acn} X$, which is one less than the minimal number of affine open sets required to cover X. The affine stratification number is bounded by $\operatorname{acn} X$, is better behaved (e.g.,

---

2000 *Mathematics Subject Classification.* Primary 14A15, Secondary 14H10.

The first author is partially supported by NSERC.

The second author is partially supported by NSF grant DMS-0228011 and NSF CAREER grant DMS-0238532.

This is the final form of the paper.

is bounded by dimension, cf. Example 4.8), and has the same topological consequences. We know of no interesting consequences of bounded acn that are not already consequences of the same bound on asn.

For equidimensional X, the definition is particularly simple.

1.1. DEFINITION. The (equidimensional) affine stratification number of an equidimensional scheme X is the minimal number $\operatorname{easn} X$ such that there is a (finite) stratification of X by locally closed affine subschemes of codimension at most $\operatorname{easn} X$.

This is the form most likely to be of interest. The appropriate generalization to arbitrary schemes is only slightly more complicated.

1.2. DEFINITION. An *affine stratification* of a scheme X is a finite decomposition $X = \bigsqcup_{k \in \mathbb{Z}^{\geq 0}, i} Y_{k,i}$ into disjoint locally closed affine subschemes $Y_{k,i}$, where for each $Y_{k,i}$,

$$\overline{Y}_{k,i} \setminus Y_{k,i} \subseteq \bigcup_{k' > k, j} Y_{k',j}. \tag{1.1}$$

The *length* of an affine stratification is the largest k such that $\cup_j Y_{k,j}$ is nonempty. The *affine stratification number* $\operatorname{asn} X$ of a scheme X is the minimum of the length over all possible affine stratifications of X.

The inclusion in (1.1) refers to the underlying set. We do not require that each $Y_{k,i}$ be irreducible. We also do not require any relation between k and the dimension or codimension of $Y_{k,i}$ in X. We will see however (Theorem 3.1) that it is always possible to assume that the stratification has a very nice form.

Strictly speaking, the term "stratification" is inappropriate, as $\overline{Y}_{k,i} \setminus Y_{k,i}$ need not be a union of $Y_{k',j}$: let X be the coordinate axes in $\mathbb{A}^2$, $Y_{0,1}$ the x-axis, and $Y_{1,1}$ the y-axis minus the origin. However, Theorem 3.1(a) shows that we may take (1.1) to be an actual stratification.

The affine stratification number has many good properties, including the following (Lemma 2.1, Propositions 4.2, 4.6, 2.10).

- $\operatorname{asn} X = 0$ if and only if X is affine.
- $\operatorname{asn} X \leq \dim X$.
- $\operatorname{asn} X \leq \operatorname{acn} X$. (Equality does not always hold.)
- $\operatorname{asn}(X \times Y) \leq \operatorname{asn} X + \operatorname{asn} Y$.
- If D is an effective Cartier divisor on X, then $\operatorname{asn}(X - D) \leq \operatorname{asn} X$.
- If $Y \to X$ is an affine morphism, then $\operatorname{asn} Y \leq \operatorname{asn} X$.

Even if one is only interested in equidimensional schemes, the more general Definition 1.2 has advantages over Definition 1.1. For example, the last property is immediate using Definition 1.2, but not obvious using Definition 1.1.

In Section 2, we establish basic properties of affine stratifications. In Section 3, we show that affine stratifications can be reorganized into a particularly good form. In particular, if X is equidimensional, then $\operatorname{easn} X = \operatorname{asn} X$ (Proposition 3.7), so the notation easn may be discarded. In Section 4, we give topological consequences of bounded asn.

Our motivation is to bound the affine stratification number of moduli spaces (in particular, of pointed curves) to obtain topological and cohomological consequences. We describe our work in progress in the form of several conjectures in Section 5. For example, the conjectures bound the homotopy type of the moduli spaces of curves

(a) of compact type (stable curves whose dual graph is a tree, or equivalently stable curves with compact Jacobian), (b) with "rational tails" (stable curves having a smooth component of genus g), and (c) with at most k rational components (a locus introduced in [**4**]), see Proposition 5.9.

The proper integral threefold with no smooth embeddings, promised in the abstract, is Example 4.9.

Acknowledgments. This note arose from our ongoing efforts to prove a conjecture of E. Looijenga, and much of what is here derives from questions, ideas, and work of his. In particular, we suspect that he is aware of most of the results given here, and that we are following in his footsteps. We thank him for inspiration. We also thank J. Starr for Example 4.8, W. Fulton for pointing out the examples of Nori [**12**] and Horrocks [**9**], and T. Graber for helpful conversations. Finally, we are grateful to the organizers of the 2003 conference *Algebraic structures and moduli spaces*, at the Centre de recherches mathématiques (CRM), which led to this work.

2. Basic properties of affine stratifications

The most basic property is that an affine stratification always exists, and hence $\operatorname{asn} X$ is defined for any scheme X: if $\cup_{i=0}^n U_i$ is a covering of X by open affine sets, then

$$U_0 \sqcup (U_1 \setminus U_0) \sqcup (U_2 \setminus (U_0 \cup U_1)) \sqcup \cdots \tag{2.1}$$

gives an affine stratification of X.

The following lemma is trivial.

2.1. LEMMA. (a) *The affine stratification number depends only on the reduced structure of X, i.e.,* $\operatorname{asn} X = \operatorname{asn} X^{red}$.

(b) *If $X \to Y$ is an affine morphism, then* $\operatorname{asn} X \leq \operatorname{asn} Y$.

(c) $\operatorname{asn}(X \times Y) \leq \operatorname{asn} X + \operatorname{asn} Y$.

(d) *If D is an effective Cartier divisor on X, then* $\operatorname{asn}(X - D) \leq \operatorname{asn} X$.

Part (d) requires the following well-known fact.

2.2. LEMMA. *Any irreducible affine scheme X, minus an effective Cartier divisor D, is affine.*

(REASON. The inclusion $X \setminus D \hookrightarrow X$ is an affine morphism, since this can be verified locally. But X is affine.)

Here is a partial converse to Lemma 2.2. A more precise converse is given in Proposition 2.6.

2.3. LEMMA. *Suppose that V is an irreducible affine scheme, and that $U \subset V$ is an open affine subset. Then the complement $Z := V \setminus U$ is a Weil divisor in V.*

PROOF. We first assume that V (and hence U) is normal. Let $Z = \cup_i Z_i$ be the decomposition of Z into irreducible components, and let $Z' = \cup_j Z_j$ be the union of those components of codimension one in V. We set $U' = V \setminus Z'$, and let $i : U \hookrightarrow U'$ be the natural open immersion. Since U' is normal, and the complement of U in U' is of codimension at least 2 in U', we have $i_* \mathcal{O}_U = \mathcal{O}_{U'}$. We will use this and the fact that both U and V are affine to see that $U = U'$.

Let $A = \Gamma(V, \mathcal{O}_V)$ and $B = \Gamma(U, \mathcal{O}_U) = \Gamma(U', \mathcal{O}_{U'})$. We have an inclusion of rings $A \hookrightarrow B$ corresponding to the opposite inclusion of open sets. Suppose that

$U \neq U'$, and let x be any point of $U' \setminus U$. Since V is affine, x corresponds to a prime ideal $\mathcal{P}_x$ of A. Since $x \in U'$, no element of $\mathcal{P}_x$ can be a unit in $\Gamma(U', \mathcal{O}_{U'})$, and hence $\mathcal{P}_x$ remains a prime ideal in B, which is a localization of A. Therefore, since U is affine, $x \in U$, contrary to assumption.

Passing to the general case, we drop the assumption that V and U are normal, and let $\widetilde{V}$ and $\widetilde{U}$ be their normalizations. We have the commutative diagram

$$\begin{array}{ccc} \widetilde{U} & \hookrightarrow & \widetilde{V} \\ \downarrow & & \downarrow \\ U & \hookrightarrow & V, \end{array}$$

where the vertical arrows are the normalization maps, and the horizontal arrows are open immersions. By the first part of the lemma, the complement $\widetilde{Z}$ of $\widetilde{U}$ in $\widetilde{V}$ is of codimension one in $\widetilde{V}$. Since $\widetilde{Z}$ maps finitely and surjectively onto Z, $\dim(Z) = \dim(\widetilde{Z})$, and hence Z is of codimension one in V. □

The next corollary follows immediately. (Note that X need not be equidimensional here.)

2.4. COROLLARY. *The complement of a dense affine open subset in any scheme is of pure codimension one.*

2.5. EXAMPLES. (a) Let X be the affine cone over an elliptic curve, embedded in $\mathbb{CP}^2$ as a cubic. Let Z be the cone over any point of the curve of infinite order in the group law. Then $X \setminus Z$ is affine, but Z is not $\mathbb{Q}$-Cartier. This shows that the complement of an affine open set in an affine scheme need not be the support of a Cartier divisor: we cannot hope to improve the conclusion of Lemma 2.3 to match the hypothesis of Lemma 2.2.

(b) Let S be $\mathbb{P}^2$ blown up at a point, and let X be the affine cone over some projective embedding of S. Let $Z \subset X$ be the affine cone over the exceptional divisor of the blowup. Then Z is of codimension one in X, but $\operatorname{cd}(X \setminus Z) = 1$, so in particular it is not affine. This shows that, conversely, the complement of a Weil divisor in an affine scheme need not be affine: we can not hope to improve the hypothesis of Lemma 2.2 to match the conclusion of Lemma 2.3.

However, there is a more precise statement giving a necessary and sufficient condition on a closed subset Z of an affine scheme V for the complement $V \setminus Z$ to be affine.

2.6. PROPOSITION. *Let V be an affine scheme (possibly reducible) and Z a closed subset of V. Then $U := V \setminus Z$ is affine if and only if $H^i_Z(\mathcal{F}) = 0$ for all quasicoherent sheaves $\mathcal{F}$ on V and all $i \geq 2$.*

Here $H^i_Z(\mathcal{F})$ is the local cohomology group. This also implies the same fact for the local cohomology sheaves $\mathcal{H}^i_Z(\mathcal{F})$, see Corollary 2.7(a) below.

PROOF. Let $\mathcal{F}$ be any quasicoherent sheaf on V. We have the long exact excision sequence of cohomology groups

$$(2.2)\quad 0 \to H^0_Z(\mathcal{F}) \to H^0(V, \mathcal{F}) \to H^0(U, \mathcal{F}|_U) \to H^1_Z(\mathcal{F}) \to H^1(V, \mathcal{F}) \\ \to H^1(U, \mathcal{F}|_U) \to H^2_Z(\mathcal{F}) \to H^2(V, \mathcal{F}) \to H^2(U, \mathcal{F}|_U) \to \cdots .$$

Since V is affine, we have $H^i(V,\mathcal{F}) = 0$ for all $i \geq 1$, so that $H^i(U, \mathcal{F}|_U) = H^{i+1}_Z(\mathcal{F})$ for all $i \geq 1$. Hence (using Serre's criterion for affineness) U is affine if and only if $H^i_Z(\mathcal{F}) = 0$ for all $i \geq 2$ and all quasicoherent sheaves $\mathcal{F}$. $\square$

2.7. COROLLARY. *Let X be a scheme (possibly reducible) and U a dense affine open subset. Let $Z := X \setminus U$. For any quasicoherent sheaf $\mathcal{F}$ on X,*

(a) $\mathcal{H}^i_Z(\mathcal{F}) = 0$ *for all* $i \geq 2$, *and*
(b) $H^i_Z(\mathcal{F}) = 0$ *for all* $i > \operatorname{cd} Z + 1$.

The notation cd denotes cohomological dimension, see Section 1.

PROOF. (a) Since the local cohomology sheaf $\mathcal{H}^i_Z(\mathcal{F})$ is the sheafification of the functor $V \mapsto H^i_{Z\cap V}(\mathcal{F}|_V)$ [**7**, Proposition 1.2], it is sufficient to check that the local cohomology group vanishes for sufficiently small V around any point of Z. But if V is any open affine set, then $V \cap U$ is nonempty (since U is dense) and also affine (by separatedness). Hence $H^i_{Z\cap V}(\mathcal{F}|_V) = 0$ for $i \geq 2$ by Proposition 2.6 and so $\mathcal{H}^i_Z(\mathcal{F}) = 0$ as well.

(b) The local cohomology sheaves $\mathcal{H}^i_Z(\mathcal{F})$ are quasicoherent and are supported on Z. The local cohomology groups can be computed by a spectral sequence with E_2^{pq} term $H^p(X, \mathcal{H}^q_Z(\mathcal{F})) = H^p(Z, \mathcal{H}^q_Z(\mathcal{F}))$. Since $H^p(Z,\cdot) = 0$ for $p > \operatorname{cd} Z$, and $\mathcal{H}^q_Z(\mathcal{F}) = 0$ for $q > 1$ by part (a), we have $H^i_Z(\mathcal{F}) = 0$ for $i > \operatorname{cd} Z + 1$. $\square$

2.8. COROLLARY. *Let X be a scheme, U a dense affine open subset, and set $Z := X \setminus U$. Then* $\operatorname{cd} X \leq \operatorname{cd} Z + 1$.

PROOF. For any any quasicoherent sheaf $\mathcal{F}$ on X, the excision sequence (2.2) and the fact that U is affine gives $H^i(X,\mathcal{F}) = H^i_Z(\mathcal{F})$ for all $i \geq 2$, and that $H^1(X,\mathcal{F})$ is a quotient of $H^1_Z(\mathcal{F})$. Hence, for any $i \geq 1$, $H^i_Z(\mathcal{F}) = 0$ implies that $H^i(X,\mathcal{F}) = 0$. Since $H^i_Z(\mathcal{F}) = 0$ for all $i > \operatorname{cd} Z + 1$ by Corollary 2.7(b), we have $\operatorname{cd} X \leq \operatorname{cd} Z + 1$. $\square$

2.9. Bounding asn by finite flat covers. The following result is useful to bound $\operatorname{asn} X$ by studying covers of X.

2.10. PROPOSITION. *Suppose $\pi \colon Y \to X$ is a surjective finite flat morphism of degree not divisible by the characteristic of the base field, and Y is affine. Then X is affine.*

PROOF. The hypothesis implies that $\pi_* \mathcal{O}_Y$ is a coherent locally free sheaf on X. The trace map gives a splitting $\pi_*\mathcal{O}_Y \cong \mathcal{O}_X \oplus E$ for some vector bundle E on X. If $\mathcal{F}$ is any coherent sheaf on X, then the flatness of π gives $\pi_*\pi^*\mathcal{F} = \mathcal{F} \oplus (E \otimes \mathcal{F})$, and it then follows from the Leray spectral sequence and the finiteness of π that $H^i(X,\mathcal{F})$ is a direct summand of $H^i(Y, \pi^*\mathcal{F})$ for all $i \geq 0$. Since Y is affine, these vanish if $i \geq 1$, hence the cohomology groups on X do as well, and therefore X is affine by Serre's criterion for affineness. $\square$

3. Reorganizing affine stratifications

We describe various ways that we can reorganize the stratification which are more convenient for analyzing X. The main results of this section are summarized in the following theorem.

3.1. THEOREM. *If X is any scheme and* $\operatorname{asn} X = m$, *then there exists an affine stratification $\{Z_0, \ldots, Z_m\}$ of X such that for any $k \leq m$:*

(a) $\overline{Z}_k = \bigcup_{k' \geq k} Z_{k'}$,
(b) *each* Z_k *is a dense open affine subset of* $\overline{Z}_k$, *and*
(c) $\overline{Z}_k$ *is of pure codimension one in* $\overline{Z}_{k-1}$.

If in addition X *is equidimensional, then we also have*

(d) *each* $\overline{Z}_{k'}$ *is of pure codimension* $k' - k$ *in* $\overline{Z}_k$ *for any* $k' \geq k$.

Even if X *is not equidimensional, if we have an affine stratification* $\{Y'_{k,l}\}$ *of length* M *such that each* $Y'_{k,i}$ *is of pure codimension* k *in* X, *then setting* $Z'_k := \bigcup_i Y'_{k,i}$ *for* $k = 0, \ldots, M$ *we have*

(e) *the affine stratification* $\{Z'_0, \ldots, Z'_M\}$ *satisfies* (a)–(d) *above.*

(*We are not guaranteed that* $M = m$, *so this stratification may not be optimal.*)

The proof is summarized in Section 3.9. In analogy with CW-complexes, we define an *affine cell decomposition* of a scheme X to be an affine stratification

$$X = \bigsqcup_k Z_k$$

where the Z_k's satisfy (a)–(c) of Theorem 3.1. The theorem guarantees that such a decomposition exists for any scheme X, with length asn X.

3.2. LEMMA. *Let* $\{Y_{k,i}\}$ *be an affine stratification of a scheme* X *and let* $Z_k := \bigcup_i Y_{k,i}$ *be the union of all the affine pieces of index* k. *Then each* Z_k *is an open dense affine subset of* $\overline{Z}_k$, *i.e.,* Z_k *is locally closed and affine.*

PROOF. By definition, Z_k is a dense subset of $\overline{Z}_k$. We will see that it is an open subset, and most importantly, affine.

Since the affine stratification is finite, we have $\overline{\cup_i Y_{k,i}} = \cup_i \overline{Y}_{k,i}$. For any distinct $Y_{k,i}$ and $Y_{k,j}$ and any point $y \in \overline{Y}_{k,i} \cap \overline{Y}_{k,j}$, the fact that the Y's are disjoint, along with the stratification condition (1.1), implies that y must be in some $Z_{k'}$ with $k' > k$. In particular, y is in neither $Y_{k,i}$ nor $Y_{k,j}$.

If we let $C_i := \bigcup_{j \neq i} \overline{Y}_{k,j}$ be the closed subset consisting of the closures of other $Y_{k,j}$'s, and $V_i := X \setminus C_i$ the open complement, then the previous remark shows that $Y_{k,i} \subseteq V_i$, and therefore that $Z_k \cap V_i = Y_{k,i}$.

Since every locally closed subset is an open subset of its closure, $Y_{k,i}$ is an open subset of $\overline{Y}_{k,i} \cap V_i = \overline{Z}_k \cap V_i$. Since $\overline{Z}_k \cap V_i$ is an open subset of $\overline{Z}_k$, we see that $Y_{k,i}$ is an open subset of $\overline{Z}_k$, and therefore that $Z_k = \bigcup_i Y_{k,i}$ is an open subset of $\overline{Z}_k$.

Let $\widetilde{Z}_k$ be the disjoint union

$$\widetilde{Z}_k = \bigsqcup_i Y_{k,i},$$

and $f \colon \widetilde{Z}_k \to X$ the natural morphism with image Z_k. The map f is one-to-one on points, and the fact that $Z_k \cap V_i = Y_{i,k}$ for each i implies that f is a homeomorphism, and in fact an immersion. Therefore, $\widetilde{Z}_k \cong Z_k$ as schemes, and so Z_k is affine since $\widetilde{Z}_k$ is. □

3.3. PROPOSITION. *Let* $\{Y_{k,i}\}$ *be an affine stratification of* X *of length* m. *Then there exists an affine stratification* $\{Y'_{k,j}\}$ *of length at most* m *such that the generic points of all components of* X *are contained in the zero stratum* $\cup_j Y'_{0,j}$ *of* $\{Y'_{k,j}\}$.

PROOF. We first set $Y'_{0,i} := Y_{0,i}$ for all valid indices i. Now let $Y_{k,i}$ be any piece of the stratification with $k \geq 1$. If $Y_{k,i}$ does not contain the generic point of any component of X then set $Y'_{k,i} := Y_{k,i}$. On the other hand, suppose that $Y_{k,i}$ contains $\eta_1, \ldots, \eta_r$ where each η_j is a generic point of X. In this case, for each $j \in \{1, \ldots, r\}$ choose an open affine subset U_j of $Y_{k,i}$ containing η_j so that U_j intersects no components of X other than $\overline{\{\eta_j\}}$. Now set $Y'_{k,j} := Y_{k,j} \setminus (\bigcup_{j=1}^{r} U_j)$, and add the U_j in as elements of the zero stratum, $Y'_{0,i_j} := U_j$, where the i_j are chosen not to conflict with previously existing indices. It is straightforward to verify that this decomposition satisfies the affine stratification condition (1.1). □

3.4. LEMMA. *Let $\{Y_{k,i}\}$ be an affine stratification of X of length m. Then there exists an affine stratification $\{Y'_{k,i}\}$ of length at most m such that if $Z'_k := \cup_i Y'_{k,i}$ is the union of all affine pieces of index k, then for any k,*

$$\overline{Z'_k} = \bigcup_{k' \geq k} Z'_{k'}.$$

PROOF. By Proposition 3.3 we may assume that all the generic points of components of X occur in the zero stratum of $\{Y_{k,i}\}$, and therefore that $\bigcup_i \overline{Y}_{0,i} = X$. We now proceed by induction on the length m of the stratification, the case $m = 0$ being trivial.

Let $U := \bigcup_i Y_{0,i}$ be the union of the pieces in the zero stratum, and $Z := \bigcup_{k\geq 1,i} Y_{k,i}$ the complement. Note that U is open and hence Z is closed by Lemma 3.2.

The $\{Y_{k,i}\}$ with $k \geq 1$ form an affine stratification of Z of length $m - 1$ (after reindexing the k's to start with zero). Therefore by induction Z has an affine stratification of length at most $m-1$ satisfying the hypothesis of the lemma. Reindexing the k's again, and adding the $Y_{0,i}$'s as the zero stratum, we end up with an affine stratification $\{Y'_{k,i}\}$ of length at most m which also satisfies the hypothesis of the lemma, completing the inductive step. □

3.5. COROLLARY. *For any scheme X, if $\operatorname{asn} X = m$ then there is an affine stratification $\{Z_0, \ldots, Z_m\}$ of X with $\overline{Z}_k = \bigcup_{k' \geq k} Z_{k'}$ for each k, and such that each Z_k is an open dense affine subset of $\overline{Z}_k$.*

PROOF. Combine Lemmas 3.2 and 3.4. □

3.6. COROLLARY. *For any scheme X, $\operatorname{asn} X \leq \dim X$.*

PROOF. Let $m = \operatorname{asn} X$ and $\{Z_0, \ldots, Z_m\}$ be a stratification as in Corollary 3.5. By Corollary 2.4, each $\overline{Z}_{k+1}$ is of pure codimension one in $\overline{Z}_k$. If $\overline{Z}'_m$ is any irreducible component of $\overline{Z}_m$, then that means we can inductively find a chain of closed irreducible subsets $\overline{Z}'_m \subset \overline{Z}'_{m-1} \subset \overline{Z}'_{m-2} \subset \cdots \subset \overline{Z}'_1 \subset \overline{Z}'_0$, with each $\overline{Z}'_k$ an irreducible component of $\overline{Z}_k$. Then $\dim X \geq \dim \overline{Z}'_m + m \geq m$. □

If we assume an additional hypothesis about X or the stratification $\{Y_{k,i}\}$, we have slightly stronger results about the stratification $\{Z_0, \ldots, Z_m\}$ of Corollary 3.5.

3.7. PROPOSITION. *If X is an equidimensional scheme, and $\{Z_0, \ldots, Z_m\}$ the stratification of Corollary 3.5, then we have in addition that $\overline{Z}_{k'}$ is of pure codimension $k' - k$ in $\overline{Z}_k$ for all $k' \geq k$. In particular,* $\operatorname{easn} X = \operatorname{asn} X$.

PROOF. By Corollary 2.4 each $\overline{Z}_{k+1}$ is of pure codimension one in $\overline{Z}_k$. If $\overline{Z}_0 = X$ is equidimensional, then it follows that each $\overline{Z}_k$ is equidimensional as well, and from this that $\overline{Z}_{k'}$ is of pure codimension $k'-k$ in $\overline{Z}_k$ for any $k' \geq k$. □

Even if X is not equidimensional, if the affine stratification $\{Y_{k,i}\}$ satisfies a suitable condition we get a similar good result about the stratification by the Z_k's.

3.8. PROPOSITION. *Let $\{Y_{k,i}\}$ be an affine stratification of a scheme X with each $Y_{k,i}$ of pure codimension k in X. Let $Z_k := \bigcup_i Y_{k,i}$ be the union of all the affine pieces of codimension k. Then*

(i) *For $k' \geq k$, $\overline{Z}_{k'}$ is of pure codimension $k'-k$ in $\overline{Z}_k$; in particular, $\overline{Z}_{k'} \subseteq \overline{Z}_k$.*

(ii) $\overline{Z}_k = \bigcup_{k' \geq k} Z_{k'}$.

PROOF. Since the decomposition is finite, the irreducible components of $\overline{Z}_m$ are all of the form $\overline{W}_m$ with W_m an irreducible component of some $Y_{m,j}$.

We prove (i) by induction on k. For $k=0$ the result is obvious, since $\overline{Z}_0 = X$, and $\overline{Z}_{k'}$ is of pure codimension k' in X. So assume that $k > 0$ and that (i) is true for $k-1$.

Let $\overline{W}_{k'}$ by any irreducible component of $\overline{Z}_{k'}$ with $k' \geq k$. By the induction hypothesis, $\overline{W}_{k'} \subset \overline{Z}_{k-1}$, and is of codimension $k'-k+1$ in $\overline{Z}_{k-1}$. Let T_{k-1} be any irreducible component of Z_{k-1} whose closure contains $\overline{W}_{k'}$ and such that $\operatorname{codim}(\overline{W}_{k'}, \overline{T}_{k-1}) = k'-k+1$. Lemma 3.2 gives us that Z_{k-1} is affine, and therefore T_{k-1} is affine also.

By Lemma 2.3, the closed set $\overline{T}_{k-1} \setminus T_{k-1}$ has codimension one in $\overline{T}_{k-1}$. Let η_k be the generic point of any component of $\overline{T}_{k-1} \setminus T_{k-1}$ containing $W_{k'}$; one exists by our choice of T_{k-1}. Since $\operatorname{codim}(\overline{W}_{k'}, \overline{T}_{k-1}) = k'-k+1$ and $\operatorname{codim}(\overline{\{\eta_k\}}, \overline{T}_{k-1}) = 1$, we have $\operatorname{codim}(W_{k'}, \overline{\{\eta_k\}}) = k'-k$.

The Z_m's partition X, and so η_k must be in exactly one Z_m. We cannot have $m \leq k-1$, since that would contradict the stratification condition. We cannot have $m \geq k+1$, since this would contradict $\operatorname{codim}(\overline{Z}_m, \overline{Z}_{k-1}) = m-k+1$, which holds by the induction hypothesis. Therefore η_k is in Z_k, and so $W_{k-1} \subset \overline{Z}_k$.

We have already seen that $\operatorname{codim}(W_{k'}, \overline{\{\eta_k\}}) = k'-k$. Since

$$\operatorname{codim}(\overline{W}_{k'}, \overline{Z}_k) = \sup_i \bigl(\operatorname{codim}(\overline{W}_{k'}, \overline{Y}_{k,i})\bigr),$$

with $\overline{Y}_{k,i}$ running over the components of $\overline{Z}_k$, we get $\operatorname{codim}(\overline{W}_{k'}, \overline{Z}_k) \geq k'-k$. But for any three closed schemes $\overline{W}$, $\overline{Z}$, and X with $\overline{W} \subseteq \overline{Z} \subseteq X$, we always have

$$\operatorname{codim}(\overline{W}, \overline{Z}) + \operatorname{codim}(\overline{Z}, X) \leq \operatorname{codim}(\overline{W}, X).$$

Since the codimensions of $\overline{W}_{k'}$ in X, and $\overline{Z}_k$ in X are k' and k by hypothesis, this gives $\operatorname{codim}(\overline{W}_{k'}, \overline{Z}_k) \leq k'-k$, and hence $\operatorname{codim}(\overline{W}_{k'}, \overline{Z}_k) = k'-k$. Therefore $\overline{Z}_{k'}$ is contained in $\overline{Z}_k$, and is of pure codimension $k'-k$, completing the inductive step for (i).

To prove (ii), the stratification condition gives $\overline{Z}_k \subseteq \cup_{k' \geq k} Z_{k'}$, while part (i) above gives the opposite inclusion. □

3.9. Proof of Theorem 3.1. (a) is Lemma 3.4, (b) Lemma 3.2, (c) Corollary 2.4, (d) Proposition 3.7, and (e) Proposition 3.8 and Lemma 3.2 again.

4. Topological consequences of bounded affine stratification number

We now describe the topological consequences of bounded asn, in particular: relation to dimension (Section 4.1), affine covering number (Section 4.5), cohomological dimension (for quasicoherent sheaves, Section 4.11, as well as constructible and ℓ-adic sheaves, Section 4.15), dimension of largest proper subscheme (Section 4.14), and homotopy type (Section 4.20).

4.1. Relation to dimension.

4.2. PROPOSITION. $\operatorname{asn} X \leq \dim X$. *If one top dimensional component of X is proper, then equality holds.*

The first statement is Corollary 3.6. The second statement follows from Proposition 4.12 ($\operatorname{cd} \leq \operatorname{asn}$) and the following theorem, first conjectured by Lichtenbaum.

4.3. THEOREM (Grothendieck [**7**, 6.9], Kleiman [**10**, Main Theorem]). *If $d = \dim X$, then $\operatorname{cd} X = d$ if and only if at least one d-dimensional component of X is proper.*

4.4. EXAMPLE (All values between 0 and $\dim X$ are possible). Let $X_k = \mathbb{P}^n \setminus \{(n-k-1)\text{-plane}\}$, for k between 0 and $n-1$. Then clearly $\operatorname{cd} X_k = k$ and $\operatorname{asn} X_k \leq k$. We will see that $\operatorname{cd} \leq \operatorname{asn}$ (Proposition 4.12), from which the result follows.

4.5. Relation to affine covering number. Recall that the affine covering number $\operatorname{acn} X$ of a scheme X is the minimal number of affine open subsets required to cover X, minus 1. The invariant acn does not obviously behave as well as asn with respect to products (cf. Lemma 2.1(c)); it also is not bounded by dimension (Example 4.8 below).

The argument of (2.1) gives the following.

4.6. PROPOSITION. $\operatorname{asn} X \leq \operatorname{acn} X$.

4.7. EXAMPLE. In general, $\operatorname{acn} X \neq \operatorname{asn} X$. As an example, let X be a complex K3 surface with Picard rank 1, minus a very general point. Then $\operatorname{acn} X = 2$: if for every point p of X, $\operatorname{acn}(X-p) = 1$, then (given the hypothesis that the Picard rank is 1) any two points of X are equivalent in $A_0(Y)$ (with $\mathbb{Q}$-coefficients), contradicting Mumford's theorem that $A_0(Y)$ is not countably generated [**11**]. Example 4.8 below gives another example (in light of Proposition 4.2).

4.8. EXAMPLE ($\operatorname{acn} X$ may be larger than $\dim X$). When X is quasiprojective, $\operatorname{acn} X \leq \dim X$. (REASON. Let $\overline{X}$ be a projective compactification such that the complement $\overline{X} \setminus X$ is a Cartier divisor D. Consider an embedding $\overline{X} \hookrightarrow \mathbb{P}^n$ and let $H_0, \ldots, H_{\dim X}$ be hypersurfaces so that $\overline{X} \cap H_0 \cap \cdots \cap H_{\dim X} = \varnothing$. Then the $\overline{U_i} := \overline{X} \setminus (\overline{X} \cap H_i)$ form an affine cover of $\overline{X}$. We conclude using Lemma 2.2.)

However, the following example, due to J. Starr, shows that $\operatorname{acn} X$ may be greater than $\dim X$. Given any n, we describe a reducible, reduced threefold that requires at least n affine open sets to cover it. Recall Hironaka's example (e.g., [**8**, Example B.3.4.1]) of a nonsingular proper nonprojective threefold X. Nonprojectivity is shown by exhibiting two curves ℓ and m whose sum is numerically trivial. Hence no affine open set can meet both ℓ and m; otherwise its complement would be a divisor (Lemma 2.3), hence Cartier (as X is nonsingular), which meets both

ℓ and m positively. Now choose points p and ℓ and q on m. Consider $\binom{n}{2}$ copies of (X, p, q), corresponding to ordered pairs (i,j) $(1 \leq i < j \leq n)$; call these copies (X_{ij}, p_{ij}, q_{ij}). Let r_1, ..., r_n be copies of a reduced point. Glue r_i to p_{ji} and q_{ik}. Then no affine open can contain both r_i and r_j for $i < j$ (by considering X_{ij}).

4.9. EXAMPLE (a family of integral threefolds with arbitrary high affine covering number (and no smooth embeddings)). This leads to an example of an integral (but singular) threefold that requires at least n affine open sets to cover it. (QUESTION. Is there a family of nonsingular irreducible varieties of fixed dimension with unbounded affine covering number?) Our example will be a blow-up of $\mathbb{P}^3$. Choose n curves C_1, ..., C_n in $\mathbb{P}^2 \subset \mathbb{P}^3$ that meet in n simple n-fold points $p_1, \ldots, p_n$ (and possibly elsewhere). Away from $p_1, \ldots, p_n$ blow up $C_1, \ldots, C_n$ in some arbitrary order. In a neighborhood of p_i (not containing any other intersection of the C_j) blow up C_i first (giving a smooth threefold) and then blow up the local complete intersection $\bigcup_{j \neq i} C_j$ (or more precisely, the proper transform thereof), giving a threefold with a single singularity (call it q_i). The preimage of p_i is the union of two $\mathbb{P}^1$'s, one arising from the exceptional divisor of C_i (call it ℓ_i), and one from the exceptional divisor of $\bigcup_{j \neq i} C_j$; they meet at q_i. By Hironaka's argument, $\ell_i + \ell_j$ is numerically trivial for all $i \neq j$. Then no affine open U can contain both q_i and q_j: the complement of U would be a divisor, meeting ℓ_i and ℓ_j properly and at smooth points of our threefold (i.e., not at q_i and q_j), and the same contradiction applies.

This is also an example of a scheme which cannot be embedded in any smooth scheme, or indeed algebraic space. (Earlier examples are the topic of papers of Horrocks [**9**] and Nori [**12**].) If $X \hookrightarrow W$ with W smooth, then for any divisor D (automatically Cartier) on W, $D \cdot \ell_i + D \cdot \ell_j = 0$ for all i, j. If $n \geq 3$, this implies that $D \cdot \ell_i = 0$ for all i. But for any affine open set U of W with $U \cap \ell_i \neq \varnothing$, the complement $D = W \setminus U$ would intersect ℓ_i properly, giving us the contradiction $D \cdot \ell_i > 0$. Hence no such embedding is possible.

By combining Proposition 4.2 with Proposition 4.6, we obtain the following.

4.10. PROPOSITION. *If X is proper, then* $\operatorname{acn} X > \dim X$.

We note that this also follows from Theorem 4.3. Example 4.7 shows that it is *not* true that $\operatorname{acn} X = \dim X$ if and only if X is proper, even for quasiprojective X.

4.11. Relation to cohomological dimension. Just as the dimensions of the cells in a CW-complex bounds the topological (co)homology, the length of the stratification into affine cells bounds the quasicoherent sheaf cohomology.

4.12. PROPOSITION. $\operatorname{cd} X \leq \operatorname{asn} X$.

PROOF. We prove the result by induction on $\operatorname{asn} X$. It is clear for $\operatorname{asn} X = 0$, so assume that $m := \operatorname{asn} X > 0$ and that the result is proved for all schemes Z with $\operatorname{asn} Z < \operatorname{asn} X$. Let $\{Z_0, \ldots, Z_m\}$ be an affine cell decomposition given by Theorem 3.1. Set $Z := X \setminus Z_0 = \overline{Z}_1 = \bigcup_{k \geq 1} Z_k$.

By Theorem 3.1(b) Z_0 is an open dense affine subset of X, so by Corollary 2.8, $\operatorname{cd} X \leq \operatorname{cd} Z + 1$. Next, $Z = \bigsqcup_{k=1}^{m-1} Z_k$ is (after reindexing) an affine stratification of Z of length $m - 1$, so $\operatorname{asn} Z \leq \operatorname{asn} X - 1$. Finally, by the inductive hypothesis, $\operatorname{cd} Z \leq \operatorname{asn} Z$. Combining these three inequalities gives $\operatorname{cd} X \leq \operatorname{asn} X$, completing the inductive step. □

We remark in passing that by combining Proposition 4.12 with Corollary 3.6 we obtain another proof of Grothendieck's dimensional vanishing theorem ([**5**, Theorem 3.6.5], [**8**, Theorem III.2.7]).

We conclude with an obvious result.

4.13. PROPOSITION. $\operatorname{cd} X = 0$ *if and only if* $\operatorname{asn} X = 0$ *if and only if* $\operatorname{acn} X = 0$.

PROOF. Each of the three is true if and only if X is affine (the first by Serre's criterion for affineness). □

4.14. Relation to dimension of largest complete subscheme. Motivated by Diaz' theorem [**3**], let $\operatorname{psv} X$ be the largest dimension of a proper closed subscheme of X. If Z is a proper closed subscheme of X (with inclusion $j\colon Z \hookrightarrow X$), and if $\mathcal{F}$ is a quasicoherent sheaf on Z, then $j_*\mathcal{F}$ is a quasicoherent sheaf on X, and $H^i(X, j_*\mathcal{F}) = H^i(Z, \mathcal{F})$ for all i. By Theorem 4.3 we can find a quasicoherent sheaf $\mathcal{F}$ on Z with $H^{\dim Z}(Z, \mathcal{F}) \neq 0$, and so this gives $\operatorname{psv} X \leq \operatorname{cd} X$. Hence by Proposition 4.12,

$$\operatorname{psv} X \leq \operatorname{asn} X.$$

4.15. Relation to cohomological vanishing for constructible and ℓ-adic sheaves. In this section all notions related to sheaves (including stalks, push-forwards, and cohomology groups) are with respect to the étale topology. For instance, "sheaf on X" means "sheaf on X in the étale topology."

To show how asn implies cohomological vanishing for constructible and ℓ-adic sheaves (Corollary 4.19), we first recall a theorem and some notation of Artin. For any (étale) sheaf F of abelian groups on X, let

$$\mathrm{d}(\mathsf{F}) := \sup\{\dim(\overline{\{x\}}) \mid x \in X, \mathsf{F}_{\overline{x}} \neq 0\}$$

be the dimension of the support of F.

4.16. ARTIN'S THEOREM ([**2**, Theorem 3.1]). *Let $f\colon X \to Y$ be an affine morphism of schemes of finite type over a field k, and F a torsion sheaf (i.e., sheaf of torsion groups) on X. Then $\mathrm{d}(R^q f_* \mathsf{F}) \leq \mathrm{d}(\mathsf{F}) - q$ for all $q \geq 0$.*

We will apply Artin's Theorem in the following form:

4.17. PROPOSITION. *Suppose that X is a scheme, U an affine open subset of X, and $Z := X \setminus U$ the complement. Then for any torsion sheaf F on X,*

$$\mathrm{d}\big(\mathcal{H}^q_{\text{ét},Z}(\mathsf{F})\big) \leq \mathrm{d}(\mathsf{F}) - q + 1.$$

Here the $\mathcal{H}^q_{\text{ét},Z}(\mathsf{F})$ are the local cohomology sheaves in the étale topology. The usual excision and spectral sequences for local cohomology remain true in the étale setting, see [**13**, Section 6].

PROOF. If $i\colon U \hookrightarrow X$ is the inclusion, then for any sheaf F of abelian groups on X we have the exact sequence [**13**, Proposition 6.5]

$$0 \to \mathcal{H}^0_{\text{ét},Z}(\mathsf{F}) \to \mathsf{F} \to i_*(\mathsf{F}|_U) \to \mathcal{H}^1_{\text{ét},Z}(\mathsf{F}) \to 0,$$

as well as isomorphisms

$$\mathcal{H}^q_{\text{ét},Z}(\mathsf{F}) \cong R^{q-1} i_*(\mathsf{F}|_U) \quad \text{for all } q \geq 2.$$

If $q \geq 2$ the proposition then follows from the above isomorphism and Artin's Theorem 4.16 applied to the inclusion morphism i, which is affine since U is.

If $q = 1$ it is enough to bound $\mathrm{d}(i_*(\mathsf{F}|_U))$, since $\mathcal{H}^1_{\text{ét},Z}(\mathsf{F})$ is a quotient of $i_*(\mathsf{F}|_U)$. The points $x \in X$ where $(i_*(\mathsf{F}|_U))_{\overline{x}} \neq 0$ are the points $x \in U$ with $\mathsf{F}_{\overline{x}} \neq 0$ and points $x \in Z$ such that there exists a point $x' \in U$, $x \in \overline{\{x'\}}$ with $\mathsf{F}_{\overline{x'}} \neq 0$. In particular, the support of $i_*(\mathsf{F}|_U)$ is contained in the support of F, so $\mathrm{d}(\mathcal{H}^1_{\text{ét},Z}(\mathsf{F})) \leq \mathrm{d}(\mathsf{F})$, which is exactly the statement of the proposition when $q = 1$.

If $q = 0$ note that $\mathrm{d}(\mathcal{H}^0_{\text{ét},Z}(\mathsf{F})) \leq \mathrm{d}(\mathsf{F})$ since $\mathcal{H}^0_{\text{ét},Z}(\mathsf{F})$ is a subsheaf of F, while the proposition only claims the weaker bound $\mathrm{d}(\mathcal{H}^0_{\text{ét},Z}(\mathsf{F})) \leq \mathrm{d}(\mathsf{F}) + 1$. □

4.18. LEMMA. *If F is a torsion sheaf on X, then $H^n_{\text{ét}}(X, \mathsf{F}) = 0$ for all $n >$* $\mathrm{d}(\mathsf{F}) + \operatorname{asn} X$.

PROOF. We show the result by induction on $\operatorname{asn} X$, the case $\operatorname{asn} X = 0$ being Artin's Theorem 4.16 again. Let $\{Z_0, \dots, Z_{\operatorname{asn} X}\}$ be an affine cell decomposition of X (as given by Theorem 3.1). Set $Z := X \setminus Z_0 = \bigcup_{k \geq 1} Z_k$.

We have $H^n_{\text{ét}}(Z_0, \mathsf{F}|_{Z_0}) = 0$ for all $n > \mathrm{d}(\mathsf{F}|_{Z_0})$ by Artin's Theorem, and since $\mathrm{d}(\mathsf{F}|_{Z_0}) \leq \mathrm{d}(\mathsf{F})$ the excision sequence ((2.2) holds in this context, [**13**, (6.5.3)]) shows that $H^n_{\text{ét}}(X, \mathsf{F})$ is a quotient of $H^n_{\text{ét},Z}(\mathsf{F})$ for all $n > \mathrm{d}(\mathsf{F})$. It is therefore enough to show that $H^n_{\text{ét},Z}(\mathsf{F}) = 0$ for $n > \mathrm{d}(\mathsf{F}) + \operatorname{asn} X$.

We can compute $H^n_{\text{ét},Z}(\mathsf{F})$ by a spectral sequence with E_2^{pq} term $H^p_{\text{ét}}(X, \mathcal{H}^q_{\text{ét},Z}(\mathsf{F}))$ ([**13**, Proposition 6.4]). We have

$$H^p_{\text{ét}}(X, \mathcal{H}^q_{\text{ét},Z}(\mathsf{F})) = H^p_{\text{ét}}(Z, \mathcal{H}^q_{\text{ét},Z}(\mathsf{F}))$$

since $\mathcal{H}^q_{\text{ét},Z}(\mathsf{F})$ is supported on Z. By Proposition 4.17 we have $\mathrm{d}(\mathcal{H}^q_{\text{ét},Z}(\mathsf{F})) \leq \mathrm{d}(\mathsf{F}) - q + 1$. Since $\operatorname{asn} Z < \operatorname{asn} X$ we can apply the inductive hypothesis to conclude that $H^p_{\text{ét}}(Z, \mathcal{H}^q_{\text{ét},Z}(\mathsf{F})) = 0$ for $p > \mathrm{d}(\mathsf{F}) - q + 1 + \operatorname{asn} Z$, or $p + q > \mathrm{d}(\mathsf{F}) + \operatorname{asn} Z + 1$. Again using $\operatorname{asn} Z < \operatorname{asn} X$, this gives $H^n_{\text{ét},Z}(\mathsf{F}) = 0$ for $n > \mathrm{d}(\mathsf{F}) + \operatorname{asn} X$. □

4.19. COROLLARY. (a) *If F is a torsion sheaf, then $H^n_{\text{ét}}(X, \mathsf{F}) = 0$ for all* $n > \dim X + \operatorname{asn} X$.

(b) *If F is a constructible sheaf, then $H^n_{\text{ét}}(X, \mathsf{F}) = 0$ for all $n >$* $\dim X + \operatorname{asn} X$.

(c) *If $\mathcal{F}_\ell$ is an ℓ-adic sheaf on X, then $H^n_{\text{ét}}(X, \mathcal{F}_\ell) = 0$ for all $n >$* $\dim X + \operatorname{asn} X$.

PROOF. (a) Clearly we have $\mathrm{d}(\mathsf{F}) \leq \dim X$. (b) A constructible sheaf is a special case of a torsion sheaf (compare [**1**, Proposition 1.2(ii)] with [**1**, Definition 2.3]). (c) follows from (a). □

4.20. Relation to homotopy type. We expect that the affine stratification number bounds the homotopy type as follows.

4.21. CONJECTURE. *If the base field is $\mathbb{C}$, then X has the homotopy type of a finite complex of dimension at most* $\operatorname{asn} X + \dim X$.

5. Applications to moduli spaces of curves

One motivation for the definition of affine stratification number is the study of the moduli space of curves, and certain geometrically important open subsets. We will use Definition 1.1 (which we may, by Proposition 3.7).

5.1. Preliminary aside: the affine stratification number of Deligne–Mumford stacks. As we have only defined the affine stratification number of schemes, throughout this section, we will work with coarse moduli space of curves. One should presumably work instead with a more general definition for Deligne–Mumford stacks. One possible definition is to replace the notion of "affine" in the definition of affine stratification number with that of a Deligne-Mumford stack that has a surjective finite flat cover by an affine scheme (see Proposition 2.10).

5.2. Recall the following question of Looijenga's.

5.3. CONJECTURE (Looijenga). *(a)* $\operatorname{acn} M_g \leq g-2$ *for* $g \geq 2$. *(b) More generally,* $\operatorname{acn} M_{g,n} \leq g-1-\delta_{n,0}$ *whenever* $g>0$, $(g,n) \neq (1,0)$.

The case $n=1$ of (b) implies the cases $n>1$, as the morphism $M_{g,n+1} \to M_{g,n}$ is affine for $n \geq 1$.

This suggests the following, weaker conjecture, which is straightforward to verify for small (g,n) (using Proposition 2.10 judiciously). We are currently pursuing a program to prove this (work in progress).

5.4. CONJECTURE (Looijenga [**6**, Problem 6.5, p. 112]). $\operatorname{asn} M_g \leq g-2$ *for* $g \geq 2$.

From this statement (and properties of asn), we obtain a number of consequences.

5.5. PROPOSITION (Looijenga [**6**, p. 112]). *Conjecture* 5.4 *implies that*

$$\operatorname{asn} M_{g,n} \leq g-1-\delta_{n,0}$$

whenever $g>0$, $(g,n) \neq (1,0)$.

PROOF. As $M_{g,n+1} \to M_{g,n}$ is affine for $n \geq 1$, it suffices to prove the result for $M_{0,3}$ and $M_{g,1}$ with $g>0$. The cases $g=0$ and $g=1$ are immediate. For $g>1$, let D be a multisection of $M_{g,1} \to M_g$ (e.g., a suitable Weierstrass divisor). Then the morphisms $D \to M_g$ and $(M_{g,1} \setminus D) \to M_g$ are affine and surjective, so pulling back the affine stratification of M_g to $M_{g,1}$ and intersecting with $(M_{g,1} \setminus D) \sqcup D$ yields the desired affine stratification of $M_{g,1}$. ☐

Examination of small genus cases suggests the following refinement of Conjecture 5.4.

5.6. CONJECTURE. *There is an affine stratification of* $M_{g',n'}$ *preserved by the symmetric group acting on the* n' *points. The induced decomposition of* $\overline{M}_{g,n}$ *is a stratification.*

This leads to a bound on the affine stratification number of the open subset $\overline{M}^{\leq k}_{g,n}$, corresponding to stable n-pointed genus g curves with at most k genus 0 components, defined in [**4**, Section 4].

5.7. PROPOSITION. *Conjecture* 5.6 *implies that* $\operatorname{asn} \overline{M}^{\leq k}_{g,n} \leq g-1+k$ *for all* $g>0$, $n>0$.

This is more evidence of the relevance of this strange filtration of the moduli space of curves. In particular, compare this to Theorem $\star$ of [**4**], that the tautological ring of $\overline{M}^{\leq k}_{g,n}$ vanishes in codimension greater than $g-1+k$. (In [**4**, p. 3], Looijenga asks precisely this question, with asn replaced by acn.)

PROOF. We show that dimension of any stratum of $\overline{M}_{g,n}$ appearing in $\overline{M}_{g,n}^{\leq k}$ is at least

$$3g-3+n-(g-1+k)=2g-2+n-k.$$

This is true for strata in $M_{g,n}$ by Proposition 5.5. Consider any other boundary stratum, say with j rational components ($j \leq k$) with $m_1, \dots, m_j$ special points respectively; and s other components, with genus $g_1, \dots, g_s$ and $n_1, \dots, n_s$ special points respectively. By Proposition 2.10, it suffices to pass to the finite étale cover that is isomorphic to

$$\prod_{i=1}^{j} M_{0,m_i} \times \prod_{i=1}^{s} M_{g_i,n_i}.$$

By Proposition 5.5 (and using M_{0,m_i} affine), we can decompose this space into affine sets of dimension at least

$$\sum_{i=1}^{j}(m_i-3)+\sum_{i=1}^{s}\big(3g_i-3+n_i-(g_i-1)\big).$$

Now $\sum_{i=1}^{j}(m_i-2)+\sum_{i=1}^{s}(2g_i-2+n_i)=2g-2+n$, so each affine set has dimension at least

$$2g-2+n-j$$

and thus codimension in $\overline{M}_{g,n}^{\geq k}$ at most $g-1+j \leq g-1+k$ as desired. □

This leads to bounds on other spaces of interest. Let $M_{g,n}^{ct}$ be the open subset of $\overline{M}_{g,n}$ corresponding to curves of compact type (i.e., with compact Jacobian, or equivalently with dual graph containing no loops). Let $M_{g,n}^{rt}$ be the open subset corresponding to curves with rational tails (i.e., with a component a smooth genus g curve, or equivalently with dual graph with a genus g vertex).

5.8. COROLLARY. *Conjecture* 5.6 *implies that* asn $M_{g,n}^{ct} \leq 2g-3+n$ *and* asn $M_{g,n}^{rt} \leq g+n-2$ *for* $g>0$, $n>0$.

PROOF. $M_{g,n}^{ct}$ is obtained by removing boundary strata from $\overline{M}_{g,n}^{\leq g+n-2}$. $M_{g,n}^{rt}$ is obtained by removing boundary strata from $\overline{M}_{g,n}^{\leq n-1}$. □

5.9. COROLLARY. *Conjectures* 4.21 *and* 5.6 *imply that* $\overline{M}_{g,n}^{\leq k}$ *(resp.* $M_{g,n}^{ct}$, $M_{g,n}^{rt}$*) has the homotopy type of a finite complex of dimension at most* $4g-4+n+k$ *(resp.* $5g-6+2n$, $4g-5+2n$*).*

References

1. M. Artin, *Faisceaux constructible. Cohomologie d'un courbe algébrique*, Théorie des topos et cohomologie étale des schémas. Tome 3, Lecture Notes in Math., vol. 305, Springer-Verlag, Berlin–New York, 1973, pp. 1–42.
2. ______, *Théorème de finitude pour un morphisme propre; dimension cohomologique des schémas algébriques affines*, Théorie des topos et cohomologie étale des schémas. Tome 3, Lecture Notes in Math., vol. 305, Springer-Verlag, Berlin–New York, 1973, pp. 145–167.
3. S. Diaz, *Complete subvarieties of the moduli space of smooth curves*,Algebraic Geometry, Bowdoin 1985 (Brunswick, ME, 1985), Proc. Sympos. Pure Math., vol. 46, Part 1, Amer. Math. Soc., Providence, RI, 1987, pp. 77–81.
4. T. Graber and R. Vakil, *Relative virtual localization and vanishing of tautological classes on moduli spaces of curves*, (submitted for publication); math.AG/0309227.

5. A. Grothendieck, *Sur quelques points d'algèbre homologique*, Tôhoku Math. J. (2) **9** (1957), 119–221.
6. R. Hain and E. Looijenga, *Mapping class groups and moduli spaces of curves*, Algebraic Geometry—Santa Cruz 1995, Proc. Sympos. Pure Math., vol. 62, Part 2, Amer. Math. Soc., Providence, RI, 1997, pp. 97–142.
7. R. Hartshorne, *Local cohomology*, Lecture Notes in Math., vol. 41, Springer-Verlag, Berlin–New York, 1967 (notes on a seminar by A. Grothendieck).
8. ———, *Algebraic geometry*, Grad. Texts in Math., vol. 52, Springer-Verlag, New York–Heidelberg, 1977.
9. G. Horrocks, *Birationally ruled surfaces without embeddings in regular schemes*, Bull. London Math. Soc. **3** (1971), 57–60.
10. S. L. Kleiman, *On the vanishing of* $H^n(X, \mathcal{F})$ *for an n-dimensional variety*, Proc. Amer. Math. Soc. **18** (1967), 940–944.
11. D. Mumford, *Rational equivalences of* 0*-cycles on surfaces*, J. Math. Kyoto Univ. **9** (1968), 195–204.
12. M. Nori, *Varieties with no smooth embeddings*, C. P. Ramanujam—A Tribute, Tata Inst. Fund. Res. Studies in Math., vol. 8, Springer, Berlin–New York, 1978, pp. 241–246.
13. J.-L. Verdier, *Cohomologie dans les topos* Théorie des topos et cohomologie étale des schémas. Tome 2., Lecture Notes in Math., vol. 270, Springer-Verlag, Berlin–New York, 1972, pp. 1–82.

Dept. of Mathematics and Statistics, Queens University, Kingston, ON K7L 3N6, Canada
E-mail address: mikeroth@mast.queensu.ca

Department of Mathematics, Stanford University, Stanford, CA 94305-2125, USA
E-mail address: vakil@math.stanford.edu

Centre de Recherches Mathématiques
CRM Proceedings and Lecture Notes
Volume **38**, 2004

Coherent Sheaves on Generic Compact Tori

Misha Verbitsky

ABSTRACT. Let T be a compact complex torus, $\dim T > 2$. We show that the category of coherent sheaves on T is independent of the choice of the complex structure, if this complex structure is generic. The proof is independent from [**19**], where the same result was proven for K3 surfaces and even-dimensional tori.

1. Introduction

Let M be a compact Kähler manifold. The category $\operatorname{Coh}(M)$ of coherent sheaves on M is an important invariant M which has deep physical meaning. For manifolds with ample or anti-ample canonical class, M can be reconstructed from the corresponding derived category $D\operatorname{Coh}(M)$ [**3, 14**]. In physics, the objects of $D\operatorname{Coh}(M)$ are interpreted as certain branes on M. Under Mirror Symmetry, this category corresponds to a triangulated category which is constructed by Fukaya from the symplectic geometry of Mirror dual manifold $\widehat{M}$.

Physicists believe that a Mirror dual of n-dimensional compact complex torus is again an n-dimensional compact complex torus. This is one of the first cases where Mirror Symmetry was known explicitly.

In [**19**], we studied the category $\operatorname{Coh}(M)$ for M a K3 surface or even-dimensional compact torus. We proved that $\operatorname{Coh}(M)$ is independent from M, provided that M is generic in its deformation class. The proof of this statement involves hyper-Kähler geometry.

Given a hyper-Kähler structure on M, we consider the corresponding twistor space $\operatorname{Tw}(M)$. To every object $F \in \operatorname{Coh}(M)$, we associate a coherent sheaf $\widehat{F}$ on $\operatorname{Tw}(M)$, in such a way that the restriction of $\widehat{F}$ to M, identified with a fiber of the twistor projection $\pi\colon \operatorname{Tw}(M) \to \mathbb{C}P^1$, gives F. Restricting $\tilde{F}$ to another fiber close to M, we obtain a coherent sheaf on a manifold which is deformationally equivalent to M. We are free in the choice of hyper-Kähler structure, and this allows us to obtain any given deformation M' of M after several iterations of this procedure.

2000 *Mathematics Subject Classification.* 32J25, 32J81, 14C30, 32G08, 32S40, 32Q15.

The author is EPSRC advanced fellow supported by CRDF grant RM1-2354-MO02 and EPSRC grant GR/R77773/01.

The author is grateful to F. Bogomolov for valuable advice and D. Kaledin for an important insight in the geometry of $\mathrm{SO}(2n)/\mathrm{U}(n)$.

This is the final form of the paper.

This construction gives a functor from $\operatorname{Coh}(M)$ to $\operatorname{Coh}(M')$. This functor is an equivalence if M' is also generic.

This proof is not very satisfactory, for several reasons. The equivalence $\operatorname{Coh}(M) \cong \operatorname{Coh}(M')$ is not canonical, because it involves many intermediate choices (such as the choice of a hyper-Kähler structure, and intermediate hyper-Kähler structures leading from M to M' as we indicated). This dependence is very difficult to estimate. Also, several important steps of the proof do not involve hyper-Kähler geometry at all, giving an impression that hyper-Kähler geometry is mostly irrelevant to the big picture. Finally, the isomorphism $\operatorname{Coh}(M) \cong \operatorname{Coh}(M')$ in [**19**] is proven only for even-dimensional tori, because odd-dimensional tori are not hyper-Kähler. By avoiding the hyper-Kähler geometry entirely, we are able to produce an isomorphism $\operatorname{Coh}(M) \cong \operatorname{Coh}(M')$ for generic tori of arbitrary dimension > 2.

We use the following notion of generic.

DEFINITION 1.1. Let T be a compact complex torus, $\dim T \geqslant 3$. We say that T is generic, if T has no nontrivial subtori, $H^{1,1}(T) \cap H^2(T, \mathbb{Z}) = 0$, and $H^{2,2}(T) \cap H^4(T, \mathbb{Z}) = 0$. As Proposition 2.1 indicates, the set of nongeneric tori is a set of measure zero.

In this paper, we prove the following theorem.

THEOREM 1.2. *Let T, T' be compact complex tori of dimension $n \geqslant 3$. Assume that T, T' are generic. Then the corresponding categories of coherent sheaves are equivalent*: $\operatorname{Coh}(T) \cong \operatorname{Coh}(T')$.

We prove Theorem 1.2 in Section 8.

Notice that in this paper all tori are assumed to have $\dim T \geqslant 3$. We have dealt with the case of 2-dimensional tori in [**19**].

The proof of Theorem 1.2 is based on a concept of "$\operatorname{SO}(2n)/\operatorname{U}(n)$-twistor space" for a torus. Given a compact torus T, with flat Kähler metric g, consider the space S of all complex structures I such that g is Kähler on (T, I). Clearly, S is isomorphic to a Kähler symmetric space $\operatorname{SO}(2n)/\operatorname{SU}(n)$ (so-called isotropic Grassmanian space). The natural twistor fibration

$$\operatorname{Tw}(T) \to S$$

is holomorphic (Proposition 4.2), and in many ways analogous to the twistor fibration known from the hyper-Kähler geometry. This analogy can be used to adapt the proof of equivalence $\operatorname{Coh}(M) \cong \operatorname{Coh}(M')$ from [**19**] to the present needs.

This paper is independent from [**19**], and can be used as introductory reading on the subject, as the proofs are significantly simplified.

We also prove the following results about generic tori.

THEOREM 1.3. *Let T be a generic torus.*[1] *Then*

(i) *All proper complex subvarieties of T have dimension 0.*
(ii) *Any holomorphic vector bundle on T admits a flat connection compatible with the holomorphic structure.*
(iii) *For any coherent sheaf F on T, the reflexization*

$$F^{**} := \operatorname{Hom}(\operatorname{Hom}(F, \mathcal{O}_T), \mathcal{O}_T)$$

is a vector bundle.

[1]Please notice that everywhere in this paper we assume that $\dim T \geqslant 3$.

PROOF. Theorem 1.3(i) is Proposition 3.1, Theorem 1.3(ii) is Corollary 6.6, and Theorem 1.3(iii) is Proposition 3.3. □

We shall use the following bit of notation, which is due to P. Deligne. Consider a complex vector space V. We consider V as a real vector space $V_{\mathbb{R}}$ with an action of $\mathbb{C}^* = \mathrm{U}(1) \times \mathbb{R}^{>0}$. Consider some tensor power W of $V_{\mathbb{R}}$ (for instance, the space $\Lambda^i(V_{\mathbb{R}})$ of real exterior differential forms), and let $W_{\mathbb{C}}$ be its complexification. The group $\mathbb{C}^*$ acts on W by multiplicativity:

$$\rho_W : \mathbb{C}^* \times W \to W. \tag{1.1}$$

We say that a vector $v \in W_{\mathbb{C}}$ has Hodge type (p,q) if $\rho(r,v) = r^{p+q}v$, and $\rho(\theta, v) = \theta^{p-q}v$, where $r \in \mathbb{R}^{>0}$ and $\theta \in \mathrm{U}(1)$ are standard generators in $\mathrm{U}(1) \times \mathbb{R}^{>0} = \mathbb{C}^*$.

This definition is compatible with the standard notion of Hodge decomposition.

2. Generic tori

In this section, we prove that Definition 1.1 is consistent, that is, the set of nongeneric tori has measure zero in the corresponding moduli space.

Let R be the space of group homomorphisms $\mathbb{Z}^{2n} \overset{\varphi}{\hookrightarrow} \mathbb{C}^n$ such that $\mathbb{C}^n/\varphi(\mathbb{Z}^{2n})$ is compact, up to $\mathrm{GL}(n,\mathbb{C})$-action. The space R admits a natural open embedding to $(\mathbb{C}^n)^{2n}/\mathrm{GL}(n,\mathbb{C})$,

$$R \to (\mathbb{C}^n)^{2n}, \quad \varphi \mapsto \big(\varphi(\alpha_1), \varphi(\alpha_2), \dots, \varphi(\alpha_{2n})\big), \tag{2.1}$$

where α_i are generators of $\mathbb{Z}^{2n}$. Clearly, R is identified with the moduli space of pairs $(T, \alpha_1, \alpha_2, \dots \alpha_{2n})$, where T is a torus, and $\alpha_1, \alpha_2, \dots, \alpha_{2n}$ a basis in $H^1(T,\mathbb{Z})$. The space R is called *the marked moduli space of complex tori.* It is a covering over the usual moduli of complex tori, with monodromy $\mathrm{GL}(2n,\mathbb{Z})$.

PROPOSITION 2.1. *Let $R_{\mathrm{sp}} \subset R$ be the set of all tori which are not generic, in the sense of Definition 1.1. Then R_{sp} is a countable union of closed complex subvarieties of positive codimension in R.*

PROOF. Given a proper sublattice $L \subset \mathbb{Z}^{2n}$, $\mathrm{rk}\, L = 2l$, let $Z_L \subset R$ be the set of all homomorphisms $\mathbb{Z}^{2n} \overset{\varphi}{\hookrightarrow} \mathbb{C}^n$ such that $\varphi(L)$ lies inside a complex space of rank l. For any $\alpha \in \Lambda^2_{\mathbb{Z}}(\mathbb{Z}^{2n})$, or in $\Lambda^4_{\mathbb{Z}}(\mathbb{Z}^{2n})$, denote by Z_α the set of all homomorphisms $\mathbb{Z}^{2n} \overset{\varphi}{\hookrightarrow} \mathbb{C}^n$ such that the element corresponding to α in the cohomology of the corresponding torus T is of Hodge type $(1,1)$ or $(2,2)$. Clearly, $R_{\mathrm{sp}} = \bigcup Z_L \bigcup Z_\alpha$, where the union is taken over all L and α. Clearly, Z_L, Z_α are closed complex subvarieties of R. To prove Proposition 2.1, it remains to show that $\mathrm{codim}\, Z_L > 0$, $\mathrm{codim}\, Z_\alpha > 0$, for all L and α. Since R is irreducible, we have $\mathrm{codim}\, Z_L > 0$, $\mathrm{codim}\, Z_\alpha > 0$ unless $Z_L, Z_\alpha = R$ for some L or α. Given a proper lattice $L \subset \mathbb{Z}^{2n}$, $\mathrm{rk}\, L = 2l$, it is easy to find a fomomorphism $\mathbb{Z}^{2n} \overset{\varphi}{\hookrightarrow} \mathbb{C}^n$ such that $\mathbb{C}\cdot\varphi(\alpha)$ generates a subspace of dimension $\min(2l, n)$. Therefore, $Z_L \neq R$, and Z_L has positive codimension. To prove that Z_α has positive codimension, we need to find a torus T with $\alpha \notin H^{p,p}(T)$ for a given α. We use the following (more general) lemma.

LEMMA 2.2. *For any n, there exists an n-dimensional compact complex torus T which satisfies*

$$H^{p,p}(T) \cap H^{2p}(T,\mathbb{Z}) = 0 \quad \textit{for all } 0 < p < n. \tag{2.2}$$

PROOF. We need to show that for each $v \in H^{2p}(T,\mathbb{R})$, there exists a complex structure I on T for which $v \notin H^{2p}(T_I,\mathbb{Z})$. Suppose that for some v this is not true. Let $\varphi \in R$ be a homomorphism $\mathbb{Z}^{2n} \overset{\varphi}{\hookrightarrow} \mathbb{C}^n = V$, and ρ the U(1)-action on $\Lambda^{2p}_{\mathbb{R}} V$ corresponding to the Hodge decomposition as in (1.1). Consider the group G generated by such U(1)-actions. It is easy to see that $G = \mathrm{SL}(V,\mathbb{R})$. Then, for v as above, v is $\mathrm{SL}(V,\mathbb{R})$-invariant. As follows from classical invariants theory, any nontrivial $\mathrm{SL}(V,\mathbb{R})$-invariant form on V is proportional to the volume form. This proves Lemma 2.2. □

REMARK 2.3. As the proof of Lemma 2.2 indicates, the set of tori satisfying (2.2) is also a complement of a countable union of positive-codimension subvarieties. We use the weaker version of generic (Definition 1.1) because it is sufficient for our purposes.

3. Subvarieties and vector bundles on a generic torus

3.1. Subvarieties of generic torus.

PROPOSITION 3.1. *Let T be a generic compact torus, and $Z \subset T$ a closed complex subvariety. Then* $\dim Z = 0$ *or* $Z = T$.

PROOF. Replacing Z with its irreducible component, we may assume that Z is irreducible. Assume that Z has smallest dimension among all proper subvarieties of T. Then the singular set $\operatorname{Sing} Z$ is 0-dimensional. Since T is generic, it has no Hodge cycles of dimension 1 and 2. Therefore $\dim_{\mathbb{C}} Z > 2$

The sheaf $\Omega^1(T)$ is globally generated; therefore, the same is true for $\Omega^1(Z)$. Pick k holomorphic 1-forms $\tau_1, \dots, \tau_k$, such that $\gamma := \tau_1 \wedge \cdots \wedge \tau_k$ is a nonzero section of the canonical class of Z. Unless the sheaf $\Omega^1(Z)$ is trivial outside of $\operatorname{Sing} Z$, γ has nontrivial zero divizor. This contradicts our assumption that Z has minimal dimension among all subvarieties. Therefore, $\tau_1, \dots, \tau_k$ are linearly independent everywhere. These sections trivialize $\Omega^1(Z)$ outside of $\operatorname{Sing} Z$.

Let $K := \ker\big(i^* : \Omega^1(T) \to \Omega^1(Z)\big)$, where i^* denotes the pullback map. Since $\Omega^1(Z)$ is trivialized, we have $\dim \Omega^1 Z = k$, and $\dim K = n - k$. Consider $K \subset \Omega^1 T$ as a distribution. Then K is integrable, and Z is its integral variety. Passing to a covering of T, we find that Z is determined as a common zero set of a system of linear equations. Therefore, Z is flat, and hence it is a subtorus. This is impossible, because T is generic. □

3.2. Stable bundles on generic tori.

We follow [**19**] (see also [**20**]).

PROPOSITION 3.2. *Let T be a generic compact Kähler torus, and F a stable reflexive*[2] *sheaf on T. Then F is a line bundle.*

PROOF. In [**2**], Bando and Siu construct a canonical Hermitian Yang–Mills connection ∇ on the nonsingular part of any stable reflexive sheaf on a compact Kähler manifold X, with L^2-integrable curvature Θ. Such a connection (called admissible Yang–Mills) is unique. In [**2**] it is also proven that the L^1-integrable form $\mathrm{Tr}(\Theta \wedge \Theta)$ (considered as a current on the Kähler manifold X) is closed and represents the cohomology class

$$\mathrm{Tr}[\Theta \wedge \Theta] = c_2(F) - \frac{r-1}{r} c_1^2(F) \tag{3.1}$$

[2] A sheaf is called *reflexive* is a natural map $F \to \mathrm{Hom}(\mathrm{Hom}(F, \mathcal{O}_T), \mathcal{O}_T)$ is isomorphism.

Consider the standard Hodge operator $\Lambda\colon \Lambda^{p,q}(T) \to \Lambda^{p-1,q-1}(T)$ on differential forms. Since ∇ is Yang–Mills, we have

$$\Lambda^2(\Theta^2) \geqslant 0, \tag{3.2}$$

and the equality is reached only if $\Theta = 0$.[3] Comparing (3.1) and (3.2), and using $c_1(F)$, $c_2(F) = 0$, we obtain that $\Theta = 0$, that is, ∇ is flat. Since F is reflexive, it is non-singular in codimension 2 (see [**13**]). Therefore, ∇ has no local monodromy around singularities of F. This implies that (F, ∇) can be extended to a flat holomorphic bundle on T. Any reflexive sheaf is normal, that is, equal to a pushforward from any open set $T \backslash Z$, with $\operatorname{codim} Z > 1$, (see [**13**]). Therefore, F is smooth.

We have proved that F is a smooth holomorphic bundle admitting a flat unitary connection. Denote by χ the corresponding representation of $\pi_1(T)$. Since F is stable, F cannot be split onto a direct sum of subsheaves. Therefore χ is irreducible. Since $\pi_1(T)$ is abelian, its irreducible representation is necessarily 1-dimensional. We have proved that F is a line bundle. □

3.3. Reflexive sheaves on generic tori. We follow [**19**].

PROPOSITION 3.3. *Let T be a generic compact Kähler torus, and E a reflexive sheaf on T. Then E is a bundle.*

PROOF. Since T has no nontrivial integer $(1,1)$-cycles, the first Chern class of all sheaves on T vanishes. Therefore, all coherent sheaves on T are semistable. Consider the Jordan-Holder filtration

$$0 = E_0 \subset E_1 \subset \cdots \subset E_n = E$$

with all the subfactors E_i/E_{i-1} stable. The reflexizations $(E_i/E_{i-1})^{**}$ are all smooth, by Proposition 3.2. Replacing E_i by its reflexization $E_i^{**} \subset E$, we may assume that all E_i are reflexive. Using induction, we may assume also that E_{n-1} is smooth.

We have an exact sequence

$$0 \to E_{n-1} \to E \to E_n/E_{n-1} \to 0$$

with E_{n-1} smooth, E reflexive, and $F = E_n/E_{n-1}$ a stable sheaf having (as we have shown above) a smooth reflexization.

Then E is given by a class $\nu \in \operatorname{Ext}^1(F, E_{n-1})$. Consider the exact sequence

$$0 \to F \to F^{**} \to C \to 0$$

where C is a torsion sheaf (cokernel of the reflexization map). This gives a long exact sequence

$$\operatorname{Ext}^1(F^{**}, E_{n-1}) \to \operatorname{Ext}^1(F, E_{n-1}) \xrightarrow{\delta} \operatorname{Ext}^2(C, E_{n-1}). \tag{3.3}$$

The kernel of δ in (3.3) corresponds to all extensions $\gamma \in \operatorname{Ext}^1(F, E_{n-1})$ with a reflexization isomorphic to an extension of F^{**} with E_{n-1}. Clearly, such extensions are reflexive only if $C = 0$. To prove that $C = 0$, it suffices to show that $\delta(\nu) = 0$. However, C is a torsion sheaf, and by Proposition 3.1 its support $\operatorname{Supp}(C)$ is a finite set. By Grothendieck's duality [**7**, Theorem 6.9], the group $\operatorname{Ext}^2(C, E_{n-1})$ vanishes, for $\operatorname{codim} \operatorname{Sup} C > 2$. Therefore, E is smooth. This proves Proposition 3.3. □

---

[3] This is the celebrated Lübke inequality [**11**], which implies flatness of stable bundles with zero Chern classes and the Bogomolov–Miyaoka–Yau inequality; see [**2**].

4. SO(2n)/ U(n)-twistor space

Let (V, g) be a Euclidean vector space, $\dim_{\mathbb{R}} V = 2n$, and S the set of all Hermitian structures on V compatible with g. Clearly, $S \cong \mathrm{SO}(2n)/\mathrm{SU}(n)$. The space S is a Kähler symmetric space, which can be seen from the following argument.

Consider the space $V_{\mathbb{C}} := V \otimes \mathbb{C}$, and let S' be the Grassmanian space of all complex subspaces $V_1 \subset V_{\mathbb{C}}$ such that $\dim_{\mathbb{C}} V_1 = n$, and $g_{\mathbb{C}}\big|_{V_1} = 0$, where $g_{\mathbb{C}}$ is the complexification of g. Since g is positive definite, $V_1 \cap V = 0$. Therefore, the natural projection $V_{\mathbb{C}} \to V$, $v \to \mathrm{Re}(v)$ induces an isomorphism $V_1 \cong V$. The space $V_{\mathbb{C}}$ is equipped with a Hermitian structure as a complexification of Euclidean space. This gives a Hermitian structure on V. It is easy to check that this Hermitian structure is compatible with the Euclidean structure on V. Conversely, any such structure defines a subspace $V^{1,0} \subset V_{\mathbb{C}}$ of vectors of Hodge type $(1,0)$, which is obviously isotropic. We obtain a bijection $S' \cong S$. However, S' is by construction a Kähler symmetric space.

DEFINITION 4.1. We call $S = \mathrm{SO}(2n)/\mathrm{U}(n)$ *the isotropic Grassman manifold.*

Given a torus $T := V/\mathbb{Z}^{2n}$, we may identify S with the space of complex structures on T compatible with the metric g. Consider the product $T \times S$, equipped with the complex structure as follows.

Let $(t, s) \in T \times S$ be a point. As we have explained, s defines a complex structure I_s on T. Let $\mathcal{I}\colon T_tT \times T_sS \to T_tT \times T_sS$ act as I_s on T_tT and as I_S on T_sS, where I_S is the standard complex structure on S. Clearly, $\mathcal{I}$ is an almost complex structure on $T \times S$.

PROPOSITION 4.2. *In the above assumptions, $\mathcal{I}$ is integrable.*

PROOF. Let B^- be the universal bundle on

$$S = \{V^{1,0} \subset V_{\mathbb{C}} \mid g\big|_{V_1} = 0\},$$

with the fiber of B^- at a point $V_1 \subset V_{\mathbb{C}}$ identified with $V^{1,0}$. By construction, $B^- \subset V_{\mathbb{C}} \otimes \mathcal{O}_S$ is holomorphic. We have an exact sequence

$$0 \to B^- \to V_{\mathbb{C}} \otimes \mathcal{O}_S \xrightarrow{\kappa} B^+ \to 0. \tag{4.1}$$

Consider the lattice $L \subset V$, $T = V/L$, and let $\alpha_1, \ldots, \alpha_{2n}$ be its generators. Consider the corresponding sections $\alpha_1 \otimes 1, \ldots, \alpha_{2n} \otimes 1$ of $V_{\mathbb{C}} \otimes \mathcal{O}_S$. Since $B^-\big|_s$ is an isotropic subspace of $V_{\mathbb{C}} = V_{\mathbb{C}} \otimes \mathcal{O}_S\big|_s$, for all $s \in S$, no nontrivial linear combination of $\alpha_i \otimes 1$ lies in B^-. Therefore, the sections $\kappa(\alpha_i \otimes 1) \in B^+$ are linearly independent over $\mathbb{R}$, and the quotient $B^+/\langle\kappa(\alpha_i \otimes 1)\rangle$ is a compact torus, at every point $s \in S$. The total space $\mathrm{Tw}(T)$ of this quotient is a holomorphic fibration over S, and its total space is naturally identified with $(T \times S, \mathcal{I})$. This proves Proposition 4.2. □

DEFINITION 4.3. Let (T, g) be a compact complex torus equipped with a flat metric. The complex manifold

$$\mathrm{Tw}(T) := (T \times S, \mathcal{I})$$

is called *the* $\mathrm{SO}(2n)/\mathrm{U}(n)$*-twistor space of* (T, g), or simply twistor space of T. Clearly, $\mathrm{Tw}(T)$ is equipped with a holomorphic projection $\pi\colon \mathrm{Tw}(T) \to S$, and its fibers are identified with (T, I_s), where $I_s, s \in S$ are complex structure operators on T compatible with g.

5. Twistor transform for $\mathrm{SO}(2n)/\mathrm{U}(n)$-twistor space

5.1. Twistor sections for $\mathrm{Tw}(T)$.

DEFINITION 5.1. Let T be a compact torus, and

$$\mathrm{Tw}(T) \xrightarrow{\pi} S$$

its $\mathrm{SO}(2n)/\mathrm{U}(n)$-twistor space. A twistor section of $\mathrm{Tw}(T)$ is a holomorphic map $S \overset{i}{\hookrightarrow} \mathrm{Tw}(T)$ such that the composition

$$S \overset{i}{\hookrightarrow} \mathrm{Tw}(T) \xrightarrow{\pi} S$$

is identity. We denote the space of twistor sections by $\mathrm{Sec}(T)$. For any point $t \in T$, the map $I \longrightarrow (I, t)$ gives a holomorphic section of π. Such sections are called *horizontal twistor sections.*

PROPOSITION 5.2. *In the above assumptions, let $\tau\colon S \longrightarrow \mathrm{Tw}(T)$ be a twistor section, and I, $J \in S$ points in*

$$S = \left\{V^{0,1} \subset V_{\mathbb{C}} \mid g\big|_{V_1} = 0\right\},$$

such that the corresponding $V^{0,1}$-spaces do not intersect:

$$V_I^{0,1} \cap V_J^{0,1} = 0. \tag{5.1}$$

Then, in a neighbourhood U of $[\tau] \in \mathrm{Sec}(T)$, every section τ' is uniquely determined by $\tau'(I)$, $\tau'(J)$. This defines an open embedding $U \hookrightarrow (T, I) \times (T, J)$.

PROOF. By construction, $\mathrm{Tw}(T)$ is a quotient of the total space $\mathrm{Tot}\, B^+$ by a lattice. Since S is simply connected, every section $\tau \in \mathrm{Sec}(T)$ is lifted to a holomorphic section of the projection

$$\mathrm{Tot}\, B^+ \xrightarrow{\tilde{\pi}} S.$$

Therefore, to prove Proposition 5.2 it suffices to show that a section of B^+ is uniquely determined by its values at I and J. This is implied by the following

LEMMA 5.3. *Let S be an isotropic Grassmanian (Definition 4.1), B^+ be the universal bundle (see (4.1)), and I, $J \in S$ points satisfying $V_I^{0,1} \cap V_J^{0,1} = 0$. Consider the restriction maps*

$$H^0(B^+) \xrightarrow{r_I} B^+\big|_I, \qquad\qquad H^0(B^+) \xrightarrow{r_J} B^+\big|_J.$$

Then

$$H^0(B^+) \xrightarrow{r_I \oplus r_J} B^+\big|_I \oplus B^+\big|_J. \tag{5.2}$$

is an isomorphism of vector spaces. In other words, a section of B^+ is uniquely determined by its values at I, J.

PROOF. Consider the exact sequence

$$0 \longrightarrow B^- \longrightarrow V_{\mathbb{C}} \otimes \mathcal{O}_S \xrightarrow{\kappa} B^+ \longrightarrow 0$$

(4.1). As Lemma 5.4 (below) implies, $H^0(B^-) = H^1(B^-) = 0$. From the corresponding long exact sequence, we obtain that the induced map

$$H^0(V_{\mathbb{C}} \otimes \mathcal{O}_S) \xrightarrow{\kappa} H^0(B^+) \tag{5.3}$$

is an isomorphism. Therefore, $H^0(B^+)$ is naturally identified with $V_{\mathbb{C}}$, and $\dim H^0(B^+) = 2\operatorname{rk} B^+$. We obtain that the spaces on the left and the right-hand side of (5.2) have the same dimension.

To prove that (5.2) is an isomorphism of vector spaces, it suffices to show that $\ker r_I \cap \ker r_J = 0$. This is equivalent to $V_I^{0,1} \cap V_J^{0,1} = 0$, because under the isomorphism (5.3) $\ker r_I$ corresponds to $V_I^{0,1}$ and $\ker r_J$ corresponds to $V_J^{0,1}$. We reduced Lemma 5.3 and Proposition 5.2 to the following algebro-geometric lemma.

LEMMA 5.4. *Let S be the isotropic Grassmanian (Definition 4.1) and*

$$0 \to B^- \to V_{\mathbb{C}} \otimes \mathcal{O}_S \xrightarrow{\kappa} B^+ \to 0 \tag{5.4}$$

the exact sequence defined above. Then $H^0(B^-) = H^1(B^-) = 0$.

PROOF. We use the Kodaira–Nakano vanishing theorem. Since B^- is a holomorphic sub-bundle of a trivial bundle, it is negative: $\Theta_{B^-} \leqslant 0$, where $\Theta_{B^-} \in \Lambda^{1,1}(S) \otimes \operatorname{End}(B^-)$ is the curvature of B^- [**6**]. The form Θ_{B^-} can be computed explicitly as follows:

$$\Theta_{B^-} = A \wedge A^\perp,$$

where

$$A \in \Lambda^{1,0}(S) \otimes \operatorname{Hom}(B^-, B^+) \tag{5.5}$$

is the second form of the exact sequence (5.4), and

$$A^\perp \in \Lambda^{0,1}(S) \otimes \operatorname{Hom}(B^+, B^-)$$

its Hermitian conjugate (see [**6**] for details). The second form A can be computed explicitly, as follows.

The tangent bundle $T(S)$ is naturally identified with $\operatorname{Hom}(B^-, B^+)$, because S is a Grassmanian, and a tangent vector to a Grassmanian of planes $V_0 \subset V_1$ is identified with $\operatorname{Hom}(V_0, V_1/V_0)$. Under the identification

$$\Lambda^{1,0}(S) \cong \operatorname{Hom}(B^+, B^-), \tag{5.6}$$

A becomes an identity operator.

$$\begin{aligned} A \in \Lambda^{1,0}(S) \otimes \operatorname{Hom}(B^-, B^+) &= \operatorname{Hom}(B^+, B^-) \otimes \operatorname{Hom}(B^-, B^+) \\ &= \operatorname{End}\bigl(\operatorname{Hom}(B^-, B^+)\bigr). \end{aligned}$$

This implies that $\Theta_{B^-} = A \wedge A^\perp$ is strictly negative. Now, by Kodaira–Nakano theorem, $H^i(B^-) = 0$ for all $i < \dim S$. This proves Lemma 5.4. We finished the proof of Proposition 5.2. □

REMARK 5.5. Consider an anticomplex involution $\iota\colon S \to S$, $I \to -I$. Then $\iota \times \operatorname{Id}$ is an anticomplex involution of $\operatorname{Tw}(T)$. This involution maps twistor sections to twistor sections. Abusing notation, we denote the corresponding involution of $\operatorname{Sec}(T)$ by ι.

There is a natural real analytic embedding $T \overset{\gamma}{\hookrightarrow} \operatorname{Sec}(T)$, $t \to \{t\} \times S$ mapping t to the corresponding holomorphic section. Clearly, $\gamma(T)$ is a fixed point set of the anticomplex involution $\iota\colon \operatorname{Sec}(T) \to \operatorname{Sec}(T)$. By Proposition 5.2, $\operatorname{Sec}(T)$ is locally in a neighbourhood of $\gamma(T)$ identified with a neighbourhood of diagonal in $(T, I) \times (T, -I)$. We shall think of $\operatorname{Sec}(T)$ as of complexification of its totally real submanifold $T = \gamma(T)$.

5.2. Twistor transform for Tw(T). Let (B, ∇) be a flat bundle on T, and $(\sigma^* B, \sigma^* \nabla)$ its pullback to $\mathrm{Tw}(T)$. Clearly, $(\sigma^* B, \sigma^* \nabla)$ is flat, and therefore holomorphic. This defines a functor from the category of flat bundles on T to the category of holomorphic bundles on $\mathrm{Tw}(T)$:

$$\mathrm{Fl}(T) \to \mathrm{HolBun}(\mathrm{Tw}(T)). \tag{5.7}$$

We call this functor *the direct twistor transform.* It turns out that (5.7) is invertible.

The following theorem is adapted from [**8**], where a similar result was proven for a twistor space of a hyper-Kähler manifold.

THEOREM 5.6. *Let T be a torus, $\mathrm{Tw}(T) \xrightarrow{\pi} S$ its $\mathrm{SO}(2n)/\mathrm{U}(n)$-twistor space, and B a holomorphic vector bundle on $\mathrm{Tw}(T)$. Assume that*

$$B\big|_{S\times\{t\}\subset S\times T=\mathrm{Tw}(T)} \text{ is trivial for all } t \in T, \tag{5.8}$$

that is, B is trivial on all horizontal sections of π. Then B is obtained as a twistor transform of a flat bundle $(B_{\mathrm{fl}}, \nabla)$ on T. Moreover, $(B_{\mathrm{fl}}, \nabla)$ is unique, and determines an equivalence

$$\mathrm{HolBun}_0\big(\mathrm{Tw}(T)\big) \to \mathrm{Fl}(T)$$

of the category $\mathrm{HolBun}_0\big(\mathrm{Tw}(T)\big)$ of holomorphic bundles on $\mathrm{Tw}(T)$ satisfying (5.8) and the category of flat bundles on T.

We prove Theorem 5.6 in Section 5.3.

5.3. Real analytic differential operators and holomorphic vector bundles on twistor spaces. We work in assumptions of Theorem 5.6. Let $B \in \mathrm{HolBun}_0\big(\mathrm{Tw}(T)\big)$ be a holomorphic vector bundle which is trivial on all horizontal sections of twistor projection. Denote by $U_B \subset \mathrm{Sec}(T)$ the space of all twistor sections $S \overset{\varphi}{\hookrightarrow} \mathrm{Tw}(T)$ for which $B\big|_{\varphi(S)}$ is trivial.

LEMMA 5.7. *In the above assumptions, U_B is open in $\mathrm{Sec}(T)$.*

PROOF. We need to show that a small deformation of a trivial bundle on S is again trivial. It is well known that deformations of vector bundles are classified by $\mathrm{Ext}^1(B,B) = H^1\big(\mathrm{End}(B)\big)$ [**9**]. If B is a trivial bundle, $\mathrm{End}(B)$ is also trivial. Therefore, Lemma 5.7 is implied by

$$H^1(\mathcal{O}_S) = 0. \tag{5.9}$$

From (5.6) and ampleness of B^+, we obtain that TS is ample (this is also clear because S is symmetric). Therefore, the canonical class K of S is negative. By Serre's duality,

$$H^1(\mathcal{O}_S) = H^{\dim S - 1}(K)^*.$$

By Kodaira–Nakano theorem, $H^i(K) = 0$ for all $i < \dim S$, because K is negative. This proves Lemma 5.7. □

Consider the evaluation map $S \times U_B \xrightarrow{\mathrm{ev}} \mathrm{Tw}(T)$, and let $\mathrm{ev}^* B$ be the pullback of B. Then $\mathrm{ev}^* B$ is trivial on all $S \times \{s\} \subset S \times U_B$, and therefore, the pushforward $B_{\mathbb{C}} = \pi_*(\mathrm{ev}^* B)$ is a holomorphic vector bundle on U_B.

One should think of U_B as of complexification $T_{\mathbb{C}}$ of T (Remark 5.5). Then $B_{\mathbb{C}}$ becomes a complexification of a real analytic bundle on T underlying B. The holomorphic structure operator on B can be interpreted as a holomorphic differential operator on $B_{\mathbb{C}}$, as follows.

Fix $I \in S$, and let $T_I \subset TU_B$ be a tangent hyperplane field to a holomorphic foliation $\mathcal{L}$ on $\operatorname{Sec}(T)$ with leaves

$$L_t := \{\varphi\colon S \to \operatorname{Tw}(T) \mid \varphi(I) = t\}$$

parametrized by $t \in T$. We can write $L_t = ev_I^{-1}(t)$. Clearly, $B_{\mathbb{C}} \cong \operatorname{ev}_I^*(B\big|_{(T,I)})$. Therefore, $B_{\mathbb{C}}$ is trivialized along the leaves of $\mathcal{L}$. This defines a trivial holomorphic conection along the leaves of $\mathcal{L}$:

$$\nabla_I\colon B_{\mathbb{C}} \to B_{\mathbb{C}} \otimes T_I^*. \tag{5.10}$$

Restricting (5.10) to $\gamma(T) \subset \operatorname{Sec}(T)$ (see Remark 5.5), we obtain that T_I^* becomes $\Lambda^{0,1}(T)$, and (5.10) becomes the holomorphic structure operator

$$\bar\partial\colon B \to B \otimes \Lambda^{0,1}(T).$$

Writing this operator as in (5.10) allows us to study the dependence of $\bar\partial$ from I in a more explicit way.

Since U_B is an open neighbourhood of its totally real submanifold $\gamma(T)$, it contains a Stein neighbourhood of $\gamma(T)$ as Grauert's theorem implies. Replacing U_B by a smaller neighbourhood if necessary, we may assume that U_B is Stein. Then $B_{\mathbb{C}}$, $B_{\mathbb{C}} \otimes T_I^*$ are globally generated, and the map (5.10) can be interpreted as a map of $\mathcal{O}_{U_B}$-modules:

$$\nabla_I\colon H^0(B_{\mathbb{C}}) \to H^0(B_{\mathbb{C}} \otimes T_I^*).$$

Clearly, (5.10) depends holomorphically from the parameter $I \in S$. Therefore, (5.10) can be interpreted as a map

$$\nabla_I\colon H^0(B_{\mathbb{C}}) \otimes_{\mathbb{C}} \mathcal{O}_S \to H^0(B_{\mathbb{C}}) \otimes_{\mathbb{C}} B^+. \tag{5.11}$$

(we use the identification of $T_I^* = \Lambda^{0,1}(T)$ with B^+ obtained earlier).

Let Γ be a vector space, possibly infinite-dimensional. Consider the exact sequence

$$0 \to B^- \otimes \Gamma \to V_{\mathbb{C}} \otimes \mathcal{O}_S \otimes \Gamma \xrightarrow{\kappa} B^+ \otimes \Gamma \to 0 \tag{5.12}$$

(this is (4.1) tensored by Γ). Since $H^0(B^-) = H^1(B^-) = 0$, the long exact sequence associated with (5.12) gives

$$\operatorname{Hom}(\Gamma_1 \otimes \mathcal{O}_S, \Gamma_2 \otimes B^+) \cong \operatorname{Hom}(\Gamma_1 \otimes \mathcal{O}_S, \Gamma_2 \otimes V_{\mathbb{C}} \otimes \mathcal{O}_S),$$

for any vector spaces Γ_1, Γ_2. Therefore, the map (5.11) lifts, uniquely, to a holomorphic map

$$\nabla_S\colon H^0(B_{\mathbb{C}}) \otimes \mathcal{O}_S \to H^0(B_{\mathbb{C}}) \otimes V_{\mathbb{C}} \otimes \mathcal{O}_S = H^0(B_{\mathbb{C}} \otimes_{\mathcal{O}_{U_B}} \Omega^1(U_B)) \tag{5.13}$$

It is easy to check that ∇_S is in fact a holomorphic connection on $B_{\mathbb{C}}$. Restricting ∇_S to $\gamma(T) \subset U_B$, we obtain a real analytic connection operator ∇ on B. By construction, the $(0,1)$-part of ∇ is the holomorphic structure operator on B.

We are going to show that ∇ is flat. Indeed, the $(0,2)$-part of the curvature Θ of ∇ vanishes for all $I \in S$, because $\bar\partial = 0$. If we replace I by $-I$, the $(2,0)$-forms become $(0,2)$-forms and vice versa. Therefore, the $(2,0)$-part of the curvature Θ also vanishes. We obtain that Θ is of type $(1,1)$ with respect to all $I \in S$. In other words, Θ is invariant under the natural $\mathrm{U}(1)$-action induced by the complex structure. It is easy to see that such $\mathrm{U}(1)$ generate $\mathrm{SO}(2n)$. Therefore, Θ is an $\mathrm{SO}(2n)$-invariant 2-form on $2n$-dimensional space. Using standard invariant theory, we infer that $\Theta = 0$. Therefore, ∇ is flat. From the construction it is clear that ∇ is unique.

We obtained a functor from $\operatorname{HolBun}_0(T)$ to the category of flat bundles, which is obviously inverse to the twistor transform. This proves Theorem 5.6. □

6. Flat connections on semistable bundles

6.1. Deformation theory and flat bundles.

DEFINITION 6.1. Let T be a compact complex torus. Given a holomorphic bundle B on T, we say that B is *polystable* if B is a direct sum of line bundles.

DEFINITION 6.2. Let T be a compact complex torus with flat Kähler metric, and $\operatorname{Tw}(T)$ the corresponding twistor space. Fix a point $I \in S$. Denote by Fl_I the category of flat bundles B on $\operatorname{Tw}(I)$ such that $B\big|_{(T,I)}$ is polystable, and every irreducible subquotient of B admits a Hermitian metric.

We are interested in the category Fl_I because of the following theorem.

THEOREM 6.3. *Let T be a compact torus, $\operatorname{Tw}(T) \xrightarrow{\pi} S$ its twistor space, and I, J points which satisfy $V_I^{0,1} \cap V_J^{0,1} = 0$ (5.1). Consider the restriction functor from Fl_I to the category $\operatorname{Bun}(T, J)$ of holomorphic bundles on $(T, J) \subset \operatorname{Tw}(T)$:*

$$\operatorname{Fl}_I \xrightarrow{\varphi} \operatorname{Bun}(T, J). \tag{6.1}$$

Then (6.1) is equivalence of categories.

We prove Theorem 6.3 in Section 6.2. In the present section, we prove a weaker form of Theorem 6.3. A similar result is proven in [**19**].

PROPOSITION 6.4. *In assumptions of Theorem 6.3, let B be a holomorphic bundle on (T, J). Then $B = \varphi(\widetilde{B})$, for some flat bundle $\widetilde{B} \in \operatorname{Fl}_I$. Moreover, φ induces a bijection on the set of isomorphism classes of the objects of corresponding categories.*

PROOF. We use the following lemma

LEMMA 6.5. *Let T be a compact complex torus, $B_1, \dots, B_n$ flat holomorphic Hermitian vector bundles and B a holomorphic vector bundle with a filtration*

$$0 = E_0 \subset E_1 \subset \cdots \subset E_n = B,$$

such that $E_i/E_{i-1} \cong B_i$. Then the holomorphic structure on B can be obtained as follows. Identify B with $B_{\mathrm{gr}} := \bigoplus B_i$ as C^∞-bundle. Then there is a cohomology class ν

$$\nu \in \bigoplus_{i>j} \operatorname{Ext}^1(B_i, B_j) \subset \operatorname{Ext}^1(B_{\mathrm{gr}}, B_{\mathrm{gr}}) \tag{6.2}$$

such that the holomorphic structure operator in B is written as

$$\bar\partial = \bar\partial_{\mathrm{gr}} + \nu_0, \tag{6.3}$$

where $\bar\partial_{\mathrm{gr}}$ is the holomorphic structure operator on B_{gr}, and

$$\nu_0 \in \Lambda^{0,1}(\operatorname{End}(B_{\mathrm{gr}}))$$

denotes the harmonic (and hence, flat) representative of ν.

PROOF. Write the holomorphic structure operator on B as $\bar{\partial} = \bar{\partial}_{\mathrm{gr}} + \widetilde{\nu}$, where $\widetilde{\nu}$ is a (0,1)-form with values in

$$\bigoplus_{i>j} \Lambda^{0,1}\big(T, \mathrm{Hom}(B_i, B_j)\big).$$

The form $\widetilde{\nu}$ satisfies the Maurer–Cartan equation

$$\bar{\partial}^2 = \bar{\partial}_{\mathrm{gr}}(\widetilde{\nu}) + 2\widetilde{\nu} \wedge \widetilde{\nu} = 0 \tag{6.4}$$

Every automorphism $g \in \mathrm{End}\, B_{\mathrm{gr}}$ acts on $\widetilde{\nu}$ as $\widetilde{\nu} \to g(\widetilde{\nu}) + \bar{\partial}_{\mathrm{gr}}(g)$ (this is the well-known gauge action). To produce ν with the properties described in Lemma 6.5, we need to find a correct gauge transform.

Consider the group $(\mathbb{C}^*)^n$ acting on B_{gr} by diagonal automorphisms, in such a way that the ith component α_i of $(\mathbb{C}^*)^n$ acts trivially on $B_j \subset B_{\mathrm{gr}}$ for $i \neq j$, and as a multiplication by α_i on B_i.

We shall write the action of $(\mathbb{C}^*)^n$ on $\Lambda^{0,1}\big(\mathrm{End}(B_{\mathrm{gr}})\big)$ as follows. Let

$$\widetilde{\nu} \in \Lambda^{0,1}(\mathrm{End}(B_{\mathrm{gr}}) = \bigoplus_{i,j} \Lambda^{0,1}(B_i, B_j), \qquad \widetilde{\nu} := \sum_{i,j} \widetilde{\nu}_{ij}, \quad \widetilde{\nu}_{ij} \in \Lambda^{0,1}(B_i, B_j)$$

If $\alpha \in (\mathbb{C}^*)^n$, $\alpha = \prod_i \alpha_i$, then

$$\alpha(\widetilde{\nu}) = \sum_{i,j} \alpha_i \alpha_j^{-1} \widetilde{\nu}_{ij}$$

The group $(\mathbb{C}^*)^n$ acts in this fashion on the solutions of Maurer-Cartan equation, and maps every solution to an equivalent one. If $\alpha_j \gg \alpha_i$ for all $i > j$, then α maps

$$\widetilde{\nu} \in \bigoplus_{i>j} \Lambda^{0,1}\big(T, \mathrm{Hom}(B_i, B_j)\big)$$

to a form which is arbitrarily small.

Consider the local deformation space $\mathrm{Def}(B_{\mathrm{gr}})$ for B_{gr}, constructed in [**16**]. The above argument implies that every neighbourhood of the point $[B_{\mathrm{gr}}] \in \mathrm{Def}(B_{\mathrm{gr}})$ contains a bundle which is isomorphic to B.

Let E be a holomorphic vector bundle over a compact Kähler manifold. The local deformation space $\mathrm{Def}(E)$ can be constructed explicitly in terms of Massey products as follows.

One can define the Massey products as obstructions to constructing a solution of the Maurer–Cartan equation (see, e.g., [**1**], or [**12,15**] for a more classical approach). Locally, $\mathrm{Def}(E)$ is embedded to the vector space $\mathrm{Ext}^1(E, E)$, and the image of this embedding is a germ of all vectors $\theta \in \mathrm{Ext}^1(E, E)$, such that $\theta \wedge \theta = 0$ and all the higher Massey products of θ with itself vanish.

Fix such a vector $\theta \in \mathrm{Ext}^1(E, E)$. We construct the corresponding vector bundle $E_\theta \in \mathrm{Def}(E)$ using the Hodge theory as follows (see, e.g., [**18**]).

Let $\theta_0 \in \mathcal{H}^{0,1}(\mathrm{Hom}(E, E))$ be the harmonic representative of θ. Using induction, we define

$$\theta_n := -\frac{1}{2} G_{\bar{\partial}} \sum_{i+j=n-1} \theta_i \wedge \theta_j$$

where $G_{\bar{\partial}}$ is the Green operator inverting the holomorphic structure operator

$$\bar{\partial} \colon \Lambda^{0,k} \otimes E \to \Lambda^{0,k+1} \otimes E.$$

on its image. The Green operator $G_{\bar\partial}$ is compact. This can be used to show that for θ sufficiently small, the series $\tilde\theta := \sum \theta_i$ converges. The vanishing of Massey products is equivalent to the following condition

$$\bar\partial\theta_n = -\frac{1}{2}\sum_{i+j=n-1}\theta_i\wedge\theta_j,$$

which is apparent from the definition given in [**1**]. In this case, we have

$$\bar\partial\tilde\theta = -\tfrac{1}{2}\tilde\theta\wedge\tilde\theta$$

and $\tilde\theta$ is a solution of the Maurer-Cartan equation (6.4). Therefore, the operator $\bar\partial_\theta = \bar\partial + \tilde\theta$ satisfies $\bar\partial_\theta^2 = 0$, and by Newlander–Nirenberg theorem [**9**, Proposition 4.17, Chapter 1] this operator defines a holomorphic structure on E. On the deformation space $\operatorname{Def}(E) \subset \operatorname{Ext}^1(E,E)$, the point θ corresponds to a bundle $(E, \bar\partial_\theta)$.

Now we return to holomorphic bundles over a compact torus and the proof of Lemma 6.5. We obtain that B is given by some $\nu \in \operatorname{Ext}^1(B_{\mathrm{gr}}, B_{\mathrm{gr}})$. Since B and $\bigoplus B_i$ have compatible filtrations, by functoriality we may assume that

$$\nu \in \bigoplus_{i>j}\operatorname{Ext}^1(B_i, B_j). \tag{6.5}$$

The higher Massey operations in $\operatorname{Ext}^*(B_{\mathrm{gr}}, B_{\mathrm{gr}})$ vanish, because the bundle B_{gr} is flat (the same proof works as was used in [**4**]; see also [**5**]). Therefore, B can be reconstructed from ν for any ν such that the cohomology class $\nu\wedge\nu$ vanishes. Pick a harmonic representative ν_0 of ν. Since B_{gr} is flat, ν_0 is parallel. Therefore $\nu_0\wedge\nu_0$ is also parallel, hence harmonic. We obtain that the cohomology class of $\nu_0\wedge\nu_0$ vanishes if and only if this form vanishes identically.

Starting from a bundle B with a filtration which satisfies the assumptions of Lemma 6.5, we have constructed a cohomology class $\nu \in \bigoplus_{i>j}\operatorname{Ext}^1(B_i, B_j)$, with $\nu\wedge\nu = 0$. The harmonic representative ν_0 of ν satisfies $\nu_0\wedge\nu_0 = 0$. Therefore, ν_0 is a solution of Maurer–Cartan equation, and $\bar\partial_{\mathrm{gr}} + \nu_0$ is equivalent to the holomorphic structure operator of B. This proves Lemma 6.5. □

Let ∇_{gr} be the standard Hermitian flat connection on B_{gr}. The bundle

$$(B_{\mathrm{gr}}, \nabla_{\mathrm{gr}} + \nu_0)$$

is also flat, because ν_0 is by construction parallel. We obtained the following corollary.

Corollary 6.6. *Let B be a holomorphic vector bundle on a compact generic torus T, $\dim_{\mathbb C} T > 2$. Then B admits a flat connection compatible with the holomorphic structure.*

Return to the proof of Proposition 6.4. We use the notation of Lemma 6.5. Let B be a holomorphic bundle on (T, J), and B_{gr} the associated graded bundle of Jordan–Holder filtration. Then B_{gr} is polystable, and hence admits a flat Hermitian connection (Proposition 3.2). Consider the bundle $\widetilde{B}_{\mathrm{gr}}$ on $\operatorname{Tw}(T)$, obtained from the flat bundle B_{gr} via twistor transform, and let $R^1\pi_*(\operatorname{End}(\widetilde{B}_{\mathrm{gr}}))$ be the corresponding direct image sheaf on S. Clearly, $R^1\pi_*(\operatorname{End}(\widetilde{B}_{\mathrm{gr}}))$ is isomorphic to a sum of several copies of the vertical tangent bundle $T_{vert}\operatorname{Tw}(T) \cong B^+$. Given a holomorphic section $\tilde\gamma \in R^1\pi_*(\operatorname{End}(\widetilde{B}_{\mathrm{gr}}))$, $\gamma^2 = 0$, we take the fiberwise harmonic (hence, flat) representative $\tilde{\gamma}_0$ of $\tilde{\gamma}$. The bundle $(\widetilde{B}_{\mathrm{gr}}, \bar{\partial}_{\mathrm{gr}} + \tilde{\gamma}_0)$ is holomorphic. Consider the 1-form $\nu_0 \in H^1((T,J), \operatorname{End}(\widetilde{B}_{\mathrm{gr}}))$ associated with B as in Lemma 6.5.

Let $\widetilde{\nu}_0$ be a section of $R^1\pi_*(\operatorname{End}(\widetilde{B}_{\mathrm{gr}}))$ which vanishes in I and is equal to ν_0 at J. Such a section exists by Lemma 5.3 (we surmise that $R^1\pi_*(\operatorname{End}(\widetilde{B}_{\mathrm{gr}}))$ is isomorphic to a sum of several copies of B^+). Clearly, $\widetilde{B} := (\widetilde{B}_{\mathrm{gr}}, \bar{\partial}_{\mathrm{gr}} + \widetilde{\nu}_0)$ is a holomorphic bundle with $\widetilde{B}\big|_{(T,I)} = B_{\mathrm{gr}}$ and $\widetilde{B}\big|_{(T,J)} = B$. Therefore, $\widetilde{B}$ belongs to Fl_I, and $\varphi(\widetilde{B}) = B$. This bundle is uniquely determined by B, because $\widetilde{\nu}_0$ is uniquely determined by the conditions $\widetilde{\nu}_0\big|_I = 0$, $\widetilde{\nu}_0\big|_J = \nu$ (Lemma 6.5). This proves Proposition 6.4. □

6.2. Abelian categories of finite length.

DEFINITION 6.7. Let $\mathcal{C}$ be an abelian category. An object $B \in \mathcal{C}$ is called *simple* if it has no proper sub-objects: for all $B' \subset B$, either $B' = B$ or $B' = 0$. An object B is called *semisimple* if B is a direct sum of simple objects.

An abelian category $\mathcal{C}$ is called *of finite length* if every object $B \in \mathcal{C}$ is a finite extension of simple objects. Equivalently, $\mathcal{C}$ is of finite length if any increasing or decreasing chain of sub-objects of B stabilizes, for all $B \in \mathcal{C}$.[4]

A *length* of $B \in \mathcal{C}$ is a minimal length of such a chain.

LEMMA 6.8. *Let $\mathcal{C} \xrightarrow{\gamma} \mathcal{C}'$ be a functor of abelian categories of finite length. Assume that γ induces equivalence on the respective subcategories of semisimple objects. Assume, moreover, that γ induces a bijection on the sets of equivalence classes of the objects of $\mathcal{C}$, $\mathcal{C}'$. Then γ is an equivalence.*

PROOF. The proof is obtained via the trivial diagram-chasing argument. We use an induction by the length of B. Let $B, B' \in \mathcal{C}$ be objects of length $\leqslant n$. Consider an exact sequence

$$0 \to B_0 \to B \to B_1 \to 0$$

with $l(B_1) = n-1$ and B_0 semisimple. There is a long exact sequence

$$0 \to \operatorname{Hom}(B', B_0) \to \operatorname{Hom}(B', B) \to \operatorname{Hom}(B', B_1) \to \operatorname{Ext}^1(B', B_0) \to \cdots$$

Applying γ, we obtain a commutative diagram with exact rows

$$\begin{array}{ccccccccc} 0 & \longrightarrow & \operatorname{Hom}(B', B_0) & \longrightarrow & \operatorname{Hom}(B', B) & \longrightarrow & \operatorname{Hom}(B', B_1) & \longrightarrow & \operatorname{Ext}^1(B', B_0) \\ & & \downarrow{\scriptstyle\gamma} & & \downarrow{\scriptstyle\gamma} & & \downarrow{\scriptstyle\gamma} & & \downarrow{\scriptstyle\gamma} \\ 0 & \longrightarrow & \operatorname{Hom}(\gamma(B'), \gamma(B_0)) & \longrightarrow & \operatorname{Hom}(\gamma(B'), \gamma(B)) & \longrightarrow & \operatorname{Hom}(\gamma(B'), \gamma(B_1)) & \longrightarrow & \operatorname{Ext}^1(\gamma(B'), \gamma(B_0)). \end{array} \tag{6.6}$$

The maps of Ext^1-groups are isomorphisms because Ext^1 classify the classes of extensions, and we know that γ induces bijection on isomorphism classes of objects. Using induction by length of B, B', we may assume that all vertical arrows of ((6.6)) are isomorphisms, except, possibly, the map

$$\operatorname{Hom}(B', B) \xrightarrow{\gamma} \operatorname{Hom}(\gamma(B'), \gamma(B)).$$

Using the snake lemma, we obtain that this map is also an isomorphism. Therefore, γ is an equivalence. □

[4]One also says $\mathcal{C}$ *satisfies the ascending and descending chain condition.*

We apply Lemma 6.8 to prove Theorem 6.3. Simple objects in $\mathrm{Bun}(T, J)$ are line bundles, as Proposition 3.2 implies. Indeed, simple objects must be stable, and stable bundles are line bundles. Clearly, line bunles on T admit flat Hermitian connections.

As a definition of Fl_I implies, simple objects here are line bundles obtained from flat Hermitian line bundles (T, J). This implies that

$$\varphi\colon \mathrm{Fl}_I \to \mathrm{Bun}(T, J)$$

is an equivalence on the subcategory of semisimple objects. By Proposition 6.4, φ induces bijection on equivalence classes of objects. Now, Lemma 6.8 implies that it is an equivalence. This proves Theorem 6.3. □

7. Singularities of coherent sheaves on a generic torus

7.1. Local geometry on $\mathrm{Tw}(T)$. Let T be a compact torus, $\dim T > 1$. Fix a point $x \in T$, and the correspoinding point in twistor space $(I, x) \in \mathrm{Tw}(T)$. We shall study the twistor sections passing through $(I, x) \in \mathrm{Tw}(T)$ in a neighbourhood of the horisontal section $s_x = S \times \{x\}$.

Let L, $L' \in S$ be complex structures satisfying $V_L^{0,1} \cap V_I^{0,1} = V_{L'}^{0,1} \cap V_I^{0,1} = 0$ (see (5.1) for details). Consider a point (t, L') close to (x, L'). By Proposition 5.2, there exists a unique twistor section $s\colon S \to \mathrm{Tw}(T)$ passing through (t, L') and (x, I) and close to $s_x = \{x\} \times S$. This section is unique in a small neighbourhood of $\{x\} \times S$. We define $\tilde{\Psi}_I(t) := s(L) \in (T, L)$. This map is holomorphic and invertible in a sufficiently small neighbourhood of (x, L'). We think of this map as an isomorphism

$$\Psi_I(L, L')\colon \mathcal{O}_{x,L} \to \mathcal{O}_{x,L'} \tag{7.1}$$

of the corresponding local rings. Clearly, $\Psi_I(L, L')$ holomorphically depends on I, L, L'.

Let now B be a bundle on $\mathrm{Tw}(T)$ obtained from the twistor transform (Theorem 5.6). This means that the restriction of B to any horizontal twistor section s_x is trivial. As we have shown, a small deformation of a trivial bundle on S is again trivial. Therefore, the restriction of B to a twistor section close to s_x is trivial as well.

This allows one to extend the map $\Psi_I(L, L')\colon \mathcal{O}_{x,L} \to \mathcal{O}_{x,L'}$ to the space of germs of holomorphic sections of B. We obtain an isomorphism

$$\Psi_I(L, L', B)\colon B_{x,L} \to B_{x,L'}, \tag{7.2}$$

where $B_{x,L}$ and $B_{x,L'}$ are spaces of germs of $B\big|_{(T,L)}$ and $B\big|_{(T,L')}$ in x.

Denote the infinitesimal neighbourhood of $s_x \backslash (I, x)$ in $\mathrm{Tw}(T) \backslash (T, I)$ by

$$\mathrm{Tw}(T)_{x,I} = \mathrm{Spec}(\mathcal{O}_{\mathrm{Tw}(T)\backslash (T,I), s_x}).$$

This space is fibered over $S \backslash I$ with fibers isomorphic to a germ of a smooth complex manifold.

The maps (7.1) produce a canonical trivialization of $\mathrm{Tw}(T)_{x,I}$ over $S \backslash I$. Denote the trivialization map by

$$\varphi\colon \mathrm{Tw}(T)_{x,I} \to \mathrm{Spec}\, \mathcal{O}_0(\mathbb{C}^n) \times (S \backslash I) \tag{7.3}$$

where $\mathrm{Spec}\, \mathcal{O}_0(\mathbb{C}^n)$ is the germ of $\mathbb{C}^n$ in 0.

Let $\xi\colon \operatorname{Tw}(T)_{x,I} \to \operatorname{Spec}\mathcal{O}_0(\mathbb{C}^n)$ be the composition of φ and the projection $\operatorname{Spec}\mathcal{O}_0(\mathbb{C}^n)\times(S\backslash I)\to\operatorname{Spec}\mathcal{O}_0(\mathbb{C}^n)$. Using the maps $\Psi_I(L,L',B)$ of (7.2), we obtain a trivialization of $B\big|_{\operatorname{Tw}(T)_{x_0,I}}$ over $S\backslash I$. More precisely, we obtain a bundle B_S over $\operatorname{Spec}\mathcal{O}_0(\mathbb{C}^n)$, and a natural isomorphism

$$B\big|_{\operatorname{Tw}(T)_{x,I}}\cong \xi^* B_S. \tag{7.4}$$

7.2. The category $\mathcal{C}_I$.

DEFINITION 7.1. Let T be a compact torus, $\operatorname{Tw}(T)$ its $\operatorname{SO}(2n)/\operatorname{U}(n)$-twistor space The category $\mathcal{C}_I$ is defined as follows.

An object of $\mathcal{C}_I$ is a coherent sheaf F on $\operatorname{Tw}(T)$ satisfying the following conditions.

(1) The reflexization $B := F^{**}$ belongs to Fl_I.
(2) The sheaf F is non-singular outside of $A\times S\subset\operatorname{Tw}(T)$, where $A\subset T$ is a finite set.
(3) Let $Z:=\{A\}\times I\subset(T,I)\subset\operatorname{Tw}(T)$ be the finite subset of $(T,I)\subset\operatorname{Tw}(T)$ corresponding to A, and $j\colon \operatorname{Tw}(T)\backslash Z\hookrightarrow\operatorname{Tw}(T)$ the natural embedding. Then the canonical homomorphism $F\to j_*j^*F$ is an isomorphism.
(4) Let $x\in A$ be a point. Consider the trivialization

$$B\big|_{\operatorname{Tw}(T)_{x,I}}\xrightarrow{\Psi}\xi^* B_S \tag{7.5}$$

constructed in (7.4), where $B=F^{**}$ is the reflexization of F. Then there is a sheaf F_S on $\operatorname{Spec}\mathcal{O}_0(\mathbb{C}^n)$, equipped with an isomorphism $F_S^{**}\cong B_S$, and a trivialization

$$F\big|_{\operatorname{Tw}(T)_{x,I}}\xrightarrow{\Psi}\xi^* F_S$$

which is compatible with (7.5).

The morphisms of $\mathcal{C}_I$ are morphisms of coherent sheaves.

THEOREM 7.2. *Let T be a compact torus, $\operatorname{Tw}(T)\xrightarrow{\pi} S$ its twistor space, and I, $L\in S$ complex structures which satisfy $V_L^{0,1}\cap V_I^{0,1}=0$. Consider the category $\mathcal{C}_I$ constructed above, and let $\mathcal{C}_I\xrightarrow{\varphi}\operatorname{Coh}(T,L)$ be the restriction map. Then φ is equivalence of categories.*

PROOF. We construct the inverse functor $\Upsilon\colon\mathcal{C}_I\to\operatorname{Coh}(T,L)$ as follows. Take $F_L\in\operatorname{Coh}(T,L)$. Let B_L be its reflexization, which is smooth by Proposition 3.3, and $B\in\mathcal{C}_I$ the corresponding bundle over the twistor space, which is defined in a canonical way by Theorem 6.3.

Let $x_0\in T$ be a singular point of F_L. Consider the infinitesimal neighbourhood of

$$\{x_0\}\times(S\backslash I)\subset\operatorname{Tw}(T)\backslash(T,I)$$

in $\operatorname{Tw}(T)\backslash(T,I)$, denoted, as in Section 7.1, by $\operatorname{Tw}(T)_{x_0,I}$.

Then B is trivialized over $\operatorname{Tw}(T)_{x_0,I}$, and we may write $B\big|_{\operatorname{Tw}(T)_{x_0,I}}=\xi^*(B_{L,x_0})$, where B_{L,x_0} is the germ of E_L in x_0. Let $F\big|_{\operatorname{Tw}(T)_{x_0,I}}:=\xi^*(F_{L,x_0})$ be the sheaf corresponding to F_L. Since F_L has isolated singularities, the sheaves $F\big|_{\operatorname{Tw}(T)_{x_0,I}}$ and $B\big|_{\operatorname{Tw}(T)_{x_0,I}}$ are canonically isomorphic outside of $\{x_0\}\times(S\backslash I)$. Gluing together $F\big|_{\operatorname{Tw}(T)_{x_0,I}}$ and B, we obtain a sheaf F_0 on $\operatorname{Tw}(T)\backslash(T,I)$. Let A be the singular set of F_L. Then B is equal to F_0 outside of $\{A\}\times S \subset \mathrm{Tw}(T)$, hence F_0 can be extended smoothly from $\mathrm{Tw}(T)\backslash(T,I)$ to a sheaf F_1 on $\mathrm{Tw}(T)\backslash Z$, where $Z = \{A\} \times \{I\}$ is the finite set defined in Definition 7.1. Consider the sheaf $F := j_*(F_1)$, where $j\colon \mathrm{Tw}(T)\backslash Z \hookrightarrow \mathrm{Tw}(T)$ is the natural embedding. The correspondence $F_L \to F$ is clearly functorial.

The sheaf F by construction belongs to $\mathcal{C}_I$, and satisfies $\varphi(F) = F_L$. Therefore, $F_L \to F$ gives an inverse functor of $\mathcal{C}_I \xrightarrow{\varphi} \mathrm{Coh}(T, L)$. We have proved Theorem 7.2. □

The following corollary immediately follows from Theorem 7.2.

COROLLARY 7.3. *Let T be a generic compact torus, equipped with flat Hermitian metric g, and S the set of complex structures compatible with g. Then, for any generic complex structures $I, J \in S$, the categories of coherent sheaves $\mathrm{Coh}(T, I)$ and $\mathrm{Coh}(T, J)$ are isomorphic.*

8. Moduli of compact tori

Let $T = \mathbb{C}^n/\mathbb{Z}^{2n}$ be a compact torus with a flat Riemannian metric g. We have shown that for all generic complex structures I, J compatible with g, the category $\mathrm{Coh}(T, I)$ is equivalent to $\mathrm{Coh}(T, J)$. To prove that $\mathrm{Coh}(T, I)$ is equivalent to $\mathrm{Coh}(T, J)$ for arbitrary generic complex structures, it suffices to prove the following proposition.

PROPOSITION 8.1. *Let T be a compact torus, $T = V/\mathbb{Z}^{2n}$, $V \cong \mathbb{C}^n$, and R its marked moduli space (Section 2). Given $I \in R$, pick a flat Riemannian metric compatible with I, and let $R_g \subset R$ be the set of all complex structures $J \in R$ compatible with g. Take a generic complex structure $I_1 \in R_g$, and repeat the same procedure, obtaining I_2. Then, after at most 6 iterations, we can obtain any generic complex structure on T from I.*

PROOF. We use the following claim. Please note that the flat metrics on T are in one-to-one correspondence with the Euclidean metrics on V.

CLAIM 8.2. *In assumptions of Proposition 8.1, let $\mathcal{G}$ be the set of all Euclidean metrics g_1 on V such that g_1 is compatible with some complex structure $I \in R_g$. Then $g_1 \in \mathcal{G}$ if and only if all eigenvalues of the symmetric matrix $g_1 g^{-1}$ occur in pairs. In other words, $g_1 \in \mathcal{G}$ if and only if there is an orthonormal basis in (V, g) such that g_1 is written in this basis as*

$$\begin{pmatrix} \alpha_1 & & & & & & \\ & \alpha_1 & & & & & \\ & & \alpha_2 & & & & \\ & & & \alpha_2 & & & \\ & & & & \ddots & & \\ & & & & & \alpha_n & \\ & & & & & & \alpha_n \end{pmatrix}. \tag{8.1}$$

PROOF. "If" part. Let $I_1 \in R_g$ be a complex structure such that g_1 is compatible with I_1. Then g, g_1 are Hermitian forms on the complex vector space (V, I_1). A standard linear-algebraic argument implies that the corresponding Hermitian metrics have a common orthogonal basis $z_1, z_2, \ldots, z_n \in (V, I_1)$. After rescaling, we may assume that $z_1, z_2, \ldots, z_n$ is orthonormal with respect to g. Consider the corresponding basis z_1, $I_1(z_1)$, z_1, $I_1(z_1), \ldots,$ in V over $\mathbb{R}$. In this basis, g is written as identity matrix, and g_1 as (8.1). This proves the "if" part of Claim 8.2.

Conversely, assume that $g^{-1}g_1$ has eigenvalues occuring in pairs. Finding a common orthogonal basis, which is orthonormal in g, we find that g_1 is written in this basis as (8.1). Let I_1 act in this basis as

$$\begin{pmatrix} 0 & 1 & & & & & \\ -1 & 0 & & & & & \\ & & 0 & 1 & & & \\ & & -1 & 0 & & & \\ & & & & \ddots & & \\ & & & & & 0 & 1 \\ & & & & & -1 & 0 \end{pmatrix}. \tag{8.2}$$

Then I_1 is a complex structure compatible with g and g_1. Therefore, $g_1 \in \mathcal{G}$. This proves Claim 8.2. □

The following elementary claim is proven in the same way as one proves Proposition 2.1.

CLAIM 8.3. *In assumptions of* 8.2, *let* R_g° *be the set of all generic complex structures* $I_1 \in R_g$. *Then* $R_g \backslash R_g^\circ$ *has measure zero.*

Using Claims 8.3 and 8.2 and Fubini theorem, we obtain

COROLLARY 8.4. *In the above assumptions, let* $\mathcal{G}^\circ$ *be the set of all Euclidean metrics* $g_1 \in \mathcal{G}$ *such that* g_1 *is compatible with* $I_1 \in R_g^\circ$. *Then* $\mathcal{G} \backslash \mathcal{G}^\circ$ *has measure zero.*

Now we can prove the following lemma.

LEMMA 8.5. *In the assumprions of* 8.1, *let* $R_3(I)$ *be the set of all* $I_3 \in R$ *which can be obtained after three iterations of the procedure defined in Proposition* 8.1. *Then* $R \backslash R_3(I)$ *has measure zero.*

PROOF. Let $\mathcal{F} \in \operatorname{End} V$ be the set of all symmetric operators on (V, g) which can be written as UAU^{-1}, where U is orthogonal, and A takes form (8.1). From Claim 8.2 it is clear that $\mathcal{G} = \mathcal{F} \cdot g$. The second iteration of this procedure gives the following trivial claim.

CLAIM 8.6. *In the above notation, let* $\mathcal{G}_1$ *be the set of all Euclidean metrics on* V *which are compatible with some* $I_2 \in \mathbb{R}$, *with* I_2 *being compatible with some* $g_1 \in \mathcal{G}$. *Then* $\mathcal{G}_1 = \mathcal{F} \cdot \mathcal{F} \cdot g$.

Clearly, every Euclidean metric can be obtained this way; even if A_1 and A_2 commute, $U_1 A_1 A_2 U_1^{-1}$ runs through the set of all symmetric matrices. Indeed, any diagonal matrix can be obtained as a product of two diagonal martices with eigenvalues occuring in pairs.

Let $\mathcal{G}_1^\circ$ be the set of all metrics compatible with a generic $I_2 \in R$, after a second iteration of the procedure defined in Proposition 8.1. Then $\mathcal{G}_1 \backslash G_1^\circ$ has measure 0, as one can easily surmise from Corollary 8.4. Then, $\mathcal{G}_1$ is the set of all Euclidean metrics except, possibly, a measure zero set, and $\mathcal{G}_1$ is compatible with all $I \in R$ except, possibly, a measure zero set. This proves Lemma 8.5. □

Return to the proof of Proposition 8.1. Let I, $J \in R$ be generic complex structures, and $R_3(I)$, $R_3(J) \subset R$ the sets associated with I, J as in Lemma 8.5. The sets $R\backslash R_3(I)$, $R\backslash R_3(J)$ have measure zero, as Lemma 8.5 implies. Therefore, $R_3(I)$ and $R_3(J)$ have nonempty intersection. Take a chain

$$I \to I_1 \to I_2 \to I_3 = J_3 \to J_2 \to J_1 \to J$$

of generic complex structures with each successive pair having a Euclidean metric compatible with both. We obtain that J can be obtained from I after 6 iterations of the procedure defined in Proposition 8.1. Proposition 8.1 is proven. □

References

1. I. K. Babenko and I. A. Taimanov, *Massey products in symplectic manifolds*, Mat. Sb. **191** (2000), no. 8, 3–44 (Russian); English transl., Sb. Math. **191** (2000), no. 7-8, 1107–1146.
2. S. Bando and Y.-T. Siu, *Stable sheaves and Einstein–Hermitian metrics*, Geometry and Analysis on Complex Manifolds, Festschrift for Professor S. Kobayashi's 60th Birthday, (T. Mabuchi, J. Noguchi, and T. Ochiai, eds.) World Scientific, Singapore, 1994, pp. 39–50.
3. A. Bondal and D. Orlov, *Semiorthogonal decomposition for algebraic varieties*, alg-geom/9506012
4. P. Deligne, Ph. Griffiths, J. Morgan, and D. Sullivan, *Real homotopy theory of Kähler manifolds*, Invent. Math. **29** (1975), no. 3, 245–274.
5. W. M. Goldman and J. J. Millson, *The deformation theory of representations of fundamental groups of compact Kähler manifolds*, Inst. Hautes Études Sci. Publ. Math. **67** (1988), 43–96.
6. Ph. Griffiths and J. Harris, *Principles of algebraic geometry*, Pure Appl. Math., Wiley-Intersience, New York, 1978.
7. R. Hartshorne, *Local cohomology*, Lecture Notes in Math., vol. 41, Springer-Verlag, Berlin–New York, 1967 (s seminar given by A. Grothendieck, Harvard University, Fall 1961).
8. D. Kaledin and M. Verbitsky, *Non-Hermitian Yang–Mills connections*, Selecta Math. (N.S.) **4** (1998), no. 2, 279–320.
9. S. Kobayashi, *Differential geometry of complex vector bundles*, Publ. Math. Soc. Japan, vol. 15, Princeton Univ. Press, Princeton, NJ; Iwanami Shoten, Tokyo, 1987.
10. M. Kontsevich, *Homological algebra of mirror symmetry*, Proceedings of the International Congress of Mathematicians, Vol. 1, 2 (Zürich, 1994), Birkhäuser, Basel, 1995, pp. 120–139.
11. M. Lübke, *Chernklassen von Hermite–Einstein–Vektorbündeln*, Math. Ann. **260** (1982), no. 1, 133–141.
12. J. P. May, *Matric Massey products*, J. Algebra **12** (1969), 533–568.
13. C. Okonek, M. Schneider, and H. Spindler, *Vector bundles on complex projective spaces*, Progr. Math., vol. 3, Birkhäuser, Boston, MA, 1980.
14. L. Positselski, *All strictly exceptional collections in* $D^b_{\text{coh}}(P^m)$ *consist of vector bundles*, alg-geom/9507014.
15. V. S. Retah, *Massey operations, obstructions to formality in Sullivan's sense and obstructions to Kodaira–Spencer deformations*, Uspekhi Mat. Nauk **35** (1980), no. 3(213), 217–220 (Russian); English transl., Russian Math. Surveys **35** (1980), no. 3, 279–283.
16. Y. T. Siu and G. Trautmann, *Deformations of coherent analytic sheaves with compact supports*, Mem. Amer. Math. Soc. **29** (1981), no. 238.
17. K. Uhlenbeck and S.-T. Yau, *On the existence of Hermitian Yang–Mills connections in stable vector bundles*, Comm. Pure Appl. Math. **39** (1986), no. S, suppl., S257–S293.
18. M. Verbitsky, *Hyperholomorphic bundles over a hyper-Kähler manifold*, J. Algebraic Geom. **5** (1996), no. 4, 633–669.
19. ———, *Coherent sheaves on general K3 surfaces and tori*, math.AG/0205210.
20. C. Voisin, *A counterexample to the Hodge conjecture extended to Kähler varieties*, Int. Math. Res. Not. (2002), no. 20, 1057–1075 math.AG/0112247.

Department of Mathematics, University of Glasgow, University Gardens, Glasgow G12 8QW, Scotland

E-mail address: verbit@maths.gla.ac.uk, verbit@mccme.ru

Centre de Recherches Mathématiques
CRM Proceedings and Lecture Notes
Volume **38**, 2004

The Cohomology Rings of Hilbert Schemes via Jack Polynomials

Wei-Ping Li, Zhenbo Qin, and Weiqiang Wang

1. Introduction

Fundamental and deep connections have been developed in recent years between the geometry of Hilbert schemes $X^{[n]}$ of points on a (quasi-)projective surface X and combinatorics of symmetric functions. Among distinguished classes of symmetric functions, let us mention the monomial symmetric functions, Schur polynomials, Jack polynomials (which depend on a *Jack parameter*), and Macdonald polynomials, etc. (cf. [**7**]). The monomial symmetric functions can be realized as certain ordinary cohomology classes of the Hilbert schemes associated to an embedded curve in a surface (cf. [**8**]). Nakajima [**9**] further showed that the Jack polynomials whose Jack parameter is a positive integer γ can be realized as certain $\mathbb{T}$-equivariant cohomology classes of the Hilbert schemes of points on the surface $X(\gamma)$ which is the total space of the line bundle $\mathcal{O}_{\mathbb{P}^1}(-\gamma)$ over the complex projective line $\mathbb{P}^1$. Here and below $\mathbb{T}$ stands for the one-dimensional complex torus. In other words, the Jack parameter is interpreted as minus the self-intersection number of the zero-section in $X(\gamma)$. With very different motivations, Haiman (cf. [**5**] and the references therein) developed connections between the Macdonald polynomials and the geometry of Hilbert schemes, and in particular realized the Macdonald polynomials as certain $\mathbb{T}$-equivariant K-homology classes of the Hilbert schemes of points on the affine plane $\mathbb{C}^2$ (a similar result has been conjectured in [**9**]).

In this note, we shall establish a link somewhat different from [**9**] between equivariant cohomology of Hilbert schemes and Jack polynomials, and then use this to describe the equivariant and ordinary cohomology rings of the Hilbert schemes of points on the surface $X(\gamma)$. We first show that the Jack polynomials can be realized in terms of certain $\mathbb{T}$-equivariant cohomology classes of the Hilbert schemes of points on *the affine plane*, and the Jack parameter comes from the ratio of the $\mathbb{T}$-weights on the two affine lines preserved by the $\mathbb{T}$-action. In our view, the present

---

2000 *Mathematics Subject Classification.* Primary: 14C05; Secondary: 05E05, 14F43, 17B69.

The first author is partially supported by the grant HKUST6114/02P.

The second author is partially supported by an NSF grant and a University of Missouri Research Board grant.

The third author is partially supported by an NSF grant.

This is the final form of the paper.

construction is conceptually simpler than the original one in [**9**]. This result is probably not very surprising however and could be well anticipated by experts (as it is done by elaborating the ideas of [**9**] with new inputs from [**10**]). But we feel it is nice to formulate it precisely and to make it accessible to the public. In the case when the ratio mentioned above is one, the Jack polynomials are the Schur polynomials, and our current construction specializes to [**10**] (also cf. [**6**]).

In addition, we study the $\mathbb{T}$-equivariant cohomology ring of $X(\gamma)^{[n]}$ with respect to a certain $\mathbb{T}$-action. The $\mathbb{T}$-action used in this note has the property that the $\mathbb{T}$-fixed points are isolated, and is quite different from the one used in [**9**]. Generalizing an idea in [**10**], we can define a ring structure on $H^{2n}_{\mathbb{T}}(X(\gamma)^{[n]})$, which is modified from and in turn encodes the $\mathbb{T}$-equivariant cohomology ring structure of $X(\gamma)^{[n]}$. We identify the classes in $H^{2n}_{\mathbb{T}}(X(\gamma)^{[n]})$ associated to the $\mathbb{T}$-fixed points in $X(\gamma)^{[n]}$ with 2-tuple Jack polynomials, by using the connections between the Hilbert schemes of points on $\mathbb{C}^2$ and Jack polynomials as formulated above. We further prove that the ring $H^{2n}_{\mathbb{T}}(X(\gamma)^{[n]})$ can be simply described by declaring these 2-tuple Jack polynomials to be (essentially) idempotents.

In both cases when X is $\mathbb{C}^2$ or $X(\gamma)$, our main technical tool is the construction of a Heisenberg algebra acting on the direct sum $\bigoplus_{n=0}^{\infty} H^{2n}_{\mathbb{T}}(X^{[n]})$ of the equivariant cohomology groups of middle degree. This construction generalizes the one in [**10**], which is in turn a modification of the construction in [**8**] (also cf. [**4**]).

Finally, we note that the ordinary cohomology ring of the Hilbert scheme $X(\gamma)^{[n]}$ can be shown to be isomorphic to the graded ring associated to a natural filtration on the ring $H^{2n}_{\mathbb{T}}(X(\gamma)^{[n]})$. In this way, we obtain an algorithm for computing the ordinary cup product of cohomology classes in $X(\gamma)^{[n]}$.

2. Equivariant cohomology rings of Hilbert schemes

2.1. Surfaces with torus actions. Let $\mathbb{T} = \mathbb{C}^*$, and θ be the 1-dimensional standard $\mathbb{T}$-module. For an algebraic variety M with a $\mathbb{T}$-action, let $H^*_{\mathbb{T}}(M)$ be the equivariant cohomology ring of M with $\mathbb{C}$ coefficients. Then $H^*_{\mathbb{T}}(M)$ is a $\mathbb{C}[t]$-module if we identify $H^*_{\mathbb{T}}(\mathrm{pt})$ and $\mathbb{C}[t]$ (t is an element of degree-2). For a $\mathbb{T}$-equivariant and proper morphism $f\colon N \to M$ of algebraic varieties, there is a Gysin homomorphism $f_!\colon H^*_{\mathbb{T}}(N) \to H^*_{\mathbb{T}}(M)$ of $\mathbb{T}$-equivariant cohomology groups. If N is a $\mathbb{T}$-equivariant codimension-k closed subvariety of M and $i\colon N \to M$ is the inclusion map, define $[N] = i_!(1_N) \in H^{2k}_{\mathbb{T}}(M)$ where $1_N \in H^0_{\mathbb{T}}(N)$ is the unit of the algebra $H^*_{\mathbb{T}}(N)$.

In this note, we shall consider two types of surfaces with $\mathbb{T}$-actions, and the Hilbert schemes of points on these surfaces. The first one is the complex plane $\mathbb{C}^2$, while the second is the total space of a line bundle over $\mathbb{P}^1$. The $\mathbb{T}$-actions on these surfaces are specified in the following two examples.

EXAMPLE 2.1. Fix two nonzero integers α and β with the same signs. Let u, v be the standard coordinate functions on $\mathbb{C}^2$. We define the action of $\mathbb{T}$ on $\mathbb{C}^2$ by

$$s \cdot (u, v) = (s^{\alpha} u, s^{-\beta} v), \quad s \in \mathbb{T}. \tag{2.1}$$

The origin of $\mathbb{C}^2$ is the only fixed point, which will be denoted by x. Let Σ and Σ' be the u-axis and v-axis respectively in $\mathbb{C}^2$. As $\mathbb{T}$-modules, we have $T_x\Sigma = \theta^{-\alpha}$ and $T_x\Sigma' = \theta^{\beta}$. By the localization theorem (see [**1**]), we get

$$[\Sigma] = -\alpha^{-1} t^{-1} [x], \qquad [\Sigma'] = \beta^{-1} t^{-1} [x]. \tag{2.2}$$

EXAMPLE 2.2. Fix an integer $\gamma > 1$. Let $X(\gamma)$ be the total space of the line bundle $\mathcal{O}_{\mathbb{P}^1}(-\gamma)$ over $\mathbb{P}^1$. The quasi-projective surface $X(\gamma)$ can be regarded as the quotient space of $\mathbb{C} \times (\mathbb{C}^2 - \{0\})$ by the $\mathbb{C}^*$-action defined by

$$s \cdot (b, b_1, b_2) = (s^{-\gamma} b, s b_1, s b_2), \quad s \in \mathbb{C}^*. \tag{2.3}$$

We use $[(b, b_1, b_2)]$ to denote the equivalence class. Define a $\mathbb{T}$-action on $X(\gamma)$ by

$$s \cdot [(b, b_1, b_2)] = [(sb, s^{-1} b_1, b_2)], \quad s \in \mathbb{T}. \tag{2.4}$$

For $i = 1, 2$, let X_i be the open subset of $X(\gamma)$ given by

$$X_1 = \{[(b, b_1, b_2)] \mid b_2 = 1\}, \qquad X_2 = \{[(b, b_1, b_2)] \mid b_1 = 1\}. \tag{2.5}$$

Then X_1 and X_2 form an affine open cover of $X(\gamma)$. Moreover, each X_i is $\mathbb{T}$-invariant. For simplicity, denote the point $[(b, b_1, 1)] \in X_1$ by (b, b_1). Similarly, denote $[(b, 1, b_2)] \in X_2$ by (b, b_2). Then $\mathbb{T}$ acts on the points of X_1 by $s \cdot (b, b_1) = (sb, s^{-1} b_1)$, i.e., $\mathbb{T}$ acts on the coordinate functions u_1 and v_1 of X_1 by

$$s \cdot (u_1, v_1) = (s^{-1} u_1, s v_1), \quad s \in \mathbb{T}. \tag{2.6}$$

Similarly, $\mathbb{T}$ acts on the coordinate functions u_2 and v_2 of X_2 by

$$s \cdot (u_2, v_2) = (s^{\gamma - 1} u_2, s^{-1} v_2), \quad s \in \mathbb{T}. \tag{2.7}$$

Let x_i be the origin of X_i. Then $X(\gamma)^{\mathbb{T}} = \{x_1, x_2\}$. Let $\rho \colon X(\gamma) \to \mathbb{P}^1$ be the projection sending $[(b, b_1, b_2)]$ to $[b_1, b_2]$. Let $\Sigma_0 \cong \mathbb{P}^1$ be the zero section of ρ, and

$$\Sigma_1 = \rho^{-1}([0, 1]), \qquad \Sigma_2 = \rho^{-1}([1, 0]). \tag{2.8}$$

Then as $\mathbb{T}$-modules, we have $T_{x_1}\Sigma_1 = \theta$, $T_{x_1}\Sigma_0 = \theta^{-1}$, $T_{x_2}\Sigma_0 = \theta$, and $T_{x_2}\Sigma_2 = \theta^{1-\gamma}$. By the localization theorem, we get

$$[\Sigma_1] = t^{-1}[x_1], \quad [\Sigma_0] = -t^{-1}[x_1] + t^{-1}[x_2], \quad [\Sigma_2] = (1-\gamma)^{-1} t^{-1}[x_2]. \tag{2.9}$$

REMARK 2.3. (i) Let $X(1)$ be the total space of $\mathcal{O}_{\mathbb{P}^1}(-1)$. Using $[(b, b_1, b_2)] \in X(1)$ to denote the equivalence class defined by (2.3), we define a $\mathbb{T}$-action on $X(1)$ by

$$s \cdot [(b, b_1, b_2)] = [(sb, s^{-2} b_1, b_2)], \quad s \in \mathbb{T}.$$

Then our methods and results below apply to $X(1)$ as well.

(ii) Other $\mathbb{T}$-actions on the surfaces $X(\gamma)$, $\gamma \geq 1$, with isolated fixed points can be treated similarly.

2.2. Distinguished equivariant cohomology classes. In the rest of this note, let X be a surface in Example 2.1 or Example 2.2. Our goal in this subsection is to define some distinguished equivariant cohomology classes for the Hilbert schemes. Let $X^{[n]}$ be the Hilbert scheme parametrizing all the 0-dimensional closed subschemes ξ of X with $\dim_{\mathbb{C}} H^0(\mathcal{O}_\xi) = n$. The $\mathbb{T}$-action on X induces a $\mathbb{T}$-action on $X^{[n]}$. The support of a $\mathbb{T}$-fixed point in $X^{[n]}$ is contained in $X^{\mathbb{T}}$. By the results in [**2**], the $\mathbb{T}$-fixed points of $X^{[n]}$ are isolated and parametrized in terms of (multi-)partitions.

Next, let $X = \mathbb{C}^2$ as in Example 2.1. The $\mathbb{T}$-fixed points of $X^{[n]}$ are supported in $X^T = \{x\}$ and indexed by partitions λ of n. We use ξ_λ to denote the fixed point in $(X^{[n]})^{\mathbb{T}}$ corresponding to a partition λ of n. Then for $\lambda \vdash n$, the tangent space of $X^{[n]}$ at the fixed point ξ_λ is $\mathbb{T}$-equivariantly isomorphic to (see [**2, 8, 9**]):

$$T_{\xi_\lambda} X^{[n]} = \bigoplus_{\square \in D_\lambda} \left(\theta^{\alpha(\ell(\square)+1)+\beta a(\square)} \oplus \theta^{-\alpha \ell(\square) - \beta(a(\square)+1)}\right) \tag{2.10}$$

where D_λ is the Young diagram associated to the partition λ, $\square$ is a cell in D_λ, $\ell(\square)$ is the leg of $\square$, and $a(\square)$ is the arm of $\square$ (see [7] for the notations). So

$$e_{\mathbb{T}}(T_{\xi_\lambda} X^{[n]}) = (-1)^n c_\lambda(\alpha,\beta) c'_\lambda(\alpha,\beta) t^{2n}, \tag{2.11}$$

where $e_{\mathbb{T}}(\cdot)$ stands for the equivariant Euler class and

$$c_\lambda(\alpha,\beta) = \prod_{\square\in D_\lambda} \big(\alpha(\ell(\square)+1) + \beta a(\square)\big), \tag{2.12}$$

$$c'_\lambda(\alpha,\beta) = \prod_{\square\in D_\lambda} \big(\alpha\ell(\square) + \beta(a(\square)+1)\big). \tag{2.13}$$

Note that $[\xi_\lambda] \in H^{4n}_{\mathbb{T}}(X^{[n]})$. We define the following distinguished class:

$$[\lambda] = \frac{(-1)^n}{c_\lambda(\alpha,\beta)} t^{-n} [\xi_\lambda]. \tag{2.14}$$

Now let $X = X(\gamma)$ as in Example 2.2. In view of (2.1) and (2.6), there is a $\mathbb{T}$-equivariant identification between X_1 and the complex plane in Example 2.1 with $\alpha = \beta = -1$; similarly for X_2 and the complex plane in Example 2.1 with $\alpha = \gamma - 1$ and $\beta = 1$. Then the $\mathbb{T}$-fixed points of $X^{[n]}$ are of the form $\xi_{\lambda^1} + \xi_{\lambda^2}$ where λ^1 and λ^2 are partitions with $|\lambda^1| + |\lambda^2| = n$, and ξ_{λ^1} and ξ_{λ^2} are defined in the previous paragraph as we identify X_1 and X_2 with $\mathbb{C}^2$. For simplicity, put

$$\xi_{\lambda^1,\lambda^2} = \xi_{\lambda^1} + \xi_{\lambda^2}.$$

We have a $\mathbb{T}$-equivariant splitting $T_{\xi_{\lambda^1,\lambda^2}} X^{[n]} \cong T_{\xi_{\lambda^1}} X_1^{[|\lambda^1|]} \oplus T_{\xi_{\lambda^2}} X_2^{[|\lambda^2|]}$. By (2.11),

$$e_{\mathbb{T}}(T_{\xi_{\lambda^1,\lambda^2}} X^{[n]}) = (-1)^n c_{\lambda^1}(-1,-1) c'_{\lambda^1}(-1,-1) c_{\lambda^2}(\gamma-1,1) c'_{\lambda^2}(\gamma-1,1) t^{2n}.$$

Also, as in (2.14), we introduce the distinguished class

$$[\lambda^1,\lambda^2] = \frac{(-1)^n}{c_{\lambda^1}(-1,-1) c_{\lambda^2}(\gamma-1,1)} t^{-n} [\xi_{\lambda^1,\lambda^2}]. \tag{2.15}$$

2.3. Bilinear forms on the equivariant cohomology. It is known from [**2, 3**] that the odd Betti numbers of $X^{[n]}$ are equal to zero and $H^k(X^{[n]}) = 0$ for $k > 2n$. Hence the spectral sequence associated with the fibration $X^{[n]} \times_{\mathbb{T}} E\mathbb{T} \to B\mathbb{T}$ degenerates at the E_2-term. We have

$$H^{2k}_{\mathbb{T}}(X^{[n]}) = t^{k-n} \cup H^{2n}_{\mathbb{T}}(X^{[n]})$$

for $k \geq n$. Therefore, the classes defined in (2.14) and (2.15) are contained in

$$\mathbb{H}_n \stackrel{\text{def}}{=} H^{2n}_{\mathbb{T}}(X^{[n]}).$$

Moreover, we can define a product structure $\star$ on $\mathbb{H}_n$ as follows (also cf. [**10**]):

$$t^n \cup (A \star B) = A \cup B \in H^{4n}_{\mathbb{T}}(X^{[n]}) \tag{2.16}$$

for A, $B \in H^{2n}_{\mathbb{T}}(X^{[n]})$. We see that $(H^{2n}_{\mathbb{T}}(X^{[n]}), \star)$ is a ring.

Let $H^*_{\mathbb{T}}(\cdot)' = H^*_{\mathbb{T}}(\cdot) \otimes_{\mathbb{C}[t]} \mathbb{C}(t)$ be the localization, and let

$$\iota\colon (X^{[n]})^{\mathbb{T}} \to X^{[n]}$$

be the inclusion map. By abusing notations, we also use $\iota_!$ to denote the induced Gysin map on the localized equivariant cohomology groups:

$$\iota_!\colon H^*_{\mathbb{T}}((X^{[n]})^{\mathbb{T}})' \to H^*_{\mathbb{T}}(X^{[n]})', \tag{2.17}$$

which is an isomorphism by the localization theorem.

Define a bilinear form $\langle -,-\rangle\colon H^*_{\mathbb{T}}(X^{[n]})' \times H^*_{\mathbb{T}}(X^{[n]})' \to \mathbb{C}(t)$:

$$\langle A, B\rangle = (-1)^n p_! \iota_!^{-1}(A \cup B) \tag{2.18}$$

where p is the projection $(X^{[n]})^{\mathbb{T}} \to \mathrm{pt}$. This induces a bilinear form $\langle -,-\rangle$ on

$$\mathbb{H}'_X \stackrel{\text{def}}{=} \bigoplus_{n=0}^{\infty} H^*_{\mathbb{T}}(X^{[n]})'. \tag{2.19}$$

Next, we study the restriction of the bilinear form $\langle -,-\rangle$ to $\mathbb{H}_n = H^{2n}_{\mathbb{T}}(X^{[n]})$. When $X = \mathbb{C}^2$, we see from the projection formula and (2.11) that

$$\begin{aligned} [\xi_\lambda] \cup [\xi_\mu] &= i_{\lambda!}(1_{\xi_\lambda}) \cup i_{\mu!}(1_{\xi_\mu}) = i_{\lambda!}\big(1_{\xi_\lambda} \cup i_\lambda^* i_{\mu!}(1_{\xi_\mu})\big) \\ &= \delta_{\lambda,\mu} e_{\mathbb{T}}(T_{\xi_\lambda} X^{[n]})[\xi_\lambda] = \delta_{\lambda,\mu}(-1)^n c_\lambda(\alpha,\beta) c'_\lambda(\alpha,\beta) t^{2n}[\xi_\lambda]. \end{aligned} \tag{2.20}$$

It follows from (2.14) and (2.18) that for λ, $\mu \vdash n$, we have

$$\langle [\lambda], [\mu]\rangle = \delta_{\lambda,\mu} \frac{c'_\lambda(\alpha,\beta)}{c_\lambda(\alpha,\beta)}. \tag{2.21}$$

By the localization theorem, we see that the classes $[\lambda]$, $\lambda \vdash n$ form a linear basis of the $\mathbb{C}$-vector space $\mathbb{H}_n$. Similarly, when $X = X(\gamma)$ is from Example 2.2,

$$\langle [\lambda^1, \lambda^2], [\mu^1, \mu^2]\rangle = \delta_{\lambda^1,\mu^1}\delta_{\lambda^2,\mu^2} \frac{c'_{\lambda^2}(1-\gamma,-1)}{c_{\lambda^2}(1-\gamma,-1)} = \langle [\lambda^1], [\mu^1]\rangle \cdot \langle [\lambda^2], [\mu^2]\rangle, \tag{2.22}$$

and the classes $[\lambda^1, \lambda^2]$, where $|\lambda^1| + |\lambda^2| = n$, form a linear basis of $\mathbb{H}_n$.

It follows that the restriction to $\mathbb{H}_n$ of the bilinear form $\langle -,-\rangle$ on $H^*_{\mathbb{T}}(X^{[n]})'$ is a nondegenerate bilinear form $\langle -,-\rangle : \mathbb{H}_n \times \mathbb{H}_n \to \mathbb{C}$. This induces a nondegenerate bilinear form $\langle -,-\rangle\colon \mathbb{H}_X \times \mathbb{H}_X \to \mathbb{C}$ where the space $\mathbb{H}_X$ is defined by

$$\mathbb{H}_X = \bigoplus_{n=0}^{\infty} \mathbb{H}_n. \tag{2.23}$$

3. Heisenberg algebras, equivariant cohomology and Jack polynomials

3.1. Heisenberg algebra actions. Let X be a surface in Example 2.1 or Example 2.2. Fix a positive integer i. For a $\mathbb{T}$-invariant closed curve $Y \subset X$, we define

$$Y_{n,i} = \{(\xi,\eta) \in X^{[n+i]} \times X^{[n]} \mid \eta \subset \xi, \mathrm{Supp}(I_\eta/I_\xi) = \{y\}, y \in Y\}$$

where I_η and I_ξ are the sheaves of ideals corresponding to η and ξ respectively. Let p_1 and p_2 be the projections of $X^{[n+i]} \times X^{[n]}$ to the two factors. As in [**10**], we define a linear operator $\mathfrak{p}_{-i}([Y]) \in \mathrm{End}(\mathbb{H}'_X)$ by

$$\mathfrak{p}_{-i}([Y])(A) = p_{1!}(p_2^* A \cup [Y_{n,i}]) \tag{3.1}$$

for $A \in H^*_{\mathbb{T}}(X^{[n]})'$. Note that the restriction of p_1 to $Y_{n,i}$ is proper. We define $\mathfrak{p}_i([Y]) \in \mathrm{End}(\mathbb{H}'_X)$ to be the adjoint operator of $\mathfrak{p}_{-i}([Y])$ with respect to the bilinear form $\langle -,-\rangle$ on $\mathbb{H}'_X$. For $A \in H^*_{\mathbb{T}}(X^{[n]})'$, we have

$$\mathfrak{p}_i([Y])(A) = (-1)^i p'_{2!}(\iota \times \mathrm{Id})_!^{-1}(p_1^* A \cup [Y_{n-i,i}]) \tag{3.2}$$

where p'_2 is the projection of $(X^{[n]})^{\mathbb{T}} \times X^{[n-i]}$ to $X^{[n-i]}$. Finally we put $\mathfrak{p}_0([Y]) = 0$.

By the definition of $\mathfrak{p}_{-i}([Y])$ for $i > 0$, its restriction to $\mathbb{H}_X$ gives a linear operator in $\mathrm{End}(\mathbb{H}_X)$, denoted by $\mathfrak{p}_{-i}([Y])$ as well. Next, we recall from Section 2.3 that there is a nondegenerate bilinear form $\langle -,-\rangle\colon \mathbb{H}_X \otimes_{\mathbb{C}} \mathbb{H}_X \to \mathbb{C}$, which is the restriction of the bilinear form $\langle -,- \rangle$ on $\mathbb{H}'_X$. Thus, the restriction of $\mathfrak{p}_i([Y])$ to $\mathbb{H}_X$ is the adjoint operator of $\mathfrak{p}_{-i}([Y])$ with respect to the bilinear form $\langle -,- \rangle$ on $\mathbb{H}_X$, and hence is an operator in $\mathrm{End}(\mathbb{H}_X)$ which will again be denoted by $\mathfrak{p}_i([Y])$.

By (2.2) and (2.9), $H^2_{\mathbb{T}}(X)$ is linearly spanned by the classes $[Y]$ where Y denotes $\mathbb{T}$-invariant closed curves in X. So we can extend the notion $\mathfrak{p}_k([Y])$ linearly to obtain the operator $\mathfrak{p}_k(\omega) \in \mathrm{End}(\mathbb{H}_X)$ for an arbitrary class $\omega \in H^2_{\mathbb{T}}(X)$. Note from Example 2.1 (respectively, Example 2.2) that Σ and Σ' (respectively, Σ_0 and Σ_i where $i = 1$ or 2) intersect transversely at one point. This (simple but crucial) observation together with an argument parallel to [**8, 10**] leads to the following.

PROPOSITION 3.1. *The operators $\mathfrak{p}_k(\omega)$, $k \in \mathbb{Z}$ and $\omega \in \mathbb{H}_1 = H^2_{\mathbb{T}}(X)$, acting on $\mathbb{H}_X$ satisfy the following Heisenberg commutation relation:*

$$[\mathfrak{p}_k(\omega_1), \mathfrak{p}_\ell(\omega_2)] = k\delta_{k,-\ell}\langle \omega_1, \omega_2 \rangle \mathrm{Id}. \tag{3.3}$$

Furthermore, $\mathbb{H}_X$ becomes the Fock space over the Heisenberg algebra modeled on $H^2_{\mathbb{T}}(X)$ with the unit $|0\rangle \in H^0_{\mathbb{T}}(X^{[0]})$ of $H^*_{\mathbb{T}}(X^{[0]})$ being a highest weight vector.

3.2. The case of the complex plane $\mathbb{C}^2$. In this subsection, we consider $X = \mathbb{C}^2$ from Example 2.1. Let $\mathfrak{p}_i = \mathfrak{p}_i([\Sigma])$ for $i \in \mathbb{Z}$. Given a partition $\lambda = (\lambda_1, \ldots, \lambda_\ell) = (1^{m_1} 2^{m_2} \cdots)$, define

$$\mathfrak{z}_\lambda = \prod_{i \geq 1} r^{m_i} m_i!, \qquad \mathfrak{p}_{-\lambda} = \frac{1}{\mathfrak{z}_\lambda} \prod_{i \geq 1} \mathfrak{p}_{-i}^{m_i}. \tag{3.4}$$

By (2.2), (2.14) and (2.21), we have $\langle [\Sigma], [\Sigma] \rangle = \beta/\alpha$. So by Proposition 3.1,

$$\langle \mathfrak{p}_{-\lambda}|0\rangle, \mathfrak{p}_{-\mu}|0\rangle \rangle = \delta_{\lambda,\mu} \frac{1}{\mathfrak{z}_\lambda} (\beta/\alpha)^{\ell(\lambda)}. \tag{3.5}$$

Let $S^n X$ (respectively, $S^n \Sigma$) be the nth symmetric product of X (respectively, of Σ). Let $\pi \colon X^{[n]} \to S^n X$ be the Hilbert–Chow morphism. For $\lambda \vdash n$, define

$$S^n_\lambda \Sigma = \left\{ \sum_{i=1}^{\ell} \lambda_i y_i \in S^n \Sigma \;\middle|\; y_i \in \Sigma, \text{ and } y_i \neq y_j \text{ if } i \neq j \right\}.$$

Let $\Sigma^{(\lambda)}$ be the closure of $\pi^{-1}(S^n_\lambda \Sigma)$ in $X^{[n]}$. An argument parallel to the proof of the Corollary 6.10 in [**9**] (also cf. [**10**]) shows that

$$[\Sigma^{(\lambda)}] = [\lambda] + \sum_{\mu < \lambda} c_{\lambda,\mu} [\mu] \tag{3.6}$$

where $c_{\lambda,\mu} \in \mathbb{C}$ and "$<$" denotes the dominance partial ordering of partitions.

Let Λ be the ring of symmetric polynomials in infinitely many variables, Λ^n be the space of degree-n symmetric polynomials, m_λ be the monomial symmetric function associated to a partition λ, and p_k be the kth power sum symmetric function. Given a partition $\lambda = (1^{m_1} 2^{m_2} \cdots)$, we define

$$p_\lambda = \frac{1}{\mathfrak{z}_\lambda} \prod_{i \geq 1} p_i^{m_i}. \tag{3.7}$$

It is known that the symmetric functions p_λ form a linear basis of the $\mathbb{C}$-vector space Λ. So we can define a bilinear form $\langle -,- \rangle$ on Λ by

$$\langle p_\lambda, p_\mu \rangle = \delta_{\lambda,\mu} \frac{1}{\mathfrak{z}_\lambda} (\beta/\alpha)^{\ell(\lambda)}. \tag{3.8}$$

Let $P_\lambda^{(\beta/\alpha)}$ denote the Jack polynomials (see [**7**, p. 379] for their characterization). We introduce a ring structure $\circ$ on Λ^n defined by

$$\frac{P_\lambda^{(\beta/\alpha)}}{c'_\lambda(\alpha,\beta)} \circ \frac{P_\mu^{(\beta/\alpha)}}{c'_\mu(\alpha,\beta)} = \delta_{\lambda,\mu} \frac{P_\lambda^{(\beta/\alpha)}}{c'_\lambda(\alpha,\beta)}. \tag{3.9}$$

THEOREM 3.2. *There exists a linear isomorphism $\phi\colon \mathbb{H}_X \to \Lambda$ preserving bilinear forms such that $\phi(\mathfrak{p}_{-\lambda}|0\rangle) = p_\lambda$, $\phi([\Sigma^{(\lambda)}]) = m_\lambda$, and $\phi([\lambda]) = P_\lambda^{(\beta/\alpha)}$. Furthermore, the restriction ϕ_n of ϕ to $\mathbb{H}_n$ is an isomorphism of rings, i.e., $\phi_n\colon (\mathbb{H}_n, \star) \cong (\Lambda^n, \circ)$.*

PROOF. The linear isomorphism $\phi\colon \mathbb{H}_X \to \Lambda$ is defined by mapping $\mathfrak{p}_{-\lambda}|0\rangle$ to p_λ. So

$$\phi(\mathfrak{p}_{-i}A) = p_i\phi(A), \quad i > 0,\ A \in \mathbb{H}_X. \tag{3.10}$$

Next, an argument similar to the proof of the Theorem 4.6 in [**9**] (also cf. [**10**]) verifies that for all $i > 0$, we have

$$\mathfrak{p}_{-i} \cdot [\Sigma^{(\lambda)}] = \sum_\mu a_{\lambda,\mu}[\Sigma^{(\mu)}] \tag{3.11}$$

where the coefficients $a_{\lambda,\mu}$ are the same as in $p_i m_\lambda = \sum_\mu a_{\lambda,\mu} m_\mu$. Therefore we conclude from an induction, (3.11) and (3.10) that $\phi([\Sigma^{(\lambda)}]) = m_\lambda$.

By (3.5) and (3.8), ϕ preserves the bilinear forms. So we see from (2.21) that

$$\langle \phi([\lambda]), \phi([\mu]) \rangle = \delta_{\lambda,\mu} \frac{c'_\lambda(\alpha,\beta)}{c_\lambda(\alpha,\beta)} = \delta_{\lambda,\mu} \frac{c'_\lambda(1,\beta/\alpha)}{c_\lambda(1,\beta/\alpha)}.$$

By (3.6), we have $[\lambda] = [\Sigma^{(\lambda)}] + \sum_{\mu<\lambda} d_{\lambda,\mu}[\Sigma^{(\mu)}]$. It follows that

$$\phi([\lambda]) = m_\lambda + \sum_{\mu<\lambda} d_{\lambda,\mu} m_\mu.$$

By the characterization of the Jack polynomials in [**7**], $\phi([\lambda]) = P_\lambda^{(\beta/\alpha)}$.

Finally, by (2.20) and the definition of $\star$, we get

$$\frac{[\lambda]}{c'_\lambda(\alpha,\beta)} \star \frac{[\mu]}{c'_\mu(\alpha,\beta)} = \delta_{\lambda,\mu} \frac{[\lambda]}{c'_\lambda(\alpha,\beta)}, \qquad \lambda, \mu \vdash n.$$

In view of (3.9), $\phi_n\colon (\mathbb{H}_n, \star) \to (\Lambda^n, \circ)$ is an isomorphism of rings. □

3.3. The case of the surface $X(\gamma)$. Let $X = X(\gamma)$ be the surface from Example 2.2. Recall that there is a $\mathbb{T}$-equivariant identification between the affine open subset X_1 of X and the complex plane in Example 2.1 with $\alpha = \beta = -1$; similarly for X_2 and the complex plane in Example 2.1 with $\alpha = \gamma - 1$ and $\beta = 1$. By the discussions in Section 3.1, $\mathbb{H}_X$ is the Fock space of the Heisenberg algebra generated by $\mathfrak{p}_i([\Sigma_1])$ and $\mathfrak{p}_i([\Sigma_2])$ with $i \in \mathbb{Z}$, where Σ_1 and Σ_2 are the two $\mathbb{T}$-equivariant fibers in X defined by (2.8). Note also that $\mathbb{H}_{X_1}$ is the Fock space of the Heisenberg algebra generated by $\mathfrak{p}_i([\Sigma_1])$ for $i \in \mathbb{Z}$, where Σ_1 is considered as a $\mathbb{T}$-equivariant closed curve in the affine open subset $X_1 \subset X$. Similarly, $\mathbb{H}_{X_2}$ is the Fock space of the Heisenberg algebra generated by $\mathfrak{p}_i([\Sigma_2])$ for $i \in \mathbb{Z}$. To avoid confusions, we use $\mathfrak{p}_i^{X_1}$ to denote the operators $\mathfrak{p}_i([\Sigma_1])$ acting on $\mathbb{H}_{X_1}$, $\mathfrak{p}_i^{X_2}$ to denote the operators $\mathfrak{p}_i([\Sigma_2])$ acting on $\mathbb{H}_{X_2}$, and $\mathfrak{p}_i([\Sigma_1])$ and $\mathfrak{p}_i([\Sigma_2])$ for the

Heisenberg operators acting on $\mathbb{H}_X$. As in (3.4), for a partition $\lambda = (1^{m_1}2^{m_2}\cdots)$ and for $j = 1$ or 2, we define

$$\mathfrak{p}_{-\lambda}([\Sigma_j]) = \frac{1}{\mathfrak{z}_\lambda}\prod_{i\geq 1}\mathfrak{p}_{-i}([\Sigma_j])^{m_i}.$$

Define a linear map $\Psi\colon \mathbb{H}_{X_1}\otimes_{\mathbb{C}}\mathbb{H}_{X_2}\to\mathbb{H}_X$ as follows:

$$\Psi([\lambda^1]\otimes[\lambda^2]) = [\lambda^1,\lambda^2]. \tag{3.12}$$

The linear map Ψ is an isomorphism since the classes $[\lambda^1]$, $[\lambda^2]$, $[\lambda^1,\lambda^2]$ defined in (2.14) and (2.15) form linear bases of $\mathbb{H}_{X_1}$, $\mathbb{H}_{X_2}$, $\mathbb{H}_X$ respectively. In addition, we see from (2.22) that Ψ preserves the bilinear forms.

LEMMA 3.3. *The linear isomorphism Ψ commutes with Heisenberg operators, i.e.,*

$$\Psi\circ(\mathfrak{p}_{-i}^{X_1}\otimes\mathrm{Id}) = \mathfrak{p}_{-i}([\Sigma_1])\circ\Psi,$$
$$\Psi\circ(\mathrm{Id}\otimes\mathfrak{p}_{-i}^{X_2}) = \mathfrak{p}_{-i}([\Sigma_2])\circ\Psi.$$

PROOF. By the symmetry between Σ_1 and Σ_2, we need only to prove the first identity. Also, since Ψ preserves the bilinear forms and $\mathfrak{p}_i$ is the adjoint operator of $\mathfrak{p}_{-i}$, it suffices to prove the first identity for $i > 0$.

Since the two fibers Σ_1 and Σ_2 do not intersect, we see that for partitions λ^1 and λ^2 with $|\lambda^1|+|\lambda^2| = n$, the $\mathbb{T}$-equivariant closed subvariety $\Sigma_1^{(\lambda^1)}\times\Sigma_2^{(\lambda^2)}\subset X^{[n]}$ is the closure of $\pi^{-1}(S_{\lambda^1}^{|\lambda^1|}\Sigma_1\times S_{\lambda^2}^{|\lambda^2|}\Sigma_2)$ in $X^{[n]}$, where $\pi\colon X^{[n]}\to S^nX$ denotes the Hilbert–Chow morphism. We conclude that

$$\Psi([\Sigma_1^{(\lambda^1)}]\otimes[\Sigma_2^{(\lambda^2)}]) = [\Sigma_1^{(\lambda^1)}\times\Sigma_2^{(\lambda^2)}]$$

by writing $[\Sigma_1^{(\lambda^1)}\times\Sigma_2^{(\lambda^2)}]$ in terms of $[\lambda^1,\lambda^2]$ and $[\mu^1,\mu^2]$ where $\mu^1<\lambda^1$ or $\mu^2<\lambda^2$, similarly as in (3.6). It implies that the classes $[\Sigma_1^{(\lambda^1)}\times\Sigma_2^{(\lambda^2)}]$ form a linear basis of $\mathbb{H}_X$. By (3.11) and a similar computation for $\mathfrak{p}_{-i}([\Sigma_1])[\Sigma_1^{(\lambda^1)}\times\Sigma_2^{(\lambda^2)}]$, we have

$$\begin{aligned}\Psi\big((\mathfrak{p}_{-i}^{X_1}\otimes\mathrm{Id})([\Sigma_1^{(\lambda^1)}]\otimes[\Sigma_2^{(\lambda^2)}])\big) &= \Psi\big((\mathfrak{p}_{-i}^{X_1}[\Sigma_1^{(\lambda^1)}])\otimes[\Sigma_2^{(\lambda^2)}]\big)\\ &= \mathfrak{p}_{-i}([\Sigma_1])[\Sigma_1^{(\lambda^1)}\times\Sigma_2^{(\lambda^2)}]\\ &= \mathfrak{p}_{-i}([\Sigma_1])\Psi([\Sigma_1^{(\lambda^1)}]\otimes[\Sigma_2^{(\lambda^2)}]).\end{aligned}$$

It follows that $\Psi\circ(\mathfrak{p}_{-i}^{X_1}\otimes\mathrm{Id}) = \mathfrak{p}_{-i}([\Sigma_1])\circ\Psi$ for $i>0$. □

Let Λ_1 be the ring Λ of symmetric functions with the bilinear form

$$\langle p_\lambda, p_\mu\rangle = \delta_{\lambda,\mu}\frac{1}{\mathfrak{z}_\lambda},$$

Λ_2 be the same ring Λ equipped with a different bilinear form

$$\langle p_\lambda, p_\mu\rangle = \delta_{\lambda,\mu}\frac{1}{\mathfrak{z}_\lambda}\left(\frac{1}{\gamma-1}\right)^{\ell(\lambda)},$$

and $\Lambda_i^{n_i}$ be the space of degree-n_i symmetric polynomials in Λ_i. The tensor product $\Lambda_1\otimes_{\mathbb{C}}\Lambda_2$ has an induced bilinear form. Let

$$(\Lambda_1\otimes_{\mathbb{C}}\Lambda_2)^n = \bigoplus_{n_1+n_2=n}\Lambda_1^{n_1}\otimes_{\mathbb{C}}\Lambda_2^{n_2}.$$

We define a ring structure on $(\Lambda_1 \otimes_{\mathbb{C}} \Lambda_2)^n$ by declaring the elements

$$\frac{P^{(1)}_{\lambda^1}}{c'_{\lambda^1}(-1,-1)} \otimes \frac{P^{(1/(\gamma-1))}_{\lambda^2}}{c'_{\lambda^2}(\gamma-1,1)} \in (\Lambda_1 \otimes_{\mathbb{C}} \Lambda_2)^n,$$

where $|\lambda^1| + |\lambda^2| = n$, to be idempotents.

THEOREM 3.4. *There exists a linear isomorphism* $\Phi\colon \mathbb{H}_X \to \Lambda_1 \otimes_{\mathbb{C}} \Lambda_2$ *preserving bilinear forms such that* $\Phi(\mathfrak{p}_{-\lambda^1}([\Sigma_1])\mathfrak{p}_{-\lambda^2}([\Sigma_2])|0\rangle) = p_{\lambda^1} \otimes p_{\lambda^2}$,

$$\Phi\big([\Sigma_1^{(\lambda^1)} \times \Sigma_2^{(\lambda^2)}]\big) = m_{\lambda^1} \otimes m_{\lambda^2},$$

and $\Phi([\lambda^1, \lambda^2]) = P^{(1)}_{\lambda^1} \otimes P^{(1/(\gamma-1))}_{\lambda^2}$. *Furthermore, the restriction of the map* Φ *to* $\mathbb{H}_n$ *is a ring isomorphism onto* $(\Lambda_1 \otimes_{\mathbb{C}} \Lambda_2)^n$.

PROOF. Follows from Lemma 3.3 and arguments similar for Theorem 3.2. □

Next, we discuss the implication of Theorem 3.4 to the ordinary cohomology ring $H^*(X^{[n]})$. With the notations for Heisenberg algebras in the ordinary cohomology setting in [**9**], a linear basis of $H^*(X^{[n]})$ consists of classes of the form

$$\mathfrak{Q}_{\lambda^1,\lambda^2} \stackrel{\text{def}}{=} \prod_{k\geq 1} P_{1_X}[-k]^{m_k(1)} \prod_{k\geq 1} P_{\Sigma_0}[-k]^{m_k(2)}|0\rangle \tag{3.13}$$

where $|\lambda^1| + |\lambda^2| = n$, $\lambda^i = (1^{m_1(i)}2^{m_2(i)}\cdots)$, $1_X \in H^0(X)$ is the fundamental cohomology class, and Σ_0 is the zero-section in $X = X(\gamma)$. Denote

$$\mathfrak{Q}^{\mathbb{T}}_{\lambda^1,\lambda^2} \stackrel{\text{def}}{=} \prod_{k\geq 1} \mathfrak{p}_{-k}(t)^{m_k(1)} \prod_{k\geq 1} \mathfrak{p}_{-k}([\Sigma_0])^{m_k(2)}|0\rangle \in \mathbb{H}_n. \tag{3.14}$$

The graded element in $H^*(X^{[n]})$ associated to $\mathfrak{Q}^{\mathbb{T}}_{\lambda^1,\lambda^2}$ is $\mathfrak{Q}_{\lambda^1,\lambda^2}$. Since $\{t, [\Sigma_0]\}$ and $\{[\Sigma_1], [\Sigma_2]\}$ are two different linear basis for $H^2_{\mathbb{T}}(X)$, it follows from the definitions that there is a simple explicit transition matrix M_1 between the basis of the equivariant cohomology $H^*_{\mathbb{T}}(X^{[n]})$ given by (3.14) and the basis given by

$$\mathfrak{R}^{\mathbb{T}}_{\lambda^1,\lambda^2} \stackrel{\text{def}}{=} \prod_{k\geq 1} \mathfrak{p}_{-k}([\Sigma_1])^{m_k(1)} \prod_{k\geq 1} \mathfrak{p}_{-k}([\Sigma_2])^{m_k(2)}|0\rangle. \tag{3.15}$$

For a fixed r, denote by $\big(g^{(r)}_{\lambda,\mu}\big)$ the transition matrix between Jack polynomials $P^{(r)}_\lambda$ and the power-sums p_μ. By Theorem 3.4, the transition matrix between the basis (3.15) and the basis $\{[\mu^1, \mu^2]\}$ is provided by $M_2 \stackrel{\text{def}}{=} \big(g^{(1)}_{\lambda^1,\mu^1}\big) \otimes \big(g^{(1/(\gamma-1))}_{\lambda^2,\mu^2}\big)$.

Now the structure of the ordinary cohomology ring $H^*(X^{[n]})$ can be described as follows in terms of the Heisenberg monomials (3.13). Given two Heisenberg monomials $\mathfrak{Q}_{\lambda^1,\lambda^2}$ and $\mathfrak{Q}_{\mu^1,\mu^2}$ in $H^*(X^{[n]})$, we consider the product of $\mathfrak{Q}^{\mathbb{T}}_{\lambda^1,\lambda^2}$ and $\mathfrak{Q}^{\mathbb{T}}_{\mu^1,\mu^2}$ in $\mathbb{H}_n$. The latter can be calculated by transferring over to the basis $\{\mathfrak{R}^{\mathbb{T}}_{\lambda^1,\lambda^2}\}$ via the transition matrix M_1, and then to the basis $\{[\mu^1, \mu^2]\}$ via the transition matrix M_2, where the product is explicitly known. Finally, we reverse the above steps and pass to the associated graded ring to express the product of $\mathfrak{Q}_{\lambda^1,\lambda^2}$ and $\mathfrak{Q}_{\mu^1,\mu^2}$ in terms of the Heisenberg monomial basis (3.13) of $H^*(X^{[n]})$.

References

1. D. A. Cox and S. Katz, *Mirror symmetry and algebraic geometry*, Math. Surveys Monogr., vol. 68, Amer. Math. Soc., Providence, RI, 1999.
2. G. Ellingsrud and S. A. Strømme, *On the homology of the Hilbert scheme of points in the plane*, Invent. Math. **87** (1987), no. 2, 343–352.
3. L. Göttsche, *The Betti numbers of the Hilbert scheme of points on a smooth projective surface*, Math. Ann. **286** (1990), no. 1-3, 193–207.
4. I. Grojnowski, *Instantons and affine algebras. I. the Hilbert scheme and vertex operators*, Math. Res. Lett. **3** (1996), no. 2, 275–291.
5. M. Haiman, *Vanishing theorems and character formulas for the Hilbert scheme of points in the plane*, Invent. Math. **149** (2002), no. 2, 371–407.
6. W.-P. Li, Z. Qin, and W. Wang, *Hilbert schemes, integrable hierarchies, and Gromov-Witten theory*, Internat. Math. Res. Notices (to appear); math.AG/0302211.
7. I. G. MacDonald, *Symmetric functions and Hall polynomials*, second edition, Oxford Math. Monogr., Oxford Univ. Press, New York, 1995.
8. H. Nakajima, *Lectures on Hilbert schemes of points on surfaces*, Univ. Lecture Ser., vol. 18, Amer. Math. Soc., Providence, RI, 1999.
9. ______, *Jack polynomials and Hilbert schemes of points on surfaces*, math.AG/9610021.
10. E. Vasserot, *Sur l'anneau de cohomologie du schéma de Hilbert de* $\mathbb{C}^2$, C. R. Acad. Sci. Paris Sér. I Math. **332** (2001), no. 1, 7–12.

Department of Mathematics, HKUST, Clear Water Bay, Kowloon, Hong Kong
E-mail address: `mawpli@ust.hk`

Department of Mathematics, University of Missouri, Columbia, MO 65211, USA
E-mail address: `zq@math.missouri.edu`

Department of Mathematics, University of Virginia, Charlottesville, VA 22904
E-mail address: `ww9c@virginia.edu`